CHEMISTRY
Third Edition

Brad R. Batdorf

Rachel Santopietro

bju press®

Greenville, South Carolina

NOTE: The fact that materials produced by other publishers may be referred to in this volume does not constitute an endorsement of the content or theological position of materials produced by such publishers. Any references and ancillary materials are listed as an aid to the student or the teacher and in an attempt to maintain the accepted academic standards of the publishing industry.

CHEMISTRY
Third Edition

Brad R. Batdorf, MAEd
Rachel Santopietro
Heather E. Cox
Thomas E. Porch, DMD
John S. Wetzel, MS

Consultants
Verne L. Biddle, PhD
Amy Corey, MS
Joshua George
Bill Harmon
Robert E. Lee, PhD
Brian S. Vogt, PhD

Bible Integration
Bryan Smith, PhD
Brian Collins, MA, MDiv

Contributing Authors
R. Terrance Egolf, CDR, USN (retired)
Lynne Woodhull

Project Editors
Adelé Hensley
Lincoln Mullen

Project Managers
Vic Ludlum
Donald Simmons

Cover Designer
Elly Kalagayan

Designer & Cover Illustrator
Aaron Dickey

Composition
MediaLynx

Illustrators
Amber Cheadle
John Cunningham
Ethan Mongin
John Roberts
Dave Schuppert
Del Thompson
Brian S. Vogt

Photo Acquisition
Rita Mitchell
Susan Perry

George Matzko (PhD), Verne Biddle (PhD), and Bill Harmon were consultants for the second edition. George L. Mulfinger (MS), Brian S. Vogt (PhD), John R. Wolsieffer (PhD), and John E. Jenkins were contributors to the first edition. Doug Rumminger and Robert Grass were the editors of the second edition.

Produced in cooperation with the Bob Jones University Division of Natural Science of the College of Arts and Science and Bob Jones Academy.

© 2009 BJU Press
Greenville, South Carolina 29614
First Edition © 1984 BJU Press
Second Edition © 2000 BJU Press

Printed in the United States of America
All rights reserved

ISBN 978-1-59166-540-3

15 14 13 12 11 10 9 8 7 6 5 4

Handcrafted.

BJU Press employs a team of experienced writers and artists whose best work goes into every book we produce. Because of our emphasis on quality, our textbooks are the top choice in Christian education. Each book is designed to give your student a learning experience that is enjoyable, academically excellent, and biblically sound.

To find out more, call **1.800.845.5731** or visit **www.bjupress.com**.

bju press
BECAUSE **IT MATTERS**

Contents

Introduction ix

CHAPTER 1 FOUNDATIONS OF CHEMISTRY 1
1A Why Study Chemistry? 2
1B What Is Chemistry? 8
1C What Do Chemists Do? 12

FACET
A Theory Goes Up in Smoke   14

CHAPTER 2 MATTER 23
2A The Classification of Matter 24
2B Energy and Matter 31
2C The States of Matter 39

FACET
Extreme Matter   40

CHAPTER 3 MEASURING AND CALCULATING 48
3A Measuring and Units 49
3B Measurements 55
3C Mathematics and Measurements 61
3D Orderly Problem Solving 63

CHAPTER 4 ATOMIC STRUCTURE 72
4A The Development of Atomic Models:
    A Historical Perspective 73
4B The Electron Structure of the Atom 82
4C Useful Element Notations 93

FACET
Spectroscopy: Fingerprinting Atoms   77

CHAPTER 5 ELEMENTS 104
5A The Periodic Table 105
5B Periodic Trends 113
5C Element Families and Their Properties 118

FACETS
Get 'em While They're Hot!   110
Stardust or Special Creation?   121

CHAPTER 6 CHEMICAL BONDS 139
6A Basics of Bonding 140
6B Types of Bonds 143
6C Properties of Compounds 153

FACET
Diamonds   154

CHAPTER 7 BOND THEORIES AND MOLECULAR GEOMETRY 160
7A Bond Theories 161
7B Molecular Geometry 166

FACET
The Nose Knows   176

CHAPTER 8 CHEMICAL COMPOSITION AND REACTIONS ... 181
- 8A Oxidation Numbers and Formulas ... 182
- 8B Nomenclature ... 187
- 8C Writing Equations ... 196
- 8D Types of Reactions ... 202

FACET
Drinking Chemicals 207

CHAPTER 9 CHEMICAL CALCULATIONS ... 215
- 9A The Mole ... 216
- 9B Stoichiometry ... 227

FACET
Clearing the Air 238

CHAPTER 10 GASES ... 243
- 10A Properties of Gases ... 244
- 10B Gas Laws ... 249
- 10C Gases and the Mole ... 258

FACET
A Breath of Fresh Air 269

CHAPTER 11 SOLIDS AND LIQUIDS ... 275
- 11A Intermolecular Forces ... 276
- 11B Solids ... 279
- 11C Liquids ... 286

FACET
Cryogenics: It's Really Cold in Here! 293

CHAPTER 12 SOLUTIONS ... 299
- 12A The Dissolving Process ... 300
- 12B Measures of Concentration ... 308
- 12C Colligative Properties ... 313
- 12D Colloids ... 319

FACET
That's One Sweet Crystal! 309

CHAPTER 13 CHEMICAL THERMODYNAMICS ... 326
- 13A Thermochemistry ... 327
- 13B Reaction Tendency ... 337

FACET
The Bible and the Second Law of Thermodynamics 347

CHAPTER 14 CHEMICAL KINETICS ... 352
- 14A Reaction Rates ... 353
- 14B Reaction Mechanisms ... 361

FACET
Spontaneous Combustion 356

CHAPTER 15   CHEMICAL EQUILIBRIUM 371
    15A   Theories of Chemical Equilibrium 372
    15B   Applications of Equilibrium Chemistry 384

> **FACET**
> Fritz Haber (1868–1934):
>   Gases for War and Peace   384

CHAPTER 16   ACIDS, BASES, AND SALTS 396
    16A   Defining Acids and Bases 397
    16B   Acid-Base Equilibria 402
    16C   Neutralization 413

> **FACET**
> Heartburn and Acid Reflux   414

CHAPTER 17   OXIDATION–REDUCTION 424
    17A   Redox Reactions 425
    17B   Electrochemical Reactions 432

> **FACETS**
> The Battle Against Corrosion   429
> Oxidation, Reduction, and the
>   Evolution of Life   433

CHAPTER 18   ORGANIC CHEMISTRY AND BIOCHEMISTRY 444
    18A   Organic Compounds 445
    18B   Hydrocarbons 447
    18C   Substituted Hydrocarbons 455
    18D   Organic Reactions 463
    18E   Biochemistry 465

> **FACETS**
> Astrobiology and Polycyclic
>   Aromatic Hydrocarbons   454
> Trans Fats   460

CHAPTER 19   MODERN MATERIALS 480
    19A   Ceramics 481
    19B   Polymers 484
    19C   Nanotechnology 492

> **FACET**
> The Calling of Charles Goodyear   487

CHAPTER 20   NUCLEAR CHEMISTRY 500
    20A   Natural Radioactivity 501
    20B   Induced Reactions 512

> **FACET**
> Radioactive Age-Dating Methods   518

APPENDIX A	PHYSICAL CONSTANTS	525
APPENDIX B	UNIT CONVERSIONS	525
APPENDIX C	COMMONLY USED ABBREVIATIONS AND SYMBOLS	527
APPENDIX D	ELEMENT DATA	529
APPENDIX E	ELECTRON CONFIGURATIONS OF THE ELEMENTS	531
APPENDIX F	NAMES AND CHARGES OF COMMON IONS	534

GLOSSARY ... 535

INDEX ... 551

PHOTOGRAPH CREDITS ... 559

SERVING GOD AS A...
- Chemistry Teacher ... 43
- Metallurgist ... 138
- Science Illustrator ... 214
- Anesthetist ... 274
- HVAC Technician ... 348
- Groundskeeper ... 443
- Patent Attorney ... 496

To the Student

As you study chemistry this year, you will study about the makeup of matter, the design of atoms, the periodic table, molecular bonds, formulas, reactions, solutions, acids, bases, and several specialized branches of chemistry—topics that make up a comprehensive high-school chemistry course. Sounds overwhelming, doesn't it? This course does cover a lot of ground, but fundamental ideas are used to explain each concept. You'll learn each concept first in words, not math. For example, verbal descriptions of the forces between positive and negative electrical charges help to explain where electrons exist in atoms, why molecules have unique three-dimensional shapes, why some substances dissolve in water while others do not, and why salt has a very high melting point.

Features of This Book

This text has been designed to help you learn. A familiarity with its built-in tools will allow you to organize your thoughts, use your time more efficiently, and perform better on tests. Take some time now to leaf through some chapters to identify the features discussed in this section.

Each chapter begins with a dominion science problem. This brief paragraph introduces a dilemma that might be solvable using chemistry. A later section, titled "Using _____ to Solve Problems," uses the chemistry taught in the chapter to propose a possible solution. This material should help you see the relevance of the chemical concepts you are learning and to appreciate how God has given humans the ability to use science to serve others.

A list of objectives appears in a margin box at the beginning of each section. Read these before the section so that you can watch for the "big ideas" of the passage. The objectives will also help you to understand what you should be able to do after reading the section.

Success in any subject requires that you understand its vocabulary. Important terms have been boldfaced in the text and defined in the glossary. A list of these terms with their page numbers is included in each chapter review under the title "Coming to Terms." Terms of secondary importance are sometimes italicized within the text. Many of these also appear in the glossary. Terms aren't always defined the first time that they appear; the text defines them wherever their meanings can be explained best. The glossary definition plus the way the term is used in the text will give you a fuller understanding of the concept. If the term is blue, look in the margin for an etymology or more information. Understanding the word's roots can help you to recall the meaning. Some less familiar terms have pronunciation guides. The key for the pronunciation system used in this book is in the margin.

Example problems, along with the steps leading to their solutions, are included in the text. These problems are not just extra work to keep you busy but will increase your understanding of the material. Pay close attention to these problems because you will be asked to solve similar problems in the section or chapter reviews. Often the chapter test will include similar problems too. The key to successful problem solving is to understand the logical steps that lead

Pronunciation Key

The pronunciations given in this text are designed to be self-evident and should give the average reader an acceptable pronunciation of the word. For precise pronunciations, consult a good dictionary. This sample pronunciation key may help those who have difficulty with the symbols used.

Stressed syllables appear in large capital letters. Syllables with secondary stress and one-syllable words appear in small capital letters. Unstressed syllables are in lowercase letters. Most consonants and combinations of consonants (*ng, sh*) make the sounds normally associated with them.

Examples:

a	cat = KAT, laugh = LAF	
a_e	cape = KAPE, reign = RANE	
ah	father = FAH thur	
ar	car = KAR	
aw	all = AWL, caught = KAWT	
ay	neigh = NAY, paint = PAYNT	
e	jet = JET	
ee	fiend = FEEND	
eh	rebel = REH bul, care = KEHR	
eye	ivory = EYE vuh ree	
i	women = WIM un	
i_e	might = MITE	
ih	pity = PIH tee	
o	potion = PO shun	
oh	own = OHN	
o_e	groan = GRONE	
oo	tune = TOON	
oo	foot = FOOT	
ow	loud = LOWD	
oy	toil = TOYL	
th	thin = THIN	
th	then = *TH*EN	
u,uh	above = uh BUV	
ur	person = PUR suhn	
wh	where = WHEHR	
y	mighty = MY tee	

to the correct answer. It is impossible to memorize the answers to every possible chemical problem. If you know how to structure the problems, however, most can be solved with ease. In some chapters, helpful problem-solving strategies appear in the margins. In a few chapters, schematic diagrams illustrate the steps in successful problem solving (e.g., Figure 9-11 on page 233).

You may notice that this book contains many colorful illustrations and photos. They aren't decoration. Each will help you better understand the concepts of the chapter or appreciate the way the chemistry is applied. Even the humorous cartoons have something to teach you. Take time to read the captions and any labels on illustrations. The answers to some review and test questions are found only in the captions.

The tables throughout this book help to organize the information or show relationships between concepts. Some contain data that you will need to solve problems. In a few cases, a table may contain information you will use again in later chapters.

Many chapters contain facets or career features. The facets provide background information on a topic from the chapter, or perhaps a historical or technical application of some concept. In a few cases, boldface terms appear in facets. The "Serving God as a _____" career features cover a wide range of occupations, all of which use chemistry. Their purpose is to demonstrate the wide impact of chemistry on our lives and to encourage you to consider how you might use chemistry to serve others if God calls you in that direction.

Review questions at the end of each section and of each chapter give you an opportunity to practice and solidify what you have learned. Some of the questions require you merely to recall information or terms you have read. Others ask you to demonstrate a deeper understanding of the material by applying the concepts. Those marked with a blue bullet (⊙) require computation or problem solving. A red-bullet (⊙) question may require outside research. Questions preceded by *DS* relate to the dominion-science material in that chapter.

The chapter summaries are concise summaries of the key ideas and terms in each chapter. These can be a helpful tool for preparing for tests or for clarifying misconceptions. Not all of the important terms are found here, however, so do not use the chapter summary as your only review.

Many helpful tools appear in the back of this textbook. Take a few minutes now to familiarize yourself with the appendixes, glossary, and index.

A Plan of Attack

Tough nuts can be cracked with a combination of force and strategy—one without the other will not work. Succeeding in chemistry is much like cracking a nut: it takes a combination of diligent study and an intelligent approach. Nothing takes the place of study, but a plan of attack can make your work more efficient. If chemistry seems like a tough nut to crack, follow these suggestions:

1. Relate new material to previously learned concepts. By doing this, you will avoid being overcome with many isolated facts. Only a few concepts will be totally new to you. You will have seen most of the material already in your physical science and biology courses.

2. Look for basic explanations of complicated ideas. What seems complicated will become more easily comprehended when you understand "why."

3. Develop an appreciation for God's creation. Seeing just how marvelous it is will whet your appetite to understand it better.

4. Expose yourself to the practical applications of chemistry. Many of the review questions illustrate how the theory from the text is applied in the real world.

5. Use the mathematical skills you have learned in the past. Most problems in chemistry rely on the math you learned in algebra. Organize your work, and do it neatly. It is critical that you use units accurately in your computation and in your answers. The example problems are good models to follow.

6. View the laboratory exercises as opportunities to review the chapter in order to prepare for the tests. Each activity is carefully crafted to apply certain concepts and terms from its chapter.

7. Keep an open mind about enjoying what you are learning. Many people miss out on enjoying delicious foods because they make up their minds ahead of time that they will not like them. Do not deny yourself the fascination of chemistry because you think you will dislike it.

May your study of chemistry bring you an increased understanding of the world around you, a solid preparation for future studies, and a heightened interest in science. More importantly, may your studies this year help you to honor the Lord Jesus Christ and to do His will.

Foundations of Chemistry

Chapter 1

Introduction

Chemistry is in the news. Are steroids safe for athletes? Is nuclear power generation the answer to America's energy needs? How can we reduce carbon dioxide emissions? How can we recycle better to keep our planet clean?

Yet there is another side to these scientific questions. Is the use of performance-enhancing drugs ethical? Should humans pursue widespread nuclear power generation? Why should we be concerned about maintaining a healthy planet? These are questions that science cannot answer. But they must be answered for science to be of any value.

1A Why Study Chemistry?	2
1B What Is Chemistry?	8
1C What Do Chemists Do?	12
Facet	
A Theory Goes Up in Smoke	14

1A Why Study Chemistry?

1.1 Chemistry in Biblical Focus

If you have experienced God's saving grace, you have insight many experienced chemists lack. We cannot properly study chemistry unless we approach it from a Christian perspective. To many people, that statement sounds controversial—even nonsensical. After all, what does Scripture have to do with chemistry? What Bible verses explain the structure of atoms or the nature of chemical bonds?

True, the Bible is not a chemistry textbook. But the Bible confronts humans with a distinct **worldview**, a perspective from which to see and interpret all of life. Like a corrective lens, the Bible brings into focus every part of the world, including chemistry. As we look at chemistry through the lens of Scripture, we find that the best reasons to study chemistry are biblical ones.

1.2 Declaring the Glory of God

Why is the universe here? Why are we here? Romans 11:36, one of the most important verses in the Bible, answers these questions: "For of him [God], and through him, and to him, are all things: to whom be glory for ever. Amen." God made the world to declare His greatness. Even the study of chemistry shows us God's glory. As we look at the most basic components of matter, we should be amazed at the intelligence and power of the Creator. It's easy to admire people who can shape clay into beautiful pottery, steel into automobiles, petroleum into chairs, or soybeans into hormones. How much more should we marvel at the One Who created our universe with simply the word of His mouth?

The whole world is God's cathedral of praise, and chemistry is one of its stained-glass windows. This window is too vast and intricate for us to comprehend or re-create, but we can and must admire its beauty and the wisdom of its Maker. If you make an A in chemistry class but never lift your heart to God in worship during your study, you have failed to really understand chemistry. All things exist to declare God's glory. If we fail to see the glory of God in chemistry, we have missed the point.

1.3 Genesis and a Proper View of Chemistry

In its first chapter, the Bible begins to give us the proper worldview for science. Genesis 1 reveals that God made man as His great masterpiece. We are the only creation made in the **image of God** (Gen. 1:26–27). Being created in God's image means that we are like God in every way that it is appropriate for a creature to be like the Creator. This image makes us precious in His sight. We are in His faithful care (1 Pet. 4:19).

How does God's image in us affect what we do and how we act? Christ closely associates the **first commandment** with the **second commandment**: "And thou shalt love the Lord thy God… this is the

---

Chapter One

1A Section Objectives

After finishing this section, you should be able to

- list five reasons a Christian should study chemistry.
- define the Creation Mandate and the image of God in man and explain their significance to studying chemistry.
- define scientific modeling and contrast science and Scripture as paths to truth.
- compare and contrast a naturalistic worldview with a Christian worldview.
- explain how a career in chemistry can be of service to God.

1-1 A biblical worldview, like these lenses, allows you to see the world in focus.

"Today biologists are beginning to understand the origins of life's complexity—the exquisite optical mechanism of the eye, the masterly engineering of the arm, the architecture of a flower or a feather, the choreography that allows trillions of cells to cooperate in a single organism. The fundamental answer is clear: In one way or another, all these wonders evolved."*

This quotation demonstrates the power of a person's worldview. The secularist and the Christian study the same marvelously complex world. But the secularist is moved to admire evolution, while the Christian praises God.

---

*Carl Zimmer, "A Fin is a Limb is a Wing," *National Geographic*, November 2006, 114.

first commandment. And the second is like, namely this, Thou shalt love thy neighbor as thyself" (Mark 12:30–31). God deserves our total devotion because He is God, our Maker, Master, and Savior. Humans deserve our love because they are image bearers of God.

Soon after creating man in His own image, God gave His first command to mankind. The **Creation Mandate** (Gen. 1:26, 28) reveals why God made humans. He wants us to exercise good and wise dominion over the earth by managing and using His world and its resources.

Both the image of God in man and the Creation Mandate should motivate and guide our study of chemistry. The world God has given us to manage is made of chemicals. As soon as we seriously consider our calling to have dominion, we will ask questions that drive the study of chemistry. As we learn about the chemical composition of the world around us, we become more able to enrich the lives of God's image bearers. Useful medication, safer automobiles, and even better food packaging all depend on applications of chemistry.

1-2 The chemicals within these capsules, the capsules themselves, and even the container that holds them are all the products of careful chemistry.

Understanding that we live in a fallen world heightens the value of chemistry. The world we have been assigned to manage is not simply wild; it is severely broken. Children develop leukemia, water often carries deadly diseases, and diabetes afflicts millions. How should Christians respond to these challenges? By imitating their Lord. The Lord Jesus had compassion on people suffering in a fallen world. Compassion did not distract Him from teaching and preaching; instead, His works of mercy empowered His message of forgiveness and eternal life. Meeting humans' spiritual needs and meeting their physical needs are linked in the Bible: "And Jesus went about all Galilee, teaching in their synagogues, and preaching the gospel of the kingdom, and healing all manner of sickness and all manner of disease among the people" (Matt. 4:23). Furthermore, it would be hypocritical for us to minister the gospel without meeting people's physical needs when doing so lies within our power. "As we have therefore opportunity, let us do good unto all men, especially unto them who are of the household of faith" (Gal. 6:10).

When we look at chemistry through the lens of a Christian worldview, we realize that chemistry gives us an opportunity to practice dominion science. **Dominion science** is scientific activity that exercises dominion for the glory of God and the benefit of other humans. To keep you thinking about this connection between Genesis and a correct view of chemistry, this textbook begins each chapter with a section titled "Dominion Science Problem." The section presents a real problem that humans have faced, and it invites you to solve the problem using the chemistry you will learn in the chapter. These problems let you practice studying chemistry for God's glory by obeying His most basic commands: exercising dominion and loving others (Gen. 1:28; Mark 12:30–31).

1.4 Science as Modeling

Having a proper view of science in general and chemistry in particular helps us put science in its proper context. How does knowledge gained from scientific pursuits compare to biblical truths? Science, unlike God's

1-3 Chemistry and the models it develops help us to recognize our God's greatness.

Richard Feynman (1918–88) was one of the greatest American physicists of the twentieth century. His work expanded our current understanding of quantum electrodynamics. A Nobel Prize winner in 1965, Feynman worked on the atomic bomb and was a member of the presidential committee investigating the Space Shuttle *Challenger* accident.

1-4 Maps represent reality just as models represent reality. They are not the reality itself.

What is faith? Read Hebrews 11:1–3. How does faith relate to one's presuppositions and view of science?

Word, does not establish truth or depict reality absolutely. Science is about modeling the world.

What comes to mind when you think of a model? You may think of a smaller physical representation of a larger object, like a model car or airplane. Some models are more than physical representations; instead, they are mental or mathematical representations. **Scientific models** are simplified representations that characterize a system or explain a phenomenon. Scientific models may be expressed as computer simulations, pictures, formulas, descriptions, and physical representations.

Richard Feynman, physicist and Nobel Prize recipient, said, "What I cannot create, I cannot understand." Scientists create models to represent reality in an effort to understand the universe. These models are simplified replicas of reality that help us understand phenomena that overwhelm our abilities to observe directly. Feynman did not use science to recreate reality but to create a visual or mathematical representation of what reality might be like.

Chemists operate under the same modeling process as physicists like Feynman. For example, the atomic model of matter, which says that all matter is composed of atoms, has been refined over the centuries to best represent experimental data. The ways in which atoms connect and interact are described by several molecular-bonding models. All these models are attempts to better understand matter and its interactions.

Modeling and Uncertainty

The modeling character of chemistry and of science in general impacts how certain one can be in science. Consider Feynman's words: "Scientific knowledge is a body of statements of varying degrees of certainty—some most unsure, some nearly sure, but none *absolutely* certain. Now, we scientists are used to this, and we take it for granted that it is perfectly consistent to be unsure—that it is possible to live and *not* know."* Feynman understood that *the goal of science is not truth but workability.* Models are not constructed to match the natural world. Such models would be too complicated for a finite mind to construct or use. Models are constructed to help scientists produce workable, useful answers to the questions scientists ask. Models help us *live* in this world, not *know* this world.

Uncertainty and Faith

Christians must realize that there is uncertainty in science. Science has become a sort of religion in our culture. Centuries ago, European and American cultures derived their beliefs and values from some kind of religious authority, usually related to the teachings of the Bible. Modern civilization, however, relies on science for certainty (infallibility) and objectivity. We deem science worthy of faith. But science is not about truth. It is about producing workable models.

---

*Richard P. Feynman, *What Do You Care What Other People Think?: Further Adventures of a Curious Character* (New York: W. W. Norton, 1998), 245.

Only God and His Word are infallible and objective. His Word deserves our trust: "Every word of God is pure: he is a shield unto them that put their trust in him" (Prov. 30:5). His Word cannot be disproved: "The scripture cannot be broken" (John 10:35). His Word is truth: "Thy word is truth" (John 17:17). Scripture's teachings, though not specifically scientific, affect how science is conducted.

As we grapple with the nature of science, we realize the foolishness of using science to disprove the Bible. Since science is concerned with modeling and workability, it cannot test a book filled with final answers and ultimate truths. To do so would be like trying to measure the height of the Empire State Building using a ruler; the tool does not fit the purpose. This world belongs to God; therefore, every part of it testifies to His existence and truthfulness. When someone claims that chemistry (or any other science) disproves the Bible, he is in fact claiming that his model contradicts biblical reality. There is always another way to interpret the evidence so that it glorifies God and is consistent with His Word.

Usefulness of Modeling

The model-making nature of science should not keep us from taking it seriously. It certainly did not keep Richard Feynman from doing science. He participated in scientific research such as the Manhattan Project, and his scientific investigation led to important answers regarding the cause of the tragic explosion of the *Challenger* space shuttle. Scientific model making is a glorious calling for the Christian. The application of science creates opportunities to glorify God as we glimpse His creative power and to use the earth's resources to help other people.

1.5 Confronting Error

Another reason to study chemistry is that it exposes error in false worldviews. All scientists approach their work with certain presuppositions, which they derive from their worldview. A **presupposition** is an idea that a person assumes to be true without proof. The most common presuppositions concern the nature of reality and the nature of good and evil. Some chemists approach their science with the presuppositions of a Christian worldview; some with the presuppositions of a naturalistic or secular worldview. Both groups live by faith, but they have different objects for their faith. A scientist's presuppositions affect his interpretation of scientific phenomena by causing him to prefer certain conclusions over other conclusions. This preference is called **bias**.

A Naturalistic Worldview

A **naturalistic worldview** assumes that matter is all that exists and that human reason informed by science is the only reliable path to truth. Sometimes this extreme faith in reason and science is called **scientism**. Carl Sagan expressed the essence of this worldview when he said, "The cosmos [the material universe] is all there is, or was, or ever will be."

Naturalists have developed the theory of evolution to explain the chance existence of elements, plants, animals, people, and ultimately the universe. Since there is no Creator in this worldview, humans are not accountable to a God. There is no absolute code of morality. People are tiny specks in a huge, impersonal world with little to elevate their

Scripture tells us...
- why we should study chemistry.
- how and when the elements came into existence.
- our substance and significance.
- how to use chemistry.

1-5 Using science in an attempt to disprove the infallible Word of God is like trying to use a ruler to determine the height of the Empire State Building.

1-6 The Manhattan Project led to the development of nuclear weapons.

A bias is a person's preference for certain conclusions over other conclusions. A bias may be good or bad depending on which presuppositions it is based on.

1-7 Carl Sagan (1934–96) was an American astronomer and the narrator of the award-winning *Cosmos* mini-series aired on PBS in 1980 and rebroadcast for years.

Making Sense of It All

Why do people put faith in a naturalistic worldview that degrades their value as humans and reduces their lives to a fight for survival?

value above other things that evolved. They are themselves nothing more than bags of chemicals with arms and legs. The things they value most—love, happiness, justice, peace, and life itself—are nothing more than chemical reactions.

Why do people with a naturalistic worldview study chemistry? To them, chemistry is a method of survival, a way to preserve or improve the current human way of life. But scientism cannot defend this way of life as good. Goodness implies a moral standard independent of science. The naturalistic worldview is thus unable to defend the work of chemistry as a good thing.

A Christian Worldview

A **Christian worldview** is based on the teachings of the Christian Scriptures. Human reason and scientific research are not irrelevant or necessarily wrong, but they are not a sufficient foundation for knowledge. God's Word is the foundation, and reason and science are important tools for building on that foundation.

The Christian sees the world through the lens of the Bible's story of God's *Creation*, man's *Fall*, and God's *Redemption* of the world and mankind. A Christian worldview gives one hope and peace, not the discontent and despair that a naturalistic worldview brings.

God made this world by speaking it into existence in an act of marvelous, supernatural power. God made humans in His own image and called them to exercise dominion over His world. Humans are made of chemicals, but that is not what we are. We are image bearers of God. Our lives are eternally significant.

But we have rebelled against God. Because of this rebellion, God has punished us by cursing our world (Gen. 3:17–19; Rom. 8:22). The world rebels against our dominion, just as we rebelled against God. Despite our best efforts, pain, disease, and death dominate our existence. God has also cursed our affections, minds, and wills. We do not feel and think about ourselves, our world, and our God as we should (Gen. 3:12; 1 Cor. 2:14; Eph. 4:17–18). This inner brokenness and self-deception is why people cannot accurately perceive reality, including chemistry, without the light of God's truth (Ps. 119:130).

These tragic truths are not the end of the story, however. God has promised to redeem this fallen world to Himself. God's Son, Jesus Christ, came to earth to die for our sins and rise again in another act of marvelous, supernatural power (Gen. 3:15; Matt. 1:21; Luke 24:46–47). In every generation He sends His people into the world to proclaim His gift of forgiveness and eternal life. We are called to proclaim the gospel to remedy how sin separates us from God. We are also to relieve suffering, another of the effects of sin (Matt. 5:16; 10:8; Gal. 2:10; 6:10).

A right-thinking Christian deems chemistry to be a high and noble activity to meet the needs of the image bearers of God. Chemistry is important because humans are important, and humans are important because God made them in His image.

1.6 Chemistry and Your Life Choices

There are several reasons that you should give yourself fully to the study of chemistry. God may call you to a chemistry-related career.

Studying chemistry could also prepare you to face important life choices. Many of those choices concern issues with which all humans must grapple.

Careers in Chemistry

Many careers require knowledge of chemistry. Obviously, chemistry teachers need to study the subject, but some other professionals that use chemistry regularly include anesthetists, dieticians, air-conditioner technicians, materials engineers, medical lab technicians, pharmacists, and even zookeepers.

A Christian can serve God in a chemistry-related career. God sent His Son "to reconcile all things unto himself... whether they be things in earth, or things in heaven" (Col. 1:20). It is not just souls that God desires to redeem. He also intends to redeem all the things that souls do and think about.

Is there anything that needs redeeming in chemistry? Much of chemistry is dominated by a naturalistic worldview. As you might suspect, most chemists do science without any concern to declare God's glory. In fact, many chemists refuse to believe the Bible because of their work in chemistry. They need to encounter believers who know both the Bible and chemistry—believers who can show them that faith in God helps a person understand the natural order. As Solomon said long ago, being right with God enables us to know His world: "The fear of the Lord is the beginning of knowledge" (Prov. 1:7).

Some applications of chemistry are dominated by self-interest and greed. A chemical-production plant that deliberately and recklessly releases toxic waste, a rogue government that develops deadly chemical weapons, or a pharmaceutical manufacturer who sets exorbitantly high drug prices are all guilty of these sins. Chemistry needs devoted Christians who can show that chemistry is more than a tool for self-advancement and money. It is a tool for loving God's image bearers. Love is a powerful motivator for scientific advancement. Living for love in such an environment is one way of glorifying God: "Let your light so shine before men, that they may see your good works, and glorify your Father which is in heaven" (Matt. 5:16).

Chemistry for Life

You might not pursue a career in chemistry. Nevertheless, the knowledge and skills you gain in this course will prove useful. This course will give you many reasons to praise God for His marvelous intelligence and His care for us. The logical thinking that this course demands of you will shape your ability to think critically about the many ideas and issues you will face throughout your life. Getting used to thinking about chemistry as a tool for loving others will prepare you to think properly about issues that many Christians dismiss as unimportant—issues such as recycling, groundwater contamination, and global warming. Gaining a proper view of modeling in chemistry will prepare you to defend the Scripture and correctly present the limits of science.

This course will also help you identify naturalistic biases. We hear about scientific issues in the news all the time that affect politics, governments, education, religion, and health. Many Christians are not able to form a biblical perspective on these issues. Many

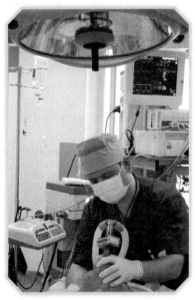

1-8 How do these people use chemistry in their occupations?

Christians and the Environment

Many Christians are rightly skeptical of the environmentalist movement. Too often environmentalists view humans as defiling intruders on the environment. Environmentalists may also oppose technological advancement. Too often their claims of looming disaster are not matched with appropriate scientific data. How would Genesis 1:26–28 encourage humans to be active in both environmental care and technological advancement?

times the core problem is the naturalistic biases of the media. We need biblical and scientific grounding to discern between truth and error.

1A Section Review

1. Why did God make all things?
2. What is the second commandment? How is it related to a Christian view of chemistry?
3. Why is the goal of science workability and not truth?
4. Is it possible for science to prove the Bible wrong? Why or why not?
5. Why does a naturalistic worldview lead to despair?
6. What are the three main components of a Christian worldview? Why does this worldview give hope?
7. Why do the media often showcase scientific theories opposed to Christianity?

1B Section Objectives

After finishing this section, you should be able to
- define the scope of the field of chemistry.
- summarize the historical development of chemistry.
- define the six major branches of chemistry.
- explain why chemistry is the central science.
- demonstrate how the different fields of chemistry interact.

1B What Is Chemistry?

1.7 Introduction

Everything in the material universe with which science deals is composed of chemicals. You are sitting on chemicals, wearing chemicals, and eating chemicals at every meal. **Chemistry**, then, is the study of matter and the changes that it undergoes. **Matter** is anything that takes up space and has mass. This definition of chemistry shows you the breadth of its field. Just like science in general, chemistry frequently uses models to describe the composition, structure, and interactions of matter.

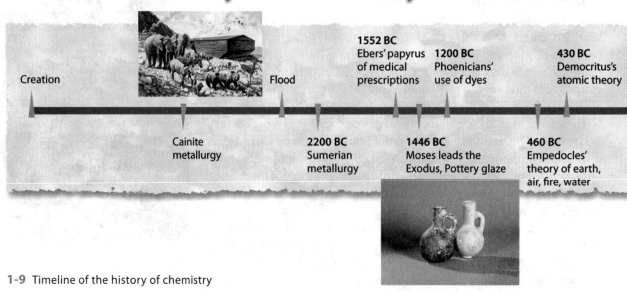

1-9 Timeline of the history of chemistry

1.8 Chemistry in the Old Testament

Chemistry developed as a practical skill during Old Testament times. Work in **metallurgy** is anticipated in one of the Bible's earliest passages. In describing how God's original creation met the needs and desires of the human race, Scripture states that God put the Garden of Eden near Havilah, a land "where there is gold; And the gold of that land is good" (Gen. 2:11–12). As the descendants of Adam gained the necessary skills to make weapons, tools, and other utensils, they laid the early foundations of chemistry. The first people known to be skilled in metallurgy were descendents of Cain. Tubal-Cain, the great-great-great-great grandson of Cain, made tools and implements from iron and bronze, a copper-tin alloy (Gen. 4:22). By the time of Abraham, man had already gained a significant degree of chemical technology. The Sumerians, for example, were highly skilled in metallurgy. Years later the Israelites demonstrated further advancements in chemistry with the building of the temple (966 BC). Solomon made extensive use of refining technology and alloy production. The amount of gold and bronze used to build the temple boggles the mind (1 Kings 6–7).

The Sumerians were not the only civilization developing chemical technology at this time. The Egyptians were also making great progress, especially in medicine. By the time Joseph was sold into slavery, the apothecary was an important part of Egyptian culture. **Apothecaries** were early pharmacists who prepared and sold a wide variety of chemicals and herbs. Apothecaries are mentioned several times in the Old Testament (Exod. 30:25, 35; 37:29; Eccl. 10:1).

chemistry: (Gk. *chemia*—black earth, perhaps the black earth along the Nile contrasted with desert sand and connected with black arts), (Gk. *chumeia*—mix together as in a solution or alloy)

metallurgy (MET al lur jee) is the science of obtaining metals from their ores.

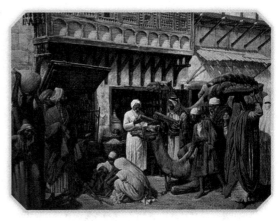

1-10 Early apothecary shops sold goods along with many chemicals used for medicinal purposes.

apothecary (ah POTH ih care ee)

1.9 Greek Influences on Chemistry

From 600 BC through the early church era, Greek philosophers debated the nature of matter. Unlike their predecessors, the Greeks

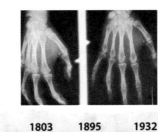

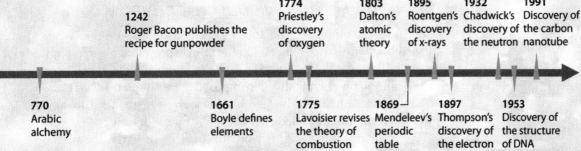

- 770 Arabic alchemy
- 1242 Roger Bacon publishes the recipe for gunpowder
- 1661 Boyle defines elements
- 1774 Priestley's discovery of oxygen
- 1775 Lavoisier revises the theory of combustion
- 1803 Dalton's atomic theory
- 1869 Mendeleev's periodic table
- 1895 Roentgen's discovery of x-rays
- 1897 Thompson's discovery of the electron
- 1932 Chadwick's discovery of the neutron
- 1953 Discovery of the structure of DNA
- 1991 Discovery of the carbon nanotube

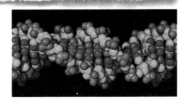

desired knowledge more than practical skills. The Greeks were the first to introduce an organized approach to chemistry.

The theories developed by the Greek philosophers were so logical that they dominated science for the next two thousand years. In some ways, however, this domination hindered the development of chemistry. The Greeks' high regard for deductive logic and their lack of technology prevented them from making observations and doing experiments. Even philosophers like Aristotle rejected Democritus's idea, formulated about 430 BC, that the atom was the building block for all matter and that the difference in materials is based on the shape, position, and arrangement of these atoms.

> Democritus (460–370 BC) was a Greek philosopher who first proposed the term *atom*. He believed all matter was made up of identical atoms.

1.10 The Chemistry of the Alchemists

During the time of Christ, a large scientific community formed in Alexandria, Egypt. Greeks who traveled to this center of learning blended their deductive logic with the ancient skills of the apothecaries. It was in ancient writings from this Greco-Egyptian city that the word *chemia*, from which the word *chemistry* is derived, first appeared.

After the decline of the Roman Empire, the Arabs took control of Alexandria. Arab scholars learned all they could from the Greek *chemia*. As *chemia* became a part of their culture, they added the Arabic prefix *al* ("the") and called it **alchemy**.

The alchemists gradually forsook other studies to pursue the secret of converting ordinary metals into gold. These attempts led to an alchemy that was interwoven with astrology and mysticism. Alchemists also believed that chemistry could cure diseases, extend one's lifespan, achieve immortality, and even create life.

Despite their failures, alchemists did provide modern chemistry with a rich legacy of laboratory techniques and equipment. But the most important contribution of the alchemists was their experimental approach. Experimentation sparked a scientific revolution in Europe. Alchemists like Tycho Brahe, Isaac Newton, Robert Boyle, and Francis Bacon made many valuable contributions to science.

'An Alchemist in his Study,' 17th century, Egbert Jaspersz van Heemskerk I (Haarlem 1610–1680), Oil on Canvas, Fisher Collection 00.01.278, Chemical Heritage Foundation Collections, Photo by Wil Brown

1-11 Though they failed to achieve their primary goal, alchemists made important contributions to the field of chemistry.

> alchemy (AHL keh mee)

1.11 The Rise of Modern Chemistry

Not until the middle of the 1600s did chemists dispute the Greek idea of the four basic elements. Robert Boyle proposed a completely new definition of elements. He said that elements are substances that cannot be chemically decomposed into simpler substances. The elements identified by Greek philosophy—earth, air, fire, and water—could not be called elements by this new definition.

One of the most significant discoveries in early chemistry was the discovery of oxygen. In 1774, Joseph Priestley heated mercuric calx (mercury [II] oxide) and obtained a gas in which substances burned easily. Shortly thereafter, the French chemist Antoine Lavoisier observed that several substances gained rather than lost weight after combustion. Lavoisier concluded

> Joseph Priestley (1733–1804) was an English minister who never took a formal science course. A friendship with Benjamin Franklin awakened his interest in science.

> Antoine-Laurent de Lavoisier (lah VWAH zee aye) (1743–94) was a French nobleman who is often called the Father of Modern Chemistry. He was guillotined during the French Revolution for being a tax collector.

that the gas Priestley had discovered was a common substance in the air that combined with other substances in combustion. He named Priestley's discovery oxygen. Lavoisier's use of the scientific method and his reliance on careful measurements served as a model for many other chemists. His method of experimentation led to a chemical revolution.

1.12 The Chemistry of Today

By 1800 chemistry had become an academic discipline. In the New World, several colleges made chemistry a part of their curriculum. Benjamin Rush of the College of Philadelphia was the first professor of chemistry in the United States. During his training at the College of New Jersey, professors attacked the old Greek ideas and used demonstrations and experiments in their teaching.

As knowledge of chemistry expanded, specialized branches of chemistry developed. Scientists soon had to concentrate in one area. For example, the investigations of Friedrich Wöhler developed the branch of chemistry called organic chemistry.

Soon other branches of chemistry developed. Inorganic chemistry, analytical chemistry, physical chemistry, nuclear chemistry, and biochemistry all expanded into separate fields. Today these branches overlap considerably. Each field will be covered to some extent in this textbook.

Just as the branches of chemistry overlap with each other, the entire subject of chemistry overlaps with other subjects. For example, the production of new elements and the study of subatomic particles blend physics with nuclear chemistry. Likewise, studies of protein structure combine biology and biochemistry. Chemistry by its very nature touches every major field of science.

> Benjamin Rush (1746–1813) was an American statesman and physician. He graduated from the College of New Jersey (now Princeton University) at fifteen before earning a medical degree from the University of Edinburgh in Scotland. He worked in hospitals in London and Paris and soon became the professor of chemistry and medicine at the College of Philadelphia. He was a surgeon general of the Continental Army during the American Revolution, and he is considered the Father of American Psychiatry.

> Friedrich Wöhler (1800–82) was a German chemist who accidentally synthesized urea while attempting to prepare ammonium cyanate from silver cyanide and ammonium chloride.

1-1	**The Major Branches of Chemistry**
Inorganic chemistry	the study of all elements (except carbon) and their components, often in the context of their ability to conduct electricity
Organic chemistry	the study of compounds containing carbon, most commonly in living things
Biochemistry	the study of the chemical processes in living things
Nuclear chemistry	the study of radioactivity, the nucleus, and the changes that the nucleus undergoes
Physical chemistry	the foundational theories of chemistry that allow detailed study of interactions between substances and the accompanying energy changes
Analytical chemistry	the techniques used in all branches of chemistry to (1) discover what substances are in a sample (qualitative) and (2) determine how much of each component it contains (quantitative)

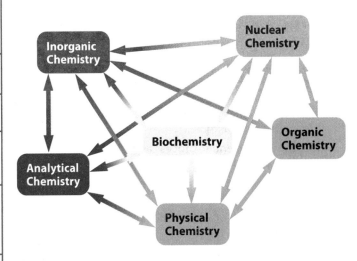

1-12 The major branches of chemistry

1-13 Modern chemistry has many practical applications.

1B Section Review

1. Look at Figure 1-9. Abraham was born in 2106 BC. What was happening in the field of chemistry in his lifetime?
2. Search for the word *iron* in a concordance to see when metallurgy first occurs in the Bible.
3. How has human history had an effect on the field of chemistry?
4. Based on the definitions in Table 1-1 and your own knowledge, place each of the six major branches of chemistry in the following Venn diagram. The diagram shows how the three major branches of science intersect. As an example, the field of biophysics has been placed in the overlap between physics and biology.

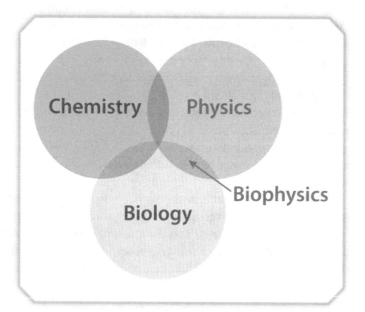

1C Section Objectives

After finishing this section, you should be able to
- define *science*.
- compare and contrast deductive and inductive reasoning.
- classify examples of logic as inductive or deductive reasoning.
- describe the scientific process as a method of inquiry.
- compare and contrast hypotheses, theories, and laws.
- give an example of when a model should be used.
- create a scientific question that would best be addressed by a survey.
- create a scientific question that would best be addressed by an experiment.
- analyze case studies of the scientific process.

1C What Do Chemists Do?

1.13 Defining Science

How would you define science? **Science** is the total collection of knowledge gained through the systematic observation of nature. Although science is an established body of knowledge, it is also an ongoing activity or process. Science does not simply involve making observations; it also involves relating, explaining, describing, predicting, refining, and testing for workability.

Scientists use different tools when trying to make sense of the world around them. They do not follow a specified, step-by-step approach but rather customize their method of investigation to what they are studying. An astronomer does not use the same process to search for new asteroids as a chemist uses to search for a more efficient meat-packaging plastic. This process of choosing and using a workable method of inquiry can be referred to as the scientific process.

Scientists have different motivations for their work. **Applied science** explores natural products and processes for specific applications, like the chemist searching for a more efficient meat-packaging plastic. **Pure science** probes nature simply to learn new things about the universe we live in, like the astronomer searching for new asteroids.

1.14 Scientific Questions

Is there a cure for HIV/AIDS? Is the earth going through a period of global warming? Are there viable energy alternatives to fossil fuels? Can stem cells be used to cure diseases? These are some of the current questions that are driving science today. **Scientific questions** or **scientific problems** are questions which direct and stimulate scientific inquiry.

1.15 Observation

Scientific inquiry attempts to answer these questions through observations. We gather **observations** of the natural world by using our senses of taste, touch, hearing, seeing, and smelling. But our senses sometimes need help. For example, our sense of sight can be greatly enhanced by a telescope, microscope, or mini-video camera such as those used in arthroscopy or colonoscopy. But observations are gathered by humans who are finite beings. We are not able to observe all that needs to be observed to give us absolute answers. Our observations are also subject to our own errors in measurement or recording of data. We therefore must interpret our observations from our own perspective, which is based on personal experiences and faith commitments. We humans are also sinners with a view of reality distorted by our own self-love. Because of these factors, observations are rarely completely **objective**, or unaffected by the observer's personal inclinations and presuppositions. Only God is truly objective, for He is infinite and His perspective is perfect.

There are two kinds of information or data that we can collect using our senses: qualitative data and quantitative data. **Quantitative data** describes observations using numerical data, such as weighing a robin's egg on a scale to obtain its weight or mass. Scientific instruments are usually necessary to collect quantitative data. In contrast, **qualitative data** describes observations using nonnumerical data and can include observations such as the texture, color, and runniness of a sample of lava.

1.16 Reasoning

Deductive reasoning proceeds from general statements to a specific conclusion. That process uses statements of assumption or fact, called *premises*. Arranged in the correct order, premises lead in a step-like pattern to a conclusion. For instance: "All matter is made of atoms. The air is matter. Therefore, the air is made of atoms." In order for the conclusion to be true, all of the premises must be true and the logic must be sound. An error in any of the premises will lead to an invalid conclusion. If the premises are arranged in the wrong order, they will lead to an invalid conclusion, even if all the premises are true.

Inductive reasoning proceeds from known data to an unknown general conclusion. Inductive reasoning uses specific facts or data to draw general conclusions. One limitation of this type of reasoning is that it does not prove its conclusions to be completely certain. Instead, it shows that a conclusion is probable on the basis of the evidence under consideration. Most scientific conclusions are based on data, so they use inductive reasoning, but deduction also plays a part.

science: (L. *scio*—to know), (L. *scientia*—knowledge)

1-14 Science involves systematically studying the natural world.

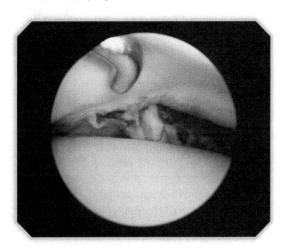

1-15 In an arthroscopy or colonoscopy, a tiny video camera is inserted into a patient's joint or intestines to diagnose potential problems and to pinpoint areas that need surgery.

Deductive reasoning proceeds from generalized premises to a specific conclusion.

Inductive reasoning proceeds from specific examples to a general conclusion. Deductive reasoning proves a conclusion to be true; inductive reasoning shows a conclusion to be probable or likely.

A Theory Goes Up in Smoke
Facets of Chemistry

Substances lose weight when they burn, right? After all, a hefty log turns into a small pile of ashes after a night in the campfire! Observations like this led many scientists to assume that burning always decreases the mass of a substance. Johann Becher and Georg Stahl theorized in the late sixteenth century that burning allowed a mysterious substance called phlogiston to escape. The loss of phlogiston supposedly accounted for the decrease in weight during burning. Yet one observant scientist saw something that threatened the widely accepted phlogiston theory. On November 1, 1772, the French chemist Antoine Lavoisier delivered a sealed note to the secretary of the French Academy of Sciences.

"About eight days ago I discovered that sulfur in burning, far from losing weight, on the contrary, gains it; it is the same with phosphorus. This increase in weight arises from a prodigious quantity of air that is consumed during combustion. The discovery, which I have established by experiment which I regard as decisive, has led me to think that what is observed in the combustion of sulfur and phosphorus may well take place in the case of all substances that gain weight by combustion."

Lavoisier had observed a significant gain in weight when phosphorus and sulfur burned. This observation contradicted the prevailing idea that all substances lost weight when they burned. Lavoisier recognized that this contradiction posed a problem well worth investigating.

Were sulfur and phosphorus isolated exceptions to the phlogiston theory, or was the entire theory faulty? After a careful study of the matter, Lavoisier proposed a daring hypothesis: substances gain something from the atmosphere when they burn. This idea went against the current theory

David, Jacques Louis (1748–1825). Antoine-Laurent Lavoisier (1743–1794) and His Wife (Marie-Anne-Pierrette Paulze, 1758–1836), 1788. Oil on canvas, $102\frac{1}{4} \times 76\frac{5}{8}$ in. (259.7 × 194.6 cm). Purchase, Mr. and Mrs. Charles Wrightsman Gift, in honor of Everett Fahy, 1977 (1977.1). The Metropolitan Museum of Art, New York, NY, U.S.A.

and would not gain acceptance without experimental proof. To prove that his hypothesis was correct, Lavoisier knew he must first identify the "something" from the atmosphere that substances gain when they burn.

At first, Lavoisier suspected that carbon dioxide caused the increase in mass. Several experiments soon proved that carbon dioxide would not support combustion of any type. The discovery of oxygen by fellow chemist Joseph Priestley gave several valuable clues. When heated, mercuric calx releases large quantities of oxygen gas and leaves silvery, elemental mercury behind. Lavoisier duplicated Priestley's procedure with this red-orange compound. The results of the experiment gave Lavoisier the necessary insight into his problem. Perhaps oxygen was the component of air that combined with burning substances and increased their masses. To test his idea, Lavoisier developed a controlled experimental procedure to produce mercuric calx from mercury and ordinary air and then from mercury and pure oxygen. Both procedures produced the same compound. Oxygen was the substance in the air that combined with substances as they burned!

Lavoisier began with a problem, made observations, researched and defined the problem, and then developed a hypothesis. He next conducted experiments in which he gathered more observations, chose his solution, and then verified it. His consistent use of the scientific process in his chemical investigation did not guarantee success, but it did keep him on the right path. With the help of other researchers and a questioning mind, Lavoisier made a discovery that changed the theoretical framework of chemistry.

EXAMPLE PROBLEM 1-1
Reasoning
Determine whether the following arguments use deductive or inductive reasoning.
a. When potassium chloride, sodium chloride, or calcium chloride is mixed in water, it dissolves readily. Because magnesium chloride and barium chloride are similar salts, they also should dissolve in water.
b. Any substance that floats on top of water has a lower density than water. Vegetable oil floats on top of water. Therefore, vegetable oil is less dense than water.

Solution
a. This argument attempts to show that the conclusion is probable or most likely based on similar situations, but it cannot prove that these other salts will dissolve in the same manner. Therefore, the argument uses inductive reasoning.
b. This argument compares premises and arrives at a conclusion that must be true if the premises are true. Therefore, the argument uses deductive reasoning.

1.17 Model Making

Scientists use scientific models to make sense of data that they gather. Models give data a unifying theme, identify causes and effects, and suggest practical applications for the data. Models make data useful by establishing connections and directing predictions. Computer modeling of weather processes makes it more possible to predict weather accurately. Models can be used to predict the behavior of the New York Stock Exchange or the side effects that a medication can cause in the human body. Models are refined as man learns more and makes technological breakthroughs.

Models are always simpler than reality for the sake of **workability**. Their simplicity limits their application in some situations. The kinetic molecular theory discussed in Chapter 2 explains most properties of matter but not the behavior of certain gases under extreme conditions. Some of the models used to describe chemical bonding are valuable in explaining the changes that occur when atoms form bonds, but exceptions show their limits and lead to alternate models.

Some significant examples of models include the model of the universe, the model of the earth's interior, the model of the atom's structure, the big bang model, the periodic table, and the DNA double-helix model. You may have recognized the unbiblical big bang model in this list. It is included as a reminder that one's presuppositions greatly affect the models one develops.

> Workability is the characteristic that makes information useful or valuable in applying to other situations.

1.18 Experimentation

When you think of a scientist, you probably picture a person in a laboratory performing experiments. You can think of an experiment as scientific "detective" work in which you are trying to solve a mystery. An **experiment** is a repeatable method that involves observing a natural process, sometimes under controlled conditions, for the purpose of analysis. Experiments are an important part of

William of Ockham (1285–1349) was an English Franciscan friar who was a controversial but influential philosopher and theologian. Although he did not originate the principle known as Ockham's razor, he used it in his treatise titled *Commentary on the Sentences*.

the model-making process, although they are often used for other purposes. In a **controlled experiment**, only one condition is varied at a time to isolate and measure its effect on the outcome. When an explanation has been determined or data gathered experimentally, it is said to be **empirical**. Experimentation makes use of inductive reasoning by beginning with specific examples and ending with a general conclusion that is probably true. The greater the amount of empirical evidence, the more certain the conclusion will be.

A scientist will often make a hypothesis about the results of his experiment. A **hypothesis** is a simple, reasonable, testable statement that tries to predict the results of an experiment from the information available about a specific scientific problem. Sometimes a hypothesis takes the form of a model. The purpose of a hypothesis is to direct scientific research. Most importantly, a hypothesis must not violate known laws of science which apply to the system under study. A good hypothesis endeavors to account for known data more efficiently and simply than all previous hypotheses. This principle of logic, known as **Ockham's razor**, states that a hypothesis must make as few assumptions and be as simple as possible. This overarching principle governs how scientists should formulate hypotheses and design models. When presented with several explanations for a phenomenon, the most logical choice is often the simplest explanation.

Hypotheses must be tested. Controlled experiments are conducted by varying one condition at a time or observing natural phenomena to come to a conclusion. Results are used to predict future outcomes. Those predictions can then be tested themselves.

Hypotheses are sometimes tested using an experiment in which the conditions cannot be controlled, or a **natural experiment**. In natural experiments, scientists usually try to determine what variables are interacting, even though they cannot control them. Research in ecology, meteorology, and astronomy often involves natural experiments.

1-16 The discovery of the planet Neptune was the result of a natural experiment.

1.19 Conducting Surveys

Sometimes the only way to test a hypothesis is to conduct a survey. A **scientific survey** is a process that involves randomly selecting representative samples from a larger population to learn about the characteristics of the population as a whole. Surveys and experiments can work together. For example, a scientist might take a sample of a population and experiment on the sample to find out characteristics of the population.

In 1993, the Mississippi and Missouri river basins flooded, resulting in $15 billion of damage and 50 deaths in the upper Midwestern states. These floodwaters appeared to have contaminated non-public water supplies such as wells, springs, and cisterns. In nine states in that region, about 18% of the water supply came from these sources. In 1994, a survey of water quality was conducted in these nine states by collecting samples from domestic

1-17 The 1993 flood of the Mississippi and Missouri river basins caused contamination of private wells. In 1994, those wells were tested in a scientific survey.

wells. The survey found that many of the water supplies were contaminated with bacteria and fertilizer chemicals.

1.20 Formulating Results

After a scientist has experimented or conducted a survey, he must formulate results and communicate the findings to the scientific world. There are many avenues for this type of communication. Often scientists publish their work in scientific journals or on the Internet in forums. In both venues, knowledgeable peers review their work. This systematic accountability is called the *peer review* process. The publication of experimental findings gets useful scientific information into the hands of those who can apply it or use it to direct further studies.

If a scientist has discovered and tested some idea which consistently explains a certain phenomenon, it may be referred to as a **theory** by the scientific community. Since theories offer explanations for what we observe, theories can be thought of as scientific models. Theories are considered sound when they can make accurate predictions and can explain several different bodies of data. Sometimes a theory may try to connect many different areas, like the theory of evolution, string theory, or the theory of relativity. Theories are modified as new intuitive breakthroughs are made. Although a scientist can say that a theory is supported by proof, he cannot say that it is true. Theories serve a different purpose than hypotheses—they are overarching explanations of a body of data. Consequently, theories often suggest areas for further study.

A scientific **law** is a statement which describes observations made under many different conditions as a recognizable, repeating pattern in nature. Laws are often stated in mathematical terms that model natural phenomena. Generally, laws are considered very reliable statements of science to which there are no known contradictions. Examples of scientific laws include Newton's laws of motion and the laws of thermodynamics.

Scientific concepts do not generally progress from hypothesis to theory to law, though each of these does involve a decreasing amount of speculation. Hypotheses, theories, and laws serve different functions in the work of science.

1.21 The Scientific Method

Scientists have modified and expanded the inductive approach of the scientific process to fit their needs. The **scientific method** is one specific process within scientific inquiry that uses a general, inductive approach to discover information about our universe. It is commonly identified with the following steps: *recognizing a problem, making a hypothesis, making observations to test the hypothesis, organizing and analyzing data, and verifying the hypothesis.* Is this list different from one you have seen before? If so, don't be surprised. There are many versions of the scientific method and no standardized form. Most of these versions can be considered correct, since they describe the scientific process.

Theories are subject to one's presuppositions. There are serious critics of the theories of evolution and the big bang among Bible-believing Christians, as well as among some unbelievers.

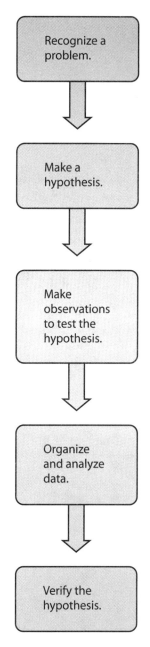

1-18 The scientific method is one tool of the scientific process that can be used to make scientific breakthroughs.

1C Section Review

1. What is the major difference between quantitative and qualitative data?

2. Determine whether the following statements use inductive or deductive reasoning.
   a. Dr. Blackford has performed a hundred cardiac bypass surgeries, and all were successful. When Dr. Blackford performs this surgery on my grandpa Harvey, it will be successful.
   b. Everything that the Bible teaches is true. The Bible teaches that man is sinful by nature. Therefore, it must be true that man is sinful by nature.
   c. If humanism is correct, then man is inherently good. Inherent goodness requires loving others; and yet war, deceit, and treachery have been continuous throughout human history. Thus, humanism must not be correct.

3. Design a survey to test if there are any gold deposits in the Grand Canyon.

4. Many people attack evolution by saying that it is "just a theory." How does that argument misunderstand the term *theory*? Why do you think that there is some confusion about the differences between hypotheses, theories, and laws?

5. (True or False) The scientific method is a set of steps that must be followed strictly in order to reach valid conclusions.

Chapter Review

Chapter Summary

- A Christian worldview gives the best reasons for studying chemistry. These reasons include declaring the glory of God, learning how to better love one's fellow humans, understanding that science is about making models, becoming aware of naturalistic biases (and how they differ from Christian biases), and preparing oneself to live for God with a knowledge of chemistry.

- Genesis 1:28 and Mark 12:30–31 reveal that all Christians should be interested in dominion science. Dominion science is scientific activity that seeks to obey the Creation Mandate for the glory of God and for the benefit of other humans.

- Scientists use models to understand the world. Scientific models are simplified representations that characterize a system or explain a phenomenon. Because science uses models, the goal of science is workability, not truth. The Bible is the only source for settled truth.

- Everyone who studies science has a worldview. A naturalistic worldview is ultimately dissatisfying and leads to despair. A Christian worldview states that humans are precious in God's sight, and it gives the human race hope.

- As a Christian studies chemistry, he prepares himself for the possibility of serving God as a chemist. Even if a Christian does not become a chemist, the study of chemistry can prepare him to live for God.

- Chemistry is the study of chemicals (matter) and the changes that they undergo. Matter is anything that takes up space and has mass.

- The field of chemistry has developed greatly throughout history. It began with metallurgy and the early apothecaries, was refined through the influence of Greek reason, and gained the experimental approach through the efforts of the alchemists. Lavoisier's scientific approach to chemistry and the work of Benjamin Rush and Friedrich Wöhler transformed the science into its modern form.

- Chemistry overlaps with the other sciences. Its six major branches are inorganic chemistry, organic chemistry, biochemistry, nuclear chemistry, physical chemistry, and analytical chemistry.

- Science is the total collection of knowledge gained through the systematic observation of nature.

- The process of science involves reasoning from particular observations to general conclusions. Scientists make observations of particular scientific problems using experiments and surveys. From their observations, they construct models, theories, and laws.

- Theories make models that try to explain scientists' observations. Scientific laws describe scientists' observations mathematically.

Coming to Terms

worldview	2
image of God	2
first commandment	2
second commandment	2
Creation Mandate	3
dominion science	3
scientific model	4
presupposition	5
bias	5
naturalistic worldview	5
scientism	5
Christian worldview	6
chemistry	8
matter	8
metallurgy	9
apothecary	9
alchemy	10
science	12
applied science	12
pure science	12
scientific questions	13
scientific problems	13
observations	13
objective	13
quantitative data	13
qualitative data	13
deductive reasoning	13
inductive reasoning	13
workability	15
experiment	15
controlled experiment	16
empirical	16
hypothesis	16
Ockham's razor	16
natural experiment	16
scientific survey	16
theory	17
law	17
scientific method	17

- The scientific method is one facet of the scientific process in which a scientist recognizes a problem, makes a hypothesis, makes observations to test the hypothesis, organizes and analyzes data, and verifies his hypothesis.

Review Questions

Concept Review

1. According to Romans 11:36, why does chemistry exist?
2. Why did the Lord Jesus closely associate the command to love others with the command to love God (Mark 12:30–31)?
3. Unlike chemistry, which gives us only useful models, Scripture gives us truth. Scripture's truth affects our study of chemistry in what four areas?
4. Identify the core problem in the following statement:
   > The Bible is often unscientific in its discussion of the natural world. For example, Numbers 11:33 states that while the Israelites were eating quail, God struck them with judgment and many of them died. There is no compelling reason to interpret this as damnation from some deity. It was most probably a case of mass food poisoning. Certain organisms or chemicals in the quail reacted with certain chemicals in the human body so that certain systems in the body were overwhelmed, and these humans expired. Their deaths have nothing to do with God; they have everything to do with unfortunate chemistry.
5. Name two ways that a Christian can serve God as a chemist.
6. By what criteria did the Greeks judge the validity of their scientific ideas? What criteria do modern scientists use?
7. Though alchemists are sometimes criticized, they developed several things that benefit modern chemists. List these benefits.
8. What is the difference between organic chemistry and inorganic chemistry?
9. Classify each of the following activities as pure chemistry, applied chemistry, or both.
    a. determining how much energy is released when iron forms rust
    b. determining how the size of silver grains on photographic film affects the resolution of the finished print
    c. isolating a chemical compound from the leaves of a newly discovered tropical plant
    d. developing a method to purify polluted well water in a small town
    e. determining what elements exist on a distant star
10. Give two examples of scientific questions.
11. What is the difference between inductive and deductive reasoning? Which type of reasoning do scientists use to develop theories and models? Which type of reasoning is used to make predictions from theories and models?

12. Design a controlled experiment to test the effect of acid rain on plants.

13. A dermatologist (scientist who studies skin) is doing a scientific survey of a population of people. His goal is to see the connection between the increase in skin cancer and the popularity of sunbathing. Which is the best hypothesis that he could propose before making observations? Why is this hypothesis the best one?
    a. If sunbathing is the largest factor in the recent increase in skin cancer, then I expect to see the most recent cases of skin cancer in younger people who spend significant amounts of time at tanning beds and outside sunning. The rate of skin cancer in elderly people, construction workers, park rangers, groundskeepers, and other people whose occupations require them to spend a great deal of time outside should be relatively stable.
    b. If sunbathing is the largest factor in the recent increase in skin cancer, then I expect to see an increasing rate of elderly people with skin cancer.
    c. If sunbathing is the largest factor in the recent increase in skin cancer, then I expect to see a direct correlation between hours spent tanning and occurrence of skin cancer.

14. Look at the example of a survey in Subsection 1.19. Why are the results of this survey somewhat questionable? What would allow the results to be more useful?

15. If scientific laws are generally reliable statements of science, can they be proved to be wrong? How?

16. Identify the five steps of the scientific method in the following example.
    NASA has been experimenting aboard the International Space Station to see how gravity influences the way that plants grow. How do plants know which way is up? In one experiment, astronauts are growing flax seed in a highly magnetic environment. The magnetic field affects certain slightly magnetic parts of the plant, called starch grains, but leaves other parts unaffected. Assume that 50 seeds are grown in the dark in orbit around the earth, free from gravitational attraction but subjected to strong magnetic fields. Suppose that scientists find that 42 of the plants seem to follow magnetic lines for a regular period of time but then change direction in a cycle-like pattern. The remainder of the seeds have no recognizable pattern in their growth, or the seeds did not germinate. They conclude that the starch grains do influence the direction of the plant growth, but there is another factor, not yet identified, that influences plant growth.

17. Choose a career involving chemistry that interests you. Consult the Internet to write a one-paragraph summary about that career.

18. Choose one of the branches of chemistry mentioned in Subsection 1.12. Write a one-paragraph summary about that

branch of chemistry, including its specific applications of the science and recent developments in that field.

True or False

19. It is possible to establish absolute truth using science.
20. All of science is about making models.
21. It is difficult for a Christian to be a chemist and not be distracted from his responsibility to share the gospel with unbelievers.
22. Chemistry is a relatively ancient science.

MATTER

2

DOMINION SCIENCE PROBLEM
Metal Fatigue
On December 21, 2005, a turboprop airplane, headed to the Bahamas from Miami Beach, plummeted into the sea after one of its wings sheared off mid-flight. All twenty people on board, including three infants, were killed. Most of the passengers had been Christmas shopping in Miami Beach. Investigations into the cause of the crash of the fifty-eight-year-old plane suggested that the wing broke off due to metal fatigue, a weakening of metal when exposed to prolonged stress. Is there some way to strengthen metal parts of aircraft so that accidents and loss of life can be avoided?

2A The Classification of Matter	
2B Energy and Matter	24
2C The States of Matter	31
Facet	39
Extreme Matter	40

2A Section Objectives

After finishing this section, you should be able to
- define matter.
- describe how to classify matter.
- explain the difference between physical and chemical properties.
- explain the difference between physical and chemical changes.
- draw and label a chart illustrating the classifications of matter.
- compare and contrast chemical symbols and formulas.
- interpret chemical formulas.
- identify the fundamental particles of elements and some compounds.

Faith or Science?

If belief in the creation of matter *ex nihilo* is a matter of faith, then what is the belief, asserted by evolutionary cosmologists, that matter is eternal or that it ultimately originated in a "big bang"?

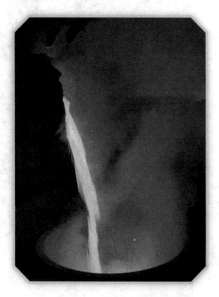

2-1 Processing iron involves both chemical changes (smelting the ore) and physical changes (solid to molten to solid).

2A The Classification of Matter

2.1 What Is Matter?

The matter that makes up the physical universe was created out of nothing (*ex nihilo*) by God. Outside of Scripture, there is no conclusive proof for or against that belief. That statement is made by faith, and it is the foundational truth of a Christian understanding of chemistry.

What is matter? **Matter** is operationally described as anything that occupies space and has mass. Volume and mass are the two measurable properties common to all matter. Matter is the basic building block of the universe; all scientifically observable things are made of matter.

You may have noticed that the definition of matter above is actually a description. It tells what matter does, not what matter is. To define matter is difficult. A *conceptual definition* of matter would require even more basic concepts, but in science there is no more basic concept than matter. For such terms scientists resort to operational descriptions.

2.2 Organizing the Study of Matter

The world that God spoke into existence contains a variety of materials. Worms, electrons, people, sand, uranium, trees, and sulfuric acid are just a few of the many forms that matter takes. Scientists have documented many millions of kinds of materials, both natural and artificial. Some of the basic materials may be in the same form as they were when created. However, nearly all accessible natural minerals have been partially or even significantly modified by the Genesis Flood. Humans also rework natural materials into other useful materials.

A study of all this matter needs the organization that a good classification scheme can provide. Although there are many possible classification systems, this text divides matter into two major categories: pure substances and mixtures. We can make distinctions between the two categories based on the physical and chemical properties of matter.

2.3 Physical and Chemical Properties

Properties are the distinguishing characteristics of matter. Scientists divide these characteristics into two classes: physical and chemical. The **physical properties** of a material are determined by how the particles that compose the material relate to one another. What does the material look like? How many particles are there? How closely are they packed together? Physical properties of matter can be measured or observed without changing the actual composition of the material. Color, shape, texture, physical state, odor, and taste are examples of properties that can be observed without making the material something different. The following are common physical properties of matter.

Density. The amount of matter packed into a given volume is density. Dense objects have a lot of matter packed into the space they occupy. Less dense objects have less matter in the same space. We know that wood is less dense than water because a piece of wood will float on water. Density will be explored further in Chapter 3.

Malleability. Materials that can be hammered easily into shapes or thin sheets are malleable. This property is possible because the connections between the particles that make up these materials are strong but allow some movement. Most metals are malleable. When Bezaleel the Hebrew followed God's orders to prepare the tabernacle, he beat gold into sheets to cover the shittim wood of the Ark of the Covenant (Exod. 37:1–2). Thirty grams of gold (approximately 1 oz) can be hammered out to cover 30 m², about the area of two standard bedrooms! Since it is such a soft metal, gold is usually mixed with another metal such as copper or silver to give it strength.

Ductility. Materials that can be drawn into long, thin wires are ductile. Ductility is another property of most metals. The same properties that make gold the most malleable metal also make it the most ductile. A single ounce of gold can be drawn into a fine wire that is 80 km (50 mi) long. Copper and aluminum are also very ductile.

Conductivity. The ability of a material to transfer heat or electricity between its particles is conductivity. For example, the handle of a metal spoon left in a hot pot on a stove becomes hot, but the handle of a wooden spoon does not. The metal conducts heat easily; the wood does not. Silver is the most conductive metal, but because of silver's high cost, copper is usually used for electrical wiring.

Matter has a second class of characteristics as well. **Chemical properties** describe how matter acts in the presence of other materials or how it changes composition when sufficient energy is added to it. In order to determine the chemical properties of a material, scientists must know the kinds of changes that the material can undergo. The reaction between gasoline and oxygen is *combustion*, or burning. Gasoline's strong reaction with oxygen is one of its chemical properties. Other types of matter break down or decompose when exposed to various forms of energy. For example, when an electrical current is passed through pure water, the water decomposes into the elements oxygen and hydrogen. Each substance has its own set of chemical properties.

The terms *physical change* and *chemical change* are closely associated with physical and chemical properties. Changes that occur in a material without changing the identity of the material are **physical changes**. Physical changes are changes in both the state and shape of a material. Boiling is a physical change in which a material changes from its liquid state to its gaseous state but in which its identity is not altered.

Chemical changes are changes in the identity of a material—changes that result in a different material with a different composition and properties. When iron rusts, it undergoes a chemical change. The iron particles combine with oxygen particles to form rust. Not only is rust a new substance totally different from oxygen or iron, but it also has a different physical appearance. Table 2-1 contains a comparison of physical and chemical changes.

malleable (MAL ee a bul)

ductile (DUK til)

A material can be in one of several physical states. Its state is determined by the proximity and motion of the particles of matter, as defined by its temperature and pressure. Under normal circumstances, a material may be a solid, liquid, or gas. There are six known states of matter. The other three states of matter are discussed in a facet in Section 2C.

2-2 The lid of a tin can undergoes a chemical change when it corrodes. Water undergoes physical changes when it changes between ice, liquid water, and vapor.

2-1	**Physical and Chemical Changes**	
	Physical change	Chemical change
Definition	a change in state or shape that does not alter the identity of the material	a change in the composition of the particles of a material that alters the identity of the material
Changes at particle level	The positions or motions of the particles may change; chemical bonds between particles remain unchanged.	The positions and chemical bonds of the particles change.
Observable changes	Shape, state of matter, or other physical property may change. Chemical properties are not altered.	New substances are formed that have different chemical and physical properties.
Examples	ice melting, wheat being ground to flour, sugar dissolving in water	metal rusting, oil burning, wood rotting, food digesting

2.4 Using Physical Changes to Solve Problems

Have you ever repeatedly bent a paper clip? What happens? The paper clip eventually breaks because of metal fatigue. Metal that experiences repeated tension can fail, and it did with the turboprop airplane that crashed on December 21, 2005.

The solution to this problem is a process called shot peening. Glass, ceramic, or steel balls only millimeters in diameter are fired at the metal object that needs to be strengthened. The balls pit the surface to form sections that are highly compressed and thus stronger. Recently, advances in technology peen metal surfaces with lasers instead of shot.

Shot peening is used to strengthen all kinds of metal parts, such as airplane wings and landing gear, automobile axles, railroad wheels, and turbine blades. If those parts weren't strengthened, their failure could cause great expense and possibly death. Lives could have been saved if the aircraft that crashed in December 2005 had been manufactured with wings that had undergone this procedure.

2.5 Pure Substances and Mixtures

Matter can be divided into two categories based on its physical and chemical properties: *pure substances* and *mixtures*. A **pure substance** consists of only one type of matter. Some pure substances can be separated into other kinds of matter, but only by a chemical change. Pure substances can be further classified into elements and compounds.

A **mixture** is a material that can be separated into two or more pure substances by physical changes. There are two types of mixtures—*heterogeneous* and *homogeneous*. **Heterogeneous mixtures**, if examined carefully, appear to consist of two or more materials segregated into distinct regions, called *phases*. Consider oil and vinegar salad dressing. When the dressing is left undisturbed, it separates into two different layers, the oil on top and the vinegar on the bottom. No matter how vigorously the dressing is shaken, a microscopic examination will always show two separate phases—oil and vinegar. The individual mineral particles of quartz, mica, and feldspar can be clearly seen as separate phases in a piece of granite—another example of a heterogeneous mixture.

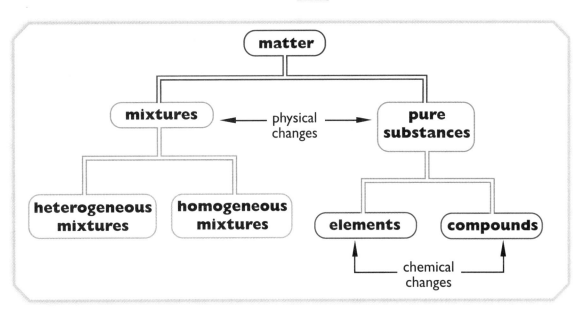

2-3 A chart showing the classification of matter

If a heterogeneous mixture consists of groups of solid particles suspended in a liquid or gaseous phase, it is called a *suspension*. We will discuss suspensions and other heterogeneous mixtures in Chapter 12.

On the other hand, a mixture that shows only a single phase is **homogeneous**. Such uniform mixtures of the particles of two or more pure substances are also known as *solutions*. The physical properties appear to be the same throughout. If you mix sugar with water and stir it until it is completely dissolved, it appears as a single liquid phase—sugar water. If gold and silver are melted together and thoroughly mixed, a single phase of white gold results. Air is an example of a solution of gases.

2-4 Oil and vinegar salad dressing is a heterogeneous mixture.

heterogeneous (HET er oh JEEN ee us): hetero- (Gk. *hetero*—different) + -gene- (Gk. *genos*—kind) + -ous

A suspension is a uniform mixture of solid particles suspended in a liquid or gas. The particles will eventually settle out due to gravity.

homogeneous (HOME oh JEEN ee us): homo- (Gk. *homo*—same, similar) + -gene- (Gk. *genos*—kind) + -ous

2.6 Elements and Their Symbols

An **element** is a pure substance that cannot be broken down into a simpler substance by ordinary chemical means. Men have known about gold, silver, sulfur, tin, and lead since biblical times. Men used gold, brass (copper and tin), and iron before the Flood. Scientists discovered many other elements shortly after the rise of modern science in the 1800s. In addition, other elements have recently been synthesized in the laboratories of nuclear physicists. You can see the

An element is a pure substance consisting of only one kind of atom.

> An **ion** is a charged particle consisting of a single atom or group of atoms bonded together that has a net electrical charge. The charge is the result of an unequal number of protons and electrons in the particle.

> monatomic (MON uh TOM ik)
> diatomic (DIE uh TOM ik)

> The subscript number in each chemical symbol shown in this paragraph indicates the number of atoms of the element that are present in the particle. If there is no subscript, it is understood that the element exists in the form of uncombined atoms.

> Atoms are the individual units of a pure substance that we call an element. In the same way, molecules are the individual units of a pure substance that we call a compound.

elements represented on any periodic table. Each element has unique chemical and nuclear properties.

An element is better defined as a pure substance that consists of only one kind of atom. For example, a lump of pure copper metal is made up of only copper atoms. **Atoms** are the basic particles that make up elements. Atoms have a nucleus containing positively charged particles called protons and neutral particles called neutrons. A cloud of negatively charged particles called electrons surrounds the nucleus. When the number of protons and electrons in an atom is balanced, the atom is neutral. When they are out of balance, the atom is charged in proportion to the imbalance of protons and electrons. Charged atoms are called **ions**.

The particles of elements in their natural state may consist of individual atoms or groups of atoms. If an element occurs naturally as individual atoms, the element is called a *monatomic element*. Monatomic elements are rare, because most atoms tend to bond with one or more other atoms. Examples of monatomic elements are neon (Ne), argon (Ar), and xenon (Xe). Elements whose atoms naturally bond into two-atom units are called *diatomic elements*. Oxygen (O_2) and hydrogen (H_2) are common examples of diatomic elements. Elements whose particles are normally composed of groups of three or more identical atoms are called *polyatomic elements*. For instance, sulfur often occurs in the form of eight atoms bonded into an S_8 unit. Oxygen can also exist as ozone (O_3) high in the atmosphere. Distinct groups of atoms bonded together are called **molecules**. Molecules may consist of atoms from one or many different elements.

Each element has a special **chemical symbol** that represents its name. The first letter of the name of the element often serves as its symbol (e.g., H for hydrogen, N for nitrogen, and O for oxygen). Frequently the names of more than one element have the same first letter. To avoid confusion in these cases, a second lowercase letter is used in the symbol. The second letter is usually related to the sound of the element's name. The single letter C stands for carbon, Ca for calcium, and Cd for cadmium.

The first letter of an element's symbol is *always* capitalized, but a second letter is *always* lower case. Careless writing of symbols can result in serious errors. Consider the symbol for the element cobalt: Co. If written carelessly as CO, this symbol would represent the compound carbon monoxide, a poisonous gas found in automobile exhaust.

Element names come from many different sources. For some elements that were known in ancient times, their Latin names serve as the basis of their symbols. For example, *cuprum*, the name for copper (Cu), comes from the Latin for "from the island of Cyprus." Elements are named after many things:

- colors: iridium (Ir) from the Latin for *rainbow*
- people: curium (Cm) for Pierre and Marie Curie, early researchers of radioactivity
- places: californium (Cf) for California
- heavenly bodies: helium (He) from *helios*, the Greek word for *sun*
- miscellaneous words: bromine (Br) from the Greek word for *stench*

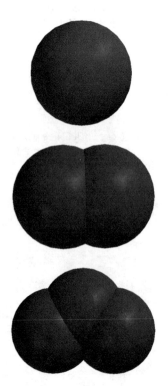

2-5 Elements in their natural states may consist of particles containing one or more atoms.

Today, it is generally accepted that whoever discovers a new element has the honor of naming it, subject to the approval of the International Union of Pure and Applied Chemistry (IUPAC). The periodic table of the elements in the back of this textbook displays the symbols for all known elements.

2.7 Compounds and Their Formulas

Pure substances that consist of two or more elements chemically combined are called **compounds**. In a compound, atoms from the different elements bond together. Some compounds form distinct particles called molecules. Other compounds form vast crystalline arrays with repeating arrangements of ions, called *formula units*. As you can imagine, there are many more compounds than elements. Just as symbols are used to represent elements, chemists use **chemical formulas** to represent the millions of kinds of molecules or formula units that make up compounds.

In compounds that form molecules, each molecule has a definite number of atoms. The molecules themselves are the smallest distinct particle of the compound. Therefore, molecular chemical formulas indicate the number of atoms of each element present in each molecule. To illustrate this, let's examine two compounds of oxygen and carbon. The formula CO represents carbon monoxide. This formula tells chemists that each molecule of this compound contains one carbon atom chemically combined with one oxygen atom. Another compound of carbon and oxygen is carbon dioxide (CO_2). This compound's molecule consists of one carbon atom combined with two oxygen atoms. Some common compounds, their formulas, and the atoms they contain are listed in Table 2-2.

> Compounds that form crystalline structures composed of ions do not have a definite number of atoms. In their chemical formulas, the element symbols and their subscripts represent the smallest ratio of the component atoms that uniquely represents the compound. These groupings of atoms are called *formula units* rather than molecules.

2-2 Some Compounds and Their Formulas		
Compound	Formula	Atoms
ammonia	NH_3	1 nitrogen, 3 hydrogens
rust	Fe_2O_3	2 irons, 3 oxygens
salt	NaCl	1 sodium, 1 chlorine
slaked lime	$Ca(OH)_2$	1 calcium, 2 oxygens, 2 hydrogens
sucrose (table sugar)	$C_{12}H_{22}O_{11}$	12 carbons, 22 hydrogens, 11 oxygens
water	H_2O	2 hydrogens, 1 oxygen

Numbers written at the lower right of a chemical symbol are called **chemical subscripts**. They indicate the number of atoms or groups of atoms in a chemical formula. A molecule of water (H_2O) contains two hydrogen atoms and a single oxygen atom (the 1 is assumed). When a subscript follows a group of symbols that is surrounded by parentheses, it refers to the entire group. A formula unit of $Ca(OH)_2$ consists of a calcium atom and two OH groups, for a total

> ### Problem-Solving Strategy 2-1
> Learning the chemical symbols of the elements begins with reviewing the periodic table of the elements frequently. As you read through the material in this chapter, look up any symbols that you do not recognize.

> ### Problem-Solving Strategy 2-2
> A subscript of a group of atoms contained in parentheses applies to all of the atoms in that grouping. In other words, each written subscript or implied subscript of 1 inside the parentheses is multiplied by the subscript outside the parentheses.

subscript: (L. *subscribere*—to write underneath)

of one calcium atom, two oxygen atoms, and two hydrogen atoms. A standard-sized number in front of a chemical formula is called a **coefficient**. A chemical coefficient applies to the entire formula immediately following it, multiplying each subscript. Thus 5 Fe_2O_3 (a form of rust) means five Fe_2O_3 formula units, for a total of ten iron atoms and fifteen oxygen atoms. You will use coefficients when you begin to write equations describing chemical reactions.

Example Problem 2-1
Deciphering Compound Formulas

How many atoms of each element are present in each of the following formulas?
a. $Na_2S_2O_3$ b. $Mg(NO_3)_2$ c. 3 $CaBr_2$

Solution
a. The subscripts show that two sodium (Na) atoms, two sulfur (S) atoms, and three oxygen (O) atoms are present.
b. No subscript after the magnesium (Mg) implies that only one atom is present. The subscript 2 after the nitrate (NO_3) grouping means that two nitrate groups are present, for a total of two nitrogen (N) atoms and six oxygen (O) atoms.
c. The coefficient 3 indicates that three of these $CaBr_2$ formula units are present. A single $CaBr_2$ formula unit has one calcium (Ca) atom and two bromine (Br) atoms. Tripling the number of atoms in one group gives a total of three calcium atoms and six bromine atoms.

Problem-Solving Strategy 2-3
Coefficients of chemical formulas multiply each single atom and the subscripts of multiple atoms and groups of atoms in the entire formula.

2-3 Elements, Compounds, and Mixtures

Elements	Compounds	Mixtures
H_2 molecules	H_2O molecules	H_2 and O_2 molecules
O_2 molecules	H_2O_2 molecules	H_2O and H_2O_2 molecules

2A Section Review

1. What is matter, and where did it come from?
2. Differentiate between physical and chemical properties of matter. Give two examples of each property.
3. What is a physical change? Give three examples of physical changes.

DS 4. What kind of change does shot peening cause in a sheet of metal? Why is it that type of change?

DS 5. What property of metal does shot peening use?

DS 6. How is shot peening an example of using technology to exercise dominion and to love your neighbor as yourself?

7. What are the two main classes of matter and what characteristics differentiate them?
8. Explain the difference between heterogeneous and homogeneous mixtures.
9. (True or False) Molecules consisting of two or more elements are the smallest identifiable particles of noncrystalline compounds.
10. For each of the following elements, give the term that best describes how its atoms exist under "normal" conditions. Be as specific as possible. For example, N_2 would be a diatomic element.
    a. oxygen (O_2)
    b. red phosphorus (P_4)
    c. sulfur (S_8)
    d. helium (He)
11. Give the number of each kind of atom in the following formula units or molecules.
    a. 5 NaCl
    b. $Ca(OH)_2$
    c. $HC_2H_3O_2$
    d. $4 Li_2O$

2B Energy and Matter

2.8 What Is Energy?

Energy is required to lift a rocket into space. A fast-moving soccer ball striking the side of your head hurts because it has a lot of energy. Sometimes when you touch a doorknob on a dry day, you receive a shock. Nuclear reactors, which generate nuclear energy, convert tiny bits of matter into immense amounts of energy. Einstein predicted this equivalence of energy and matter with his famous equation, $E = mc^2$, which relates energy (E) to the mass of matter (m) by the speed of light (c) squared. With these and so many other examples of energy, how can we define this all-important concept?

The concept of **energy**, like matter, is so fundamental to science that there are no terms that are more basic to define it. Scientists operationally describe energy as the "ability to do work." According to this description, energy is something that matter has—an ability—just like having a wallet full of cash gives you the ability to buy things. Energy that is expended doing something is called work. The value of what you purchase with cash should be equivalent to the amount of money you spend. Similarly, there is an equivalence between

2B Section Objectives

After finishing this section, you should be able to

- describe energy.
- list and give examples of the six common forms of energy.
- state the first law of thermodynamics in your own words.
- state the second law of thermodynamics in your own words.
- discuss several examples of how the second law of thermodynamics applies to nature and human activities.
- define thermal energy.
- discuss the similarities and differences in heat, thermal energy, internal energy, and temperature.
- define the thermodynamic terms *entropy*, *exothermic*, and *endothermic*.

energy and work done. In fact, scientists use the same dimensional unit to measure them—the joule (J). All forms of work involve a force moving matter through a distance.

It may seem frustrating to not be able to precisely define such an important concept, but such is the nature of science. In fulfilling the Dominion Mandate, we must often make operational use of science, realizing that ultimate answers come only from the Creator.

2.9 Forms of Energy

Energy is related to the force that matter generates and the resulting actions of that force. One of the most obvious forms of energy is *mechanical energy*. This kind of energy is commonly possessed by objects that are moving or have the potential to move. Mechanical energy can be either *kinetic energy* (the energy of motion) or *potential energy* (the energy of position). Our bodies and their various parts possess both kinds of mechanical energy.

The kinetic energy of atom-sized particles as they move in random directions is called *thermal energy*. We can perceive the transfer of thermal energy as heat and chill. If particles move in a periodic way like waves, they possess *sound energy*, or *acoustic energy*. Some sounds are loud (energetic) enough that our skin and other organs can sense the energy (e.g., earthquakes and explosions).

The forces that move electrically charged particles involve *electrical energy*. The movement of charged particles creates magnetic fields, which in turn affect other charged particles or magnetic fields. Because electricity and magnetism are so closely related, the energy associated with these two kinds of forces is called *electromagnetic energy*. Visible light is one form of electromagnetic energy.

When charged particles (electrons) move between atoms, atoms form and break bonds with other atoms. The energy responsible for these changes is *chemical energy*—a very important concept in this textbook. And finally, when the nucleus of an atom breaks apart or when particles are added to or removed from a nucleus, *nuclear energy* is involved in these processes.

As you can see, every form of energy can change the motion of matter to do work. When energy is absorbed by matter, it has done some kind of work. More importantly, energy can be converted into other forms. Read the box on Energy Conversions to see some common examples.

2-6 Tiny amounts of matter are converted to energy in nuclear reactors to propel nuclear warships for thirty or more years without refueling.

The six most common forms of energy:
- mechanical
- thermal
- acoustic
- electromagnetic
- chemical
- nuclear

2.10 The Conservation of Energy

As you can see from Figure 2-7, energy is continuously transferred and converted from one form to another. In the early nineteenth century, scientists developed the science of **thermodynamics** to study the movement and conversion of energy, especially thermal energy. This new field of study was a direct consequence of the development of the steam engine during the Industrial Revolution. The laws of thermodynamics seem to govern nearly all areas of science. No exception to these laws has ever been observed in natural processes; they apply to every field, at all times, in every instance.

Earlier we saw that energy can be converted from one form to another. During these conversions, energy is conserved in every process.

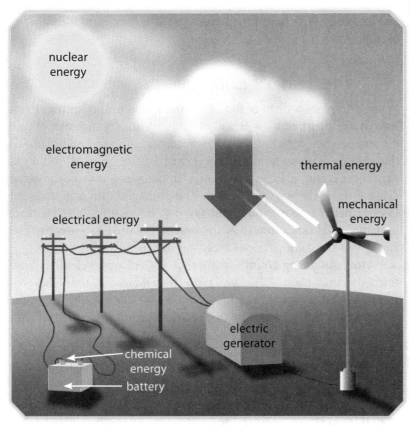

2-7 Energy conversions occur all around us.

Energy Conversions

Atomic nuclei deep in the sun undergo a nuclear reaction, releasing nuclear energy that is converted to thermal and electromagnetic energy. At the surface of the sun, electromagnetic energy is radiated into space. When it reaches the earth, that energy is absorbed by the atmosphere, ground, and water, which convert the radiant energy into thermal energy. The atmosphere is unevenly heated, causing density and pressure differences. Winds, which have kinetic energy, blow from high to low pressure areas. As wind blows through a wind turbine, one form of kinetic energy is converted into another, causing the wind turbine to spin. The turbine turns an electrical generator, rotating magnets inside a coil of wires and producing an electrical current (converting mechanical energy to electromagnetic energy). The current travels through wires to your house, releasing heat in the wires along the way due to electrical resistance. In your house, among other things, the electricity charges the battery in your cell phone, and is converted into chemical energy. The electrical current also is converted into visible light (electromagnetic energy) in your house lights and mechanical energy in your ceiling fan.

In other words, in a conversion the sum of all kinds of energy produced always equals the total energy consumed. Many careful measurements have shown that energy can be neither created nor destroyed but can only change from one form to another. This principle is known as the *law of energy conservation*.

2-8 These windmills transform wind energy into electrical energy.

Since Einstein's equation reveals the equivalence of matter and energy, the law of energy conservation is sometimes called the **law of conservation of mass-energy**. *Matter and energy can neither be created nor destroyed, only converted from one form into another.* This principle is also known as the **first law of thermodynamics**. Apart from divine intervention through miracles, the total amount of mass and energy has remained constant since the end of the sixth day of Creation (Gen. 2:1–3). All energy changes that have occurred since that time, either by natural processes or by man, have merely changed one form of energy or matter into another form of energy or matter.

The first law of thermodynamics is problematic for the evolutionist regarding the origins of the universe. He cannot explain the original source of matter and energy from a purely naturalistic perspective.

thermodynamics (THER mo dye NAM iks): thermo- (Gk. *therme*—heat) + -dynamics (Gk. *dunamikos*—powerful)

Can you give some examples of Bible miracles in which the total amount of mass or energy in the world has been changed since Creation?

Naturalism is the belief that all observations can be explained on the basis of natural cause and effect relationships. It leaves no room for the supernatural.

A creationist has no difficulty with this idea. Because both creation and scientific laws have their origin in God, they do not contradict one another.

2.11 The Law of Entropy

When you place a drop of food coloring into a beaker of water and leave it undisturbed for a period of time, eventually the color evenly disperses. The particles of color spread out and are less organized than before. This natural process is called *diffusion*. You would not expect the color particles to organize on their own into a drop-sized volume somewhere in the beaker. The probability of such an event is extremely remote—you could say it is impossible. The measure of disorder or randomness in the beaker is called its **entropy**. The entropy of the color particles increases over time.

After observing countless instances of entropy, scientists have formed a conclusion. *All natural processes tend toward the highest entropy (greatest disorder) and minimum usable energy.* This principle is known as the **second law of thermodynamics**. One significant consequence of the second law pertains directly to chemistry. During any energy conversion, some of the original energy is changed to an unusable form. For example, car engines produce mechanical energy from fuel, but they also produce heat and exhaust as waste. The thermal energy eventually ends up in the atmosphere and becomes unusable. At the same time, the combustion products released as exhaust have very little usable chemical energy remaining, and their particles are randomly dispersed throughout the air. Thus, they have high entropy and are minimally usable according to the second law of thermodynamics.

In order to simplify the analysis of a phenomenon, scientists define the boundaries within which the analysis is limited. Everything inside the boundaries is called the system of interest, or just the **system**. A system could be a beaker on a laboratory bench, a rocket engine, or the entire rocket. A system can be a distinct part of an object, such as the burner of a stove. It can also be a small portion identified by imaginary boundaries within a larger volume of a substance, such as a volume of air within the atmosphere. Throughout this discussion of energy, the system is assumed to be something with distinct boundaries, whether an object or a sample of a substance.

Every natural and artificial system is subject to the second law. Mountains wear away, buildings crumble, clothing wears out, and living organisms age and die. True, for short periods construction or growth might exceed decay. However, any decrease in entropy in one part of the universe comes at the expense of increased entropy elsewhere. For example, the growth of our bodies produces more organization in the form of complex organic molecules and highly organized structures, but it also produces unusable thermal energy and a greater number of simpler, more disordered waste products. In every case, the second law prevails.

Long before scientists formulated the second law of thermodynamics, the Psalmist recognized that the creation is wearing out and will not last forever: The heavens and earth "shall perish . . . yea, all of them shall wax old like a garment" (Ps. 102:26). He contrasts this

2-9 Scientists expect to find causes for the effects (phenomena, processes, objects) that they observe. What was the ultimate cause of everything?

Copyright © Boeing

2-10 The entropy of the fuel materials in the rocket engine dramatically increases as they burn and are discharged from the rocket.

A **system** is a portion of the universe under study.

fleeting creation with its eternal Creator. When the majestic mountains, quiet forests, resplendent star-filled sky, and the fascinating world of microorganisms are cast away like a worn-out shirt and replaced with a new heaven and a new earth, the Creator will be just as He has always been. Hebrews 1:11 quotes Psalm 102, applying the comparison of the decaying creation with the everlasting God to our Lord Jesus Christ.

2.12 Thermal Energy, Temperature, and Heat

All matter is constructed from particles, whether atoms, molecules, or combinations of both. This *particle model of matter* has been proposed in various forms since at least the fifth century BC. With the advances of quantitative chemistry early in the 1800s, scientists accepted the particle model of matter on experimental rather than philosophical grounds.

There is strong evidence for the continuous movement of particles in matter. For example, the odor of perfume diffuses from an open bottle, and steam droplets evaporate. Under a high-powered light microscope, you can see minute bits of matter in water being jostled by extremely small water particles. Botanist Robert Brown observed this in 1827. The random jiggling of microscopic matter is now known as *Brownian motion*.

In the nineteenth century, thermodynamics proposed a new model of heat. Scientists agreed that particles of matter are in constant motion. At the particle level, this motion may occur in straight lines (translational motion), as small, random vibrations (vibrational motion), or as a spin (rotational motion). The sum of all these motions determines the total **kinetic energy (KE)** of the particle.

The kinetic energy of a system is proportional to the mass (m) of the system, which is assumed to be constant, and the square of its velocity (v). Symbolically, this relationship is

$$KE = \tfrac{1}{2}mv^2.$$

You can see that for a given particle, its kinetic energy depends only on its velocity. Because the velocity is squared, doubling its speed quadruples its kinetic energy.

It is very difficult to measure the energy found in each particle of an object or system. It is much easier to measure the energy of the whole object or system. The **internal energy** of an object is the sum of the potential and kinetic energies of its component particles. The combined energy of a system's particles contributes to a dimension that *can* be measured—its **temperature**. A system's temperature is proportional to the *average* kinetic energy of its particles. Temperature is a measure of "hotness" or "coldness."

When hot soup is placed in a cold bowl, the soup cools and the bowl grows warmer. Obviously, something was taken away from the soup and added to the bowl. Since temperature is related to internal energy, it is clear that energy was transferred from the soup to the bowl. This form of energy is called thermal energy. **Thermal energy** is the sum of the kinetic energy of an object's particles. Finding out the kinetic energy of particles is difficult, so measuring thermal

2-11 The increased order represented by the structure of this ship's hull comes at the expense of the much greater decrease in order of the trees from which the lumber came as well as the energy expended to build the hull.

Robert Brown (1773–1858) was an English botanist who first described Brownian motion while studying pollen grains under a microscope.

The thermal energy of a system is the sum of the kinetic energies of all particles in the system. It is far more readily measured as the energy that is absorbed or released by a system that changes the kinetic energies of the particles in the system.

2-12 Thermal energy always flows from a higher temperature to a lower one.

Heat can correctly be used as a verb to describe a process, such as "heating a beaker." In this sense, it is the opposite of cooling and means to add thermal energy to something.

exothermic: exo- (Gk. *exo*—without, outside) + -thermic (Gk. *therme*—heat)

endothermic: endo- (Gk. *endon*—within, inside) + -thermic (Gk. *therme*—heat)

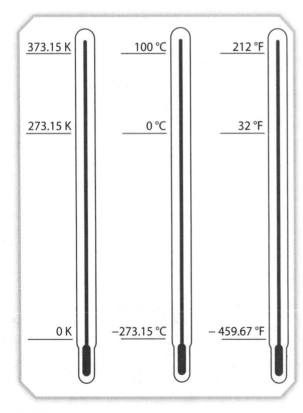

2-13 Comparison of temperature scales

energy is difficult as well. It is far easier to measure changes in thermal energy. When thermal energy is added to a system, its temperature rises. When thermal energy is removed from a system, its temperature falls. Thermal energy can flow only when there is a difference in temperature. It always flows from the higher temperature to the lower, never the other way.

The transfer of thermal energy between two objects or substances is called *heat transfer*. Within a system, the change in thermal energy is sensed as a temperature change and is called **heat**. The word *heat* is often used interchangeably with "temperature" and even "thermal energy." In science, this usage is incorrect. Heat is the quantity of thermal energy transferred from one object to another.

The amount of heat transferred to or from a system is related not only to the temperature difference that causes the heat transfer, but also to the mass of the system. A thimble full of hot water and a bathtub full of water at the same temperature contain different amounts of thermal energy because the bathtub contains more particles of water than the thimble. Spilling the thimble of hot water on yourself might cause slight discomfort. Immersing yourself in the bathtub of hot water could cause serious injury.

Every chemical change either releases or absorbs some form of energy. Chemical changes that liberate energy, such as burning natural gas or the explosion of a firecracker, are called **exothermic**. Changes that absorb energy are called **endothermic**. Chemical cold packs used to treat sprains feel cold because the chemical change absorbs thermal energy from its surroundings. We will discuss exothermic and endothermic reactions in Chapter 13.

2.13 Measuring Temperature and Thermal Energy

All matter is moving and has some kinetic energy. Instruments convert this kinetic energy into a measurement using a property of matter directly related to temperature. One such property is volume. In Chapter 11 you will study how the volume of liquids and solids generally increases proportionally with temperature. If a liquid is enclosed in a thin rod of glass, a small temperature change produces a measurable change in the height of the liquid column. This change in height can be compared to a calibrated scale. Within the measuring limits of the instrument, called a *thermometer*, the change in the height of the liquid column is proportional to the temperature change.

The two temperature scales used most often in chemistry are the Celsius scale and the Kelvin scale. The **Celsius scale** uses two reference points to anchor the temperature scale. The freezing point of pure water is its zero point, and the boiling point of pure water is 100 °C. These two reference points can be reproduced in any laboratory in the world, which makes the temperature scale especially useful for scientific work. The Celsius temperature scale contains both positive and negative temperatures, which is not a problem for observing temperatures. But if a calculation involves temperature, the negative numbers can often cause problems.

For this reason, a temperature scale was proposed in the 1800s that consisted of only positive temperature values, called the *absolute* or **Kelvin scale**. This scale was named after the prominent Scottish physicist Lord Kelvin. A temperature change of one Kelvin is equal to one Celsius degree. The Kelvin scale has two reference points. The first is the triple point of water, equivalent to 0.01 °C or 273.16 K. The triple point of water is readily measurable. The second is called **absolute zero**. Absolute zero, or 0 K, is equivalent to −273.15 °C. Absolute zero is the theoretical temperature at which all molecular and atomic movement ceases. The **third law of thermodynamics** states that it is impossible to reach absolute zero, so this reference point is unattainable.

As you can see from the relationships between the Kelvin and the Celsius scales in Figure 2-13, the numerical value of a Kelvin temperature is always 273.15 degrees more than the equivalent Celsius temperature. Note that Kelvin temperatures are *never* negative, and that their numerical values are *always* greater than the corresponding Celsius temperature. This observation should help you when working conversion problems. When required, you can make temperature measurements in degrees Celsius, then convert them to kelvins. The equations for conversion between Celsius and Kelvin temperatures are

$$T = t_C + 273.15° \text{ (in K) and}$$

$$t_C = T - 273.15° \text{ (in °C),}$$

where T is the Kelvin temperature and t_C is the Celsius temperature.

Although the Fahrenheit scale is not usually used for scientific purposes, it is the scale most Americans know best. Choose from these formulas to convert between Celsius and Fahrenheit readings.

$$t_F = (1.8 \times t_C) + 32$$

$$t_C = \frac{t_F - 32}{1.8}$$

2-14 Various instruments are used to indicate temperature.

Note that the degree symbol is not used when writing Kelvin temperatures. Degrees in the Kelvin scale are referred to only as "kelvins," not "degrees Kelvin."

EXAMPLE PROBLEM 2-2
Temperature Conversions

a. While performing an experiment, you collected a sample of gas in a bottle immersed in 32.50 °C water. What is this temperature in kelvins?
b. The temperature of seawater jetting from a "black smoker" vent on the bottom of the ocean on the Mid-Atlantic Ridge was measured at 671.50 K. What is this temperature in degrees Celsius?

Solution
a. Recall that the Kelvin temperature is always higher than the equivalent Celsius temperature. Therefore, use this formula.
$T = t_C + 273.15°$
$T = 32.50 °C + 273.15°$
$T = 305.65$ K
b. The Celsius temperature is always less than the equivalent Kelvin temperature. Therefore, use this formula.
$t_C = T - 273.15°$
$t_C = 671.50$ K $- 273.15°$
$t_C = 398.35 °C$

Problem-Solving Strategy 2-4
If required to convert from Celsius to Fahrenheit temperatures or from Fahrenheit to Celsius, use the appropriate conversion formula. If converting from degrees Fahrenheit to kelvins, first convert the Fahrenheit temperature to Celsius, then convert the Celsius temperature to Kelvin temperature.

Problem-Solving Strategy 2-5
When converting between Celsius and Kelvin temperature scales, round the final result to the precision of the original measurement. The "sizes" of the degrees on both scales are exactly the same. The conversion is just shifting the zero point, so the measured precision does not change during the conversion.

The modern metric system, called the *Système International d'Unités*, which is French for the International System of Units, is commonly abbreviated as simply the SI.

The joule (JOOL) is the SI unit of work and of any kind of energy. The joule is named for the English physicist James Prescott Joule, who distinguished himself by his scientific integrity and his precise measurements.

Calories and joules are related by this definition: 1 cal = 4.184 J (exactly).

The standard metric system (SI) unit of measurement for energy and work is the **joule (J)**. Another energy unit that is used less today than in the past is the **calorie (cal)**—the amount of heat transfer required to change the temperature of one gram of water one degree Celsius. Larger units called kilocalories (kcal), which are a thousand times as large, were used for most chemical applications before the adoption of the SI. The kilocalorie is equivalent to the Calorie (note the capital C), which is still used in reference to the energy content of foods.

2B Section Review

1. Is energy something a system does or something it has? Explain.
2. Name the principal type of energy involved in each of the following:
   a. vibrations from a loudspeaker
   b. a pot of near-boiling water
   c. a camera flash
   d. the source of an explosion of dynamite
   e. changes occurring at the center of a star
   f. a rolling ball
3. What specifically does thermodynamics investigate? What invention spurred its development?
4. State the first law of thermodynamics. What does this law imply for the origins of matter and energy?
5. Compare the entropy of a firecracker before it is lighted to its entropy after it explodes. Explain the difference.
6. State three evidences that the particles of substances are in continuous motion.
7. Explain the difference between temperature and thermal energy.
8. Since heat transfers only take place between systems with different temperatures, explain why the third law of thermodynamics is supported by our current understanding of matter.
9. Convert the following temperatures from Celsius to Kelvin scales. Express your answers to the nearest hundredth of a degree.
   a. 0.00 °C (freezing point of water)
   b. 22.22 °C (room temperature)
   c. 100.00 °C (boiling point of water)
   d. −218.30 °C (melting point of oxygen)
10. Convert the following temperatures from the Kelvin to the Celsius scales.
    a. 496 K (temperature of rock at 5.9 km depth)
    b. 2.726 K (average temperature of deep space)
    c. 1783 K (melting point of steel)
    d. 273.16 K (triple point of water)

2C The States of Matter

2.14 Common States of Matter

According to the particle model of matter, all matter is composed of submicroscopic particles (atoms, molecules, or ions) that are in constant motion. If energy is added to these particles, their motions increase. The greater the amount of energy added, the faster the resulting motions will be. If particle motions were not in some way limited, matter would fly apart. Instead, forces between atoms limit the movement of the particles toward and away from each other.

Scientists call the model described by these ideas the **kinetic-molecular theory**, since it describes the motion of particles in matter. According to the kinetic-molecular theory, particles in a **solid** have relatively little kinetic energy compared to the attractive forces that are present between particles. These attractive forces overpower the movements of the particles. The forces keep the particles of a solid in relatively fixed positions with set distances between them. Solids thus have several distinct properties. They tend to have a fixed volume and a rigid, fixed shape. In addition, solids are relatively incompressible because the structure of atoms makes them resist being squeezed more compactly. We will discuss the structure of atoms in Chapter 4.

When sufficient thermal energy is absorbed by a solid substance, its particles gain enough energy to partially overcome the attractive forces between them. The particles are no longer held rigidly in fixed positions, so they can move about somewhat. But because the attractive forces have not vanished, the particles remain close to each other. The solid has become a **liquid**. The melting of candle wax demonstrates this change. Because the particles in liquids are still close to each other, the volume of a liquid is relatively constant—it is nearly incompressible. The shape of a liquid, however, is determined by the container that holds it. In addition, liquids can flow when a force is applied. For example, gravity causes liquids to flow.

2-15 Can you see water in all three of its states in this picture?

When particles of a liquid acquire additional thermal energy, they are eventually able to completely overcome the attractive forces between them. When liquid particles have enough velocity to escape from their partners, they form a **gas**. The particles of a gas possess a large amount of kinetic energy. They move rapidly and randomly across great distances. At 25 °C, the average velocity of an oxygen molecule is 444 m/s (993 mi/h). Gases are highly compressible, have no definite shape, and eventually diffuse to

2C Section Objectives

After finishing this section, you should be able to

- list the key features of the kinetic-molecular theory of matter.
- relate a system's temperature and the distance between its particles to the state or phase of the system.
- describe the four most common states of matter and their general properties.
- properly use the terms describing changes of state and the temperature points where these changes occur.

Descriptive Terminology

- *molecular*—that which pertains to molecules
- *atomic*—that which pertains to whole atoms
- *subatomic*—that which pertains to particles making up atoms (e.g., protons, neutrons, and electrons)
- *nuclear*—that which pertains only to the nucleus and its major components (e.g., protons and neutrons)

There are two instances where the shape of a liquid is not determined by its container. Both are caused by the surface tension of the liquid. Very small drops of a liquid tend to form nearly spherical balls when resting on a surface. In weightless conditions, even large volumes of liquids form perfect spheres if not constrained by a container.

Technically, a gas diffuses quickly to fill a closed container if it has been evacuated of other gases first. In a container with multiple gases, the effect of gravity on their different densities greatly affects the rate at which the gas diffuses.

completely occupy a closed container. Since gas particles are highly mobile, not only do they flow when forced to, but their unrestrained movement allows them to move from one point to another even in the absence of a force.

Under certain extremely energetic conditions, so much energy is absorbed by an atom that some or all of its electrons are stripped away, forming a positive ion. These free, energetic ions and a seething cloud of electrons form a state of matter called **plasma**. Plasma exists on the surface and in the atmosphere of the sun and other stars. In fact, because of the immensity of stars compared to their planets, the most common state of matter in the visible universe could be the plasma state, even though it seems unusual to us. Closer to home, plasmas are found in plasma screen televisions, fluorescent lights, and neon signs, and they are used in certain industrial metal-cutting processes. Plasmas are being used in research into nuclear fusion, where temperatures of over 100 million degrees are required to sustain the plasma.

Extreme Matter
Facets of Chemistry

The three everyday states of matter—solid, liquid, and gas—are familiar to all. The fourth, plasma, is the most abundant state of matter in the universe, though there are only a few common examples on earth.

Bose-Einstein condensation at 400, 200, and 50 nano-kelvins

As our technology grows, scientists are also beginning to explore more states of matter at opposite ends of the energy spectrum: Bose-Einstein condensates and quark-gluon plasma.

In Section 2B we discussed the concept of absolute zero. What happens to matter when it approaches that temperature? In the late 1920s and early 1930s Albert Einstein, building on the work of Indian physicist Satyendra Nath Bose, predicted that a strange kind of matter would exist at temperatures that approach absolute zero. He theorized that at that temperature, all atoms in a system would exist as a "super-atom." In other words, they would all act as one single atom.

In 1995, researchers at the University of Colorado in Boulder, led by Carl Wieman and Eric Cornell, were the first to succeed in supercooling rubidium atoms to just a few billionths of a kelvin above absolute zero, and they observed the phenomenon that Einstein had predicted. They named this phase of matter the **Bose-Einstein condensate**. To cool the rubidium atoms, a special set of laser beams held the atoms in place. The light particles (photons) glancing off the atoms cooled the atoms by absorbing energy. In the final stages, magnetic evaporative cooling brought the rubidium atoms to near absolute zero.

Because only a few elements have been converted to this extreme

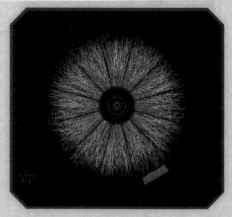

Courtesy of Brookhaven National Laboratory

Collision pathways from RHIC

state and because only a few million atoms are supercooled at a single time, progress in research is slow. Researchers in this field are continuing to study Bose-Einstein condensates to determine their physical and chemical properties.

2-4 The States of Matter

State	Particle Positions	Characteristics
solid		definite shape and volume, difficult to compress, packed close together, rigidly held
liquid		definite volume, assumes shape of container, packed close together, difficult to compress
gas		restricted only by its container as far as shape and volume; particles are easy to compress
plasma		gas-like state at very high temperatures; particles are ions, electrons, and neutral atoms traveling at extremely high speeds, no definite shape or volume, can be affected by magnetic field

More recently, physicists have theorized that if a substance absorbed enough energy, its protons and neutrons would shred and reveal the building blocks of matter. In 2005, scientists announced that they had accomplished this feat with the formation of a **quark-gluon plasma** in the Brookhaven Relativistic Heavy Ion Collider (RHIC)—an "atom smasher."

Each proton or neutron is theorized to be made up of three quarks. The quarks are the "elementary particles" that build protons and neutrons. Gluons are other particles that physicists theorize hold or "glue" quarks together. At the RHIC, scientists succeeded in accelerating two beams of heavy gold ions to almost the speed of light and then caused the ion beams to collide. The beams are shot in opposite directions around a 2.4 mi ring, and they collide six different times as their paths cross. The tubes carrying the ion beams are surrounded by electromagnets. Ironically, the electromagnets are supercooled to just 4.5 K above absolute zero to reduce electrical resistance and maximize the magnet strength.

The resulting collisions release thermal energy up to 150,000 times hotter than the center of the sun. Fortunately, the actual amount of energy is small and lasts only millionths of a second. Because the collisions happen so quickly, scientists are limited to studying some of the end products that are left in the collision area as it cools. Although the scientists originally announced that they had created quark-gluon plasma, further studies revealed that the new state of matter was not gaseous as predicted. Instead, it seemed to behave more like a special type of liquid. This high-energy liquid contained particles that did not move randomly but were coordinated like a school of fish responding as a single organism. For this reason, scientists call their discovery a "perfect" liquid.

In yet another example of how a researcher's worldview affects his work, this field is motivated by evolutionary assumptions. The main purpose for this research is to investigate the nature of matter just microseconds after the hypothesized big bang. According to the beliefs of many evolutionary scientists, this event was the origin of our universe.

Courtesy of Brookhaven National Laboratory

RHIC tunnel

2.15 Changes of Common States of Matter

As a gas cools by transferring thermal energy or converting it into other forms of energy, its particles slow down. Lacking the necessary energy to resist the attractive forces between them, the particles begin clumping and then finally draw together into drops recognizable as the liquid state of the substance. This process occurs when fog droplets or dew forms. The change of state, or phase change, from gas to liquid is called **condensation**. Condensation occurs frequently in our homes during the winter months when the warm, moist air inside the house comes into contact with a cold windowpane. The opposite process to condensation is **vaporization**, in which a liquid changes into a vapor or gas.

When particles in a liquid lose thermal energy, their thermal motion slows to the point that particles begin sticking to each other as attractive forces overcome their kinetic energy. More particles accumulate in this manner as their average kinetic energy drops. Eventually, large numbers of particles are held rigidly in place. The substance **freezes** as it enters the solid state. When the opposite process occurs, a solid **melts** into a liquid. Melting and freezing normally take place at the same temperature.

Under certain conditions, a solid may change directly into a gas, or a gas may change directly into a solid, without passing through the liquid phase. The change from a solid to a gas is **sublimation**. The change from a gas directly to the solid state is called **deposition**. Sublimation from a solid to a gas occurs even at temperatures well below the melting point of most solids. Recall that the temperature of a system is related to the average kinetic energy of its particles. Some particles can individually gain kinetic energy well above the average. They then break away from the attractive forces at the surface of the solid, entering the gaseous state. This process removes material from the solid. Examples of common substances that sublime directly from a solid to a gas are moth balls, solid air fresheners, and even ice cubes after an extended period in a freezer. These solids gradually convert to a gas and eventually "disappear." The deposition of gaseous water vapor to solid ice is the principal process in the formation of snowflakes and frost. The deposition of water vapor can cause rapid ice buildup on cold aircraft wings during flight. The ice buildup can be hazardous if the ice is not removed.

2-16 Three common forms of plasma: sun and stars (center), neon signs (left), and the medium through which lightning passes (right).

A vapor is not normally visible. The steam "vapor" that one sees emitting from a tea kettle or on the surface of a pond in winter is not true vapor but fog-like droplets of liquid water suspended in air.

sublimation (sub lih MA shun): sub- (L. *sub*—up, under, up to) + -lima- (L. *limen*—lintel) + -tion. The word was originally used in the sense of something rising to the highest levels. Sublimated crystals were often found high up on the walls of the crucible in which a substance was heated.

2-17 Dry ice is frozen carbon dioxide that sublimes at room temperature.

2C Section Review

1. List the known states of matter in order from the most energetic to the least.
2. Discuss why temperature is the most important factor in determining the state of a substance.

3. Compare and contrast gases and plasmas.
4. Name the changes of state that are described by the following processes.
    a. Pond water solidifies on a cold winter day.
    b. The fragrance of an open perfume bottle is smelled from across the room.
    c. Water leaves a teakettle that is sitting on a hot stove.
    d. Frost patterns appear on a cold window pane in winter.
    e. Water mist collects on the bathroom mirror after a shower.
    f. You smell the odor of moth balls when you open a clothes box.
    g. As a candle burns, wax drips down the sides of the candle.
5. (True or False) Liquids flow only when a force acts on them.

Serving God as a Chemistry Teacher

Job Description
A chemistry teacher works both inside and outside the classroom. In class, the teacher explains the concepts of chemistry, directs student lab work, and experiences the thrill of watching students mature in their understanding of the order in God's world. Most of a teacher's work comes outside the classroom, though. Teachers prepare course objectives, outlines, and syllabi, and they grade labs, quizzes, tests, and other assignments.

To teach effectively, teachers must know chemistry well. They must provide hands-on learning opportunities through regular labs and experiments for students. Other activities could include school science projects, technical fairs and local and national competitions, as well as training programs and field trips. A commitment to excellence is a must.

Possible Workplaces
Chemistry teachers teach in universities, public high schools, charter schools, and Christian or private high schools. Some work as private tutors, and others train employees of business and industry.

Education
A teacher should possess the minimum of a bachelor's degree in chemistry, general science, physics, science education or another science-related field. Those teaching on the college level usually have a PhD. For most jobs, teachers also need training or experience in instructional techniques and public speaking. Good writing and organization skills are critical, and knowledge of a foreign language is a plus. Public schools and many private schools require state teaching certification.

Dominion Opportunities
Chemistry teachers can encourage and help young people to better understand God's creation and develop a Christian worldview. An enthusiastic chemistry teacher may influence some students to become further involved in research. That research may well develop solutions to medical, environmental, and other problems that would improve the quality of life for humans and would help man know how to take better care of God's creation.

Coming to Terms

Term	Page
matter	24
physical property	24
chemical property	25
physical change	25
chemical change	25
pure substance	26
mixture	27
heterogeneous mixture	27
homogeneous mixture	27
element	27
atom	28
ion	28
molecule	28
chemical symbol	28
compound	29
chemical formula	29
chemical subscript	29
coefficient	30
energy	31
thermodynamics	32
law of conservation of mass-energy	33
first law of thermodynamics	33
entropy	34
second law of thermodynamics	34
system	34
kinetic energy (KE)	35
internal energy	35
temperature	35
thermal energy	35
heat	36
exothermic	36
endothermic	36
Celsius scale	36
Kelvin scale	37
absolute zero	37
third law of thermodynamics	37
joule (J)	38
calorie (cal)	38
kinetic-molecular theory	39
solid	39
liquid	39
gas	39
plasma	40
Bose-Einstein condensate (BEC)	40
quark-gluon plasma	41
condensation	42
freezing	42
melting	42
sublimation	42
deposition	42

Chapter Review

Chapter Summary

- Matter is operationally described as anything that occupies space and has mass.
- Matter is classified by its physical and chemical properties.
- Physical changes do not change matter's composition, just its appearance or arrangement of particles. Chemical changes alter the composition of matter by rearranging its particles.
- Matter can be classified as pure substances (elements or compounds) or mixtures (homogeneous or heterogeneous combinations).
- Atoms are the fundamental particles of elements. Molecules are the fundamental particles of some compounds. Other compounds do not have distinct molecules.
- Chemical symbols are used to represent the names of elements in chemical expressions. Chemical formulas represent the composition of compounds.
- Matter and energy are two fundamental quantities in the universe. Energy is the ability to do work. There are six forms of energy commonly recognized: mechanical, acoustic, thermal, electromagnetic, chemical, and nuclear.
- Thermodynamics is the study of the movement and conversion of all forms of energy. The first and second laws of thermodynamics establish some basic principles regarding the origin and destiny of energy and matter in creation.
- Matter consists of tiny particles in constant motion. This principle is called the kinetic-molecular theory. The motions of particles in matter contribute to thermal energy. The average kinetic energy of particles in a sample of matter is proportional to the temperature of the matter.
- Thermal energy is transferred between two points at different temperatures. Such a movement of energy is called a heat transfer. Heat always moves from hotter to colder temperatures.
- Temperature is measured by thermometers. Temperature scales are anchored to specific temperature reference points.
- The Celsius temperature scale and the Kelvin (absolute) temperature scale are commonly used in chemistry.
- The SI unit for all kinds of energy and work is the joule (J).
- There are four states of matter that occur commonly in nature: solid, liquid, gas, and (ionic) plasma.
- Changes of state occur as energy is added to or removed from matter.

Review Questions

Concept Review

1. Why is Creation *ex nihilo* a religious belief and not a scientific fact?
2. State the operational descriptions of matter and energy.
3. What must occur in a substance in order to determine a chemical property?
4. Tell whether each of the following properties is a physical property or a chemical property.
   a. color
   b. density
   c. insulating ability
   d. acidity
   e. magnetism
   f. flammability
5. Tell whether each of the following processes involves a physical change, a chemical change, or both.
   a. a wafer of silicon being scored and cut into chips to be used in computer microprocessors
   b. a glacier melting
   c. small water droplets forming on the outside surface of a cold can of soft drink
   d. dynamite exploding and the resulting collapse of an old building
   e. a burning candlewick melting wax
   f. the rusting of spokes on a bicycle's wheels
   g. the growth of a child
   h. the erosion of a hillside
6. Describe the two kinds of pure substances. How do these differ from mixtures?
7. How does an ion differ from an atom?
8. Classify each of the following as elements or compounds. Assuming these substances all consist of particles, describe the particles as completely as possible.
   a. oxygen (O_2)
   b. carbon monoxide (CO)
   c. ozone (O_3)
   d. helium (He)
   e. methane (CH_4)
   f. hydrogen peroxide (H_2O_2)
9. Why do some chemical symbols for elements consist of two letters and others of only one?
10. How many of each kind of atom are present in each of the following formula expressions? Example: A single $CaCl_2$ unit has one Ca atom and two Cl atoms.
    a. CsBr
    b. $NaNO_3$
    c. $2\ LiH_2PO_4$
    d. $KC_2H_3O_2$
    e. $K_2Cr_2O_7$
    f. $Al(C_2H_3O_2)_3$
    g. $2\ H_3PO_4$
    h. $2\ Ba_3(PO_4)_2$

11. Use the classification scheme of matter presented in this chapter to classify the following substances as elements, compounds, homogeneous mixtures, or heterogeneous mixtures. Assume the substances are uncontaminated. Example: Magnesium sulfate ($MgSO_4$) is a compound.
    a. vegetable soup
    b. air
    c. oxygen (O_2)
    d. gasoline
    e. sulfuric acid (H_2SO_4)
    f. granite

12. What principal form of energy is involved in the following processes?
    a. energy that heats the water in the core of a nuclear power plant reactor
    b. energy that activates the digital display of a cell phone
    c. energy possessed by a bat striking a baseball
    d. energy received by a digital camera through its lens
    e. energy released during the digestion process
    f. energy used by refrigerator magnets

13. The ideas of men are not always in harmony with the way the universe operates. Identify the law that says that each of the following is impossible.
    a. A perpetual motion machine, once started with an initial input of energy, continues running indefinitely.
    b. A machine produces more energy than it consumes.
    c. The universe spontaneously came into existence.
    d. Organisms with new structural features descended from those without the genetic information for those features.

14. When you grasp a hot cup of cocoa with cold hands, which way does the heat transfer occur? Explain.

15. Why is the joule the preferred unit of energy in scientific studies? What part of the definition of the calorie could make it more convenient for use in non-scientific studies?

16. Why are solids and liquids relatively incompressible?

17. What condition distinguishes plasmas from more "normal" states of matter?

18. During what meteorological (weather) changes does sublimation occur on earth?

True or False

19. The origin of matter and energy in the universe by naturalistic processes is an established fact beyond scientific debate.

20. A chemical property of a substance cannot be determined without causing or attempting to cause a chemical change to occur in the substance.

21. A solution, such as filtered air, is a homogeneous mixture.

22. The modern system of chemical symbols was developed by John Dalton in the early 1800s.

23. The thermal energy of a system is determined by the formula $E = mc^2$.

24. In order to be useful, energy must often be converted from one form to another.

25. One of the difficulties with the big bang origin of the universe is that many evidences predicted by the theory are not seen in the universe today or are contradicted by other evidence.
26. The belief that space dust can spontaneously come together to form a planet is a good example of the second law of thermodynamics.
27. Thermal energy, unlike internal energy, does not account for the potential energies of the particles in a system.
28. Absolute zero cannot be reached at a given location because there cannot be a colder place where thermal energy can flow to in order to cool the location.
29. The melting point is identical to the freezing point of a substance.

Application

30. Convert the following temperatures from the scale given to the scale indicated in parentheses. Report your answers to the nearest tenth of a degree.
    a. 300.0 K (°C)
    b. −118.53 °C (K)
    c. 100.0 K (°C)
    d. 37.5 °C (K)
    e. 212.0 °F (K)

31. Convert the following quantities of energy from the unit given to the unit indicated in parentheses. You can consult Appendix B for the necessary conversion factors.
    a. 157 J (cal)
    b. 3700 BTU (J)
    c. 153 Cal (J)
    d. 7.89×10^8 J (BTU)

32. Research and present a report on some aspect of the creation-evolution debate where the second law of thermodynamics has been used by either side to justify its position. State in your report which argument seems more persuasive.

Measuring and Calculating

CHAPTER 3

3A Measuring and Units	49
3B Measurements	55
3C Mathematics and Measurements	61
3D Orderly Problem Solving	63

Dominion Science Problem

Allergies

Springtime brings budding trees, blooming flowers, warm sun, melodic bird songs—and sneezing! You or someone in your family probably suffers from allergies, along with fifty million other Americans. Allergies affect one in five Americans, and they are on the rise worldwide. The number of Americans that experience asthma and allergies has more than doubled since 1980. Some people experience severe food and environmental allergies, causing extreme skin, digestive, and breathing problems, and sometimes even death. How can millions of people get relief from this modern disease?

3A Measuring and Units

3.1 Measuring and the Creation Mandate

How fast can a moth fly? How hot is the water spewing from Old Faithful? What is the distance separating the sodium and chlorine ions in a salt crystal? The answers to these questions can be found only by measuring.

In the beginning, God ordered man to subdue the earth and have dominion over the creatures living there. This charge—what we have called the Creation Mandate in Chapter 1—does not imply that humans are only to occupy the earth. Rather, we are to actively pursue knowledge about the earth and the living things that inhabit it. Gaining knowledge about God's creation so that we will be able to benefit others is the purpose underlying the study of science from a Christian worldview. To gain knowledge about creation, we must make measurements of many different aspects of nature.

> **3A Section Objectives**
> After finishing this section, you should be able to
> - describe measuring and define the terms associated with it.
> - describe the properties of a metric system.
> - list the base units and symbols of the SI.
> - differentiate between base and derived SI units.
> - correctly use metric prefixes to create different-sized SI units.
> - convert between metric units.
> - use the bridge notation in unit conversions.

3.2 What Is Measuring?

You have probably measured things before. If you enjoy woodworking or carpentry, you have measured the length, width, or thickness of a board. If you cook, you have measured dry and liquid volumes of ingredients in measuring cups. But have you ever considered what measuring actually involves? Could you define the act of measuring in your own words?

Measurement is the act of comparing an unknown quantity to a standard unit. This simplified definition implies several concepts that you must understand.

Objects and systems are measurable. For example, scientists measure quantities such as length, mass, volume, time, and electrical charge. Each of these properties is called a **dimension**. Before a dimension can be measured, there must be a **unit** of measure defined for that dimension. For example, you are probably familiar with the meter as a unit of measure for the dimension of length. Other units for length include the mile, kilometer, and inch. The word *unit* means "one of something." Dimensional units are lined up on a **scale** to provide a means of measuring. The scale is **calibrated**, or accurately subdivided into measurement units. A calibrated scale is essential to every instrument designed to measure anything. An **instrument** is any artificial device made for the purpose of refining, extending, or substituting for the human senses when measuring.

A measurement consists of two parts—a number and the unit in which the measurement was taken. Only like units can be added or subtracted, but any unit can be multiplied or divided, when appropriate.

3-1 Accurate measurement is important when building a woodworking project.

> unit: L. *unus*—one

3.3 Metric Systems of Measurement

In the eighteenth century, the need for standardized units spurred the development of what are called *metric* systems of measurement. A metric system is based on a decimal scale, that is, a scale that consists of base units multiplied by or subdivided by powers of ten.

> A metric system is any measuring system in which different-sized units are related to each other by multiples of 10.

The kilogram was originally defined as the mass of 1 cubic decimeter (1000 cm³) of water. The platinum-iridium alloy cylinder fabricated in 1889 established a stable, corrosion-proof reference that is still the standard for the kilogram today.

The SI is the standard metric system used by research scientists worldwide.

Measurement and God

God uses the concept of measurement to illustrate His greatness. In ancient times, people made measurements with parts of their bodies, such as their hands or arms. So God asks, "Who hath measured the waters in the hollow of his hand?" (Isa. 40:12). Scientists estimate that the oceans alone contain 1.37 billion cubic kilometers of water. By appealing to measurement, God shows that He is far greater than man can even conceive.

3-2 The kilogram is the only SI base unit that is still defined by a man-made object. The National Institute of Standards and Technology (NIST) administers the SI in the United States.

In 1799, France became the first European nation to adopt a national metric measurement system consisting of the kilogram and the meter. These standards were manufactured from platinum metal in the form of a cylinder and bar, respectively.

In 1960, many nations adopted the modern metric system, called the *Système International d'Unités*, which is French for the International System of Units. It is commonly abbreviated as simply the SI.

3.4 SI Units

There are seven **base units** in the SI, each of which measures a basic dimension. These units and dimensions are summarized in Table 3-1, along with their symbols.

3-1 SI Base Units

Dimension	Dimensional symbol	SI unit	Unit symbol
length	ℓ	meter	m
mass	m	kilogram	kg
time	t	second	s
temperature (absolute)	T	kelvin	K
number of particles	n	mole	mol
electrical current	I	ampere	A
light intensity	I_L	candela	cd

The beauty of the SI is that nearly every other unit of measurement used in science can be expressed in terms of these base units. For example, the unit of thermal energy can be expressed in terms of kilograms, meters, and seconds.

$$1 \frac{\text{kg} \cdot \text{m}^2}{\text{s}^2}$$

In many cases, expressing a measurement in terms of several base units is cumbersome. When these measurements are expressed in SI base units, they can become complicated and hard to remember. To avoid this problem, special units have been established that are derived from various combinations of the base units. There are twenty-two **derived units** approved for use in the SI. For example, the unit of thermal energy mentioned above is called the joule (J).

$$1 \text{ J} = 1 \frac{\text{kg} \cdot \text{m}^2}{\text{s}^2}$$

Some of the derived SI units used commonly in chemistry are listed in Table 3-2.

3-2 Common Derived Units Accepted in the SI			
Dimension	Derived unit	Unit symbol	Expressed as SI base units
area	square meter	m^2	m^2
volume	cubic meter	m^3	m^3
mass density	kilogram per cubic meter	kg/m^3	kg/m^3
frequency	hertz	Hz	s^{-1} (cycles/s)
force	newton	N	m · kg/s
pressure	pascal	Pa	$kg/(m \cdot s^2)$
energy/work/heat	joule	J	$kg \cdot m^2/s^2$
Celsius temperature	degree Celsius	°C	K
radioactive decay	becquerel	Bq	s^{-1} (disintegrations/s)

In this course, you will not use light intensity. It is included in the table for completeness. Note that the SI uses the kilogram rather than the gram as the base unit for mass. Scientists can measure the larger quantity with greater relative precision than the gram.

The small raised dot (·) signifies multiplication.

There are also some non-metric units that are used with SI units because of their convenience or longevity. Some of these are listed in Table 3-3.

3-3 Selected Non-SI Units			
Dimension	Unit	Unit symbol	Value in SI units
time	minute	min	1 min = 60 s
time	hour	h	1 h = 3600 s
time	day	d	1 d = 86 400 s
volume	liter	L*	1 L = 1 dm^3 = 10^{-3} m^3
mass	metric ton	t	1 t = 10^3 kg
mass	atomic mass unit	u**	1 u ≈ 1.66 × 10^{-27} kg

*Note that the symbol for liter is a capital *L*.
**1 u is $\frac{1}{12}$ of the mass of one carbon-12 atom. The value must be determined by experiment, so it is not exact.

3.5 SI Unit Prefixes

The SI allows scientists to size units so that they are convenient to measure and express quantities of different sizes. To change the size of a unit, the SI adds prefixes that represent factors of powers of ten to the names of the base units. For example, the meter is the base unit of length, but a meter is not always the best size for a measurement.

For Good Measure

The ability to construct a system of measurement is a blessing from God given to man when He created him and granted him dominion over the earth. Because of the Fall, man often uses these abilities for evil. For instance, measurement can be used to produce illegal drugs. Christians ought to use their abilities to measure to exercise wise dominion for the benefit of others. Exercising wise dominion may involve Christian scientists' researching new drugs to cure diseases.

To measure the length of your shoe, it is more manageable to use the *centi*meter, a unit that is $\frac{1}{100}$ of a meter (centi- = × 10^{-2} or $\frac{1}{100}$). At the same time, to measure the distance from your house to the school, it is usually more convenient to use the *kilo*meter, a unit that is one-thousand times as long as a meter (kilo- = × 10^3 or 1000).

The prefixes used in the SI are based on foreign words for the related power of ten. As the word implies, prefixes are always written in front of the base unit they modify. Table 3-4 lists the SI prefixes, their symbols, meanings, and exponential forms. You should memorize the colored prefixes, which you will use repeatedly in this course.

> The SI unit prefixes are derived from Greek, Latin, Danish, Norwegian, and Spanish roots. They are descriptive of the corresponding powers of ten.

3-4 SI Unit Prefixes

Prefix	Prefix symbol	Meaning	Factor	Exponential form
peta-	P-	quadrillion	1 000 000 000 000 000	10^{15}
tera-	T-	trillion	1 000 000 000 000	10^{12}
giga-	G-	billion	1 000 000 000	10^9
mega-	M-	million	1 000 000	10^6
kilo-	k-	thousand	1000	10^3
hect-/hecto-	h-	hundred	100	10^2
deka-	da-	ten	10	10^1
—	(none)	base	1	10^0
deci-	d-	tenth	0.1	10^{-1}
centi-	c-	hundredth	0.01	10^{-2}
milli-	m-	thousandth	0.001	10^{-3}
micro-	µ-	millionth	0.000 001	10^{-6}
nano-	n-	billionth	0.000 000 001	10^{-9}
pico-	p-	trillionth	0.000 000 000 001	10^{-12}
femto-	f-	quadrillionth	0.000 000 000 000 001	10^{-15}
atto-	a-	quintillionth	0.000 000 000 000 000 001	10^{-18}

> In this textbook, very large and very small numbers will be written with thin spaces instead of commas in between every three digits. This practice follows the SI rules for writing numbers. You should insert commas for numbers 10 000 and larger in your written work.

3.6 Conversion Factors

Units in the SI can be adjusted by multiplying the base unit by an appropriate factor of ten. That factor of ten is represented by a special prefix added to the unit name. Equivalent units of measure can be compared in an equation form. For example, look at the equivalence of 1000 millimeters and 1 meter.

$$1000 \text{ millimeters} = 1 \text{ meter}$$
$$1000 \times (10^{-3} \times \text{meter}) = 1 \text{ meter} \quad \text{Note: milli- = } 10^{-3}$$
$$(1000 \times 10^{-3}) \times \text{meter} = 1 \text{ meter}$$
$$(1) \times \text{meter} = 1 \text{ meter}$$
$$1 \text{ meter} = 1 \text{ meter}$$

Some other equivalent quantities include the following:

60 min = 1 h   1000 m = 1 km   1000 mL = 1 L   1 kg = 1000 g.

Often, calculations you do in this course may contain measurements of the same dimension using different SI units, such as kilograms and grams, or non-SI units, such as ounces. When possible, it is best to convert all measurements to the same units before calculating the answer.

Unit conversion consists of multiplying the measurement by a conversion factor. A **conversion factor** is a fraction that contains both the original unit and its equivalent value in a new unit. Mathematically, the fraction is equal to one. Since multiplying a quantity by one gives the original quantity, unit conversion does not change the amount of the quantity, just the number and units in which it is expressed.

Suppose we want to convert 2.350 liters to milliliters. The conversion factor will contain the old and new units, L and mL. The equivalence expression is

$$1 \text{ L} = 1000 \text{ mL}.$$

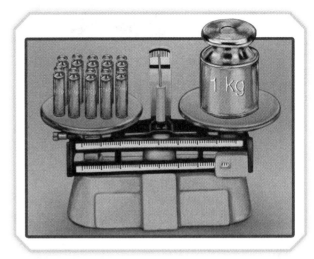

3-3 A balance demonstrates equivalence between different units of the same dimension.

Next, we have to create the conversion factor ratio. There are two possibilities in our example, obtained by dividing one side or the other by each quantity:

$$\frac{1 \text{ L}}{1000 \text{ mL}} = 1 \quad \text{or} \quad \frac{1000 \text{ mL}}{1 \text{ L}} = 1.$$

Both of these ratios are equivalent—they are both equal to 1—but which is the correct one to use? The basic rule for using conversion factors is this: *In a conversion factor ratio, the new unit is in the numerator and the old unit is in the denominator.*

Formatting the conversion factor this way allows standard cancellation to eliminate the original unit, leaving the required unit. The proper conversion factor is

$$\frac{1000 \text{ mL}}{1 \text{ L}}.$$

Problem-Solving Strategy 3-1
If the measurement you are converting is in the denominator of a rational expression, the conversion factor must be inverted. The unit that cancels with the original unit must always be placed on the opposite side of the fraction bar from the original unit.

3.7 Bridge Notation

Bridge notation is a special notation for multiplying and dividing several measurements together at the same time. (The notation has this name because it looks like a trestle bridge.) Bridge notation can be used to do any calculation, but it is almost always used when converting units. The bridge method applies only to multiplication and division; it cannot be used for addition and subtraction. It requires that every measurement be written as a fraction with a numerator and a denominator. For example, the density of a substance might be 1.50 g/cm³. As a factor in the bridge, this density would be expressed as

$$\frac{1.50 \text{ g}}{\text{cm}^3}.$$

Problem-Solving Strategy 3-2
When doing unit conversions, check that your results are reasonable. As you increase the size of the unit (including its prefix), the numerical factor gets smaller. As the unit grows smaller, the numerical factor grows proportionately larger.

Measurements and units that are not naturally ratios are written as a fraction with a 1 in the denominator.

When two such factors are multiplied together, the bridge notation replaces the multiplication sign with a vertical line and joins the fraction bars into a single horizontal line:

$$\frac{256 \text{ cm}^3}{1} \times \frac{1.50 \text{ g}}{\text{cm}^3} = \frac{256 \text{ cm}^3}{1} \left| \frac{1.50 \text{ g}}{\text{cm}^3} \right.$$

If you set up the bridge correctly, like units on opposite sides of the fraction bar will easily cancel. Multiply all the remaining numerator values, including the units, and all the denominator values, including the units. Then divide the resulting numerator by the resulting denominator to obtain the answer. If the expression was set up correctly, the numerical value and the remaining unit(s) should be correct. For example,

$$\frac{256 \cancel{\text{cm}^3}}{1} \left| \frac{1.50 \text{ g}}{\cancel{\text{cm}^3}} \right. = \frac{256 \times 1.50 \text{ g}}{1} = 384 \text{ g}.$$

3-4 A bridge connects the place you start to the place you want to go. Bridge notation performs the same function.

EXAMPLE PROBLEM 3-1
Unit Conversions

Convert the following units using bridge notation:
a. 132 540 cm to kilometers
b. 350 mg to kilograms

Solution

a. Start with the given information—132 540 cm. Arrange the necessary conversion factors to obtain the required units. Multiply numerators and denominators of conversion factors. Divide the numerator by the denominator to obtain the final answer.

$$\frac{132540 \cancel{\text{cm}}}{1} \left| \frac{1 \cancel{\text{m}}}{100 \cancel{\text{cm}}} \right| \frac{1 \text{ km}}{1000 \cancel{\text{m}}} = \frac{132540 \text{ km}}{100000}$$

$$= 1.32540 \text{ km}$$

b. Follow the steps outlined in part a. You may not know the direct conversion factor from milligrams to kilograms. It is more convenient to convert milligrams (mg) to grams (g), and then grams (g) to kilograms (kg).

$$\frac{350 \cancel{\text{mg}}}{1} \left| \frac{1 \cancel{\text{g}}}{1000 \cancel{\text{mg}}} \right| \frac{1 \text{ kg}}{1000 \cancel{\text{g}}} = \frac{350 \text{ kg}}{1000000}$$

$$= 0.000350 \text{ kg}$$
$$= 3.5 \times 10^{-4} \text{ kg}$$

Problem-Solving Strategy 3-3
When using bridge notation, the first number should be the value given in the problem. You should always proceed from the given information to the required information.

3A Section Review

1. In your own words, describe measuring.
2. In chemistry, what is a dimension?
3. What is a metric system of measurement? What metric system is currently used worldwide?
4. What is the base unit that scientists use for mass? length? time? Is there a base unit for volume?
5. What is the unit, including its prefix, for 10^{-6} of the derived unit for energy?

6. Convert the following measurements to the indicated units:
   a. 238.4 mm to m
   b. 0.003 55 L to mL
   c. 48.6 s to h
   d. 6.50 oz to mL

3B Measurements

3.8 The Limitations of Measurements

Have you ever heard someone say, "Man is the measure of all things"? According to that statement, man is the determiner of all reality. Right and wrong, good and evil are determined by man rather than an absolute standard (such as God).

If humans were correct in every life decision and absolutely accurate when it comes to mundane things like measurement, there might be some reason to trust man's abilities. But men aren't able to measure anything with complete accuracy. Since the Fall, man's abilities have been permanently affected. Even secular scientists recognize there is always some *uncertainty* involved with any measurement.

This uncertainty poses a grave problem for worldviews in which man is the measure of all things. If man cannot achieve absolute accuracy in his observations and thoughts, how can he be sure of anything? Man's finitude does not pose the same problem for the Christian. The Christian worldview rests not on human thought but upon God's revelation. Furthermore, the Christian understands that God has not called man to exhaustively understand all things. God has called us to manage the world. It is possible to manage the world without achieving absolute accuracy.

3B Section Objectives

After finishing this section, you should be able to

- identify sources of uncertainty and error in measurements.
- compare and contrast accuracy and precision.
- compute the percent error of measurements.
- describe the factors that affect precision in measurements.
- determine the precision of measuring instruments in your laboratory.
- summarize the importance of significant digits in scientific measurements.
- explain how to determine the significant digits in a measurement.
- identify the significant digits in a measurement when recording data.

3.9 Accuracy

The **accuracy** of a measurement or instrument is a numerical evaluation of how close the measured value is to the actual or accepted value of the dimension measured. For example, if you measured the height of this book and you reported 27.6 cm, you would have made an accurate measurement. But what would you say if the ruler indicated a height of 27.9 cm? Since the accepted height of the book is 27.6 cm, there is an error in your measurement. There are two possible reasons for this error (assuming the book is the correct size): either you read the ruler wrong, or the ruler is faulty. It may also be a combination of these two factors.

Accuracy is a *quantitative* term that describes how close a measurement is to the actual or accepted value of the dimension being measured.

Photo provided by NIST

3-5 Scientists use extremely precise instruments to reduce the uncertainties of their measurements.

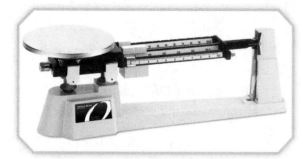

3-6 If a balance is not adjusted to zero, it will introduce error into all measurements made with it.

Scientists try to quantify errors in measurements using various statistical techniques. In this course, you will instead use a relatively simple approach to error analysis that compares the magnitude of the measurement error with the size of the measurement. This quantity is called **percent error**. This technique assumes that you know what the measurement *should* be or is calculated to be. Mathematically, percent error can be expressed in the following equation.

$$\text{percent error} = \frac{|\text{observed value} - \text{accepted value}|}{\text{accepted value}} \times 100\%$$

$$= \frac{|O - A|}{A} \times 100\%$$

Notice that the numerator of the fraction lies within absolute value signs. Written this way, the percent error is always positive and is more properly called the *absolute percent error*.

If the percent error is small, then the measurement is considered to be accurate. If the percent error is large, then the measurement is inaccurate.

> Absolute error = |observed − actual|

Example Problem 3-2
Calculating Percent Error

You measure the mass of the product of a chemical reaction to be 3.80 g. According to your theoretical calculations, you should have obtained 3.92 g. What is the percent error of your work?

Solution
Accepted value: 3.92 g
Observed value: 3.80 g

$$\text{percent error} = \frac{|O - A|}{A} \times 100\% = \frac{|3.80\text{ g} - 3.92\text{ g}|}{3.92\text{ g}} \times 100\%$$
$$= 0.0306 \times 100\%$$
$$= 3.06\%$$

Percent and Percentage

In correct usage, *percent* is a unit, meaning 1 part in 100. It can be replaced by the "%" symbol in numerical expressions.
Ex: 50 percent = $\frac{50}{100}$ = 50%. *Percent* should only be used in numerical expressions, not in place of *percentage*. *Percentage* is used to describe a portion of a whole expressed as a ratio when the ratio's denominator is understood to be 100. Ex: What percentage of 12 is 6? Six is 50 percent of 12.

3.10 Precision

Accuracy and **precision** are two terms that are often incorrectly used interchangeably. While accuracy indicates how close a measurement is to the actual value, precision indicates how repeatable a measurement is or how exactly one can make a measurement. Precision cannot say how correct a measurement is.

Certain numbers can be known exactly. For example, the number of beakers on a shelf can be counted. Counted numbers are exact. Some dimensions used in science are known exactly because they are defined. The speed of light, the second, and the meter are known exactly for this reason. Each of these is defined in relationship to some naturally occurring phenomenon.

Measurements do not have unlimited precision. Whenever we use an instrument to make a measurement, we convert the dimension we are measuring into a number by using the scale on the instrument. The digits in the measurement are based on how many large and small

> Precision is a *quantitative* evaluation of the exactness of a measurement or a measuring instrument.

scale markings are known for certain. But measurements almost never fall exactly on a scale mark, so we must estimate the measurement's location between marks. When you use an analog (nondigital) metric instrument, you should estimate the measurement to the nearest tenth of the smallest *decimal* subdivision on the instrument's scale. For most measurements in a high school laboratory, this estimated number will be the last or right-most digit in the reported value. With finer scale graduations, more decimal places can be known for certain. Thus, the reported measurement can have greater precision because it is known more exactly. We will assume for the purposes of this course that the maximum precision of an instrument is limited by its finest scale subdivisions.

3-5	Precision of Lab Instruments
Instrument	Typical precision
platform balance	0.1 g
triple-beam balance	0.01 g
analytical balance	0.0001 g
100 mL graduated cylinder	0.1 mL
10 mL graduated cylinder	0.1 mL
50 mL burette	0.01 mL
thermometer (10 °C – 110 °C)	0.1 °C

Many instruments may be classified by how they display their measurements. One that measures by comparing the object being measured directly to a scale (e.g., a ruler) or displays the reading using a pointer against a scale (e.g., a balance) is called an **analog instrument**. The size of the reading in proportion to the scale is *analogous* to the actual dimension being measured. An instrument that has a display consisting only of numerical digits (e.g., a digital thermometer) is called a **digital instrument**.

3.11 Using Measurements to Solve Problems

Can measurements really make a difference in how we live? Think about the chapter opener and springtime allergies. When you experience sneezing, coughing, watery eyes, skin rash, or difficulty breathing, your immune system is overreacting to a foreign object, called an allergen. Scientists are puzzled by the explosive increase in the number of people who experience allergies and asthma. There is no known cure for allergies, but they can be controlled.

One very effective treatment for allergies is called *immunotherapy*. The idea is that the immune system can be taught to react

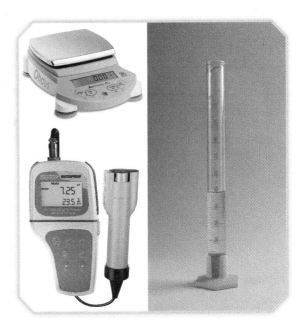

3-7 A digital analytical balance (top); a digital pH meter (bottom); an analog 100 mL graduated cylinder (right)

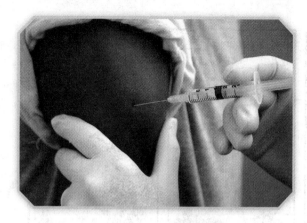

3-8 Immunotherapy uses precise measurements to give people relief from allergic symptoms.

Data (plural of *datum* but used as both a singular and a plural) is any recorded observation that is a detailed description or measurement. Scientists analyze data and draw conclusions from it.

Telling the Truth

Scientists are continually faced with moral choices. Sometimes these are issues that make the news, like embryonic stem-cell research. Other choices may arise when the scientist is alone in his laboratory. He may be tempted to falsify data on an experiment. Some scientists have done this to convince government agencies that they deserve grant money. Others have falsified data to gain prestige by writing groundbreaking articles in respected scientific journals. However good an excuse the scientist may think he has, God says, "A false balance is an abomination to the Lord: but a just weight is his delight" (Proverbs 11:1). Deception through falsifying measurement is clearly displeasing to God.

properly to allergens so that a patient has fewer and less severe allergic reactions. Immunotherapy is thus like a vaccination. In allergic immunotherapy, a patient is tested to determine his allergic triggers. Then a serum of sterile liquid extracts of these allergens is custom-made, consisting of substances like pollen, dust mites, and pet dander.

Allergy shots are then carefully administered. Usually a patient is given a shot once or twice a week. The first dose is usually 0.05 mL of the serum. Doses are increased in 0.05 mL increments depending on how the patient's body reacts to each dose, until he reaches a maintenance dose of 0.5 mL. Allergy sufferers receive shots for three to five years before any permanent improvement is observed.

3.12 Integrity in Measurements

You may wonder, "Why do I need to concern myself with measurement anyway?" One purpose of this course is to train you to think like scientists as much as possible—to become student chemists. Scientists are acutely aware of their limitations, especially when it comes to measuring. For this reason, you need to be familiar with the errors associated with measuring **data**.

Scientists try to report their quantitative results in the most accurate manner possible. To do this, they follow a set of rules for determining which digits of their answer are reliable or significant. Minimizing errors in measurement is an important part of experiment design. Estimating the uncertainty of measurements is important when evaluating the validity of conclusions taken from measured data.

Scientists also must report measurements correctly because it is a matter of integrity. Reporting measurements with too great a precision is dishonest. Reporting with too little precision is sloppy and calls into question the validity of their work. In either case scientists may lose the respect of their colleagues and perhaps even their jobs. Similarly, you must make the correct reporting of measured data a matter of personal honesty in your work as a means of imitating Christ in your life.

3.13 Significant Digits When Recording Data

Read the temperature indicated on the thermometer in Figure 3-9. The major divisions are labeled every 10 °C and the minor divisions (unlabeled) mark every degree. The fluid column is slightly higher than the mark indicating 98 °C. You are to report measurements to one-tenth of the smallest decimal increment, so you should estimate the temperature to the nearest 0.1 °C. The temperature could be legitimately reported as 98.1 °C, 98.0 °C, or 98.2 °C. The uncertainty of the tenths digit is why the last digit is called the *estimated, uncertain,* or *least significant* digit. It would be improper to report more or fewer decimal places for this temperature.

You can see from the preceding discussion that significant digits serve two purposes—they determine the important digits you need to record when making a measurement, and they indicate the precision of your measurement. The **significant digits** (SDs) in a measurement are only those that are known for certain plus one estimated digit. The estimated digit is given in tenths of the smallest subdivision on the instrument's scale. Therefore, the right-most significant digit will always be the estimated digit. Zeros may be added as placeholders after the SDs of a number greater than one or before the SDs of a number less than one in order to locate the decimal point.

3.14 Significant Digits in Recorded Data

You must be able to quickly determine which digits in a measurement are significant, especially if you are going to perform calculations with them. We have developed some easy-to-use rules that will make solutions to chemistry problems accurate and straightforward.

SD Rule 1: Significant digits apply only to measured data. Significant digits do not apply to counted or pure numbers or to exact definitions of certain quantities. These kinds of quantities do not normally have uncertainties associated with them. The following kinds of numbers are not measured data but are considered to be exact:

A. Significant digits do not apply to counted or pure numbers.

    500   one pair $\sqrt{3}$ π

B. Significant digits do not apply to ratios that are exactly 1 by definition.

$$\frac{60 \text{ s}}{1 \text{ min}} \qquad \frac{1000 \text{ g}}{1 \text{ kg}} \qquad \frac{1 \text{ °C}}{1 \text{ K}} \qquad \frac{12 \text{ in.}}{1 \text{ ft}}$$

SD Rule 2: All nonzero digits are significant.

    25.4 mL   (3 SDs)

    13.78 g   (4 SDs)

SD Rule 3: All zeros between nonzero digits are significant.

    100.5 °C   (4 SDs)

    1.09 g   (3 SDs)

SD Rule 4: Decimal points define significant zeros.

A. If a decimal point is present, all zeros to the right of the last nonzero digit are significant.

    20.0 s   (3 SDs)

    250.00 L   (5 SDs)

B. If a decimal point is not present, no trailing zeros are significant.

    2500 mL   (2 SDs)

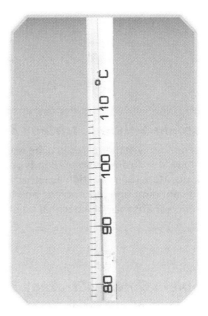

3-9 This thermometer is read to one-tenth of the smallest decimal graduation.

The significant digits of a measurement consist of all the certain digits plus one estimated digit.

Numbers without units are assumed to be counted numbers or pure numbers.

3-10 What is the precision of the weight limit? Should the manager be worried?

In 2500 mL, we cannot tell from the written data if the measurement was to the nearest 100 mL, to the nearest 10 mL, or to the nearest 1 mL. We can only be sure that the first two nonzero digits are significant (SD Rule 2). To avoid this problem with more precise measurements, we have SD Rule 5.

C. *In decimal numbers, all zeros to the left of the first nonzero digit are not significant.* They are placeholders for the decimal point of a measurement smaller than one whole unit.

0.075 kg	(2 SDs)
0.0010 s	(2 SDs)

SD Rule 5: *Significant zeros in the one's place are followed by a decimal point.*

2500. mL	(4 SDs)

This measurement was estimated to the nearest 1 mL, so all four digits are SDs.

SD Rule 6: *The decimal factor of scientific notation contains only significant digits.*

2.50×10^3 mL	(3 SDs)

This measurement indicates that the data was estimated to the nearest 10 mL.

> **Problem-Solving Strategy 3-4**
> Determining whether zeros in a measurement are significant or not depends on their position, their function as either a significant digit or a placeholder, and the presence or absence of a decimal point.

> **Problem-Solving Strategy 3-5**
> Use scientific notation when trailing zeros make the precision of the measurement ambiguous.

EXAMPLE PROBLEM 3-3
Significant Digits in Data

Determine the number of significant digits in the following quantities and the precision of each instrument's smallest scale increments:

a. 9.370 kg
b. 63 000 g
c. 705.06 mL
d. 12 cookies
e. 0.0001 s
f. 2300.000

Solution

a. 4 SDs (SD Rules 2 and 4A); the scale was calibrated in 100 g (0.01 kg) increments.
b. 2 SDs (SD Rules 2 and 4B); the scale was calibrated in 1000 g increments at least; however, the actual precision is indeterminate.
c. 5 SDs (SD Rules 2 and 3); the scale was calibrated in 0.1 mL increments.
d. This is an exact or counted number; therefore, significant digits are not applicable (SD Rule 1A).
e. 1 SD (SD Rules 2 and 4C); scale was calibrated in 1 ms (0.001 s) increments.
f. This is not a measurement, since there are no units associated with the number. Significant digits do not apply to pure numbers (SD Rule 1A).

3B Section Review

1. What is error in a measurement?
2. Compare and contrast accuracy and precision.

DS 3. Based on the dosages a patient receives, what level of precision should a syringe have in order to ensure that the patient receives the proper dosage? Why is this important?

DS 4. About four thousand Americans die each year from allergies. However, this is only about 0.2% of the total deaths in the United States per year. Is the medical research and expense for allergy and asthma treatment really necessary?

5. What are the two purposes for using significant digits in measurements?

⊙ 6. List in order which digits are significant in the following numbers:
   a. 230 mL
   b. 0.005 40 s
   c. 1440
   d. 50 °C
   e. 0.9405 g/cm³
   f. 700 jellybeans
   g. 1.05×10^6 cm
   h. π

3C Mathematics and Measurements

3.15 Introduction

Most scientists perform mathematical calculations in the normal course of their work. Math is an essential tool of science, because creating scientific models often involves mathematical formulas. Those formulas are workable descriptions of natural processes. You will learn that the ordinances established by the Creator that govern the behavior of matter nearly always involve mathematical relationships.

3C Section Objectives

After finishing this section, you should be able to

- add and subtract measured data so that the correct precision is preserved in the result.
- multiply and divide measured data so that the correct precision is preserved in the result.
- preserve the correct precision of the data in a calculation involving pure numbers.
- assign the allowed significant digits to the result of compound mathematical calculations.

3.16 Calculations with Measured Data

When using a calculator to compute the result of a scientific equation, be careful which numbers you use. Let's compute the density of a sample of the mineral galena, a lead-bearing ore. We measure the sample's volume to be 5.8 cm³ and its mass to be 44.03 g. We calculate the density by dividing the mass by the volume.

$$\text{density} = \frac{\text{mass}}{\text{volume}} = \frac{44.03 \text{ g}}{5.8 \text{ cm}^3} = 7.591\,379\,31 \text{ g/cm}^3$$

The result on the calculator indicates that the density has a precision of at least one-hundred millionth (10^{-8}) of a gram per cubic centimeter. However, the mass was measured only to a hundredth (10^{-2}) of a gram, and the volume was measured to only one tenth (10^{-1}) of a cubic centimeter. The calculated density is about one million times more precise than either the mass or volume! In scientific calculations, the precision of a computed result should have approximately the same relative precision as the measured data that were used to calculate it. As with the original measurements, scientists try to avoid adding or losing precision.

Below are some rules to follow for calculations using measured data. The rules provided here are adequate for high-school subjects, but they do not always yield answers that would be valid in scientific research. Scientists doing research must determine the precision for each calculation based on approved statistical methods. Those methods take into account the types of calculations and the sources and kinds of measured data.

3-11 Calculators will be used for solving most mathematical problems in this course.

Rules for Adding and Subtracting Measured Data

Math Rule 1: Measured data must be the same kind of dimension and have the same units before they can be added or subtracted. You cannot add apples and oranges!

Math Rule 2: The sum or difference of measured data cannot have greater precision than the least precise quantity in the sum or difference. When adding or subtracting, round the result to the place value of the least significant digit of the least precise quantity.

> **EXAMPLE PROBLEM 3-4**
> *Adding or Subtracting Measured Data*
> Add 1.008 g and 16.00 g.
>
> **Solution**
> Add the masses in a column.
>
> $$\begin{array}{r} 1.00\underline{8} \text{ g} \\ + 16.0\underline{0}\text{ g} \\ \hline 17.0\underline{08} \text{ g} \cong 17.0\underline{1} \text{ g} \end{array}$$
>
> Estimated digits are underlined. Note that the sum has two estimated digits, but only one is allowed. The sum must be rounded to the estimated digit with the largest place value.

> In this textbook, the "approximately equal to" symbol (≅) means that a calculated result has been rounded to the allowed significant digits.

Rules for Multiplying and Dividing Measured Data

Math Rule 3: The product or quotient of measured data cannot have more SDs than the quantity with the fewest SDs. Round results as necessary and add zeros as placeholders to locate the decimal point if required. The density we calculated at the beginning of this subsection should be rounded to 7.6 g/cm³ because the volume measurement contained only two significant digits.

> **EXAMPLE PROBLEM 3-5**
> *Multiplying or Dividing Measured Data*
> What is the area of a piece of paper that is 11.5 cm long and 5.5 cm wide?
>
> **Solution**
> Area is calculated by the product of length (ℓ) and width (w).
>
> $$A = \ell w = (11.5 \text{ cm})(5.5 \text{ cm})$$
> $$A = 6\underline{3}.25 \text{ cm}^2$$
>
> Since 5.5 cm has only two SDs, only 2 SDs are allowed in the final result.
>
> $$A \cong 63 \text{ cm}^2$$
>
> Study the explanation provided in the margin for this rule.

> **Problem-Solving Strategy 3-6**
> When adding or subtracting measured data in a column, draw a vertical line to the *right* of the estimated digit with the largest place value (i.e., the *least* precise measurement). In the answer, round to the digit's place to the *left* of this line (SD Rule 1A).

> When multiplying or dividing by irrational numbers (e.g., π, √2), the significant digits in the final result must be determined by the number of SDs in the measured quantities given in the problem. Irrational numbers in their symbolic forms are considered to have an infinite number of SDs.

> $$\begin{array}{r} 11.\underline{5} \text{ cm} \\ \times 5.\underline{5} \text{ cm} \\ \hline 5\ \underline{75} \\ 5\underline{7}\ 5 \\ \hline 6\underline{3}.\underline{25} \text{ cm}^2 \cong 63 \text{ cm}^2 \end{array}$$
>
> Estimated digits or the results of operations with estimated digits are underlined. As with the original measurements, the final answer can have only one estimated digit, so it is rounded to 2 SDs as shown.

Math Rule 4: The product or quotient of a measurement and a counted number, conversion factor, or defined value has the same number of decimal places, or same precision, as the original measurement.

$$3 \times 3.64 \text{ g} = 10.90 \text{ g (measured to 0.01 g)}$$

$$25.0 \text{ mL} \div 5 \cong 5.0 \text{ mL (measured to 0.1 mL)}$$

Multiplication is just repetitive addition. Therefore, multiplying a measurement by a pure number should not contradict Math Rule 2. Division is the inverse operation of multiplication, so the same principle applies.

Rule for Compound Calculations
Math Rule 5: In compound calculations, do not round off at the intermediate steps. Later in this course you will be required to calculate answers that are the result of two or more mathematical operations. Multiple operations are called compound calculations. When doing compound calculations, do not round off at the intermediate steps. Keep one extra digit while calculating, and then adjust your final answer to the proper number of significant digits.

> **Problem-Solving Strategy 3-7**
> For solutions requiring more than one calculation, do not round after each step. Instead, note the least significant digit for each result as if you were rounding. Leave one or more extra digits. Round only the final answer, taking into account the SDs of the intermediate answers.

3C Section Review

1. What restrictions apply to units when adding or subtracting measurements? when multiplying or dividing?
2. What determines the number of allowed significant digits when adding or subtracting measurements? when multiplying or dividing?
3. Why do scientists control the number of significant digits in calculations involving measurements?
4. Perform the following operations and express your answers with the allowed significant digits:
   a. 33.153 g + 3.2 g + 8.70 g
   b. 270 mL − 1.5 mL
   c. 10.10 g ÷ 4.04 mL
   d. 6.00 cm × 3.228 cm × 0.37 cm

3D Orderly Problem Solving

3.17 Problem Solving and the Creation Mandate

Problem-solving skills are necessary for daily living. Buying a car, building a house, balancing a checkbook, and even baking cookies all require problem-solving skills. As humans fulfill the Creation Mandate of Genesis 1:28, we solve the problems one by one that God has placed before us. In doing so, we learn more about the creation around us that we may glorify God and serve our fellow human beings.

Chemistry provides many opportunities to gain experience in problem solving. You can think of chemists as trained problem-solvers. Chemists may be called upon to identify unknown chemicals found in a river, to calculate the ingredients to make the most effective drug, or to create a plastic that can withstand exposure to extreme temperatures.

Problems in chemistry are solved best by an orderly, reasoned approach with a correct application of the mathematical rules you learned in the previous two sections. This section describes a proven, systematic method for solving the chemistry problems you will face.

> **3D Section Objectives**
> After finishing this section, you should be able to
> - explain why orderly problem solving is important in chemistry.
> - list each of the steps in chemical problem solving.
> - demonstrate how to use formulas in problem solving.
> - describe how to check the reasonableness of an answer.
> - use the problem-solving methods discussed in this section to obtain a correct answer to a mathematical problem.

3.18 Steps in Problem Solving

Step 1: Read the statement of the problem. Reading the problem is the most important step, but it is too often overlooked. Read the problem statement through completely to get an idea of what is being asked. Then reread the problem slowly and carefully, identifying the knowns and unknowns.

Identify what information is given. Make a list of numerical data and their units. Mark terms that set conditions or boundaries for the problem. Often, making a list of the data and their units will help you see the relationships among them.

Step 2: Determine the method of solution. After jotting down the given information, look for key words that tell you what to do. Some terms tell you what calculations to perform to solve the problem. For example, *of* implies multiplication, and *ratio* implies division.

Other terms tell you what sort of solution is required. For example, the terms *find, how many, what is the,* and *balance* direct you toward the answer. The solution may involve a mathematical calculation, writing out special symbols or formulas associated with chemical elements or compounds, drawing a sketch, making a graph, or writing one or more sentences of descriptions or reasons for a conclusion.

Step 3: Choose the specific tools to use. For mathematical problems, the kind of questions and the required method of solution usually suggest the tools needed to arrive at an answer. Each kind of problem involves the same type of data and a similar path to the solution. Most problems involve one or more mathematical operations requiring the use of bridge notation, mathematical formulas, chemical formulas, or any combination of these. You will recognize patterns as you gain experience with each kind of problem. However, you may need to be creative and link ideas together in a pattern which you have not seen before. Don't assume that every problem will be solved in the same way.

Formulas are mathematical models of natural phenomena. Often, they are statements of scientific laws. They relate quantities, whether variables or constants, in *literal equations*. You should identify all constants and variables and list them in your solution. Usually, the given variables and what is requested in the problem will alert you to which formula is needed. You may have to look up the value of any constants in the appendixes.

Step 4: Set up the problem, estimate, and calculate. Once you have identified the formula you are going to use, write it down in its standard form. Replace all of the variables with constants or with the known quantities that you have measured or been given. Write multiple products or ratios in bridge notation. Cancel units to check that the result has the dimension you are solving for. If the variable you are solving for was not isolated in the original equation, you will need to use algebra to solve for the answer.

When scientists design experiments or attempt to predict the effects of a certain phenomenon, they often try to estimate the answer they will obtain. Make your estimations using round numbers and mental calculation.

3-12 What assumptions did the scientist make regarding the ape's problem-solving abilities? What could you conjecture about his (the scientist's, not the ape's) presuppositions?

Estimating the approximate answer can help uncover errors you might make using your calculator. You can even discover mistakes in setting up the solution of a problem. For example, you know you've made a mistake if you estimate that you will end up with ten times more matter than you started with!

You can quickly find out if your answer is reasonable by comparing your calculated result with the estimated answer. The ratio of the answer to its estimate should be close to 1. If it is not, check your solution and the estimate for errors.

Step 5: Check and format. Students are often so eager to get an answer that they fail to check their answer. Did you solve for the required quantity? Did you answer the question? Go back and re-read the problem to be sure you found the answer that the problem asks for. Make sure the result has the required dimensional units.

Once you are satisfied that the problem was solved correctly, round the answer to the proper significant digits, applying the rules for rounding. For multiplication and division involving measured data, keep track of the number of SDs. For addition and subtraction, keep track of measured precision.

For very large or very small numbers, express the final answer in scientific notation. This step is particularly important if the answer contains ambiguous trailing zeros when written in standard notation.

> Literal equations are those written so that most or all quantities are represented by letters rather than numbers. One example of a literal equation you will use in chemistry is $PV = nRT$, the ideal-gas law.

> ### Problem-Solving Strategy 3-8
> When comparing large numbers, convert your answer and your estimation to scientific notation. Round the decimal factors to the nearest whole number. Find the ratio of the decimal factors and the difference of the exponents. Convert the result back to standard notation to check its reasonability.

3.19 An Example of Problem Solving

One of the physical properties of matter is density (D). Density is defined as the ratio of a substance's mass (m) to its volume (V), and it can be calculated using the following formula.

$$D = \frac{m}{V}$$

Mass is expressed in units such as grams or kilograms. When the mass is given in grams, cubic centimeters are used as the unit for volume; when the mass is given in kilograms, cubic meters are used.

People sometimes incorrectly compare the densities of two substances by saying that one is "heavier" than the other. What they actually mean is that one is denser than the other. Suppose someone says that lead is heavier than water. Actually, 1 kg of water is just as "heavy" as 1 kg of lead. The person means that lead is denser than water, or more compact. Thus, lead has more matter in a given volume than does water.

In the following example problems, density will be used to illustrate how to use the problem-solving steps discussed in this section. It is important to understand that there is a direct relationship between the mass of a substance and its volume. Thus, in questions involving the mass and volume of a given substance, density serves as a "conversion factor" between the two dimensions.

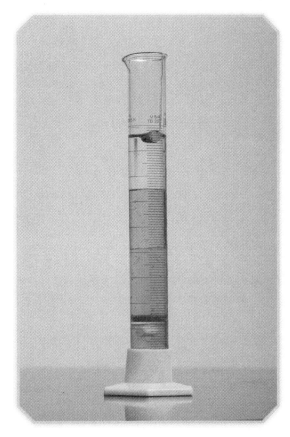

3-13 Relative density. Shown from the top down (increasing density) are a cork, isopropyl alcohol, corn oil (yellow), water (blue), a rubber stopper, and glycerol.

EXAMPLE PROBLEM 3-6
Finding Volume from Mass and Density

Use the systematic approach to problem solving described in this section to answer the following question. A lead fishing sinker has a mass of 51.0 g. If the density of lead is 11.34 g/cm³, what is the volume of the sinker?

Solution

Step 1. Read the problem statement. List the given information with the appropriate variables.
Given:

$$m = 51.0 \text{ g}$$
$$D = 11.34 \text{ g/cm}^3$$
$$V = ?$$

Step 2. Determine the method of solution. The problem requires that we calculate the volume of the lead sinker. Since we know that mass and volume are directly proportional to each other for a substance of constant density, and density is a ratio of these two dimensions, we know that the answer can be found by an equation relating these three quantities.

Step 3. We will use the density formula,

$$D = \frac{m}{V}$$

which relates density, mass, and volume.

Step 4. Set up the problem by substituting known values into the formula.

$$D = \frac{m}{V}$$

$$\frac{11.34 \text{ g}}{\text{cm}^3} = \frac{51.0 \text{ g}}{V}$$

After doing this, rearrange to isolate the unknown. You should recognize that by doing so you are essentially using bridge notation. Check that resulting units represent volume.

$$V = \frac{51.0 \text{ g}}{1} \cdot \frac{\text{cm}^3}{11.34 \text{ g}} = \frac{51.0}{11.34} \text{ cm}^3 = 4.497 \text{ cm}^3$$

Step 5. Estimate the result, and then calculate the answer. In round numbers, the ratio is about 50/10, or about 5 cm³. The calculated volume is 4.4_9_7 cm³.

Step 6. Check for reasonableness. The answer contains the proper dimensional units, and the numerical value appears to be correct based on agreement with the estimation. Round the calculated result to the correct number of significant digits. Three SDs are allowed, since the mass was expressed to 3 SDs.

$$V \cong 4.50 \text{ cm}^3$$

While this example involves only a simple density calculation, it serves as a model for how to solve more complex problems. This textbook will demonstrate how to solve new types of problems as you encounter them. As new techniques are suggested, we will describe them in numbered problem-solving strategies in the body of the chapter. Nothing, however, can replace thoughtful practice.

3D Section Review

1. Taking into consideration Genesis 1:28, what is the value of methodical problem solving?
2. Summarize the steps to orderly problem solving when dealing with a mathematical chemistry problem.
3. Describe a literal equation. Why are such formulas important in science? Give an example of a formula that you have used before.
⊙ 4. An experiment requires 45.6 g of isopropyl alcohol as a reagent. Instead of measuring the mass of the liquid on a balance, you decide to determine the mass by using an equivalent volume measured in a graduated cylinder. The density of isopropyl alcohol is 0.785 g/mL. What volume of the alcohol is required for the experiment?

Coming to Terms

Term	Page
measurement	49
dimension	49
unit	49
scale	49
calibrate	49
instrument	49
SI	50
base unit	50
derived unit	50
unit conversion	53
conversion factor	53
accuracy	55
percent error	56
precision	56
analog instrument	57
digital instrument	57
data	58
significant digit (SD)	59

Chapter Review

Chapter Summary

- Measuring is an essential part of studying the universe. It involves comparing a measurable quantity, called a dimension, to a scale calibrated in units appropriate to the dimension. A measurement is the product of a number and the dimensional unit.

- The universe seems to have three fundamental characteristics—space, matter, and time.

- Scientists have developed a metric system of units called the Système International d'Unités (SI). The SI has seven base units, from which all other metric dimensional units are derived.

- A small number of non-metric units are approved for use with the SI, including units of time and the atomic mass unit.

- Metric units that are smaller or larger than the base metric units can be created by adding special prefixes that stand for a multiple of a power of ten. Thus, unit conversions between different-sized metric units are a simple matter compared to the English system, for example.

- Accuracy is a quantitative evaluation of how close a measurement is to the actual or accepted value of the dimension being measured. Precision is a quantitative evaluation of how exact and reproducible a measurement is.

- Because measurements have uncertainties, scientists use the technique of significant digits to indicate the precision of measurements and the relative uncertainty of the data.

- For this course, simple rules determine the significant digits in measured data and in mathematical results from calculations involving measured data. In multiplication and division, the number of significant digits is more important; in addition and subtraction, the precision of the data is more important.

Review Questions

Concept Review

1. What is it called when scientists quantify their observations of their surroundings?
2. What are two required components of every measurement?
3. What is the first step in creating a unit conversion factor?
4. Why should you not use the term *percent* to indicate a portion of a larger quantity?
5. A liquid thermometer is an example of what kind of instrument display?
6. What are the significant digits in a measurement?
7. What is the only reliable way to report the correct significant digits in a measurement that contains trailing zeros?

True or False

8. A scientific measuring instrument requires some kind of calibrated scale.
9. The main function of bridge notation is to check that you accurately key the calculation into your calculator.
10. Percent error is sometimes used to quantify the accuracy of a measurement or answer.
11. The precision of an instrument is the measure of its accuracy.
12. The quantity 10 050 g has two significant digits because there are only two nonzero digits and there is no decimal point.
13. It is a good problem-solving technique to go back to the problem statement several times while working to the solution to be sure you are solving for the right quantity.

Application

14. Rename the following prefixes with their name and decimal equivalents. (Example: deka- = ten [10])
    a. deci-
    b. milli-
    c. kilo-
    d. nano-

⊙ 15. Write an equation that states the relationship between the two given units, and then write the conversion factor(s) that could be used to convert the first unit to the second. Example: Millimeters and meters are related by the equation 1000 mm = 1 m. The conversion factor 1 m/1000 mm should be used.
    a. grams and micrograms
    b. nanometers and meters
    c. kilograms and grams
    d. liters and milliliters
    e. megahertz and hertz

⊙ 16. Convert the following measurements from the units given to the units requested with the use of conversion factors. Use bridge notation.
    a. How many grams are in 5280 mg?
    b. The wavelength of blue light is about 475 nm. What is the wavelength of blue light measured in meters?
    c. What is the mass in grams of a 72.6 kg man?
    d. How many milliliters of ginger ale are contained in a 2 L bottle?

⊙ 17. Use bridge notation to convert the following measurements from the given units to the requested units. You may need some information from Appendix B to answer some of these.
    a. The earth's mass is estimated to be 5.9763×10^{27} g. What is its mass in mg?
    b. The average distance from the earth to the sun is an astronomical unit (ua) and is approximately 149 599 000 km. How many nanometers is this?
    c. The volume of the earth is approximately $1.083\,26 \times 10^{12}$ km³. How many liters is this?
    d. A box is 0.2 m high, 0.050 m wide, and 0.100 m long. How many liters of air does this box contain?

18. Evaluate the precision and accuracy of the measurements in the following situations:
    a. During three consecutive measurement trials, you obtained 45.2 °C, 44.9 °C, and 45.1 °C. The actual temperature was 45.0 °C.
    b. In three separate trials, you measured the density of a gas to be 2.865 kg/m³, 2.852 kg/m³, and 2.860 kg/m³ under the same conditions. The actual density of the gas was 3.214 kg/m³.

19. Determine the percent error for the following measurements:
    a. The density of mercury is 13.534 g/mL. What would be your percent error if you experimentally determined the density to be 13.000 g/mL?
    b. The density of oxygen gas is 1.429×10^{-3} g/cm³ at standard conditions. What would be the percent error of the average of two separate experimental determinations measuring 1.112×10^{-3} g/cm³ and 1.069×10^{-3} g/cm³? (To solve this problem, average the two readings before calculating the percent error.)

20. Name the significant digits in each of the following quantities. Also state the place value of the estimated digit.
    a. 2.40 cm
    b. 0.010 g
    c. 200 km
    d. 102 students
    e. 130.012 00 m
    f. 300. mm

21. Perform the indicated operations with the following measurements and give your answer with correct units and the allowed number of significant digits. Round off answers when necessary.
    a. 5.8 g + 0.032 g + 6.19 g
    b. 4.718 cm − 3.94 cm
    c. (6.98 cm − 2.83 cm) × 1.7 cm²
    d. $(5.45 \times 10^6 \text{ km}) \times (3.22 \times 10^3 \text{ km})$
    e. $(2.1 \text{ m})^2 \times \pi \times 100. \text{ m} \div 6.83 \text{ s}$
    f. 3.0 m × 2.54 m
    g. 1.75 cm ÷ 2
    h. 4.020 m + 0.23 m + 40.2 m + 10.00 m

22. Write the following numbers in proper scientific notation, including the allowed significant digits.
    a. 6001 kg
    b. 0.0300 g
    c. 452×10^4 cm³
    d. 50 300 000 people

23. Perform the following operations; then assign the allowed significant digits.
    a. $(4.08 \times 10^{-3} \text{ g}) - (7.2 \times 10^{-4} \text{ g})$
    b. $(7.04 \times 10^{-4} \text{ m}) \times (4.1 \times 10^4 \text{ m})$
    c. $(3.40 \times 10^{-2} \text{ cm}^3) \div (5.08 \times 10^{-3} \text{ cm}^3)$
    d. $(8.95 \times 10^3 \text{ kg}) + (9.8 \times 10^1 \text{ kg})$

⊙ 24. Use a systematic approach to solve the following density problems. Answers should include the correct units and the allowed significant digits. When appropriate, use bridge notation to solve the problem.

a. Mercury, which is often used in thermometers, has a density of 13.534 g/mL at room temperature. What volume of mercury contains 10.0 g?

b. Copper has a density of 8.920 g/cm^3. What mass of copper will occupy 45 cm^3? What volume of copper will have a mass of 1.0 kg?

c. Ethyl alcohol has a density of 0.789 g/mL. What volume of ethyl alcohol must be poured into a graduated cylinder to obtain 17.60 g of alcohol?

d. The volume of a quartz mineral sample was determined by placing the weighed sample in a graduated cylinder containing 56.3 mL water. With the mineral sample completely submerged, the water level was 68.2 mL. This measuring method, known as water displacement, uses the difference in the level of the water to determine the volume of an irregular object. The dry mass of the quartz sample was 31.18 g. What was the density of the quartz in g/cm^3?

e. A solid will float on any liquid that is denser than the solid. The volume of a piece of calcite, which is often used in the production of chalk, is 12.5 cm^3, and the mass is 33.93 g. On which of the following liquids will the calcite float: acetone (D = 0.785 g/mL), water (D = 0.998 g/mL), methylene bromide (D = 2.497 g/mL), bromine (D = 3.119 g/mL), mercury (D = 13.534 g/mL)?

⊙ 25. Research the origin and meaning of the metric prefixes listed in Table 3-4 on page 52.

DS ⊙ 26. Do some research to see what scientists think might be causing the sudden rise in allergies and asthma.

DS 27. How do you think that the recent explosion in allergy diagnoses has the potential to affect our society?

Atomic Structure

CHAPTER 4

4A The Development of Atomic Models: A Historical Perspective	73
4B The Electron Structure of the Atom	82
4C Useful Element Notations	93
Facet Spectroscopy: Fingerprinting Atoms	77

DOMINION SCIENCE PROBLEM

Detecting Stroke

On December 13, 2006, Senator Tim Johnson, a Democrat from South Dakota, was doing a phone interview with reporters. His speech suddenly became disorganized and confused. He was able to regain his composure but was later taken to the hospital with strokelike symptoms.

Strokes happen when the blood supply to a person's brain is disrupted, causing brain cells to die from lack of oxygen. They can be thought of as "brain attacks," similar to heart attacks. Strokes disable more people in the United States than any other medical problem, and they are the third most common killer. If treated early enough, however, a stroke victim's condition can be greatly improved. According to the National Institute of Neurological Disorders and Stroke (NINDS), in the first three hours after a stroke, a patient can be treated with "clot-busting therapy" that reopens the blood supply to the brain. One out of four stroke victims treated within three hours suffers no brain damage. How can we accurately diagnose and treat strokes, or even prevent them from happening?

4A The Development of Atomic Models: A Historical Perspective

4.1 Introduction

Only recently have scientists been able to "see" atoms using a technique known as scanning tunneling microscopy (STM). Before the invention of STM, most of the evidence about atoms was collected indirectly. Scientists ingeniously used fragments of evidence acquired over nearly two centuries to piece together the theories that are now called atomic models. These models are foundational to what we know about chemical bonding, chemical reactions, and the other chemical and physical properties of matter. As scientists continue to study matter, they continue to revise their models. The current model is itself open to further revision.

Christians are not concerned that science produces models rather than ultimate truth because they realize that the purpose of science is not to provide the ultimate answers for life. But Christians are not unconcerned about research or about improving existing scientific models. God has called some Christians to use chemistry to manage His creation wisely. These scientists are interested in working with the most accurate models to better perform their work.

4A Section Objectives

After finishing this section, you should be able to

- summarize the development of the atomic model as an example of scientific model making.
- explain why the development of the atomic theory of matter was not continuous between the time of the Greeks and the nineteenth century.
- sketch or otherwise describe each of the historical atomic models, emphasizing their differences and similarities.
- explain why the discovery of the wave-particle dual nature of matter was important to the development of the atomic model.
- name the key scientists who contributed to the development of the modern atomic model and their discoveries.

4.2 Early Thoughts on the Structure of Matter

The Greek philosopher Democritus was the first to state that matter is made of separate, discrete particles. He said that matter contained definite particles and that it could *not* be divided infinitely without losing its properties. Democritus used the term *atomos* to name those smallest particles. Today we use the term **atom** for the smallest particle capable of chemical interactions.

Belief in philosophical atomism continued throughout the ancient and medieval eras, but it was not widespread. It was not until the seventeenth century that men started to use science to prove that matter is composed of atoms, or particles. Philosophical atomism was thus replaced by scientific atomism.

Many scientists during the seventeenth and eighteenth centuries grappled with the particle theory of matter. Famous men of science such as Isaac Newton, Robert Boyle, and John Dalton disagreed on whether matter is continuous. The first real experimental support for the existence of atoms was finally obtained in the late 1700s. However, no one recognized the full significance of this data for several decades.

Eighteenth-century chemists observed that compounds always contained a set ratio for the masses of elements in the compound. For instance, decomposing a 9.0 g sample of water always produced 8.0 g of oxygen and 1.0 g of hydrogen, a ratio of 8:1. Decomposing an 18.0 g water sample produced 16.0 g of oxygen and 2.0 g of hydrogen—the same ratio as the smaller sample. Any sample of water always contains 8 g of oxygen for every 1 g of hydrogen.

atom: (Gk. *atomos*—indivisible)

4-1 Democritus (c. 460–c. 370 BC) was a Thracian philosopher who was a contemporary of Plato and Aristotle. He believed that not only matter but also thoughts were constructed of atoms.

The *continuous theory of matter* held that matter could be continuously subdivided without end. This idea was an offshoot of the Greeks' purely mathematical view of the world, where dividing something in half always resulted in two new wholes that could be divided in half again. This view of nature was held to be more pure and consistent with philosophy than atomism.

Further experimentation with many compounds confirmed that the ratios of composition by mass were unique and constant for each compound. This discovery led to the **law of definite composition**. This law states that *every compound is formed of elements combined in specific ratios by mass that are unique for that compound.*

4.3 The First Experimental Model: Dalton's Atomic Model

It was not until 1803 that an English schoolteacher named John Dalton first framed an atomic model based on experimental evidence instead of philosophy. With incisive logic and the support of the law of definite composition and the law of the conservation of matter, Dalton formed his remarkable theory of matter.

Dalton's theory was based on two premises. First, combinations of atoms from different elements form compounds; second, atoms of different elements have different masses. Dalton assigned relative masses to atoms of various elements so he could determine the masses of the atoms in relation to each other. Through painstaking analysis, Dalton and other chemists of his day determined that oxygen was much more massive than hydrogen and slightly more massive than carbon. He made the same calculation for other atoms as well.

The exact values in Dalton's table were not accurate. In some cases, even the order of elements from smallest to largest was incorrect. Nevertheless, Dalton's accomplishments were notable, given the lack of information available to him. His theory inspired new experimentation that brought new insights in the nature of matter. Within fifty years, nearly all scientists had accepted the atomic theory, and chemistry advanced significantly. Dalton's model is still the foundation of present atomic theories.

4-2 John Dalton (1766–1844) was originally an English schoolteacher who eventually developed a modern particle model of the atom based on experimental evidence. He also formulated the first credible table of atomic weights and established the mathematical basis for chemistry.

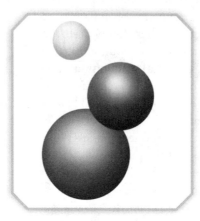

4-3 Dalton's model of the atom was based on differences in mass and the definite composition of compounds.

Effective vacuum pumps were invented as early as 1650. The mercury pump, which permitted experimentation with gas discharge tubes, was invented in 1855.

4.4 Discovery of the Electron: Thomson's Model

The invention of the electrical battery by Alessandro Volta in 1800—about the same time as Dalton's discovery—allowed scientists to study how matter behaves in the presence of an electrical current. They found that a gas sealed under very low pressure in a glass tube can carry an electrical current between two electrodes. The glass tube was called a gas discharge tube.

Scientists noticed another strange phenomenon. When most of the gas molecules were removed by a vacuum pump, the current decreased. However, when

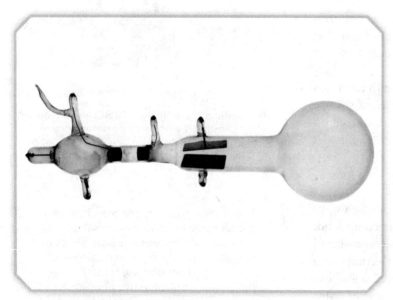

4-4 The use of cathode-ray tubes opened the way for the study of subatomic particles.

even more gas was removed, the current gradually began to increase, and the tube glowed with an eerie green light. Later, scientists determined that the color of the light depended on what gas was in the tube. Removing all gas molecules caused the glow to disappear, but current continued to flow. Obviously, something other than the gas carried the current. In a series of experiments during the late 1800s, physicists observed that current through the discharge tube came from the negative electrode, called the cathode. Emissions from the cathode were called **cathode rays**.

The English physicist J. J. Thomson finally explained cathode rays. He noted that cathode rays travel in straight lines, unaffected by gravity. That phenomenon could be explained only if the cathode rays were waves, similar to light waves, or particles moving at incredibly fast speeds. He also observed that a magnet could deflect cathode rays. Thus, the rays could possibly be tiny charged particles. When Thomson passed a cathode ray between two electrically charged plates, the ray bent toward the positively charged plate. Accordingly, Thomson concluded that the cathode-ray particles were negatively charged. Another scientist, George Johnstone Stoney, named these particles **electrons** (e^-) in 1894.

In 1897, Thomson determined the charge-to-mass ratio (e/m) of the particle. He found that this ratio was surprisingly large. Compared to the charge, the mass of the electron was almost nothing. He concluded that he was working with a particle fundamentally different from atoms and molecules, the ordinary particles of matter. Thomson concluded that every atom contained smaller, negatively charged particles.

Thomson further proved that the same type of particle was emitted from atoms of every element tested. Regardless of the gas in the tube or the type of cathode material, he obtained the same results. Gold, silver, iron, and copper all gave off the same particle. Dalton's atomic model could not explain how electrically charged particles could exist in an atom, why the atom was neutral, and how negative charges could leave an atom. It was time for a new atomic model.

Thomson accounted for these observations by postulating a new atomic model. In his model, negatively charged electrons are embedded in a positively charged substance that completely surrounds them. The charge on the positive material balances out the negative charges on the electrons so that the atom is neutral. Under certain conditions, electrons could be removed from the atom.

That model looked like English plum pudding. The electrons were negatively charged "plums" in a positively charged "pudding." Thus, Thomson's model is often known as the plum-pudding model of the atom. An analogy you might be more familiar with is a lump of chocolate chip cookie dough—the chocolate chips represent the electrons, just like the plums in Thomson's plum pudding.

4.5 Discovery of the Nucleus: Rutherford's Model

Wilhelm Roentgen's discovery of x-rays in 1896 triggered further experiments that led to the discovery of nuclear radiation. In the twenty years before World War I, the brightest minds in the scientific world pioneered the fascinating field of radioactive particles.

4-5 Joseph John Thomson (1856–1940) was an English physicist who was knighted for his groundbreaking work in discovering the structure of the atom. For his work with cathode rays, he was awarded the Nobel Prize in Physics in 1906.

Scientists have since defined the charge on an electron as −1 and determined the mass of an electron to be a scant 9.11×10^{-31} kilogram.

The fundamental electrical charge (e) is equal to 1.602×10^{-19} coulomb (C), the SI unit of electrical charge. An electron carries a charge of $-1\ e$, while a proton carries a charge of $+1\ e$.

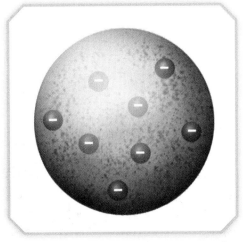

4-6 Thomson's plum-pudding (chocolate-chip-cookie-dough) model of the atom

Wilhelm Roentgen (1845–1923) was a German physicist best known for his discovery of x-rays. For his work, he won the Nobel Prize in Physics in 1901.

Ernest Rutherford (1871–1937) was an English physicist and laboratory director who was born in New Zealand. In his laboratories, many future Nobel Prize winners had their start. In 1908, he himself won the Nobel Prize in Physics for his studies in nuclear chemistry.

One of these brilliant scientists was Ernest Rutherford, a professor at Cambridge University in England. His experiments with a form of radiation known as **alpha particles** led to new insights about the inner parts of the atom.

4-7 Cambridge professor Ernest Rutherford

Rutherford discovered that alpha particles are positively charged ions with a mass of about 7300 electrons. These particles are emitted at very high speeds from some radioactive elements. Their large mass and high speed give alpha particles a great amount of energy. They also are strongly charged, having a +2 electrical charge. When a particle strikes a screen coated with zinc sulfide, its energy converts into a brief flash of light that allows it to be observed.

In 1909 Hans Geiger, one of Rutherford's assistants, had one of his students design an experiment in which a beam of alpha particles was aimed at a sheet of thin gold foil. As they struck the foil, most of the particles went straight through. A few, as expected, were slightly deflected and were detected by the flashes of light on a curved zinc sulfide screen surrounding the gold foil. Unexpectedly, however, about one in 20,000 was deflected at a large angle or even ricocheted back toward the source. Rutherford was astounded. He compared the deflected alpha particles to a cannon shell bouncing off tissue paper. Thomson's model could not explain these observations. It was time for a new atomic model.

Hans Geiger (1882–1945) was a German nuclear physicist who was involved in the attempt to build an atomic bomb for Nazi Germany during WWII. He is best remembered as the inventor of the Geiger counter, an instrument used to detect nuclear radiation.

4-8 Hans Geiger

Rutherford reasoned that atoms must be mostly empty space, otherwise most of the alpha particles should have been deflected. Because such a small percentage of the alpha particles were greatly deflected, there must be little likelihood that an alpha particle would collide with a central, dense region of an atom. Rutherford reasoned that this region, the **nucleus**, as he called it, must be very small. Subsequent calculations indicated that the diameter of a nucleus was only 1/100 000 the size of an entire atom. Rutherford's model is sometimes known as the nuclear model of the atom.

4-9 Schematic of Hans Geiger's experiment with alpha particles and gold foil

SPECTROSCOPY: FINGERPRINTING ATOMS
Facets of Chemistry

Where do you think helium was first found? If you guessed that it was isolated from air, found in some mineral, or detected anywhere else on earth, you missed the mark by some 150 million kilometers. Surprisingly, scientists discovered helium by looking at the sun. Even more surprisingly, the discovery occurred in 1868. Scientists made this remarkable discovery using one of the most diverse tools known to science: spectroscopy.

Spectroscopy is the study of how matter produces and interacts with electromagnetic radiation. When atoms are highly energized, they release light and other forms of radiation. In other instances, atoms absorb electromagnetic radiation. Because the atoms of every element have a unique arrangement of electrons, every element has its own characteristic pattern of energy interactions. The types of light that elements emit and absorb serve as their "fingerprints."

Atoms can absorb and emit all kinds of electromagnetic radiation. When an atom receives a small amount of thermal energy, electricity, or light of the proper wavelength, one or more of its electrons becomes more energetic, or "excited." It jumps from its original position to a higher energy level. Electrons in higher levels are extremely unstable, so within a fraction of a second, they fall back to some lower energy level. When they fall, they give off a burst of electromagnetic energy whose frequency (or color) depends on the energy difference between the higher and the lower levels.

Robert Bunsen

Normally these individual colors of light cannot be observed because they are all mixed together. A prism, however, can separate the colors of light so that they can be detected. When light from atoms in an incandescent material is analyzed, distinct bands of colors appear at specific locations. The set of bright lines is called a bright-line spectrum.

Visible light is the easiest to observe, so that branch of spectroscopy developed first. The simplest device for studying spectra, called the prism spectroscope, was invented in the 1850s by Gustav Kirchhoff and Robert Bunsen at the University of Heidelberg. It consisted of three parts—an arrangement of lenses; a prism; and a small, movable telescope. A flame energized the atoms in the sample to be tested. The lenses directed the light rays from the flame into a parallel beam. The prism took the unified beam of light and broke it up into its component wavelengths. The telescope swiveled around and detected the colors of light that came from the prism at various angles.

Time has brought numerous improvements in the basic spectroscope. Instead of prisms, scientists now use diffraction gratings to separate the beams of light into their separate components. More sensitive optical detection systems have been developed, and it is now possible to study the spectral lines in great detail. In many cases what was thought to be one line has turned out to be a compilation of many lines. Cameras have replaced the telescope of the old spectroscope, and it is now possible to photograph an entire spectrum instantaneously. New instruments have allowed scientists to study infrared, microwave, ultraviolet, x-ray, and even gamma-ray emissions. Armed with these new tools and the basic theory of spectroscopy, scientists can quickly determine what kinds of elements are in a sample of water, in clay from an archaeological relic, or in the plasma of a distant star.

Gustav Kirchhoff

Because spectroscopy is useful in both qualitative and quantitative analysis, it is applied to everything from quality-control tests in industry to crime-scene forensics. Even our time-keeping system is based on this technique, because the second (s) is defined by microwave spectroscopy of cesium-133 atoms.

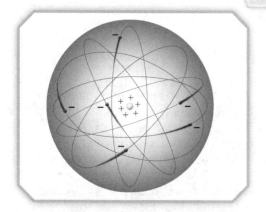

4-10 The nuclear model of the atom

4-11 James Chadwick (1891–1974) was a highly decorated English nuclear physicist best remembered as the discoverer of the neutron. He won the Nobel Prize in Physics in 1935 and was knighted in 1945.

4-12 Danish physicist Niels Bohr (1885–1962) was one of the key theoreticians who contributed to the development of the quantum model of the atom early in the twentieth century. He won the Nobel Prize in Physics in 1922.

spectrum: (L. *specere*—to see or look)

At first, Rutherford did not investigate the charge of the nucleus. Further experimentation, however, determined that the nucleus had to be positive, and so the electrons somehow had to be arranged around the nucleus. A decade after discovering the nucleus, Rutherford identified the positive particles in the nucleus, which he called **protons**. A proton has an electrical charge of +1 e, exactly opposite the electron's charge. The mass of a proton is approximately 1836 times that of an electron, or 1.673×10^{-27} kilogram.

Shortly after the discovery of the proton, scientists recognized a relationship between the number of protons in the nuclei of atoms and their chemical and physical properties. The number of protons in the nucleus is called an atom's **atomic number (Z)**.

4.6 Discovery of the Neutron: Chadwick's Model

Rutherford and others realized that atoms have much more mass than the sum of the masses of protons and electrons. Rutherford suggested that neutral particles supplied the missing mass without changing an atom's charge. He thought that the particle might consist of a closely joined proton and electron. This hypothetical particle eluded detection for many years. In 1932 James Chadwick observed neutral particles radiating from a sample of metal bombarded by alpha particles. Appropriately enough, these particles were named **neutrons** (N). A neutron has a mass of 1.675×10^{-27} kilogram, 1839 times the mass of an electron—just slightly greater than the mass of a proton. Neutrons and protons make up the nucleus and contain almost all of the mass of an atom.

4.7 Discovery of Electron Energy Levels: Bohr's Model

Rutherford's experiments supplied much information about the nucleus. But what about the electrons? Many questions were still unanswered. How were the various electrons arranged? Were they moving? If so, how?

Niels Bohr, a young Danish physicist working in Rutherford's laboratory, took up the task of answering some of these nagging questions. He did not understand why electrons did not fall into the nucleus. Physicists in the first decades of the twentieth century conjectured that some unknown force was holding electrons at a distance.

The science of *spectroscopy* (the analysis of light emitted or absorbed by matter) helped Bohr devise a model that describes the movement of the electrons around the nucleus. When white light passes through a wedge of glass, known as a prism, it is dispersed to form a **continuous spectrum** of all the colors of a rainbow. Early in the 1800s, scientists discovered that when most materials were heated until they glowed, they produced spectra that contained bright lines. Each element had its own pattern of bright lines that was characteristic to it. These spectra were called **line** spectra. The explanation of these bright lines in emission spectra eluded scientists for several decades.

In 1913 Niels Bohr devised a mathematical atomic model that accounted for the line spectra. He suggested that electrons exist

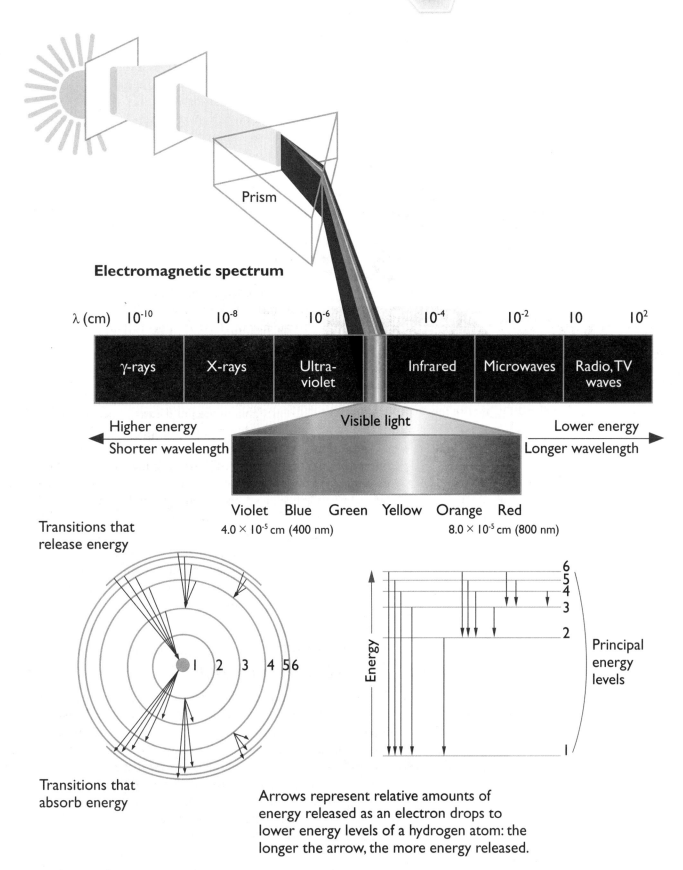

4-13 Production of a light spectrum

Quantize is the verb form of the physics term quantum: (L. *quantus*—how much, how large).

4-14 A continuous spectrum and a bright-line spectrum of hydrogen

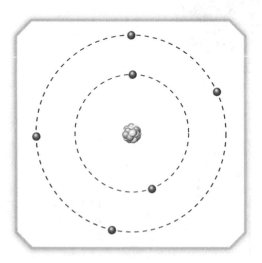

4-15 Bohr's "planetary" model of the atom

More than Atoms

When an atheist peers into the workings of subatomic particles, he is looking at the foundations of all existence according to his worldview. For him, all of life can be reduced to the random movements of these particles. If this is all the universe is, why do ethics matter?

only in certain energy levels outside the nucleus. Bohr's electron energy levels are called **principal energy levels.** The distance of the electron's orbit from the nucleus corresponds to a particular energy level. Electrons at greater distances from the nucleus possess higher potential energy levels.

Bohr suggested that electrons normally orbit the nucleus in the lowest allowed energy level. When the right amount of energy "excites" the atom, the electrons absorb the exact amount of energy required to jump up to a higher energy level. This change in energy levels is distinct, with no stopping between levels. Physicists describe these energy levels as **quantized,** meaning that energy emissions come only in certain amounts or quantities.

The movement of electrons between the various energy levels thus accounts for the bright-line spectra. Once the electrons are at their higher energy levels, they immediately fall back to a lower energy level and shed their excess energy. This excess energy is radiated as electromagnetic energy. The difference in potential energy between the higher and lower levels determines the frequency of the energy emitted. If the frequency of electromagnetic energy emitted falls in the range visible to human eyes, the color observed changes with the frequency. These colors in bright-line spectra give valuable clues about the energy levels in atoms. Using his mathematical model, Bohr was actually able to predict the exact locations of the lines in the hydrogen emission spectrum.

Theoretically, many principal energy levels exist, depending on how energetic the atom is. An atom's lowest energy state, or its **ground state**, contains six or seven levels. Figure 4-13 shows that outer energy levels are spaced so closely together that measuring the specific energy differences between them is very difficult. Therefore, electrons jumping between these levels emit low-energy radiation, such as invisible infrared light. On the other hand, electrons dropping from an outer energy level to the innermost levels release a large amount of energy, corresponding to the ultraviolet region of the electromagnetic spectrum.

4.8 Modern Physics and the Quantum Model

Bohr's model worked well for elements with one electron: neutral hydrogen and ionized helium and lithium. Bohr had developed the model by studying hydrogen. However, it did not accurately predict the spectra of elements with more electrons. Although Bohr's model was a great accomplishment, he soon saw that modifications to his theory were necessary. Scientists of the early twentieth century, including Bohr himself, set out to accomplish this task.

For more than a century scientists had known that light had properties of a wave—it could be refracted through a lens and it could be diffracted around a sharp edge. Light also could reflect off a flat surface like a particle. At the beginning of the twentieth century, Einstein's theories of matter and energy suggested that light consisted of massless particles called **photons.** During the 1920s, a young French physicist named Louis de Broglie suggested that if waves could behave like particles, then particles could behave like waves. This concept is known as **de Broglie's hypothesis**. Whether an object acts as a particle or as a wave depends on how the object is observed.

4-16 Louis de Broglie (1892–1987) was a French physicist who proposed that all matter acts like waves.

According to de Broglie's hypothesis, all forms of matter act like waves, including baseballs, pizzas, chemistry students, and smaller objects like electrons and protons. De Broglie waves increase in wavelength as the object's mass decreases and its speed increases. Such wavelengths are only significant for atomic and subatomic particles. The de Broglie wavelengths of large objects are so small as to be negligible, even at atomic sizes. On the other hand, for an electron, small whole numbers of de Broglie wavelengths can fit around the circumference of the electron's orbit in Bohr's model of the atom.

So, is an electron a particle, or is it a wave? It's both! The electron has a dual nature. The *wave-particle duality* of electrons explains the movement of electrons which seems to defy other scientific laws.

Scientists were further perplexed when they tried to locate electrons more precisely within the atom. In 1927, Werner Heisenberg determined that it is impossible to know both the energy or momentum and the exact position of an electron at the same time. This concept, known as the **Heisenberg uncertainty principle**, is a fundamental property of all submicroscopic systems. All information about atomic matter can be obtained only by photons, electrons, or other subatomic particles. Trying to find an electron through particle collisions changes its speed and location.

4-18 Werner Heisenberg (HI zen berg) (1901–76) was a German nuclear physicist who developed the principle of indeterminacy. In 1932, he received the Nobel Prize in Physics for his contributions to the development of quantum mechanics.

The uncertainty principle gave scientists a whole new view of the atom. Figure 4-19 shows a graph of probability versus distance for an electron in an atom's three lowest energy levels. The distance of highest probability corresponded very closely to Bohr's orbital radius. Bohr's precise, planetlike orbits were replaced by **orbitals**, or three-dimensional regions of probable position. An orbital is actually a four-dimensional map because it contains location information in three geometrical directions (x, y, and z), as well as the probability of finding the electron in time, as shown in Figure 4-19.

One physicist compared trying to locate an electron with a photon to trying to locate a flying seagull by throwing a baseball. You may indeed hit the seagull, but the bird's speed and direction after being struck will not be the same as before its collision with the ball!

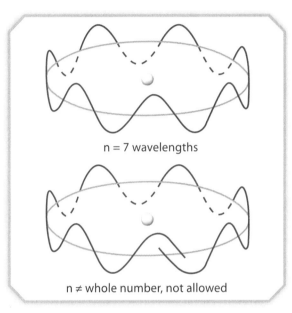

n = 7 wavelengths

n ≠ whole number, not allowed

4-17 De Broglie's electrons are described as waves that fit around the nucleus in whole numbers. In other words, an electron in the second principal energy level is exactly two waves long.

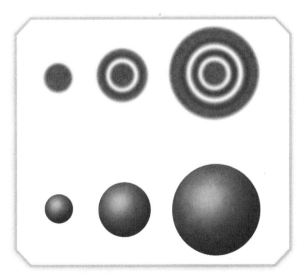

4-19 Probability versus distance graph for the first three s sublevels in an atom

In conceptual diagrams, orbitals look like fuzzy clouds without definite boundaries. The areas where electrons are most likely to be found are shaded the darkest. Away from the main part of the cloud, the shading becomes lighter as the chances of finding an electron decrease. As in the Bohr model, principal energy levels are arranged around the nucleus. Orbitals with low energies predict that the electron's average distance from the nucleus is small. Orbitals with high energies predict that the electron's average distance from the nucleus is great.

4A Section Review

1. The law of definite composition states that every compound has a definite composition by mass. What does that mean?
2. Upon what evidence was Dalton's model of the atom based? Compare that evidence with the evidence the Greeks used to support their model of the atom. Are these two models directly related?
3. What inventions permitted the investigation of the nature of atoms?
4. What four characteristics did Thomson and others determine about cathode rays that resulted in the discovery of the electron?
5. How did Thomson's discovery change the model of the atom?
6. After considering the evidence of his assistant's experiment, how did Rutherford describe the atom and on what evidence was his model based?
7. Why was the discovery of the neutron difficult? Why was it important for the development of the atomic model?
8. What does it mean to say that an electron's energy within an atom is *quantized*?
9. Describe the main differences between the standard Bohr model of the electron and the quantum model.
10. (True or False) One of the most revolutionary aspects of Dalton's model of the atom was that the atoms of each element had masses that were different from the masses of atoms of other elements.

4B The Electron Structure of the Atom

4.9 Introduction to Quantum Numbers

As evidence about the structure of electrons in the atom accumulated, scientists came to realize that every electron has its own energy-determined location within an atom. To identify each electron when the atom is in its ground state, physicists have developed an address system based on the electron's quantum mechanical properties. This addressing system uses a series of four quantum numbers. **Quantum numbers** are solutions to the various wave equations scientists use to describe the energy, momentum, and probable location of an electron.

4B Section Objectives

After finishing this section, you should be able to

- describe how the four quantum numbers give an electron's "address."
- given one quantum number, derive the other possible quantum numbers that depend on its value.
- relate quantum numbers, electron energy levels, sublevels, and orbitals.
- write the ground-state electron configuration and orbital notation for any element using only the periodic table as a reference.

4.10 The Principal Quantum Number (n)

The **principal quantum number** (*n*) identifies the principal energy level. It is the first quantum number. In essence, the principal quantum number indicates the average or most likely distance from the nucleus where an electron resides. The numbers are positive integers, beginning at 1.

$$n = 1, 2, 3, \ldots, 7$$

The highest value of *n* for large atoms in the ground state is 7, although it can be much higher in excited atoms. These numbers are the energy level numbers shown in Figure 4-13. All electrons in a given energy level have the same principal quantum number. The number of total electrons in a principal energy level equals $2n^2$, where *n* is the principal quantum number.

> The principal quantum number indicates the main energy level in which an electron is found. Thus, it is also called the *energy-level number*.

> The maximum number of electrons that can be found in an energy level *n* is $2n^2$.

4.11 The Azimuthal Quantum Number (l)

Within each principal energy level, electrons have greater probability of being found in certain regions, called sublevels. A series of small positive integers, called **azimuthal quantum numbers** (*l*), are solutions to angular momentum equations that describe these sublevels. The sublevels are usually symmetrical shapes arranged around the nucleus in orientations determined by the azimuthal quantum number. These sublevels are the orbitals described earlier in this chapter. They have distinctly shaped probability regions identified by the letters *s*, *p*, *d*, and *f*.

For each value of *n*, the principal quantum number, there are *n* possible values of the azimuthal numbers, ranging from zero to *n* − 1. The letters *s*, *p*, *d*, and *f* correspond to values for *l* of 0, 1, 2, and 3, respectively. Electrons in ground state atoms may have the following combinations of first and second quantum numbers:

> The azimuthal quantum number is also called the *sublevel quantum number*. The word *azimuth* means "a direction" in more general usage.

> For any *n*, *l* can have any integer value from 0 to *n* − 1.

4-1 Allowed Azimuthal Quantum Numbers for Each Principal Energy Level	
n	Possible azimuthal quantum numbers (*l*)
1	0
2	0, 1
3	0, 1, 2
4	0, 1, 2, 3
5	0, 1, 2, 3, 4
6	0, 1, 2, 3, 4, 5
7	0, 1, 2, 3, 4, 5, 6

> The blue values of *l* exist only for excited atoms, and the black represent possible numbers for an atom in its ground state.

4.12 The Magnetic Quantum Number (m)

The way that atoms respond to external magnetic fields reveals that magnetic sensitivity and the specific orientations of their electrons are related. The equations that define these orientations can be solved using small positive and negative integers. These numbers contribute the third quantum number for locating electrons—the **magnetic quantum number** (*m*). The magnetic quantum numbers describe the spatial orientation of the orbitals within the atom.

> The magnetic quantum number is also called the *orbital quantum number*.

For each sublevel, there are $2l + 1$ possible orbitals. For instance, in the third principal energy level ($n = 3$), and in the third sublevel ($l = 2$), there are ($2 \times 2 + 1$) or 5 orbitals where electrons can reside. Table 4-2 summarizes the possible orbitals for the first three principal quantum numbers.

> For each l number, the possible m numbers are 0 and $\pm 1, \pm 2, \pm 3, \ldots, \pm l$.

4-2	Allowed Quantum Numbers for the First Three Principal Quantum Numbers		
n	l	m	Total of possible orbitals per energy level
1	0	0	1
2	0	0	4
	1	−1	
		0	
		1	
3	0	0	9
	1	−1	
		0	
		1	
	2	−2	
		−1	
		0	
		1	
		2	

> As we have mentioned before, these orbitals may exist in the ground state of an atom.

Summary of Quantum Number Rules

1. The highest value of n for large atoms in the ground state is 7, although that value can be much higher in excited atoms.
2. There are n values for l, ranging from 0 to $n - 1$.
3. The number of values for m are given by $2l + 1$. The values of m are given by 0 and $\pm 1, \pm 2, \pm 3, \ldots, \pm l$.
4. Each orbital holds a maximum of two electrons with opposite spin, one having an m_s value of $+\frac{1}{2}$ and the other $-\frac{1}{2}$.
5. The maximum number of electrons that can fit into the nth energy level is given by $2n^2$.
6. No two electrons in an atom can have the same set of four quantum numbers.

4.13 The Electron-Spin Quantum Number (m_s)

Each orbital can hold only two electrons. These two electrons have opposite reactions to a magnetic field. To associate this property with common experience, scientists call it electron "spin," even though wave-like electrons do not actually spin inside the atom. The quantum numbers that describe this property are therefore the **electron-spin quantum numbers (m_s)**. The two possible values for the electron-spin quantum number are $+\frac{1}{2}$, indicated by a half arrow that points up, and $-\frac{1}{2}$, indicated by a half arrow that points down.

Further research confirmed that each orbital can contain only *one* electron with each kind of spin. If two electrons are in the same energy level, same sublevel, and same orbital, then they have the same n, l, and m values. Since they must have different m_s values, no two electrons in an atom can have the same set of four quantum numbers. This principle is known as the **Pauli exclusion principle**.

The four quantum numbers that we have just discussed uniquely identify each electron in the atom. Just as zip codes uniquely identify post offices, so quantum numbers uniquely identify electrons.

4.14 Sublevels and Orbitals

Scientists have developed models that describe the spatial arrangement of electrons. We will discuss those models in the following subsections, and at the same time we will see the orderly way in which God designed atoms. These theoretical models have been extensively confirmed by analysis of emission spectra and other kinds of tests.

Unlike the Bohr model, the quantum model subdivides all but the first principal energy level into **sublevels**, assigned l values. These sublevels are subdivided into orbitals assigned m values, and the orbitals may contain up to two electrons each differentiated by their spin (m_s) values. The four energy sublevels with l values of 0 through 3 have been assigned the letters s, p, d, and f respectively to avoid confusion when describing an atom's electron structure as it relates to chemical behavior. Because quantum theory predicts where an electron is *most likely* to be found, it is most accurate to call an illustration of a sublevel a probability plot.

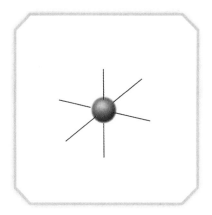

4-20 A representative s sublevel

The s sublevel

The s sublevel is the simplest of all. A probability plot of an s sublevel has a spherical shape. Unlike the other sublevels, it contains only one orbital, so the orbital is often called the s orbital, as well. Every energy level has one s sublevel that contains one s orbital. Remember that an orbital can hold only two electrons.

The p sublevel

The probability plot of a p sublevel consists of three dumbbell-shaped orbitals arranged symmetrically around the origin of a three-dimensional x-y-z graph. One lies on the x-axis, one on the y-axis, and one on the z-axis. Because of this, p orbitals are sometimes designated p_x, p_y, and p_z. Two electrons of opposite spin can occupy each of these sublevels. There are no p orbitals in the first main energy level, since it can have only one sublevel. Every neutral atom in its ground state with two or more main energy levels has a p sublevel. Since each orbital can hold a maximum of two electrons, a p sublevel can hold a maximum of six electrons.

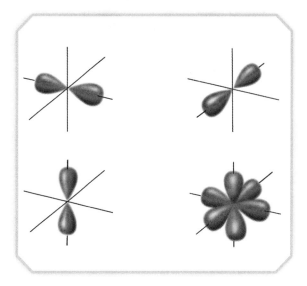

4-21 Representative p orbitals separate and combined into a p sublevel

The d sublevel

A d sublevel probability plot has a more complicated shape. The five orbitals in the d sublevel are oriented as shown in Figure 4-22. Atoms in their ground state containing three or more main energy levels may have d sublevels. A d sublevel may contain up to ten electrons.

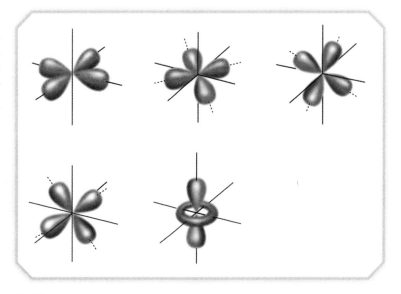

4-22 Representative d orbitals

The f sublevel

An *f* sublevel probability plot is even more complicated than the *d* sublevel plot. Figure 4-23 shows the seven possible orbitals. Only ground state atoms with four or more main energy levels may have an *f* sublevel. An *f* sublevel may contain up to fourteen electrons.

You may have noticed that some orbitals overlap one another. The orbitals overlap because they are probability regions for the locations of electrons, not definite locations.

The combination of all orbitals produces a roughly spherical region around the atom called the electron cloud. This cloud functions similarly to Dalton's spherical, BB-like atomic model, since the electron cloud of one atom tends to prevent other atoms from intruding into its space.

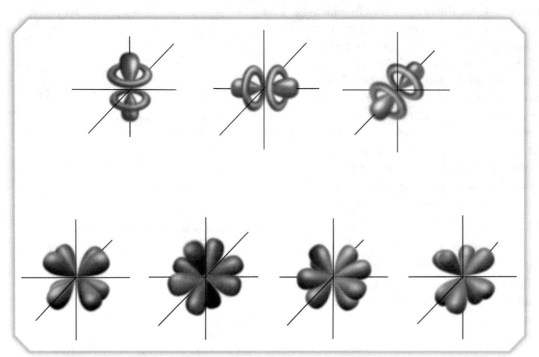

4-23 Representative *f* orbitals

As previously stated, this table applies to all known 117 elements in the ground state.

4-3 Energy Levels and Occupied Sublevels

Principal energy level	Sublevels that can be occupied
1	s
2	s, p
3	s, p, d
4	s, p, d, f
5	s, p, d, f
6	s, p, d
7	s, p

The first energy level contains one *s* sublevel containing only one orbital, which can hold two electrons. The total capacity of the first principal energy level is thus two electrons. Note that the *s* sublevel in any principal energy level can contain a maximum of two electrons. The second energy level has an *s* and a *p* sublevel. Since every *p* sublevel has three orbitals that can hold a maximum of six electrons, the total capacity of the second principal energy level is the sum of the two electrons from the *s* sublevel and the six electrons from the *p* sublevel—a total of eight electrons. The total capacities of the other principal energy levels can be determined by a similar procedure. Table 4-4 lists the total allowed electrons in each of the seven ground-state principal energy levels.

4-4 Allowed Electron Capacities of Energy Levels

Principal energy level	Allowed sublevels	Orbitals in each sublevel	Electron capacity of each sublevel	Total possible electron capacity
1	s	1	2	2
2	s	1	2	8
	p	3	6	
3	s	1	2	18
	p	3	6	
	d	5	10	
4	s	1	2	32
	p	3	6	
	d	5	10	
	f	7	14	
5	s	1	2	32
	p	3	6	
	d	5	10	
	f	7	14	
6	s	1	2	18
	p	3	6	
	d	5	10	
7	s	1	2	8
	p	3	6	

Compare the third column of Table 4-4 to the third column of Table 4-2 on page 84.

We will identify specific sublevels by adding their principal energy level number as a prefix. For example, the p sublevel of the third principal energy level is designated the 3p sublevel. We will use this notation throughout the textbook.

4.15 Relative Energies of Sublevels

In Bohr's model, as the energy level of an electron increases, the distance of the electron from the nucleus increases. In the quantum model, this simpler view was replaced by a more complex relationship. Principal energy levels become larger in size as their energy level number increases, but their sublevels too are ranked by their relative energies.

Beginning with the innermost energy level, the 1s sublevel has the least energy. The probability plot for the 1s orbital is the closest to the nucleus of any other orbital. The 2s and 2p sublevels have more energy since they are in the second principal energy level—one that is farther from the nucleus. Note that all three 2p orbitals have the same relative energy, and the electrons in these orbitals have more energy than those in the 2s sublevel. Farther out and higher in energy are the 3s and the 3p sublevels.

4-24 The sections of seats are like sublevels. The farther out you are from the center, the more sections there are, and the greater the capacity of that level.

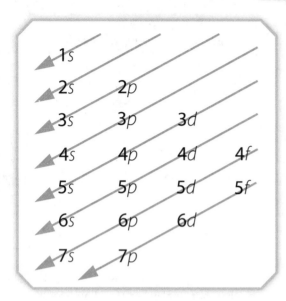

4-25 The diagonal rule helps you remember in what order the sublevels are filled.

From Table 4-4 you would expect that the next higher energy sublevel should be 3d. However, this is not the case in a ground-state atom; it is actually 4s. This break in order may seem confusing to you, but there is a reason for this apparent inconsistency. In the next subsection, you will learn how the periodic table reflects this filling order, but there is also a simple diagram that will help you. Figure 4-25 is sometimes called the diagonal rule because the parallel diagonal lines help you remember the filling order. If you start with the top arrow and work your way down, you can see the order in which the sublevels fill.

4.16 Electron Configuration and the Periodic Table

The periodic table is organized by the electron configuration of the elements. Atomic numbers increase from left to right across the rows. Each row number corresponds to the highest principal energy level for any given element. Certain regions, or blocks, of the table indicate the sublevel to which electrons are added.

For example, the first electrons to be added to a new energy level are always in the s sublevel, which can hold only two electrons. Therefore, the first two elements of each row are in what we call the s block of the periodic table. The first row can contain only two elements because the first energy level can hold only two electrons. (Helium, however, is placed in the right-hand column, which indicates that its energy level is full.) Similarly, the last six electrons to occupy the outermost energy level fill the p sublevels, so those elements occupy the p block of the table, which is six blocks wide.

Notice that the energy level numbers of the rows in the middle of the table with d sublevels are offset one number from the row in which they appear. This is because in these elements, electrons are added to the d sublevels after the next higher energy level has been established with an s sublevel. These elements occupy the d block of the periodic table. The two separate rows at the bottom of the table contain the elements in which f sublevels are filled. This is called the f block. The energy levels of the sublevels these elements fill are offset by two compared to the row in which they appear. This is because

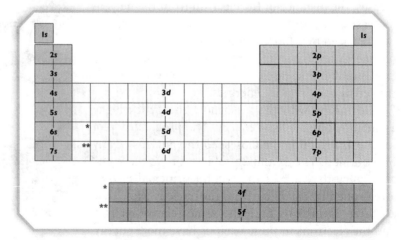

4-26 The periodic table reveals the order in which electron sublevels are filled as the atomic number increases.

they are filled only after the corresponding *d* orbitals are filled in the previous row.

After learning the significance of the sublevel blocks in the periodic table, you can now quickly determine the arrangement of electrons for any element. This method is the first of many ways the periodic table will be useful to you. The arrangement and features of the table will be discussed fully in Chapter 5.

4.17 Electron Configurations

You can now understand how electrons are arranged in a neutral atom in its ground state. In neutral atoms, the numbers of electrons and protons are exactly equal. Since the periodic table is arranged by increasing atomic number and since the atomic number is equal to the number of protons in an atom's nucleus, we can use the atomic number to determine the total number of electrons in a neutral atom.

> The number of electrons in a neutral atom in its ground state is equal to its atomic number.

The **Aufbau principle** states that the arrangement of electrons in an atom is determined by adding electrons to an atom with a lower atomic number, that is, one with fewer electrons. Progressing from hydrogen through the higher-numbered elements, electrons add to the least energetic orbital possible. Generally, lower-energy orbitals must be filled before higher-energy orbitals.

> Aufbau (OUF bough): (Ger. *Aufbau*—structure, architecture)

Hydrogen ($Z = 1$) has one electron that resides in the 1s sublevel. Hydrogen's **electron configuration**, or arrangement of electrons, is written as $1s^1$. The coefficient 1 tells us that hydrogen's electron is in the first energy level. The letter *s* indicates that the electron is in the *s* sublevel of the first energy level. The superscript above the 1s tells how many electrons occupy that sublevel. In this case, there is one electron in the 1s sublevel, which is also the only orbital in the atom.

Physicists and chemists also use a special notation, called **orbital notation**, to illustrate the electron configuration of an atom. Orbital notation consists of a horizontal line representing the orbital above which the orbital name is written. The two electrons occupying the orbital are written as up or down half arrows on the baseline. The up arrow represents the spin quantum number $m_s = +\frac{1}{2}$, and a down arrow represents $m_s = -\frac{1}{2}$. The orbital notation for hydrogen is written as follows.

$$\text{H}: \underset{1s}{\uparrow}$$

Helium ($Z = 2$) has two electrons when neutral. The second electron joins the first to fill the 1s sublevel. Helium's electron configuration is thus $1s^2$. The 1s sublevel is now full.

$$\text{He}: \underset{1s}{\uparrow\downarrow}$$

Lithium ($Z = 3$) has three electrons. Two electrons occupy the lower energy sublevel, the 1s. The third electron can only reside in the next higher energy sublevel, the 2s. Its electron configuration is written as $1s^2 2s^1$. Notice that lithium is on the second row of the periodic table. Similarly, the electron configuration of beryllium ($Z = 4$) is $1s^2 2s^2$.

> Note that hydrogen and helium are in the first row, or period, of the table. The numbers along the left of the table indicate the period. Also note that with the beginning of the second energy level, elements line up in the second period in increasing atomic number.

As shorthand for electron configurations, chemists often write the symbol of the element that fills the previous energy level in brackets before adding the next sublevels. Doing so shortens the notation and emphasizes the highest energy electrons. Because helium filled the first energy level, lithium's abbreviated configuration can be written [He] $2s^1$. Because this last column of the periodic table is a group called the noble gases, this abbreviated form is sometimes called noble gas notation. Review Table 4–5 to see the orbital notations and electron configurations of the next eight elements after helium.

4-5 Electron Locations for Elements 3–10

Atomic number	Name	1s	2s	2p			Electron configuration
3	lithium (Li)	↑↓	↑				[He] $2s^1$
4	beryllium (Be)	↑↓	↑↓				[He] $2s^2$
5	boron (B)	↑↓	↑↓	↑			[He] $2s^2 2p^1$
6	carbon (C)	↑↓	↑↓	↑	↑		[He] $2s^2 2p^2$
7	nitrogen (N)	↑↓	↑↓	↑	↑	↑	[He] $2s^2 2p^3$
8	oxygen (O)	↑↓	↑↓	↑↓	↑	↑	[He] $2s^2 2p^4$
9	fluorine (F)	↑↓	↑↓	↑↓	↑↓	↑	[He] $2s^2 2p^5$
10	neon (Ne)	↑↓	↑↓	↑↓	↑↓	↑↓	[He] $2s^2 2p^6$

When determining the electron configuration of carbon ($Z = 6$), you have a choice in which blank to place the half arrows. Its electron configuration is [He] $2s^2 2p^2$. But what about the orbital notation? Are the two electrons in the $2p$ sublevel in the same orbital, or are they in different orbitals?

$$\text{C: [He]} \quad \underset{2s}{\underline{\uparrow\downarrow}} \quad \underset{2p}{\underline{\uparrow\downarrow} \, \underline{} \, \underline{}} \quad \text{or}$$

$$\text{[He]} \quad \underset{2s}{\underline{\uparrow\downarrow}} \quad \underset{2p}{\underline{\uparrow} \, \underline{\uparrow} \, \underline{}}$$

This question was answered by Friedrich Hund after he carefully studied the movements of electrons within the orbitals of a single sublevel. He determined that for orbitals of the same sublevel, the lowest energy state exists when the number of unpaired electrons with the same spin is maximized. In other words, **Hund's rule** states that as electrons fill a sublevel, all orbitals receive one electron with the same spin before they begin to pair up. Therefore, by Hund's

---

Friedrich Hermann Hund (1896–1997) was a German physicist best known for his work on the structure of atoms and molecules. He was also a brilliant university lecturer.

rule, carbon's orbital notation is the second option. Notice also that there are three blanks for the *p* sublevel. These blanks stand for the three orbitals in the *p* sublevel, p_x, p_y, and p_z.

$$\text{C:[He]} \quad \underset{2s}{\underline{\uparrow\downarrow}} \quad \underset{2p_x}{\underline{\uparrow}} \; \underset{p_y}{\underline{\uparrow}} \; \underset{p_z}{\underline{}}$$

Using the Aufbau principle and Hund's rule, you can determine the electron configuration and orbital notation of any element.

EXAMPLE PROBLEM 4-1
Ground State Electron Configuration and Orbital Notation

Give the ground-state electron configuration of manganese ($Z = 25$) and draw its orbital notation.

Solution
Referring to Figures 4-25 and 4-26 for the order in which sublevels are filled, we find that the twenty-five electrons in a manganese atom fill the 1s, 2s, 2p, 3s, 3p, 4s, and 3d sublevels in that order. However, we write the sublevels in order of their principal energy levels.

$$1s^2\,2s^2\,2p^6\,3s^2\,3p^6\,3d^5\,4s^2$$

When writing the full form, always check to make sure that the sum of the superscripts equals the atomic number.

Since the electron configuration for argon is $1s^2\,2s^2\,2p^6\,3s^2\,3p^6$, we can substitute argon's symbol into this long expression to shorten it.

$$[Ar]\,3d^5\,4s^2$$

All sublevels prior to the final one are filled, but the 3d sublevel is only partly filled. Since there are 5 electrons in the 3d sublevel and 5 orbitals in any *d* sublevel, each orbital is occupied by one electron; no electron pairing occurs (Hund's rule). The following is the shorthand orbital notation.

$$\text{Mn:[Ar]} \quad \underset{3d}{\underline{\uparrow}\;\underline{\uparrow}\;\underline{\uparrow}\;\underline{\uparrow}\;\underline{\uparrow}} \quad \underset{4s}{\underline{\uparrow\downarrow}}$$

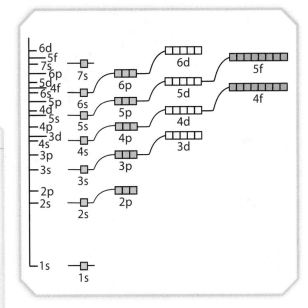

4-27 Energy increases as you move up the vertical axis. Each box is one orbital within a sublevel.

4.18 Using Atoms to Solve Problems

To effectively prevent, diagnose, and treat strokes, doctors need to be able to see inside the brain. A technology called magnetic resonance imaging (MRI), developed by Raymond Damadian and others in the 1970s, allows just that.

The MRI technique is based on the magnetic fields produced by atoms. All atomic nuclei spin, creating a magnetic field. When a person undergoes an MRI, his body is placed in the center of a stable and very strong magnetic field. The magnetic field aligns the spinning hydrogen nuclei in the body's water molecules. The machine also emits a radio frequency that makes those spinning hydrogen nuclei wobble like a spinning top. This wobbling is called *magnetic precession*. When the radio frequency is switched off, the hydrogen nuclei relax and realign in the magnetic field. This relaxation releases energy in the form of a radio signal that can be interpreted by a computer to form an image. Different tissues in the body

Raymond Damadian (1936–) pioneered magnetic resonance imaging. In 1977, he built his first machine, named "Indomitable." He received several awards for his invention but no Nobel Prize, though other scientists who improved on his work received one in 2003. Though many believe he was unfairly excluded because he was a creationist, Damadian is nevertheless an example of a Christian who used his scientific expertise to help others.

respond differently to the magnetic field and radio frequency, so different tissues appear distinct in the image.

MRI technology can be used to diagnose and direct the treatment of all kinds of diseases and injuries, such as cysts, cancer, tumors, infections, joint injuries, tendonitis, and multiple sclerosis. As for Senator Tim Johnson, doctors found that he had not suffered a stroke but instead had a congenital abnormality that caused the blood vessels in his brain to become tangled and bleed. They were able to perform emergency surgery to correct the problem. An MRI revealed that the language center of his brain had escaped injury, and he began to speak again. This action may have prevented a stroke for Senator Johnson.

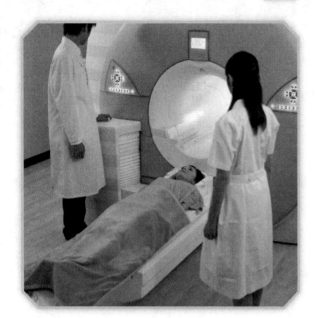

4-28 Modern MRI machine

4B Section Review

1. List the formal names and symbols for the four quantum numbers, their alternative names (if applicable), and all possible values in the ground state of an atom.

2. To which quantum numbers do the sublevel designations s, p, d, and f correspond? Give the maximum electron capacity of each.

3. What do the Aufbau principle and Hund's rule tell us about the electron structure of an atom in its ground state?

4. (True or False) The electron structure of an atom is determined by direct observation using x-rays.

5. (True or False) Only an odd number of orbitals is possible for any electron sublevel.

⊙ 6. For each of the following atoms, (1) give the abbreviated ground-state electron configuration, (2) draw the abbreviated orbital notation, and (3) find the set of four quantum numbers for the last electron added.
   a. silicon ($Z = 14$)   b. nickel ($Z = 28$)

⊙ 7. Tell whether each of the following sets of quantum numbers, given in the order n, l, m, m_s, is possible or impossible. If impossible, tell why.
   a. $(0, 1, 0, +\frac{1}{2})$
   b. $(4, 1, 0, -\frac{1}{2})$
   c. $(2, 0, 1, +\frac{1}{2})$

DS 8. Moving electrical charges create magnetic fields. Why is a hydrogen nucleus charged? Do you think electrons also generate magnetic fields?

DS ⊙ 9. Do some research on Raymond Damadian, the inventor of the MRI machine. Do you think that he was overlooked for the Nobel Prize because of his beliefs? Why or why not?

DS 10. MRI machines have their disadvantages. They weigh several tons, are extremely expensive, are very noisy, and can cause claustrophobia. Toward what goals should research in MRI development be directed?

4C Useful Element Notations

4.19 Introduction

Even though the quantum model might one day be modified or replaced, scientists generally agree on the basic subatomic particles—protons, neutrons, and electrons. Chemists and physicists have developed notations to easily identify the numbers and arrangement of these particles within the atom. These symbols will help you understand the various properties of each element.

4.20 Atomic Number

We have already mentioned that the atomic number of an element is the number of protons in the nucleus. The atoms of each element contain a unique number of protons, which determines the element. If the atomic number of an element changes, its identity also changes.

It is very important to know the atomic number of an element (Z). The size of the positive nuclear charge determines the number of electrons surrounding the nucleus, which in turn determines the chemical properties of the atom.

4.21 Mass Number and Isotopes

After Chadwick discovered the neutron in 1932, scientists could explain an observation that had puzzled them. When ionized atoms of a single gaseous element were "shot" between electrically charged plates, the electrostatic force deflected the atoms. However, not all the atoms were deflected the same amount. Some seemed to be heavier and were deflected less, while others seemed to be lighter and were deflected more. Since these atoms all had the same number of protons, the differences in mass were attributed to differences in the number of neutrons in their nuclei. Repeated experiments confirmed this observation.

If the number of neutrons can vary from atom to atom in an element, then the atoms of a given element are not truly identical, contrary to what Dalton and other nineteenth-century scientists thought. Physicists and chemists call atoms with the same number of protons but different numbers of neutrons **isotopes**. Isotopes of an element have essentially the same chemical properties. However, certain physical properties that depend on mass, such as density, evaporation, and diffusion rate, vary between isotopes. In order to identify the isotopes of a given element, scientists established the **mass number** (A). The mass number is a whole number that is equal to the sum of the protons and neutrons in an atom.

$$A = Z + N$$

Isotopes of an element are named by writing the element's name followed by a hyphen and the isotope's mass number. For example, the most common isotope of oxygen ($Z = 8$) is oxygen-16. An atom of oxygen-16 contains 8 protons and 8 neutrons.

When chemists want to indicate the specific isotope in a chemical expression, they use a special notation called **isotopic notation**. This notation is based on the element's symbol and includes the atomic number as a leading subscript and the mass number as

4C Section Objectives

After finishing this section, you should be able to

- determine an element's atomic number from its nuclear charge and electron configuration.
- define the term *isotope* and describe various notations to represent isotopes.
- state the number of protons, neutrons, and electrons present in an atom of the isotope when given an element written in isotopic notation.
- compare and contrast mass number and atomic mass.
- calculate the weighted average atomic mass of an element when given the percent abundances of its natural isotopes.
- define valence electrons and identify them in an atom based on its electron configuration.
- write the electron-dot notation for any atom when given its electron configuration.
- use electron-dot notation to describe how atom ionization occurs.

4-29 Standard format for isotopic notation. The letter X represents the element symbol. The isotopic mass number is A. Note that the atomic number (Z) is often omitted.

The number of neutrons for a given isotope can be easily determined by subtracting the atomic number from the mass number.

The name of an isotope is sometimes shortened to just the element's symbol followed by a hyphen and the mass number. For example, oxygen-16 can be written O-16 and carbon-14 can be written C-14.

a leading superscript. The generalized isotopic notation is shown in Figure 4-29. The isotopic notation for oxygen-16 is written

$$^{16}_{8}O.$$

Carbon-14 is written

$$^{14}_{6}C.$$

The identity of a particular isotope does make a difference. For example, the element hydrogen has three known isotopes: ^1H; ^2H (or D), called deuterium; and ^3H (or T), called tritium. The most important compound of deuterium is D_2O, or "heavy" water, which in large quantities is poisonous to human beings. Its only large-scale use is in nuclear reactors. Tritium is radioactive and highly unstable. It is an extremely toxic radiation source, even in small quantities. Tritium is used as a "trigger" in nuclear fusion bombs, known more commonly as hydrogen bombs.

Example Problem 4-2
Writing Isotopic Notation

Write the isotopic notations for the following elements using atomic numbers from the periodic table:
a. sodium-24 c. lead-208
b. hydrogen-2 (deuterium) d. uranium-235

Solution
a. sodium-24 (Z = 11): $^{24}_{11}Na$
b. hydrogen-2 (Z = 1): $^{2}_{1}H$
c. lead-208 (Z = 82): $^{208}_{82}Pb$
d. uranium-235 (Z = 92): $^{235}_{92}U$

Example Problem 4-3
Counting Neutrons in Isotopes

Determine the numbers of neutrons in the nuclei of the isotopes listed in Example Problem 4-2.

Solution
Use $N = A - Z$.
a. sodium-24: $N = 24 - 11 = 13$ neutrons
b. hydrogen-2: $N = 2 - 1 = 1$ neutron
c. lead-208: $N = 208 - 82 = 126$ neutrons
d. uranium-235: $N = 235 - 92 = 143$ neutrons

4.22 Atomic Mass

Once the chemists of Dalton's day started to analyze chemical compounds, they tried to determine the masses of individual atoms. They knew that water was made of hydrogen and oxygen, containing 8 g of oxygen for every 1 g of hydrogen. Because no one knew the real formula for water, Dalton assumed that there must be one atom of each. Thus, water's formula would have been OH. Therefore, he assigned the smallest atom, hydrogen, a mass of one and oxygen a mass of 8. We now know that the real formula is H_2O, so oxygen's mass is really 16.

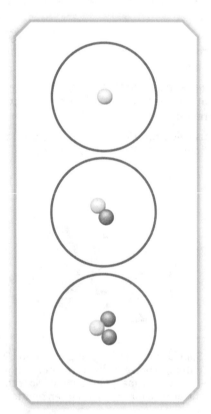

4-30 Isotopes of hydrogen, deuterium, and tritium

Chemists express these masses in **unified atomic mass units** (**u**) (often abbreviated amu). An atomic mass unit is currently defined as $\frac{1}{12}$ the mass of a carbon-12 atom, or as approximately 1.66054×10^{-27} kg. It is about the same size as a proton (1.0073 u) or a neutron (1.0087 u). The electron is much smaller (0.00055 u).

4-6	*Relative Atomic Masses of Some Common Elements*		
Element	Atomic number (Z)	Mass number (A)	Atomic mass* (u)
H	1	1	1.008
C	6	12	12.01
O	8	16	16.00
Na	11	23	22.99
S	16	32	32.07
Ca	20	40	40.08
U	92	238	238.0

*given to four significant digits

> The unit for the unified atomic mass unit is u. Though this unit is not an SI unit, it is approved for use with the SI. The symbol u will be used for atomic mass units when they appear with measurements.

> The atomic masses used in this book will have four significant digits. A few common elements showing their relative atomic masses are listed in Table 4-6.

Why are atomic masses not whole numbers? In nature, samples of even pure elements are mixtures of the element's isotopes, which have different masses. Atomic masses as printed in references such as the periodic table are the *weighted averages* of the element's isotopes as they occur in nature. A weighted average is an average that gives more importance to certain values than to others. In the case of atomic mass, more common isotopes are given more weight than less common isotopes. Thus, the listed atomic mass is closer to the mass of a typical atom than a simple average would be. For example, a sample of chlorine is composed of two chlorine isotopes, 75.77% Cl-35 (34.969 u) and 24.23% Cl-37 (36.966 u). The simple average mass of these isotopes is 35.968 u. However, because Cl-35 is three times as common, it is much more likely that a randomly chosen atom would have the smaller mass. Thus the weighted average atomic mass of a sample containing innumerable atoms will be closer to 35 u than to 37 u.

> Simple averages give equal representation to all quantities being averaged. Weighted averages take into account the relative number of each quantity being averaged. Some teachers use weighted averages to calculate grades.

Let's calculate the average atomic mass of a chlorine atom. In nature, 75.77% of chlorine atoms have a mass of 34.969 u per atom (Cl-35) and 24.23% of chlorine atoms have a mass of 36.966 u per atom (Cl-37). To find the weighted average of a sample of chlorine:

1. Convert the percents to decimals.
2. Multiply the decimal percentage by the mass of each isotope to find the total mass of each isotope in the sample.

   mass of Cl-35 atoms in sample: 0.7577×34.969 u $= 26.50$ u

   mass of Cl-37 atoms in sample: 0.2423×36.966 u $= 8.957$ u

3. Add the results to get the weighted average mass of the sample.

   26.50 u $+ 8.957$ u $\cong 35.46$ u

weighted average mass of a typical chlorine atom: $\cong 35.46$ u

> Note that the results of weighted average calculations often vary from published values in the place of the least significant digit. The isotopic masses are known to much higher precision than the values used in this textbook, so rounding errors may occur.

Remember also that since your final answer is a *sum*, it can have no more decimal places than the least precise value being added. Though there are three decimal places in 8.957 u, there are only two decimal places in 26.50 u; therefore, the answer can have only two decimal places.

> **EXAMPLE PROBLEM 4-4**
> *Computing a Weighted Average Atomic Mass*
>
> The element magnesium (Mg) has three naturally occurring isotopes: Mg-24 (23.985 u) with an abundance of 78.99%, Mg-25 (24.986 u) with an abundance of 10.00%, and Mg-26 (25.983 u) with an abundance of 11.01%. Determine the average atomic mass of magnesium.
>
> **Solution**
> First, express each percentage as a decimal; then multiply each by the atomic mass of the corresponding atom. Finally, add the results.
>
> $(23.985 \text{ u} \times 0.7899) + (24.986 \text{ u} \times 0.1000) + (25.983 \text{ u} \times 0.1101)$
> $18.945 \text{ u} + 2.4986 \text{ u} + 2.8607 \text{ u}$
> 24.3043 u
> $\cong 24.30 \text{ u}$
>
> As expected, the weighted average is closest to the mass of the magnesium-24 atom because it has the highest natural abundance.

> As you perform these kinds of calculations, be sure to observe the rules for significant digits provided in Chapter 2.

4.23 Valence Electrons

Another type of notation describes the electrons that are responsible for an atom's chemical behavior. The electrons in the outermost energy level are the most important electrons to study, because they are the ones most likely to be involved in chemical bonding. They are also the ones that give elements many of their physical properties. These special electrons are called **valence electrons**.

Valence electrons are not necessarily the last electrons to fill an orbital. Rather, they are the electrons that have the highest principal quantum number, *n*. These electrons occupy the highest numbered *s* and *p* sublevels, even in atoms that have higher-energy electrons occupying inner *d* and *f* sublevels. Study the following example to see how this concept works.

> **valence** (VAY lents): (L. *valentia*—strength or capacity). The valence of an atom is a measure of its capacity to form bonds.

> Valence electrons can *never* be in the *d* or *f* orbitals because these orbitals are always filled after filling the *s* orbital of the higher energy level.

EXAMPLE PROBLEM 4-5

Identifying Valence Electrons

How many valence electrons do neutral atoms of these elements have?
a. argon   b. nickel

Solution
a. The abbreviated electron configuration for argon ($Z = 18$) is [Ne] $3s^2 3p^6$. The outermost energy level—the one with the largest value for n—contains 2 s and 6 p electrons. The total number of valence electrons for argon is eight.
b. The electron configuration for nickel ($Z = 28$) is [Ar] $3d^8 4s^2$. Even though the electron configuration obtained by using the periodic table shows the $3d$ orbitals are occupied after the $4s$ sublevel, valence electrons are defined as those which occupy the outermost principal energy level—the fourth in the case of nickel. Therefore, nickel has two valence electrons. This number of valence electrons is typical for most of the metals in the middle of the periodic table.

4.24 Electron-Dot Notation

Sometimes scientists represent the valence electrons of an atom using **electron-dot notation**. Dots that represent valence electrons are placed around the element's symbol. Because valence electrons usually involve only the outer *s* and *p* electrons (a maximum of eight electrons), no more than eight dots are used.

There is no standard way of arranging the dots. In this textbook, we will adopt the following convention. The first two dots, which represent the *s* valence electrons, are placed in a pair to the right of the element symbol. For the next element in atomic number sequence, the third dot is placed at the top of the symbol, the next to the left, and the next under the symbol. These three unpaired electrons occupy the three valence *p* orbitals according to Hund's rule. The remaining three electrons pair with the single electrons in the

4-7 Representative Electron-Dot Symbols for a Series of Elements

Valence electrons	Electron-dot notation
1	Na·
2	Mg:
3	·Al:
4	·Si:
5	·P:
6	·S:
7	:Cl:
8	:Ar:

same order until the *p* sublevel is filled. Table 4-7 illustrates the electron dot notation for a series of eight elements.

In later chapters, we will discuss bonding between atoms. In bonding, the positions of the *s* and *p* electron dots shift around the element symbol to best illustrate bonds formed with adjacent atoms. However, the number of valence electrons does not change.

4.25 Ionized Atoms

Electrons move between energy levels within an atom when they gain or lose energy. Given enough energy, an electron can escape or be pulled from an atom entirely. Other atoms may acquire electrons to attain a more stable electron configuration. Losing or gaining an electron causes an imbalance in the number of protons and electrons, thus giving the atom an electrical charge. A charged atom is an *ion*.

Ionized atoms may have positive or negative charges. If an atom loses an electron, it will have a net positive charge because there will be one less negative charge compared to the positive nuclear charges of the protons. Positive ions are called **cations**. If a neutral atom gains an electron, it will have a net negative charge when compared to the positive nuclear charges of the protons. Negative ions are called **anions**.

Electron-dot notation can also be used to represent isolated ions. For example, if a neutral fluorine atom gains an electron, it acquires a charge of negative one. We indicate this by placing a – superscript to the right of the symbol to represent ionized fluorine.

$$:\ddot{F}: + 1e^- \rightarrow :\ddot{F}:^-$$

Similarly, a magnesium atom that lost two electrons would have a 2+ charge.

$$Mg: - 2e^- \rightarrow Mg^{2+}$$

> **cation** (KAT eye un). In early electrochemical experiments, positive ions were attracted to the negative electrode, or cathode. Thus, positive ions are named cations.
>
> **anion** (AN eye un). Similarly, negative ions, or anions, were attracted to the positive electrode, or anode, in the same experiments.

> ### Example Problem 4-6
> *Electron-Dot Notation*
>
> Write the electron-dot symbols for the atoms of the following elements:
> a. sulfur   b. boron   c. chlorine anion (Cl⁻)
>
> **Solution**
> a. Sulfur ($Z = 16$) has the electron configuration [Ne] $3s^2 3p^4$, and therefore has 6 valence electrons. Its electron-dot symbol is $\cdot\ddot{S}:\cdot$
> b. Boron ($Z = 5$) has the electron configuration [He] $2s^2 2p^1$ and therefore has three valence electrons. Its electron-dot symbol is $\dot{B}$.
> c. Chlorine ($Z = 17$) has the electron configuration [Ne] $3s^2 3p^4$ and therefore has seven valence electrons. As an anion, it acquires an additional electron, for a total of eight valence electrons. Its electron-dot symbol is $:\ddot{Cl}:^-$.

4C Section Review

1. Identify the symbol or notation that is used to represent the following information:
   a. atomic number
   b. mass number
   c. number of neutrons
   d. identification of different isotopes
   e. configuration of valence electrons

2. Compare and contrast an element's mass number and its atomic mass.

3. Determine the number of protons, neutrons, and electrons in the neutral atoms of the following elements.
   a. silicon-28
   b. calcium-44

4. How is the current atomic mass unit defined?

5. Which electrons in an atom are its valence electrons? Why are these electrons significant?

6. Write the abbreviated (noble gas) electron configurations for the elements listed below. How many valence electrons does each have?
   a. carbon
   b. fluorine
   c. potassium
   d. krypton

7. Explain the difference between a positive ion and a negative ion.

8. (True or False) Isotopes of a given element are chemically identical but may have different physical or nuclear properties (e.g., radioactivity).

⊙ 9. Write the electron-dot notations for all the elements in the following columns of the periodic table. (See pp. 108–109.)
   a. column 2
   b. column 15

   What generalization seems to be true about the electron-dot notation for elements in a column of the periodic table?

Chapter Review

Coming to Terms

Term	Page
atom	73
law of definite composition	74
cathode rays	75
electron	75
alpha particle	76
nucleus	76
proton	78
atomic number (Z)	78
neutron	78
continuous spectrum	78
line spectrum	78
principal energy level	80
quantize	80
ground state	80
photon	80
de Broglie's hypothesis	81
Heisenberg uncertainty principle	81
orbital	81
quantum number	82
principal quantum number (n)	83
azimuthal quantum number (l)	83
magnetic quantum number (m)	83
electron spin quantum number (m_s)	84
Pauli exclusion principle	84
sublevel (s, p, d, f)	85
Aufbau principle	89
electron configuration	89
orbital notation	89
Hund's rule	90
isotope	93
mass number (A)	93
isotopic notation	93
unified atomic mass unit (u)	95
valence electron	96
electron-dot notation	97
cation	98
anion	98

Chapter Summary

- Democritus, a Greek philosopher, was the first to propose that matter was made of particles. Atomism was not investigated scientifically until the seventeenth century.

- Quantitative chemistry began with John Dalton's particle model of the atom. He suggested that atoms of different elements were unique, with different masses and other properties. They were also indivisible and indestructible.

- J. J. Thomson discovered the electron in the late nineteenth century, after analyzing cathode rays in gas discharge tubes. He discovered that the atom could be subdivided. His plum-pudding model had negative electrons embedded in a positive substance.

- After the discovery of alpha particle radiation, Ernest Rutherford and his assistants discovered the atomic nucleus. At first, his nuclear model assumed that electrons were arranged in the space around the nucleus. Later he discovered that positively charged protons resided in the nucleus.

- Nearly three decades after the discovery of the proton, James Chadwick discovered and named the neutron.

- Niels Bohr determined that electrons could have only certain amounts of energy and that their energy corresponded to their distance from the nucleus.

- Bohr and many others refined the electron model of the atom, producing the quantum model of the atom in the first half of the twentieth century.

- Electrons and all matter behave both as particles and as waves, depending on how they are observed.

- Every electron in an atom has a unique set of four quantum numbers: the principal quantum number (n), azimuthal quantum number (l), magnetic quantum number (m), and the electron-spin quantum number (m_s).

- The quantum numbers define energy levels, sublevels, electron orbitals, and electron spin. As atomic number increases, electrons occupy those regions that increase in relative energy and average distance from the nucleus. In larger atoms, the higher principal energy levels overlap, and some inner sublevels are filled after outer sublevels.

- Electron configuration can be illustrated in three ways: electron-configuration notation, electron-orbital notation, and electron-dot notation.

- The atoms of an element can differ according to the number of neutrons in their nuclei. These different species of atoms are called isotopes. Isotopic notation is used to identify isotopes.

- Atoms that gain or lose electrons through any process become ionized. Losing electrons produces positive atoms called cations. Gaining electrons produces negative ions called anions.

Review Questions

Concept Review

1. Briefly describe each of the following atomic models:
   a. Dalton's model
   b. plum-pudding model
   c. nuclear model
   d. Bohr's model
   e. quantum model

2. Describe the discoveries or advancements that made each of these atomic models obsolete:
   a. Dalton's model
   b. plum-pudding model
   c. nuclear model
   d. planetary model

3. Explain how atoms can be electrically neutral even though they contain charged particles.

4. How did observations of emission line spectra lead to conclusions that the energy given off by excited atoms is quantized?

5. What unusual and unexpected aspect of electrons and, indeed, of all matter was discovered during the development of the quantum model?

6. Certain numbers are used as the values for quantum numbers. In general, why were those particular numbers chosen?

7. What is the electron capacity of the following?
   a. an orbital
   b. an s sublevel
   c. a p sublevel
   d. the second principal energy level
   e. the fourth principal energy level

8. How many orbitals are in
   a. an s sublevel?
   b. a d sublevel?
   c. the second principal energy level?
   d. the fourth principal energy level?

9. What is the main difference between electron-configuration notation and electron-orbital notation?

10. How can we know the number of electrons in a neutral atom? What is the symbol for the atomic number?

11. Describe an isotope of an element. Give two ways it can be represented.

12. Natural boron contains both B-10 and B-11 isotopes. Considering that boron's listed atomic mass is 10.81 u, which isotope is more common? Explain.

13. Compare and contrast valence electrons and the highest energy electrons in an atom.

14. Which electrons in an atom are represented by the dots in electron-dot notation?

15. Describe how an ion forms. What does *not* change when an ion forms?

True or False

16. The particle model of matter evolved continuously from Democritus around 400 BC until the time of Dalton in the early 1800s and on through the nineteenth and twentieth centuries.

17. The larger the object, the longer the wavelength of its de Broglie waves is.

18. The 3d sublevel electrons are at a higher energy than the 4s sublevel electrons that are in the next higher principal energy level.

19. When writing standard isotopic notation, the atomic mass is entered in the superscript position and the atomic number is entered in the subscript position.

Application

20. For each of the following atoms,
    i. write out the ground-state electron configuration;
    ii. write out the orbital notation;
    iii. show the number of electrons in each principal energy level.

    Example: carbon, C ($Z = 6$)
    i. [He] $2s^2 2p^2$

    ii. [He] $\underline{\uparrow\downarrow}$ $\underline{\uparrow}$ $\underline{\uparrow}$ $\underline{}$
          2s   2p

    iii. 2, 4

    a. oxygen, O ($Z = 8$)
    b. sulfur, S ($Z = 16$)
    c. potassium, K ($Z = 19$)
    d. titanium, Ti ($Z = 22$)
    e. bromine, Br ($Z = 35$)
    f. barium, Ba ($Z = 56$))

21. Draw a blank orbital notation chart that contains all seven energy levels (through the 7p sublevel). Instead of drawing in arrows, number the blanks according to the filling order. Follow the Aufbau principal and Hund's rule. Example: The first two levels should be labeled as follows:

    1s   2s   2p
    $\underline{1, 2}$ $\underline{3, 4}$ $\underline{5, 8}$ $\underline{6, 9}$ $\underline{7, 10}$, and so-on.

22. What are the possible values for the third quantum number (m) for an electron in the following sublevels?
    a. 1s
    b. 3p
    c. 4f
    d. 3d

23. If the first quantum number (n) for an electron is 2, what possible values may the other three quantum numbers have?

24. For each of the following atoms, tell how many protons, neutrons, and electrons are present.
    a. $^{9}_{4}Be$
    b. $^{45}_{21}Sc$
    c. $^{127}_{53}I$
    d. $^{133}_{55}Cs^+$

25. Fill in the blanks in the following table. Assume each is a neutral atom of the most common isotope. You may need to reference a periodic table.

Element	Symbol	Atomic number	Mass number	Electrons	Protons	Neutrons
hydrogen	H	___	___	___	___	0
___	___	___	20	___	10	___
___	___	29	65	___	___	___
___	___	___	104	44	___	___

26. Write isotopic notation for the atoms or ions that have these compositions.
    a. 14 protons, 14 neutrons
    b. 1 proton, 1 neutron
    c. 80 protons, 120 neutrons
    d. 26 protons, 30 neutrons, 23 electrons
    e. 52 protons, 78 neutrons, 54 electrons

27. A naturally occurring sample of the element gallium, (Ga) is a mixture of two isotopes. In a sample of gallium, 60.11% of the atoms have a mass of 68.9256 u and 39.89% of the atoms have a mass of 70.9247 u. Calculate the atomic mass of gallium.

28. State the number of valence electrons in the neutral atoms of the following elements. (*Hint*: Check Figure 4-26 to see in which sublevel block and column each element lies.)
    a. oxygen
    b. sulfur
    c. potassium
    d. bromine
    e. barium
    f. lead

29. Draw electron-dot notation symbols for each of the elements in the previous question.

30. State the charge of an ion if a neutral atom were to
    a. gain one electron.
    b. lose two electrons.

Elements

CHAPTER 5

5A The Periodic Table	105
5B Periodic Trends	113
5C Element Families and Their Properties	118
Facet	
Get 'em While They're Hot!	110
Stardust or Special Creation?	121

DOMINION SCIENCE PROBLEM

Making Ends Meet

In 1987, England and France began work on a monumental tunnel connecting England and France under the English Channel. The project was named the Chunnel. Many challenges confronted planners. Good ventilation, structural support, and funding for the planned 31 mi of tunnel were essential. Before the two train tunnels could be constructed, the crews had to build a service tunnel to be used for maintenance and emergency services. One crew began from the British side and one from the French side using two monstrous tunnel-boring machines. These machines weighed 1.36 million kg (1,499 tn) and measured 5 m in diameter and two football fields in length. They used a circular shield embedded with blades to cut a cylindrical hole through the rock at the rate of 4.6 m (15 ft) per hour, moving forward like an earthworm. It was critical that the two machines meet within a span of 250 cm. If they were off by more than that, the project would be in jeopardy, especially since the machines could not reverse direction.

5A The Periodic Table

5.1 Early Organizational Attempts

A description of how elements differ from one another is found in an unexpectedly simple form—the **periodic table**. The periodic table organizes essential information in a manageable format. The history of its development demonstrates the process of model making in science.

Alchemists in the Middle Ages suspected that some of the substances they worked with were pure, in the sense that they were not composed of other substances. However, it was not until 1793 that the first list of elements was published. The great French chemist Antoine Lavoisier presented a list of thirty substances that he declared could not be broken down into simpler substances. He used the word *element* to describe such materials. Unknown to him, several of his "elements" were actually compounds of oxygen with other elements. Lavoisier had not yet been able to decompose those compounds.

Shortly after John Dalton formulated his atomic theory in 1803, he produced his own list of elements. His list too contained many substances that were actually compounds, and he revised his list several times, eventually documenting as many as sixty elements. Dalton identified his elements and compounds with a complex symbol system that was not readily accepted by scientists of his time.

5.2 The Discovery of Element Periodicity

Early in the 1800s, chemists began to arrange the known elements systematically. When arranged by atomic mass, groups of elements with similar properties emerged. One of the first scientists to discover such common properties was Johann Döbereiner. In 1829, he announced that among the known elements he had observed several *triads*, or groups of three elements with similar properties. One of Döbereiner's triads was chlorine, bromine, and iodine. Each of those elements occurs as a gas with a distinctive color and properties similar to the others.

As more elements were discovered, Döbereiner's triads did not hold up. Elements with similar properties joined the triads to form quartets and quintets. However, Döbereiner's practice of grouping chemical elements into families with similar chemical properties was an important step in developing the concept of *periodicity* of the elements.

In 1864, John Newlands presented a classification scheme that, like Döbereiner's, organized elements by atomic mass. He noticed that similar elements fell into groups when arranged into seven columns. Having both musical and scientific training,

5-2 John Newlands (1837–98) worked to organize the elements and suggested the law of octaves.

5A Section Objectives

After finishing this section, you should be able to

- summarize early attempts to organize the elements.
- explain how the properties of the elements exhibit periodicity.
- describe the development of the modern periodic table.
- state the periodic law.
- describe the information that is provided in the cells of a periodic table.
- identify element periods, families, and series.
- recognize the format styles of the major types of periodic tables.
- identify the regions of the periodic table occupied by metals, nonmetals, and metalloids.

Complex Made Simple

As different from one another as they are, the elements can be neatly ordered in the periodic table. How is the combination of such complexity with such simplicity evidence of an intelligent Designer?

Johann Wolfgang Döbereiner (1780–1849) was a German chemist who investigated the properties of chemically similar elements.

5-1 Döbereiner's triads. The middle element's mass was approximately the average of the masses of the first and third elements in each triad.

Periodicity exists when some measurable property repeats on a regular basis in a sequential list or time sequence.

Newlands saw a correlation between the two fields. He labeled this observed relationship the law of octaves. His ideas, however, were not well received, and his parallel to music was ridiculed. Newlands' idea that atomic mass and chemical properties might be related was correct, but he did not undertake a detailed study of each element's mass and characteristics.

5.3 Mendeleev's Periodic Table

Credit for the development of the modern periodic table goes mostly to the Russian chemist **Dmitri Mendeleev**. Like Döbereiner and Newlands, Mendeleev arranged elements by their atomic masses. Elements with similar chemical properties appeared in the same rows. But unlike earlier chemists, Mendeleev did not just organize the elements that were already known; he also organized elements that had not yet been discovered.

Mendeleev noted that in many cases the atomic masses increased more than usual from those of the preceding elements. He reasoned that undiscovered elements belonged in the gaps, so he left blanks in his chart. Using information about the physical and chemical characteristics of nearby elements, he predicted the properties of the elements that would fit into the blanks. After its publication in 1869, Mendeleev's table was dismissed by most chemists until one of the missing elements, gallium, was discovered in 1875.

Another innovation of Mendeleev's table involved elements called *transition metals*. These elements did not fit into the major families of the chart, but they all had similar chemical and physical characteristics. Those similarities made them difficult to purify, and mixtures of two or more of these metals were incorrectly identified as elements during the latter 1800s and early 1900s. Mendeleev put them in the chart but did not let them interfere with the groupings of the other elements.

Mendeleev summarized his chart by formulating the *periodic law*. That law stated that the properties of elements vary with their atomic masses in a periodic way.

> Dmitri Ivanovich Mendeleev (men duh LAE uhf) (1834–1907) was the famous Russian chemist who established the structure of the modern periodic table. He spent most of his life as a chemistry professor and researcher in St. Petersburg.

> By 1870, sixty-one of the ninety-four naturally occurring elements were known.

> **Understanding the Universe**
> Even before protons were part of the atomic model, scientists were making advances which led to our current periodic table. Though we do not completely understand the universe, we can develop basic ideas about how the world works. What does this testify about God's nature as opposed to our own?

Tabelle II

Reihen	Gruppe I — R^2O	Gruppe II — RO	Gruppe III — R^2O^3	Gruppe IV RH^4 RO^2	Gruppe V RH^3 R^2O^5	Gruppe VI RH^2 RO^3	Gruppe VII RH R^2O^7	Gruppe VIII — RO^4
1	H = 1							
2	Li = 7	Be = 9,4	B = 11	C = 12	N = 14	O = 16	F = 19	
3	Na = 23	Mg = 24	Ai = 27,3	Si = 28	P = 31	S = 32	Cl = 35,5	
4	K = 39	Ca = 40	— = 44	Ti = 48	V = 51	Cr = 52	Mn = 55	Fe = 56, Co = 59, Ni = 59, Cu = 63
5	(Cu = 63)	Zn = 65	— = 68	— = 72	As = 75	Se = 75	Br = 80	
6	Rb = 85	Sr = 87	?Yt = 88	Zr = 90	Nb = 94	Mo = 98	— = 100	Ru = 104, Rh = 104, Pd = 106, Ag = 108
7	(Ag = 108)	Cd = 112	In = 113	Sn = 118	Sb = 122	Te = 125	J = 127	
8	CS = 133	Ba = 137	?Di = 138	?Ce = 140	—	—	—	— — — —
9	(—)							
10	—	—	?Er = 178	?La = 160	Ta = 182	W = 184	—	Os = 195, Ir = 197, Pt = 198, Au = 199
11	(Au = 199)	Hg = 200	Ti = 204	Pb = 207	Bi = 208	—	—	
12	—	—	—	Th = 231	—	U = 240	—	— — — —

5-3 Mendeleev's table contained gaps for undiscovered elements.

However, even after most of Mendeleev's "missing" elements were found, the table still had some problems. Arranging the elements in order of increasing atomic masses did not always produce a table that grouped similar elements. The periodic table needed refining.

5.4 The Modern Periodic Law

The solution to this apparent problem was suggested by one of Ernest Rutherford's assistants, a young Englishman named Henry Moseley. Even though Rutherford discouraged him from doing so at the time, Moseley started working in the relatively new field of the interactions of x-rays with matter. Moseley finally convinced Rutherford that this line of research would reveal information about the structure of the atom. He developed an experimental technique for counting the protons in an atom's nucleus. In 1912, he found that if the elements were arranged in order of increasing atomic numbers, problems in the periodic table disappeared. Moseley's work led to a revision of the periodic law. The modern **periodic law** states that *the properties of elements vary with their atomic numbers in a periodic way.*

During the early twentieth century, elements continued to be added to the periodic table. As noted earlier, the transition elements were problems for chemists because of their similar properties. With the invention of the mass spectrograph, a device that uses electrical charges to sort particles by their mass, these elements were finally isolated. With the exception of promethium ($Z = 61$), all elements with atomic numbers up through 92 (uranium) had found their place on the periodic table by the beginning of World War II.

5.5 Artificial Elements

The only major additions to the periodic table occurred during and after World War II. Experiments with nuclear energy produced very heavy elements in man-made nuclear reactions. These elements were collectively called the **transuranium elements** because their atomic numbers are greater than 92, uranium's atomic number. Just two of these elements, neptunium and plutonium, are found in trace amounts in nature. All the rest occur only as products of artificial processes in nuclear reactors or bomb explosions. Currently, scientists have produced the elements through 118, with the exception of 117. Several of these new elements have been produced by only one laboratory, and so their discoveries have not yet been confirmed.

5.6 The Structure of the Periodic Table

The periodic table has a cell for each element. In this textbook, the following basic information is included for each element.

- *Atomic number.* The element's atomic number appears at the top of the cell. The atomic number is the number of protons.
- *Name.* The name of the element appears above its symbol.
- *Symbol.* The one-, two-, or three-letter symbol is the most prominent item in the cell.
- *Atomic mass.* Located under the symbol is the weighted average atomic mass. The mass is given in atomic mass units (u). In our

Henry Gwyn Jeffreys Moseley (1887–1915) was an English physicist who found that the frequencies of x-rays scattered by the elements are related to the number of protons in the nucleus.

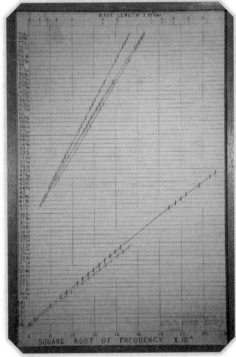

5-4 Henry Moseley's x-ray scattering data revealed that the electron structure of the atom was related to atomic number.

Because the elements 112 and beyond decay so rapidly into lighter elements, they do not have formal names assigned by the International Union of Pure and Applied Chemistry (IUPAC). They are referred to by names derived from the Latin or Greek roots for their atomic numbers.

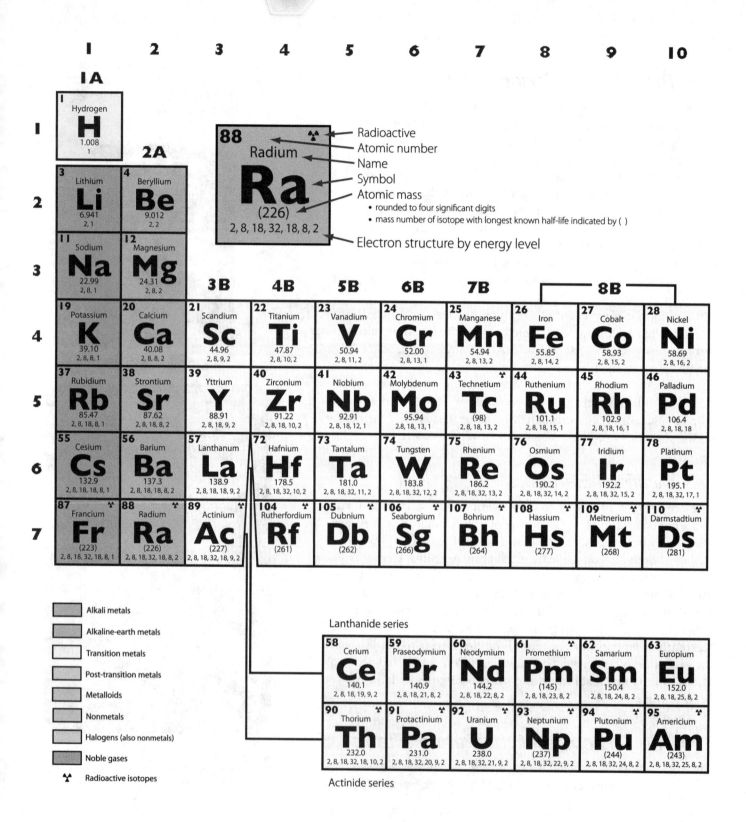

11	12	13	14	15	16	17	18
		3A	4A	5A	6A	7A	8A
							2 Helium **He** 4.003 2
		5 Boron **B** 10.81 2,3	6 Carbon **C** 12.01 2,4	7 Nitrogen **N** 14.01 2,5	8 Oxygen **O** 16.00 2,6	9 Fluorine **F** 19.00 2,7	10 Neon **Ne** 20.18 2,8
1B	2B	13 Aluminum **Al** 26.98 2,8,3	14 Silicon **Si** 28.09 2,8,4	15 Phosphorus **P** 30.97 2,8,5	16 Sulfur **S** 32.07 2,8,6	17 Chlorine **Cl** 35.45 2,8,7	18 Argon **Ar** 39.95 2,8,8
29 Copper **Cu** 63.55 2,8,18,1	30 Zinc **Zn** 65.41 2,8,18,2	31 Gallium **Ga** 69.72 2,8,18,3	32 Germanium **Ge** 72.64 2,8,18,4	33 Arsenic **As** 74.92 2,8,18,5	34 Selenium **Se** 78.96 2,8,18,6	35 Bromine **Br** 79.90 2,8,18,7	36 Krypton **Kr** 83.80 2,8,18,8
47 Silver **Ag** 107.9 2,8,18,18,1	48 Cadmium **Cd** 112.4 2,8,18,18,2	49 Indium **In** 114.8 2,8,18,18,3	50 Tin **Sn** 118.7 2,8,18,18,4	51 Antimony **Sb** 121.8 2,8,18,18,5	52 Tellurium **Te** 127.6 2,8,18,18,6	53 Iodine **I** 126.9 2,8,18,18,7	54 Xenon **Xe** 131.3 2,8,18,18,8
79 Gold **Au** 197.0 2,8,18,32,18,1	80 Mercury **Hg** 200.6 2,8,18,32,18,2	81 Thallium **Tl** 204.4 2,8,18,32,18,3	82 Lead **Pb** 207.2 2,8,18,32,18,4	83 Bismuth **Bi** 209.0 2,8,18,32,18,5	84 Polonium **Po** (209) 2,8,18,32,18,6	85 Astatine **At** (210) 2,8,18,32,18,7	86 Radon **Rn** (222) 2,8,18,32,18,8
111 Roentgenium **Rg** (272)	112 Ununbium **Uub** (285)	113 Ununtrium **Uut** (284)	114 Ununquadium **Uuq** (289)	115 Ununpentium **Uup** (288)	116 Ununhexium **Uuh** (291)		118 Ununoctium **Uuo** (294)

The names given to elements 112–116 and 118 represent the Latin and Greek names for their Arabic numbers.

64 Gadolinium **Gd** 157.3 2,8,18,25,9,2	65 Terbium **Tb** 158.9 2,8,18,27,8,2	66 Dysprosium **Dy** 162.5 2,8,18,28,8,2	67 Holmium **Ho** 164.9 2,8,18,29,8,2	68 Erbium **Er** 167.3 2,8,18,30,8,2	69 Thulium **Tm** 168.9 2,8,18,31,8,2	70 Ytterbium **Yb** 173.0 2,8,18,32,8,2	71 Lutetium **Lu** 175.0 2,8,18,32,9,2
96 Curium **Cm** (247) 2,8,18,32,25,9,2	97 Berkelium **Bk** (247) 2,8,18,32,27,8,2	98 Californium **Cf** (251) 2,8,18,32,28,8,2	99 Einsteinium **Es** (252) 2,8,18,32,29,8,2	100 Fermium **Fm** (257) 2,8,18,32,30,8,2	101 Mendelevium **Md** (258) 2,8,18,32,31,8,2	102 Nobelium **No** (259) 2,8,18,32,32,8,2	103 Lawrencium **Lr** (262) 2,8,18,32,32,9,2

Get 'em While They're Hot!
Facets of Chemistry

In the early days of the periodic table, scientists toiled over lab samples to isolate elements from compounds. They discovered many elements through these separation procedures. Eventually, nearly all of the naturally occurring elements were discovered and named.

Today, the periodic table is growing through a very different process. Nuclear physicists have learned how to artificially create new elements by smashing atoms of natural elements together. Only a small percentage of the collisions form new atoms, and those that do form in these reactions are radioactive. To further complicate the research, most exist for just fractions of a second before protons and electrons escape and the elements are transformed into other elements.

The most recently formed elements are called the *super-heavy elements*. They begin with rutherfordium ($Z = 104$) and currently extend to element 118. None of the labs currently involved in this research have yet announced the discovery of element 117, but it could happen anytime.

As you can imagine, this type of research is highly specialized, and only a few labs participate. The process involves highly complicated equipment. A particle accelerator energizes the atoms and drives them together with enough force to cause fusion. Then, before the new atoms decay, special detectors gather data about the few new atoms that form.

The discovery of element 118, ununoctium, is a good example of how a super-heavy element is created. A team of scientists from both the Lawrence Livermore National Laboratory in Berkeley, California, and the Joint Institute for Nuclear Research in Dubna, Russia, bombarded californium ions with large calcium ions. Three different times in 2002 and 2005 they were successful in forming a single new atom. This new atom contained 118 protons and 176 neutrons. These new atoms did not hang around long; within a thousandth of a second they decayed to element 116, then decayed to 114 and 112, then split in two.

Ununoctium (Uuo) has a special, three-letter symbol while it awaits a formal name. When a laboratory discovers a new element, they publish their results in a scientific journal. If the research is deemed valid by their peers, a temporary name and symbol are granted by the International Union of Pure and Applied Chemistry (IUPAC). This simple naming system assigns Latin or Greek roots to the three digits of the element's atomic number. The element name consists of these three roots combined with an *-ium* ending. The symbol is composed of a single letter for each root. (See the table below.)

To get a permanent name and a two-letter symbol, research must be verified. Only after verification will the IUPAC consider a naming proposal. You might have noticed that all of the elements following uranium have names based on scientists or places. These are all scientists or cities associated with research on radioactive elements.

Roots for New Element Names

Digit	Root	Symbol
0	nil	n
1	un	u
2	b(i)	b
3	tr(i)	t
4	quad	q
5	pent	p
6	hex	h
7	sept	s
8	oct	o
9	en(n)	e

So, why are scientists working so hard to create new elements which exist only in labs for fractions of a second and will probably never be used to make anything? The prestige of being the first to create something is certainly part of the motivation, but there are also loftier goals. This type of research can reveal why some nuclei are more stable than others and even support or refute our current models of nuclear structure. Ultimately, it may help us discover if and why nature contains only a limited number of elements.

table this number is always given to four significant digits. If the atomic mass of an element is not known to that precision, the number is surrounded by parentheses.

- *Electron configuration.* The number and location of electrons in each occupied energy level in the neutral atom are listed below the atomic mass.
- *Physical properties.* Colors and icons give information about the physical properties of the element, such as whether it is a metal or nonmetal and whether it is radioactive.

Together the various cells form columns and rows in the periodic table. A vertical column of elements is called a **group** or **family** because they usually have similar physical and chemical properties due to their similar electron configurations. Horizontal rows of elements are called **periods** or **series**. The period number on the left side of the table indicates the highest principal energy level n that electrons occupy in ground-state atoms of that period.

We have adopted the **International Union of Pure and Applied Chemistry (IUPAC)** convention of numbering the columns from 1 through 18. The advantage of this scheme is that each family is uniquely identified. However, we have also included the **North American Convention Periodic Table** notation below the IUPAC numbers. In this convention, a combination of Arabic numerals and letters are placed above each column to indicate (to some extent) the valence electron structure of the elements in each column. For example, a 1 above a column indicates that all neutral atoms of the elements in the family normally have one valence electron. Likewise, a 2 shows that each member of the family normally possesses two valence electrons.

In the North American convention, the groups' Arabic numerals are followed by an *A* or *B*. Elements in the *s* and *p* blocks of the table are members of the main or **representative groups**. The Arabic numerals above these groups give a reasonably accurate indication of valence electron configuration. Arabic numerals above representative groups are followed by an *A*. Many transition metals, however, are exceptions to almost every electron-configuration rule. Arabic numerals above transition metal groups are followed by a *B*. *Inner transition metals*, those in the lowest two rows, are not numbered in either convention.

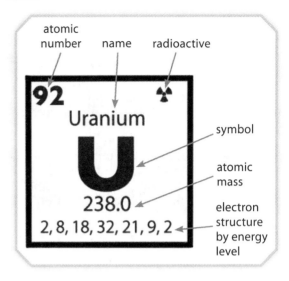

5-5 A typical cell of a periodic table and the information it contains

> In the North American Convention, numerals followed by an *A* indicate representative groups of elements and their normal number of valence electrons. Numerals followed by a *B* are transition metals. These numerals are less useful for predicting valence electron numbers or other chemical properties.

Example Problem 5-1

Getting Around the Periodic Table

a. What element in Period 2 is a member of Group 15 (5A)?
b. To what period and group does the element with the atomic number 10 belong?

Solution
a. Nitrogen
b. Neon (Z = 10) appears in the second row and in the far right column. It is in Period 2 and Group 18 (8A).

The left side and the middle of the table (Groups 1–12) contain metals. **Metals** are usually solid, lustrous (shiny), malleable, and ductile, and they are good conductors of heat and electricity. However, there are exceptions to these characteristics. For example, manganese is a brittle solid, and mercury is a liquid.

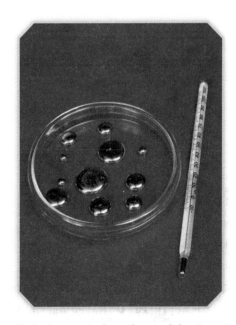

5-6 Mercury is the only metal that is liquid at room temperature.

5-7 Although most people associate the metalloid arsenic with poisoning, it is actually quite useful in manufacturing glass and transistors.

metalloid: metall- (Gk. *metallon*—metal, mine) + -oid (Gk. *oiedes*—like)

The right side of the table contains the nonmetals. The heavy, stair-step line toward the right side of the table marks the boundary between the metals and the **nonmetals**. Nonmetals are in Groups 13–18 above the line. The nonmetals are generally gases or soft, crumbly solids. Some exceptions include bromine, which is a liquid, and diamond (a form of carbon), which is the hardest known natural substance. Nonmetals in their natural forms are generally poor conductors of thermal energy and electrical current.

The **metalloids** are the elements immediately adjacent to the stair-step line. They share properties of both metals and nonmetals. Metalloids have metallic luster but tend to be crumbly, brittle solids. They can conduct electricity better than nonmetals but not as well as metals.

Two rows of elements, called the lanthanide series and the actinide series, are placed at the bottom of the table. These are *f* block elements where interior *f* sublevel electrons are added with increasing atomic number. The **lanthanide series** fits into the table immediately after lanthanum, and the **actinide series** fits into the table after actinium. If those elements were in their proper places within the periodic table, the table would be expanded into an oversized, unmanageable shape. Figure 5-8 shows this layout.

5A Section Review

1. What did scientists discover when they began to organize the elements?
2. What convinced the scientific community that Mendeleev's periodic table worked? How does this illustrate one aspect of a good scientific model?
3. State the modern periodic law.
4. Name the element that appears at the following address in the periodic table:
    a. a member of both Period 5 and Group 13 (3A)
    b. a member of both Period 4 and Group 2 (2A)
    c. a member of both Period 2 and Group 16 (6A)
5. To what period and family do the following elements belong?
    a. the element with the atomic number 17
    b. potassium
    c. the element with an atomic mass of 32.07 u

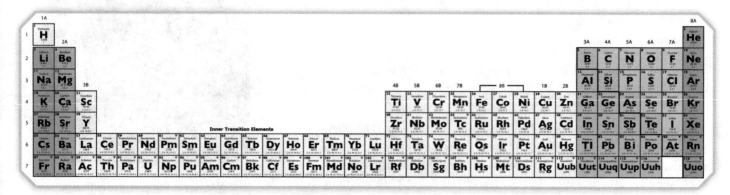

5-8 A cumbersome shape results when the inner transition metals are inserted into the main body of the periodic table.

6. Classify the following elements according to the region of the periodic table they occupy:
   a. calcium
   b. radon
   c. tungsten
   d. aluminum
   e. silicon
   f. oxygen

7. In general, how do the electron configurations of the elements in a vertical column in the periodic table compare?

8. What happens to the electron configuration of the outermost energy level as atomic number increases across a horizontal row?

9. (True or False) The shape and arrangement of the periodic table directly reflects the electron structure of the atoms composing the elements.

5B Periodic Trends

5.7 Introduction

In the last section you learned that the arrangement of the periodic table directly reflects the electron structure of atoms. As atomic number increases from left to right across a period, the elements become less metallic until they are nonmetals at the right side of the table. In this section we will examine various measurable properties that vary periodically with atomic number. These trends are important because they impact how elements interact with one another to form compounds or formula units. We will focus on how electron structure affects these properties.

5.8 Atomic Radii

The distance from the center of an atom's nucleus to its outermost electron is its **atomic radius**. Most atomic radii are measured using x-ray diffraction, in which a beam of x-rays with very short wavelengths is directed at a solid sample of an element. The resulting diffraction pattern of x-rays allows scientists to measure the distance between adjacent nuclei using simple geometry.

An atom's radius is determined by the electron cloud that surrounds it. The electrons cannot travel too far from the nucleus because of the positive attraction of the protons. The electrons of adjacent atoms repel each other, holding them at a distance. Since the radius of an atom depends on its electron cloud, one property of an atom that you would expect to show periodic variation is its atomic radius.

In general, the radii of atoms decrease in size as you move from left to right across a period of the periodic table. At first, this pattern

> **5B Section Objectives**
> After finishing this section, you should be able to
> - discuss the factors that affect atomic radius as atomic number increases.
> - compare ionic radii to the radii of neutral atoms.
> - discuss the significance and periodicity of first ionization energy, electron affinity, and electronegativity.
> - compare and contrast ionization energy, electron affinity, and electronegativity.
> - compare the relative intensities of each periodic trend for two or more elements, using a periodic table.

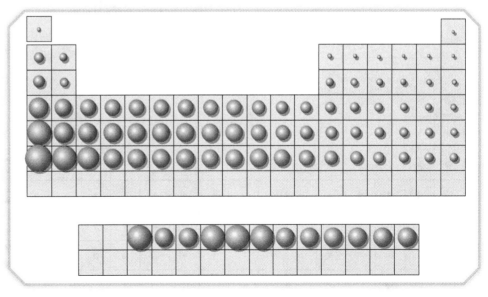

5-9 Periodicity of atomic radii. Radii decrease from left to right and increase from top to bottom.

may seem strange since each successive atom in a period has an additional electron. As atomic number increases across a period, electrons are added to the outermost shell (*s* and *p* sublevels) or to shells inside the outer shell (*d* or *f* sublevels). Shouldn't that extra electron increase the size of its cloud? Actually, as atomic number increases from left to right across a period, the positive nuclear attraction for the negative valence electrons also increases disproportionately. Remember that each element also has one more *proton* than the previous element on the table. Each additional positive charge is nearly twice as effective in attracting valence electrons as the additional electron is effective in shielding its fellow valence electrons from the additional nuclear charge. Thus, as the atomic number increases in a period, the outer-shell electrons are held progressively more tightly, and the average distance from the nucleus to the outer-shell electrons decreases.

The shell where an electron is added also affects how the atom's radius changes with increasing atomic number. When electrons are added to the outermost shell, the negative charge of the inner electron shells is unchanged, so the change in nuclear charge (number of protons) affects the outermost electron shell the most. However, if electrons are added to inner energy levels, such as the *d* or *f* sublevels, the outer electrons are shielded from the additional nuclear charge. Thus, atomic radius is affected less in the transition-metal section of the periodic table.

Atomic radius increases when moving down a group or column in the periodic table because another principal energy level (an electron shell) is added with each subsequent period. Higher energy generally corresponds to an electron with a greater average distance from the nucleus. Therefore, atomic radius increases as energy levels are added.

> Atomic radii generally decrease from left to right within a period and increase from top to bottom within a group. The largest atom is cesium.

5.9 Ionic Radii

We can extend our discussion of atomic radii to ions of single atoms. Remember, atomic ions are atoms that have lost or gained electrons. Cations—positive ions—are smaller than their parent atoms because they have fewer electrons than the fixed positive nuclear charge. The resulting charge imbalance more strongly attracts the remaining electrons, making the electron cloud smaller. In many cations, the entire outer electron shell or principal energy level is lost. This loss markedly reduces the ion's size as the excess nuclear charge draws in the electrons of the next lower energy level. Metal atoms form cations more easily than other types of atoms.

On the other hand, ions that have resulted from a gain of electrons (anions) are larger than their neutral atoms. The additional electrons are generally added to the outer main energy level, resulting

5-10 Periodicity of ionic radii for common ionic forms. Compare with Figure 5-9.

in a larger cloud. The cloud is larger for two reasons. First, the excess negative charge is more effective in shielding the other valence electrons from the positive nuclear charge, so the electrons are not as strongly held. Second, the added negative charge tends to repel the negative electrons in the same shell, forcing them apart and increasing the size of the shell.

5.10 Ionization Energy

Some atoms lose their electrons easily, while others stubbornly hold on to theirs. Atoms that form compounds either lose or take electrons to some extent. The ease with which atoms either acquire or donate electrons is a very important property for predicting chemical reactivity.

For a single neutral atom in the gaseous state, the minimum energy required to remove the first electron from its outermost shell to make it a cation is called its **first ionization energy**. Ionization energy values generally increase from left to right within a period and decrease from top to bottom within a group. This trend is the opposite of the trend in atomic radii. Figure 5-11 shows the periodic trend for first ionization energies.

Ionization energies generally increase from left to right across a period because the strength with which protons in the nucleus attract the outer-shell electrons increases. It thus becomes more difficult to remove an electron closer to and more strongly attracted to an increasingly positive nucleus. (See Figure 5-9.) First ionization energies decrease down a group for two reasons. First, the outer electrons are in increasingly higher energy levels and are thus farther away from the nucleus. The nuclear-charge attraction drops off quickly with distance. Second, the outer electrons are significantly shielded from the positive charge in the nucleus by the inner electron shells. This effect becomes more pronounced with more energy levels. You can see that ionization energy is clearly related to atomic radius.

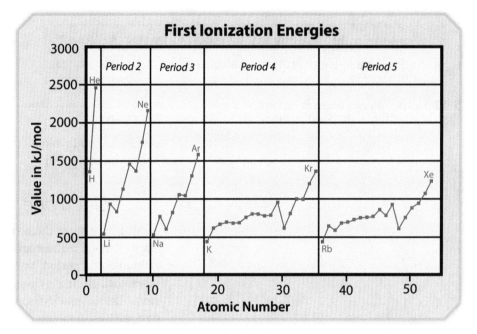

5-11 Periodicity of first ionization energies. The first and last elements of each period are indicated.

Notice that the noble gases have the largest ionization energies because they have a full outer energy level and do not gain or give up electrons easily. Because of this, they rarely react with other elements to form compounds. On the other hand, metals have the smallest ionization energies. Metals with the lowest first ionization energies tend to be the most chemically reactive. Their low ionization energies explain why they react so energetically with nonmetals.

A larger atomic radius means a smaller ionization energy, and a smaller atomic radius means a larger ionization energy. First ionization energies generally increase from left to right within a period and decrease from top to bottom within a group. Helium has the highest ionization energy.

5.11 Electron Affinity

Electron affinity is the amount of energy required to add an electron to a neutral atom to form a negative ion, or anion. Electron affinity is the opposite of ionization energy. Electron affinity measures how

Ionization energy measures how easily electrons are taken away from atoms to form cations, and electron affinity measures how easily electrons are added to atoms to form anions.

strongly an atom attracts additional electrons. This property is most affected by the fullness of an atom's highest energy sublevel. For incomplete sublevels, most elements will easily accommodate the additional electron, releasing energy. For elements where the highest energy subshells are full, such as Group 2, Group 12, and Group 18 elements, the additional electron is repelled unless energy is expended to insert the electron to the next higher energy subshell.

In general, electron affinity energies become larger from left to right along a period. As an energy level continues to fill, it has a stronger attraction for the electrons. The trend from top to bottom within a group varies depending on location in the table. For representative elements (*s* and *p* blocks), energies tend to decrease slightly from top to bottom in a group.

5-12 Electron affinity energies depend on the size of the atom and the fullness of the highest energy sublevels. (Shading indicates increasing electron affinity.)

5.12 Electronegativity

As you have seen, the attraction of the nucleus for the surrounding electron cloud influences periodic trends. In the elements toward the right end of a period, the relatively stronger nuclear charge pulls the outer-shell electrons closer to the nucleus and decreases the size of the atom. Thus, it makes the removal of electrons more difficult (increases first ionization energy) and encourages the addition of electrons (increases electron affinity).

The measure of the attraction between the nucleus and valence electrons is called **electronegativity**. In most cases, electronegativity is determined for atoms bonded *in molecules* which share electrons with other atoms. Thus, electronegativity is really a measure of an atom's ability to attract and hold electrons in a molecule.

The values of element electronegativities were determined mathematically rather than experimentally. The first chemist to quantify electronegativity was the brilliant chemist Linus Pauling. After examining the strengths of chemical bonds for many different compounds, he noted that fluorine held its electrons the

5-13 Linus Pauling developed the electronegativity scale of the elements.

Linus Carl Pauling (1901–94) was an American chemist. He was the only person to win two unshared Nobel Prizes, one for chemistry in 1954 and one for peace in 1962. Pauling, also considered the pioneer of molecular biology, believed in the integration of physics, chemistry, biology, and mathematics.

strongest and cesium held its the weakest. Using a fairly simple mathematical relationship, he arbitrarily assigned fluorine an electronegativity of 4 Pauling units and computed the other elements' values in relation to fluorine. Pauling's electronegativity values are shown in Figure 5-14.

Notice that Figure 5-14 provides values for only two of the noble gases. Other noble gases were not known to form compounds with any other elements in Pauling's time. Scientists later believed that electronegativities could be computed for all the elements using atomic properties distinct from bonding properties. Electronegativity values have now been determined for all the noble gases. These newer scales have been adjusted to agree with Pauling's electronegativity scale.

5-14 Pauling's electronegativities of the elements

EXAMPLE PROBLEM 5-2

Using Periodic Trends to Predict Electronegativity

For each of the following pairs of elements, use a periodic table and your knowledge of trends in atomic radius to predict which element in each pair has the larger electronegativity. Explain how you made your decision.
a. nitrogen (N) and phosphorus (P)
b. rubidium (Rb) and iodine (I)

Solution
a. Nitrogen (N) and phosphorus (P) are in the same group. Atomic radii tend to increase from top to bottom in a group because of increasing numbers of electron shells. As a result:
- Phosphorus is a *p* block element that has the same number of valence electrons as nitrogen. However, because $n_P = 3$ and $n_N = 2$, its valence electrons are farther from the nucleus than in nitrogen.
- Removing a valence electron from phosphorus will be easier and require less energy than removing one from nitrogen, which is smaller and so holds its valence electrons more tightly. Thus, phosphorus's first ionization energy is less than nitrogen's.
- The larger atom, phosphorus, has a smaller attraction for additional electrons than nitrogen does. Thus, phosphorus has a smaller electron affinity.

Therefore, nitrogen has the greater attraction for electrons in a molecule and so has the larger electronegativity.
b. Rubidium (Rb) and iodine (I) are in the same period ($n_{Rb} = n_I$). Atomic radii generally decrease across a period because the

Of the three periodic properties that measure the attraction between valence electrons and the nucleus of an atom, electronegativity has the widest use. It plays a central role in predicting how atoms chemically combine with each other.

increasing nuclear charge is twice as effective in attracting valence electrons as valence electrons are effective in shielding each other from the nuclear charge. As a result:

- Rubidium has a larger atomic radius than iodine. Rubidium's single valence electron is on average farther from the nucleus.
- It is more difficult to remove an outer electron from the more compact iodine atom, whose nuclear charge is more effective at attracting its valence electron. Thus, iodine has the larger first ionization energy.
- The larger atom, rubidium, has less attraction for additional electrons. Thus, it has a smaller electron affinity.

Therefore, iodine has a greater attraction for electrons in a molecule and so has the larger electronegativity.

5B Section Review

1. Write the term that matches each definition.
   a. the energy change that occurs when an atomic anion is formed
   b. a measure of an atom's size
   c. the energy required to give a neutral atom a +1 charge by removing an electron
   d. a measure of an atom's ability to attract or hold valence electrons in a molecule

2. How do the sizes of cations and anions compare to their corresponding neutral atoms?

3. In which direction in a period and a group do the following trends decrease?
   a. atomic radius   c. electron affinity
   b. ionization energy   d. electronegativity

4. Of the elements in Group 13 (3A), predict which has (a) the largest atomic radius, (b) the largest electronegativity, and (c) the smallest ionization energy. Explain your predictions.

5. If an element has a large first ionization energy, what are the trends for its atomic radius, electron affinity, and electronegativity?

5C Element Families and Their Properties

5.13 The Importance of Descriptive Chemistry

You may be thinking, "How does the periodic table affect me?" In plenty of ways! This section on descriptive chemistry will show many ways that elements can be used. **Descriptive chemistry** is the study of elements and the compounds they form. In this section, we will explore the physical and chemical properties of groups of elements, commonly called element families. You will see how the periodicity of atomic properties has been observed in the laboratory.

As you read this section, think about how descriptive chemistry can contribute to dominion science. Technologies such as detergents,

---

5C Section Objectives

After finishing this section, you should be able to

- name the families and series that define the different regions of the periodic table.
- state the general physical and chemical properties of each family or series.
- be familiar with some of the uses or occurrences for the significant elements in each family or series.

lasers, artificial joints, medicine, fuel cells, nuclear power, fertilizer, and even pest control can be beneficial to people. Science is one way to obey the great command to love our neighbors.

5.14 Hydrogen: A Family of Its Own

Hydrogen, the lightest element and the most abundant in the entire universe, has an electron configuration similar to that of the Group 1 (1A) metals. Yet hydrogen is often considered a family by itself because it is a gas at room temperature, which is a property of only nonmetals. This property is why its cell in the periodic table is colored differently and separated from the rest of the Group 1 elements.

Although other scientists probably produced hydrogen earlier, an Englishman named Henry Cavendish was the first to systematically collect and study it. In 1766, he prepared the gas by reacting a metal with an acid. The gas burned rapidly, even explosively, so Cavendish called it "inflammable air." Lavoisier later renamed it hydrogen, which means "water-former," because it produces water when combusted in air.

Physical Properties. Hydrogen is a colorless, odorless, and tasteless gas. It occurs naturally as a diatomic molecule (H_2). Because it has the least mass of all gases, its molecules move at high speeds and diffuse more quickly than other gases at the same temperature. Hydrogen molecules have little attraction for each other, so they remain in the gaseous state down to temperatures as low as –253 °C and solidify at –259 °C. Hydrogen has been compressed into a metallic form at pressures greater than 2.8 million atmospheres (2.8 Mbar).

Chemical Properties. Hydrogen has unique chemical properties. On the one hand, it has a single electron in its only occupied energy level, so it chemically reacts like other Group 1 elements. On the other hand, it is only one electron shy of filling its first and only energy level, so it is chemically similar to Group 17 (7A) elements. It is sometimes shown as a member of both families on the periodic table.

At room temperature, hydrogen molecules are very stable. But at high temperatures and pressures, molecular hydrogen can easily split apart and combine, sometimes explosively, with other elements. A mixture of hydrogen and oxygen molecules sparked by a flame or electrical discharge burns to form water vapor. When properly controlled, combustion of liquid hydrogen with liquid oxygen as rocket fuel provides a powerful and economical boost to spacecraft.

Hydrogen combines with many elements to form common but important compounds. It forms ammonia (NH_3) when it combines with nitrogen atoms. When combined with the elements of Group 17 (7A), hydrogen forms strong acids, such as hydrochloric acid (HCl). Occasionally hydrogen combines with the reactive metals of Groups 1 and 2 to form ionic compounds. In addition, hydrogen combines with carbon to form millions of compounds called **hydrocarbons**. We will study these important compounds when we survey organic chemistry and biochemistry in Chapter 18.

Commercial Sources. Hydrogen is found in only trace amounts in the earth's atmosphere. Large quantities of hydrogen are contained in gaseous byproducts of fuel and other refining processes. It can be liberated from these gaseous byproducts through a process called electrolysis, in which electricity is used to force

Henry Cavendish (1731–1810) was an English chemist and physicist who studied gases and electricity.

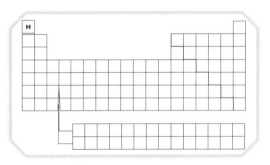

5-15 The hydrogen "family"

5-16 Most of the ammonia produced worldwide is used in agriculture as a fertilizer.

alkali (AL kah LIE): (Arab. *al-qily*—the salty ashes of certain marine plants used as fuel for cooking)

Sir Humphrey Davy (1778–1829) was an English chemist whose success with alkali metals came from separating potassium from molten potash, and sodium from common salt. In 1808 Davy was the first to isolate pure forms of magnesium, calcium, strontium, and barium. He pioneered the science of electrochemistry.

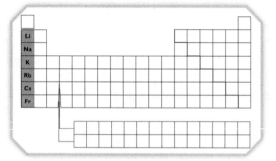

5-17 The alkali metals (Group 1)

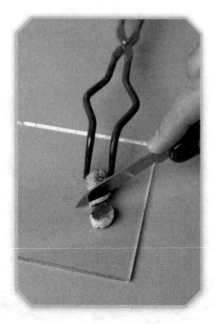

5-18 Solid alkali metals are soft enough to be cut with a knife.

a non-spontaneous reaction. Hydrogen is also obtained by passing steam over heated coke (impure carbon obtained from coal).

Uses. Ammonia manufacturers consume most of the hydrogen produced in the United States. Much of this ammonia is used in crop fertilizers. The space program is the next largest consumer, because it uses large amounts of hydrogen as rocket fuel. Producers of fats and oils combine hydrogen with their raw ingredients to form semi-solid fats in a process called hydrogenation. Fuel-cell technologies now under development will probably use hydrogen as a fuel.

5.15 Group 1: The Alkali Metal Family

Group 1 (1A) elements are metals that are very chemically reactive. They are called the **alkali metals**. Elements of this family were first isolated in 1807 by Sir Humphrey Davy. They typically form cations with a +1 charge. They consist of lithium, sodium, potassium, rubidium, cesium, and francium. Sodium is the most abundant alkali metal. It is found in star spectra and is the sixth most common element in the earth's crust.

Physical Properties. Like most metals, alkali metals conduct heat and electricity well and have a bright metallic luster on freshly cut surfaces. However, they have low densities, some less dense than water, and are very soft at room temperature.

Chemical Properties. A solitary, loosely held electron in the outermost energy level makes this family of elements very reactive. Alkali metals donate their electrons readily to attain the same stable electron configuration as the noble gases. Generally, these metals react readily with water. This reaction liberates hydrogen in the process, producing enough heat to ignite the hydrogen in an explosion. Because these metals are so reactive, they must be stored in oil so that they will not combine with atmospheric oxygen or water vapor. None are naturally found in their pure metallic form.

Uses. The most visible application of elemental sodium is in sodium-vapor streetlights, characterized by their yellow glow. Sodium is also used to manufacture baking soda, soap, rayon, and paper. Commercial salt substitutes are actually a potassium salt instead of the more common sodium salt. Potassium compounds serve as main ingredients for many industrial processes such as the production of fertilizers, soaps, glass, explosives, and fireworks. Rubidium is found in small quantities in tea and coffee. Cesium is used in atomic clocks. Lithium compounds make modern lubricants water-resistant and able to withstand extreme temperatures. Lithium battery cells have been extensively developed for a wide variety of uses. Francium has no known uses due to its scarcity and radioactivity.

5.16 Group 2: The Alkaline-Earth Metal Family

The chemical term *earth* originally applied to metal-oxygen compounds which are mostly geologic minerals that dissolve slightly in water. Some of these compounds were similar to compounds of alkali metals, so they were given the more specialized name **alkaline-earth metals**. Today the term applies to the metals in Group 2 (2A). These metals are all solid at room temperature and have typical metallic properties.

Stardust or Special Creation?
Facets of Chemistry

Evolution is a term most often associated with secular explanations for the origin of life. In a more basic sense, however, the word is also used to explain the origin of the chemical elements. All matter is composed of a limited number of distinct elements that must have come from somewhere.

To an evolutionary astronomer or chemist, the source of all naturally occurring elements is the stars. Scientists postulate this chain of events. Three minutes after the **big bang**, the giant explosion that supposedly started our universe some 13.7 billion years ago, the only elements that existed were hydrogen and a small amount of helium formed by the fusion of hydrogen molecules. Stars supposedly formed over time, and as those stars matured, hydrogen gradually converted to helium through nuclear changes. As the hydrogen was exhausted, the stars became hotter and denser. Those conditions allowed carbon, nitrogen, and oxygen to form from the helium through similar nuclear changes. At increasingly higher temperatures and pressures, protons were added to atoms to form elements as large as iron. Elements with atomic numbers beyond iron are thought to have developed through a related process called neutron capture. This theoretical chain of events is assumed to have started at the big bang and to be going on today as new stars come into existence, mature, and diminish.

So, for the past 13.7 billion years, stars have been serving as huge chemical cauldrons, producing and spewing out all of the elements that compose matter. That is supposedly enough time to have allowed some of the stardust to form other stars, planetary systems, our own planet, and eventually man. In essence, man is nothing more than a collection of cosmic dirt.

This theory seems to be supported by scientific data on the relative abundances of the elements in the universe, and there's a good reason for this correlation. The scientists who modeled this chain of events began with data on the relative abundance of each element today, and then worked backward to speculate what series of reactions and conditions might have contributed to their formation. Once they formulated their model, they discovered that their results exactly matched the current abundances. Imagine that! It's a little like guessing at the ingredients of a cake you just baked.

As far-fetched as this theory sounds, it really is the most plausible explanation for an origin of the elements by completely natural means. Thankfully, we do not have to put our faith in an explosion, space dust, and billions of years. Instead, "through faith we understand that the worlds were framed by the word of God" (Heb. 11:3).

God's Word is trustworthy. In Genesis 1 we read that God spoke the world, including its elements and the matter of which it consists, into existence. This same Creator Who formed man in His image from the dust of the ground and breathed into his nostrils the breath of life also fashioned the very atoms from which that dust was made. Such a creation makes man much more valuable and special than a collection of stardust. We are more precious. We are made in the image of God.

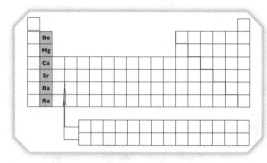

5-19 The alkaline-earth metals (Group 2)

They are denser, harder, and have higher melting points than the alkali metals. They usually form cations with a +2 charge, losing two electrons in order to have the same electron configuration as noble gases. Listed by increasing atomic number, the alkaline-earth metals of Group 2 are beryllium, magnesium, calcium, strontium, barium, and radium.

Beryllium is an important element in many minerals, including emeralds. Magnesium is the eighth most common element in the earth's crust and is an important component of the chlorophyll molecule. Calcium is the fifth most abundant substance in the earth's crust. It is a major component of bones and teeth. Together, magnesium and calcium compounds are the major constituents of hard water. Strontium's chemical reactivity is similar to alkali metals. Barium is chemically very similar to calcium. Radium is a naturally radioactive metal and exhibits luminescence.

Physical Properties. In pure samples of the alkaline-earth metals, freshly cut surfaces range from bright silvery to white in appearance. The metals quickly oxidize to a dull gray or yellow color. Densities are slightly higher than those of the alkali metals, but these metals are much harder. All the alkaline-earth metals are malleable.

Chemical Properties. Each neutral alkaline-earth metal atom has two electrons in its outermost *s* sublevel. The alkaline-earth metals typically donate the two electrons when they combine with nonmetal elements. The elements toward the bottom of the group are the most reactive because the electrons are the most loosely held. An example of this trend is how these elements react with water. Beryllium does not chemically react with water, magnesium reacts with steam, and calcium reacts vigorously with even cold water.

Uses. Elemental beryllium is used in x-ray tubes because it is transparent to low-energy x-rays. It can be mixed with other metals to produce many alloys critical in modern technologies. Similarly, magnesium is often used to make lightweight alloys for airplanes, automobiles, and boats. Military flares and some fireworks burn magnesium because it produces a brilliant, piercing white light. Epsom salt is a medically and industrially significant magnesium compound that was first discovered in the mineral water of Epsom, England. Magnesium hydroxide gives milk of magnesia its chalky taste and stomach-soothing ability. Calcium in the form of limestone is used commercially for building materials. Strontium is used for fireworks and flares since it emits a brilliant red flame. Barium is used in fireworks, rat poison, rubber, and linoleum, and for medical x-ray examinations of the gastrointestinal tract. Radium is used to treat specific forms of cancer.

Aircraft photo copyright © Boeing

5-20 Magnesium alloys reduce aircraft weight (bottom). Stalactites are formed from calcium carbonates (top left). Barium produces brilliant green fireworks (top right).

5.17 Groups 3-12: The Transition Metals

All the *d* block groups on the periodic table belong to the group of elements called **transition metals**. This name traces back to an earlier theory that the properties of these elements changed gradually but predictably into the properties of nonmetals. Today it is known that these elements are a distinct assembly of elements with their own chemical characteristics. The term *transition* still retains some significance, however, as a label for the physical location of these elements in the center of the periodic table. This region of the periodic table contains some of the most commonly used metals, such as iron and copper, as well as the precious metals silver, gold, and platinum.

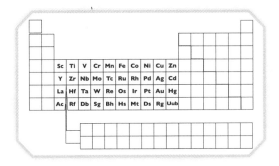

5-21 The transition metals (Groups 3–12)

Physical Properties. Transition metals exhibit the qualities we typically associate with metals. Unlike the first two metal families, most of these metals have high densities, are reasonably hard, and have considerable toughness or strength. They typically have a shiny luster on freshly cut surfaces, conduct heat and electricity well, and are more or less ductile and malleable. Many of these metals exhibit various magnetic properties. All except mercury are solids at room temperature.

Chemical Properties. Because their highest energy electrons occupy interior *d* sublevels, transition metals generally have similar chemical properties. However, there are also many exceptions, which are caused by the specific quantum mechanical relationships between the electron structure and nuclear charges. Metals showing little chemical reactivity, such as gold, silver, and platinum, can resist corrosion for centuries. Other more reactive metals, like iron and copper, corrode quickly when exposed to moist air. These chemical properties can be explained by the small energy difference between the outer *s* sublevel electrons and the highest energy *d* sublevel. Because of this, electrons have little difficulty jumping from one sublevel to another. As a result, the same atoms can combine with other atoms in a variety of ways. Nearly two-thirds of all naturally occurring transition metals occur as native minerals.

> Elements that are found as pure solids in nature are said to be native minerals. Familiar native minerals are gold, copper, silver, carbon (as diamond), and sulfur.

Uses. Modern societies use transition metals in countless ways. Most vehicles contain metals in frames, shells, and engines. Buildings and bridges are framed in iron supports. Exotic alloys compose artificial hip joints. Coins and jewelry are made of precious metals. Electrical wires use the conductivity of metal. In addition, most forms of life require trace amounts of iron, chromium, cobalt, copper, manganese, vanadium, and many other metal elements to function normally. Hemoglobin, an iron-containing complex in red blood cells, is responsible for transporting oxygen in the blood. However, excessively high concentrations of most metals are harmful to life.

5-22 Artificial hip joints of titanium alloy (top left), copper-wire armor (bottom left), and US coins (right) use many of the transition metals.

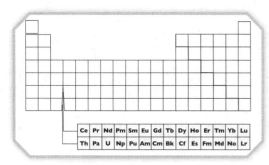

5-23 The inner transition metals

Very slight trace amounts of neptunium ($Z = 93$) and plutonium ($Z = 94$) have been discovered in nature due to nuclear changes in natural uranium.

The names *lanthanide* and *actinide* literally mean "lanthanum-like" and "actinium-like."

5.18 The Inner Transition Metals

The lanthanide and actinide series of elements are usually displayed below the main periodic table in two rows of fourteen cells. These are called series rather than periods because they actually fit into the sixth and seventh periods between the *s* block and *d* block elements for those periods, respectively. These two series are called the **inner transition metals**.

The lanthanide series extends from cerium ($Z = 58$) through lutetium ($Z = 71$). The lanthanide series elements were once called the rare-earth elements because they were difficult to identify and even more difficult to purify from their ores. They are now found in high concentrations in a considerable number of minerals and in low concentrations throughout the earth's crust.

The actinide series extends from thorium ($Z = 90$) through lawrencium ($Z = 103$). Only the first five elements in the actinide series have been found in nature. The rest have been formed only in man-made nuclear reactions. The first of the transuranium elements, elements with atomic numbers greater than 92, was discovered in 1940. These elements consist of many radioactive isotopes that are produced artificially and so are very unstable.

The transition metals scandium and yttrium are also sometimes included in the inner transition metals because of similar properties and close natural association with these elements.

Physical Properties. The majority of the lanthanide series is strongly paramagnetic. **Paramagnetism** is a property of materials that are weakly attracted by a magnetic field because of unpaired electrons. All the elements in the lanthanide series occur naturally except promethium. Freshly cut surfaces of pure forms of these metals are bright and silvery.

Chemical Properties. The highest energy electrons of the inner transition metals occupy the 4*f* and 5*f* sublevels. Although the atoms of the elements lanthanum and actinium contain no 4*f* or 5*f* electrons themselves, their outer electron structures resemble those of the lanthanide and actinide elements closely.

The lanthanide elements display great uniformity in their chemical behavior. Thus, they are difficult to purify from samples containing the other elements of the series. Unlike other metals, they usually give up their two valence electrons plus one or more of their highest energy 4*f* electrons when combining with other elements. Since these elements are relatively rare in the environment, they do not present a serious health hazard compared to some of the more common "heavy metals" such as mercury and lead.

Similar to the lanthanide series, the actinide series elements often give up most or all of their highest energy electrons, including their 5*f* sublevel electrons, when combining with other elements. The actinide series compounds are dangerous

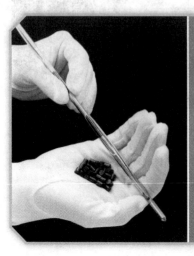 

5-24 Uranium, a familiar element of the inner transition metals, is used as "fuel" for nuclear power plants (left). Americium is used in some smoke detectors (right).

because they are radioactive. They emit tissue-destroying and cancer-producing particles and rays. Once ingested, these compounds can remain in the body indefinitely, and less than a microgram of a highly radioactive isotope can be fatal.

Uses. The lanthanides are used in the glass and television-tube industries and as catalysts in chemical reactions. Actinide series elements are valuable for use in nuclear reactions. Both uranium and plutonium have been used in atomic bombs because they produce devastating explosions. They can also be used as non-explosive fuel in nuclear power plants.

5.19 Groups 13–16: The Post-Transition Metals and the Metalloids

Post-transition metals, as their name implies, follow the transition metals on the periodic table. They include Groups 13–16 (3A–6A). This group of metals contains well-known elements such as aluminum, tin, and lead, as well as obscure elements such as thallium, indium, and gallium.

Metalloids, as we noted earlier, congregate around the stair-step line between the cells in Groups 13–17 (3A–7A). Their characteristics—luster, hardness, conductivity, and chemical reactivity—lie somewhere between those of the metals and of the nonmetals. Boron, silicon, arsenic, antimony, germanium, and tellurium are included in the metalloid family. The element astatine is considered a metalloid as well, although it is radioactive. There is probably less than 30 g of astatine in the entire earth's crust at any given time. Metalloids have a somewhat metallic luster, often showing a variety of colors beyond the typical silvery or gray metallic color. However, they tend to be more crystalline and brittle, with lower ductility and malleability than most metals. They can conduct electricity, but only under certain conditions. Consequently, they are called **semiconductors**. The post-transition metals and metalloids will be discussed later within their specific groups.

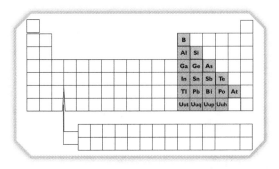

5-25 The post-transition metals and the metalloids (Groups 13–16)

5.20 Group 13: The Boron Family

Group 13 consists solely of post-transition metals and a metalloid. There are significant differences between the chemical properties of boron, a metalloid, at the top of the column and the five metals below it. These five metals are aluminum, gallium, indium, thallium, and ununtrium. Boron has some metalloid properties, but the characteristics we associate with metalloids are more recognizable in other groups.

Physical Properties. The two most economically important elements of Group 13 (3A) are boron and aluminum. Boron, in its impure form, is a crumbly black solid. It can transmit portions of the infrared spectrum (most materials absorb infrared) and it conducts electricity well only at high temperatures. Aluminum, on the other hand, is a very important metal because it combines high strength and low density, especially in its alloys. Unlike the metalloid boron, aluminum displays a silvery metallic luster, is easily machined, and is highly conductive. Aluminum is the most common metal in the earth's crust, found mostly in clays.

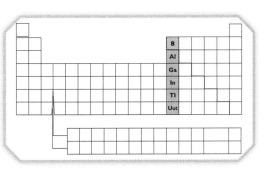

5-26 The boron family (Group 13)

The other Group 13 metals are increasingly metal-like as their atomic numbers increase. They have a silvery metallic luster, are conductive, and are malleable and ductile. Gallium is a metal that melts in your hand. Thallium is soft enough to be cut by a knife and is similar in other respects to the alkaline-earth metals.

Chemical Properties. Boron has the electron configuration of the elements in Group 13 (3A) but has several properties of silicon, which is in Group 14 (4A). Its small size allows the nucleus to hold electrons as though the nucleus had a greater positive charge than it actually does.

Aluminum is too chemically reactive to be found as a native mineral. It is usually bonded to oxygen atoms in aluminum ores, the most common of which is called bauxite. Like iron, aluminum corrodes in the presence of atmospheric oxygen. How then did aluminum get its reputation for being a durable, corrosion-resistant metal? The difference lies in the nature of the corrosion products. Iron oxide is porous and allows new oxygen molecules to penetrate deeper into the metal. Aluminum oxide, however, forms an impenetrable shield against further oxidation. Although fairly soft when pure, aluminum can be alloyed with many other elements to form many useful metals.

Uses. An oxide compound of boron is commonly used to make fiberglass. Another compound called borax softens water and helps clean clothes. Boron minimizes thermal expansion to reduce breakage in laboratory glassware. One isotope of boron is used to absorb neutrons in nuclear reactors as part of the reactor control system.

5-27 Aluminum is used for many power transmission lines (left). Laboratory glassware contains boron (right).

Aluminum and its alloys are commonly used, for everything from kitchen utensils and lawn chairs to aircraft and subway cars. Aluminum oxide forms a corrosion-proof reflective coating on telescope mirrors and other optical components. Although less conductive than copper, it weighs less and is used for long, high-voltage transmission wires in the power grid to reduce weight. Aluminum silicates form some of the most valuable natural gemstones. Aluminum oxide, known as corundum, is a very effective abrasive material used in sandpaper.

5.21 Group 14: The Carbon Family

As in Group 13, a wide range of properties is present in Group 14. The only true nonmetal, carbon, is at the top of the column in the table. It is followed by two metalloids, silicon and germanium. Two metals appear toward the bottom of the column—tin and lead. Little is known about a third metal at the bottom, ununquadium. It was identified by Russian physicists in 1999.

Carbon is more important than all the other elements in its family. Carbon compounds are the basis for life. So many biological carbon-based compounds are known that an entire area of chemistry, called biochemistry, is devoted to their study. In addition, there are literally millions of other carbon compounds that are studied in another area

5-28 The carbon family (Group 14)

of chemistry called organic chemistry. As noted in Chapter 4, an isotope of carbon, carbon-12, is used to define the atomic mass unit.

Physical Properties. Elements in the carbon family are all solids at room temperature. Carbon itself takes several forms. It can occur as an amorphous solid that is very soft and dull black in color. As graphite, it is a soft, black solid that feels slippery because it is made of sheets of carbon atoms that slide easily across each other. As diamond, carbon is a clear-to-slightly-colored, extremely hard, crystalline solid. Scientists are also engineering a whole new generation of materials using carbon nanotubes. These will be discussed in Chapter 19.

The nonmetals silicon and germanium are brittle solids with metallic luster. Both exhibit typical metalloid properties. Tin and lead are fairly common. Tin is a white or gray metal, malleable, and somewhat ductile. Lead is a soft gray metal, highly malleable and ductile.

Chemical Properties. All members of the carbon family have four valence electrons. However, the distance and strength of attraction of the valence-shell electrons and the nucleus seem to determine the basic chemical properties of these elements. Carbon shares its four valence electrons, and depending on the kind of atom it combines with, may form single, double, or even triple bonds. Carbon generally has a low chemical reactivity, but it combusts readily in oxygen to form carbon dioxide or carbon monoxide. Silicon does not react with air, water, or acids at low temperatures. Tin is relatively inactive and is often used to plate other metals to reduce corrosion. Lead is very inactive, especially in the presence of acids.

Uses. Amorphous carbon is used for electrodes in many different industrial electrolytic-decomposition processes. Graphite is used as a lubricant, as a binder in paints, and for drawing and writing instruments. Diamonds are useful as gemstones, although a much greater mass of diamond material is used as industrial abrasives than as gems. Charcoal, another form of carbon, is used in water and air filters to absorb organic impurities that cause objectionable odors and tastes.

> The natural forms of carbon depend on the bonds between carbon atoms.
> - Amorphous carbon: no particular pattern in atom linkages; non-crystalline
> - Graphite: separate layers of linked carbon atoms one atom thick that can slide across each other
> - Diamond: atoms that are cross-linked to four other atoms, forming an extremely rigid structure
> - "White carbon": sheets of cross-linked atoms many atoms thick that form crystals that can split light beams

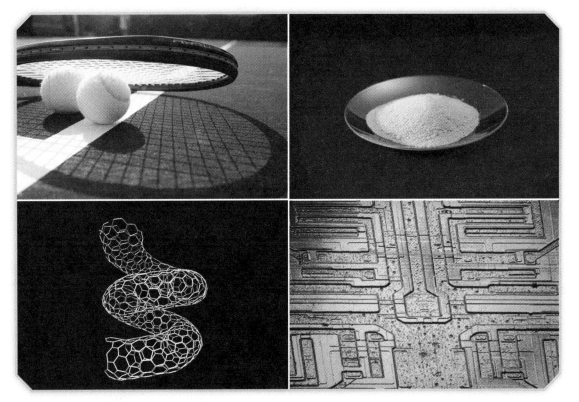

5-29 Graphite is often used to make sporting equipment (top left). Carbon structures called nanotubes are important in new developing technologies (bottom left). Sand is mainly silicon dioxide (top right). Silicon and germanium are key elements in microprocessors (bottom right).

The semiconducting properties of silicon and germanium are vital to the microprocessor industry. Silica, an oxide of silicon that makes up sand, is the basis for the glass industry and is a principal

nitrogen: nitro- (Gk. *nitron*—native soda, a form of sodium carbonate) + -gen (Gk. *genes*—forming)

Daniel Rutherford (1749–1819) was a Scottish chemist and physician who discovered nitrogen as a student and later described the properties of oxygen.

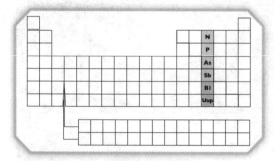

5-30 The nitrogen family (Group 15)

construction material. Silicates are the largest class of rock-forming minerals in the crust. Germanium in trace amounts is combined with semiconductor materials to enhance the function of microprocessor components.

Tin, when combined with copper, forms the alloy known as bronze. Bronze is used mainly in corrosion-proof plumbing and other hardware. Bronze is also a popular material for statues. Lead and its alloys are used to manufacture electrical storage batteries and ammunition. It is also used in radiation shielding for medical x-rays and around nuclear power reactors and weapons.

5.22 Group 15: The Nitrogen Family

Group 15 exhibits a dramatic range of properties from top to bottom. First is the element nitrogen, a gas, followed by four solids: phosphorus, arsenic, antimony, and bismuth. Nitrogen and phosphorus are nonmetals, arsenic and antimony are metalloids, and bismuth and ununpentium are post-transition metals. Neutral forms of these elements have five valence electrons, but they have an unusual ability to transfer or acquire electrons as needed when combining with other elements.

Elements in this group have an interesting history of discovery. In 1772, a Scottish physician named Daniel Rutherford first recognized nitrogen as an element. Rutherford also showed that nitrogen gas is not life-sustaining like oxygen is for animals or carbon dioxide is for plants. Nitrogen was originally named *azote*, meaning "lifeless." A German alchemist discovered phosphorus about one hundred years before the discovery of nitrogen. During his experiments, he distilled a substance that glowed in the dark. In later years, Lavoisier recognized this substance as an element and named it *phosphorus*, from a Greek word meaning "light bearer." Arsenic may have been discovered as early as 1250. It and its compounds are very poisonous, and it can be found as a native mineral as well as in many mineral compounds.

Physical Properties. Nitrogen normally exists as diatomic N_2 molecules in the gaseous state. It has no taste, no color, and no odor, and it accounts for approximately 78% of the volume of the earth's atmosphere.

Phosphorus does not occur as a native element because it is too chemically reactive. It exists in one of four or more forms that result from different arrangements of its atoms when synthesized as a pure compound. All are solid forms with a variety of colors. Some common forms of this element are white (or yellow), red, and black (or violet) phosphorus. White phosphorus is phosphorescent and is the original source for this term.

Chemical Properties. Nitrogen as an elemental gas is essentially inert. The atoms in these molecules are tightly bonded together and are difficult to split up. Only under relatively high temperatures can the element be forced to combine with other elements.

In contrast, white phosphorus is extremely reactive and will burn spontaneously in atmospheric oxygen. For this reason it must be stored under water and handled with tongs to avoid burns. It is poisonous even in small amounts.

Uses. Gaseous nitrogen is used in large quantities to create inert gas environments for applications such as manufacturing electronic components, welding, forcing oil to the surface in oil wells, and eliminating explosion hazards. Its largest use is in producing ammonia.

Phosphorus is used to manufacture matches, fireworks, signal flares, and other pyrotechnics, and it is becoming more important as a fertilizer. It is the basis for phosphoric acid, from which many other compounds are made. Phosphorus is also used in some alloying applications.

Despite its reputation as a poison and an agricultural pesticide, arsenic is finding more use as an additive to microprocessor chips. Its compounds are used to preserve animal skins and to manufacture glass. Antimony is also used in many semiconductor applications and can act as an alloying agent to harden metals. It also is combined with glass and other substances to make them heat resistant.

5.23 Group 16: The Oxygen Family

The oxygen family includes several reactive and important elements. Oxygen is the most easily recognized nonmetal element in this group, although sulfur and selenium are also nonmetals. Tellurium is a highly reactive metalloid with few industrial uses. Polonium is believed to be a reactive post-transition metal. The artificial element ununhexium is thought to be metallic and was first identified at the Lawrence Livermore National Laboratory in 1999.

In 1774, an English clergyman named Joseph Priestley discovered oxygen during a series of experiments in which he decomposed substances with the aid of the sun and a strong lens. When he focused the rays of the sun on mercury oxide (HgO), it decomposed into mercury and a gas that made flames burn brighter. Lavoisier found that when either sulfur or phosphorus was burned in this gas, it produced a gaseous compound that formed acids in water. He eventually gave the new element the name *oxygine*, which means "acid producer."

Men of ancient times knew of sulfur because it occurs in its native form and is produced as a byproduct when some ores are smelted for their metals. Selenium was discovered in 1817 by the great chemist Jöns Jakob Berzelius. It is often found as a natural trace element in copper ores.

The only naturally occurring metal in the oxygen family, polonium, was the first element discovered by Marie Curie in her research into radioactivity in 1898. It is radioactive and very rare in nature.

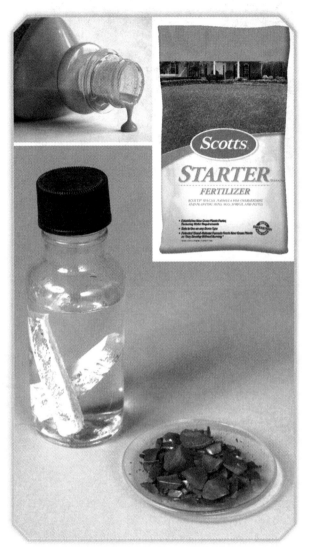

5-31 Nitrogen-containing substances are key ingredients in many fertilizers (top right). White phosphorus is used to form other phosphorus compounds, and red phosphorus is found in fireworks and some pesticides (bottom). Bismuth is found in some stomach-soothing medicines (top left).

5-33 Two forms of sulfur: a crystalline form consisting of interconnected S_8 rings (left) and the amorphous form chemists believe consists of helical chains of sulfur atoms (right)

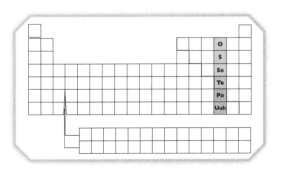

5-32 The oxygen family (Group 16)

Physical Properties. Oxygen, which is a colorless, odorless, tasteless gas, forms about 21% of the earth's atmosphere. It is slightly soluble in water and is the most abundant element by mass in the earth's crust. Through God's design, enough oxygen dissolves in lakes, rivers, and oceans to sustain fish and aquatic plants. Atmospheric oxygen exists in two forms: gas (O_2) and ozone (O_3). Lightning can convert odorless oxygen (O_2) into pungent ozone (O_3), which you can often smell after an electrical storm. The atmosphere is designed with a naturally sustained layer of ozone high in the stratosphere that screens out most of the harmful forms of ultraviolet radiation from the sun.

Sulfur exists in a variety of forms based on the arrangement of its atoms. Native sulfur is a brittle, yellow, crystalline solid. When sulfur is heated to 115 °C, it melts into a straw-colored liquid that can crystallize into another form. If the molten sulfur is quickly cooled by being poured into water, it forms amorphous globs with a plastic-like consistency. Eventually, the amorphous form transforms into the crystalline form.

Chemical Properties. While sulfur and oxygen do not have similar physical properties, they do have similar chemical properties. Oxygen is one of the most chemically active elements, having an electronegativity second only to fluorine. Its strong attraction for electrons accounts for its ability to combine with nearly every other element to form compounds called **oxides**.

5-34 Oxygen forms many important oxides with other elements, for example, cuprite (Cu_2O) (top) and amethyst, which is a type of quartz (SiO_2) (bottom).

Sulfur is reactive at room temperatures, but it does not match the reactivity of oxygen. Metals such as zinc, calcium, and iron combine with sulfur to form compounds called **sulfides**. Sulfur combines with nonmetals such as oxygen and Group 17 (halogen family) elements to form compounds such as sulfur dioxide (SO_2), sulfur dichloride (SCl_2), and sulfur dibromide (SBr_2).

Uses. As Lavoisier discovered, oxygen supports nearly all forms of combustion and is necessary for all life except anaerobic bacteria. It is a component of hundreds of thousands of compounds. In its liquid form, it is used as an oxidizer in liquid-fuel rockets.

5-35 Important sulfides: galena, a lead sulfide (PbS) (top); and pyrite, an iron sulfide (FeS_2) (bottom)

> The Bible refers to sulfur fifteen times when it describes brimstone, or "burning stone." These references associate sulfur with hell, the lake of fire, and God's judgment. The acrid fumes and searing blue flame of burning sulfur give a sobering picture of what separation from God for eternity will be like.

Sulfur atoms bonded into rubber make it supple, strong, and pliable in a wide range of temperatures. Like oxygen, sulfur forms a vast number of compounds. The majority of the sulfur used in industrial countries goes into the production of sulfuric acid, the most important industrial chemical.

Selenium is used in photocopying and in treating dandruff, acne, eczema, and other skin diseases. It is also useful in semiconductor components and photovoltaic cells.

Polonium is used primarily for its radioactivity. As it decays it emits unusually high-energy alpha particles that heat the immediate surroundings efficiently. Polonium is used in devices that ionize the air to eliminate accumulation of electrostatic charges in printing and photography equipment. Polonium is also a power source used in space probes and satellites.

5.24 Group 17: The Halogen Family

The **halogen** family is probably the most chemically uniform group of elements other than the noble gases. These elements are called halogens because they form salts when they react with reactive metals. They are so reactive that they are difficult to obtain and keep in their elemental forms. In order of increasing mass, the halogens are fluorine, chlorine, bromine, iodine, and astatine. With the exception of astatine, the halogens are nonmetals.

Fluorine's electronegativity is greater than that of any other element, making it very difficult to separate from other elements. Identified in 1886, it was one of the last halogens to be discovered. Chlorine, the second halogen on the periodic table, was first recognized in 1771 when a scientist combined hydrochloric acid (HCl) and manganese oxide (MnO_2). The greenish yellow gas that escaped was named chlorine, from the Greek word for "green." The next element in the family, bromine, stands alone among the elements as the only nonmetallic liquid at room temperature. Its irritating and even poisonous vapor has a pungent odor. For this reason it was named after the Greek word *bromos* ("stench") when it was isolated in 1826. Iodine was discovered in 1811 when seaweed treated with concentrated sulfuric acid gave off a violet-colored vapor that crystallized when cooled. Joseph Gay-Lussac soon identified the substance as an element and named it after the Greek word for "violet." Astatine, whose name means "unstable," is a highly radioactive element with no known stable isotopes. It is the only metalloid in the halogen family.

Physical Properties. Halogens show a definite trend in their physical properties. As their atomic numbers increase, their densities, melting points, and boiling points increase, and their colors exhibit increasingly darker hues. Fluorine is a pale yellow gas with a low density; chlorine is a denser, greenish yellow gas; bromine is a deep, reddish brown liquid; and iodine is a grayish black crystalline solid.

Chemical Properties. Halogens have relatively high chemical reactivities because of their large electronegativities. These elements exist in their pure forms as diatomic molecules and form acids when combined with hydrogen. They form a class of compounds called *salts* when combined with metals.

Halogens are particularly adept at taking electrons because of their high electronegativities, so they form compounds easily. Fluorine is extremely reactive and ignites when exposed to many substances, including water. The resulting chemical changes often take the form of spectacular releases of heat and light. When fluorine

halogen: halo- (Gk. *halo*—salt) + -gen (Gk. *genes*—producer, former)

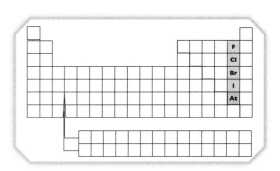

5-36 The halogen family (Group 17)

combines with metals, it often forms a protective layer of metallic fluoride that prevents all the metal from reacting.

As might be expected, the chemical properties of the other halogens are similar to those of fluorine, though less reactive. Fluorine, chlorine, bromine, and iodine are poisonous to humans in any but trace quantities.

Uses. The halogens have a variety of uses. Since World War II, fluorine has been incorporated into many products, including nonstick cookware and acids for etching glass. One common use of fluorine is as an additive to water supplies to reinforce tooth enamel and reduce cavities. Laundry bleach consists of a chlorine compound, and chlorine is also used in swimming pools to kill algae and sanitize the water. Chlorine dioxide is used to bleach wood pulp in paper production, and chlorine is also a component of table salt (NaCl). Bromine is used in photographic compounds and in producing natural gas and oil. It is also important in fumigants and sanitizer products. Iodine is important physiologically, because a lack of iodine stunts growth and causes an enlarged thyroid, or goiter. A radioactive isotope of iodine is used to treat cancer in tissues that normally use iodine, such as the thyroid. Iodine is also used with alcohol as a disinfectant and as an oxidizing agent. Potassium iodide is used in photography. Astatine has no significant uses, since it cannot be produced in any quantity and since it quickly decays. Its radioactivity makes it very hazardous to handle.

5-37 Fluoride toothpaste reinforces dental enamel (top left). The nonstick coating on some cookware contains fluorine (bottom left). Table salt (NaCl) is a compound of poisonous chlorine (center). Tincture of iodine is an effective antiseptic (right).

5.25 Group 18: The Noble Gas Family

The Group 18 elements are called **noble gases** because they do not react with other elements except under unusual conditions of pressure and temperature. Lord Rayleigh and Sir William Ramsay discovered one of these elements by carefully measuring the constituents of air. They noticed that nitrogen isolated from the atmosphere had more mass than nitrogen separated from pure ammonia. Suspecting that the atmospheric nitrogen contained some unknown substance, they separated out the residual gas and found that it would not react with other elements. Because of its chemical sluggishness, this gas was called argon, meaning "the lazy one." This discovery in 1894 was the first of the noble gases to be isolated.

More noble gases were discovered, some of which were isolated in measurable amounts from the atmosphere. Helium was discovered in a spectrogram of the sun in 1868, but was not identified on

5-38 The noble gas family (Group 18)

the earth until 1895 by Ramsay. Neon was discovered during a purification of oxygen. Scientists also isolated and identified krypton, meaning "hidden element," xenon, meaning "stranger," and radon. Radon is a gaseous byproduct of radioactive decay. Ununoctium, the heaviest of the noble gases, is a synthetic radioactive compound that lasts for less than one millisecond.

Physical Properties. All noble gases are colorless, odorless, and tasteless. Extremely low boiling points and freezing points indicate that the individual atoms of these gases have little attraction for each other.

Chemical Properties. As their name implies, noble gases tend to avoid combining with other elements. After many attempts to force noble gases to show some chemical reactivity, scientists were finally able to force all the noble gases except neon and argon to combine with other elements, including other noble gases. These elements have formed several short-lived, unstable compounds, such as helium difluoride (HeF_2), xenon difluoride (XeF_2), krypton tetrafluoride (KrF_4), xenon trioxide (XeO_3), and radon fluoride (RnF).

Uses. Balloonists prefer helium as their source of lifting power because it has an extremely low density and is not flammable like hydrogen. Helium is also used to replace nitrogen in the breathing gas mixture of deep-sea divers. The high pressures at great depths can cause nitrogen in a normal air mixture to produce a narcotic effect on the central nervous system, which can be very dangerous to a diver.

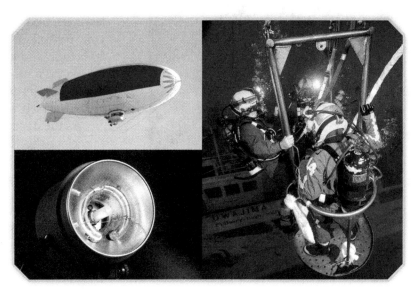

5-39 Modern dirigibles (top left) and deep-sea saturation divers (right) both depend on helium. Xenon is used in high-energy photographic strobe lights (bottom left).

Several of the noble gases are used to produce the many-colored "neon" lights in signs that advertise along busy streets. When an electric current passes through the low-pressure gases, they glow with a variety of colors depending on the gas. Fluorescent lights contain a mixture of argon and mercury vapor. Incandescent light bulbs contain either 100% argon or a mixture of argon and other noble gases at low pressure to keep the fragile metal filament from reacting with oxygen. Liquid neon is used as a refrigerant. Uses of krypton are limited because it is costly to obtain. Xenon is used primarily in lighting devices such as high-speed photographic tubes, high-intensity lamps, and lasers. Some cellular phone cameras contain xenon flash units.

5.26 Using Noble Gases to Solve Problems

Can you think of how engineers could be assured that the two tunnel-boring machines in the Chunnel project could meet with satisfactory accuracy? Obviously, technology that would cut down on machine-operator error or allow some kind of imaging of the excavation would really help.

The tunnel-boring machines used a computerized laser guidance system to keep them on the planned route. The laser in this system used helium and neon. The helium-neon laser transmitter was located toward the back of the machine and transmitted to targets at both the front and back of the machine. A computer measured the

5-40 The laser guidance system on a tunnel-boring machine

horizontal and vertical position every four feet to generate images of the excavation and pinpoint the machine's position on the plan line. The laser guidance system used by the tunnel-boring machines virtually eliminated operator error.

But how do helium-neon lasers work? *Laser* is an acronym that stands for *light amplification by stimulated emission of radiation*. The helium and neon gas in these lasers is excited, causing their valence electrons to jump to an excited energy level. As they relax, they emit packets of light energy, called *photons*. The photons emitted by helium and neon correspond to red visible light. This light is reflected off a mirror so that it comes out the end of the laser. These kinds of lasers are also used to scan barcodes at grocery stores. They are also widely used in engineering and surveying.

As the distance between the two teams narrowed in the service tunnel of the Chunnel, two men, Englishman Graham Fagg and Frenchman Philippe Cozette, were chosen to break through the thin wall of rock separating them. On December 1, 1990, England and France were connected amidst a flurry of camera flashes and international attention. The two tunnels met within 36 cm horizontally and 6 cm vertically, well within the 250 cm allowance. The guidance system had worked.

5C Section Review

1. Explain why hydrogen is often considered a family by itself.
2. How does the hardness of the alkali metals compare to that of other metals?
3. What is unusual about most of the inner transition metals in the actinide series?
4. Which Group 14 element is prominent in chemistry as well other branches of science? Explain.
5. Which element
    a. is the most abundant in the earth's atmosphere by volume?
    b. is the most abundant in the earth's crust by mass?
6. What is one major chemical property of all the noble gases that makes them useful? Why?
7. Of the elements Ag, Al, Ar, Au, C, Ca, Cl, Cu, Fe, Hg, O, Pb, U, and W, which
    a. is the principal metal found in human bones?
    b. is used as an energy source for generating electricity in nuclear power plants?
    c. forms a corrosion-proof reflective coating for telescope mirrors?
    d. can exist as diamond, graphite, or charcoal?
    e. is used to disinfect the water in swimming pools?
    f. fills fluorescent lamp tubes?
       (*Note*: More than one element may be used to answer a question.)

DS 8. How do you think xenon lamps and neon lights are similar in principle to helium-neon lasers? How are they different?

DS ⊙ 9. Research other uses of helium-neon lasers.

Chapter Review

Chapter Summary

- Humans have attempted to classify matter into elemental materials since before the time of Christ.
- The first ordered table of elements that attempted to account for similar properties was developed by the Russian chemist Dmitri Mendeleev in 1869. Mendeleev established the original periodic law when he ordered the elements by atomic mass.
- The English physicist Henry Moseley developed the modern periodic law, which states that the properties of the elements vary with their atomic numbers in a periodic way.
- The periodic table is divided into regions by the elements' properties. The three types of elements are the metals, the nonmetals, and the metalloids.
- The periodic table can be used to predict electron structure based on the location of an element on the table.
- Periodic trends can be identified on the periodic table. Those trends are attributable to the relative size of the nuclear charge compared to the size and arrangement of the negative charge in the electron energy levels.
- Atomic radius generally decreases within a period. Atomic radius normally increases with atomic number in a group.
- Electron affinity, first ionization energy, and electronegativity generally increase with increasing atomic number across a period and decrease with atomic number in a group.
- Hydrogen is often set apart from other elements because of its unique properties. It is the lightest and most plentiful element in the universe.
- The most "metallic" metals are on the left side of the periodic table. Metallic character is defined by how loosely an atom holds its valence electrons. Metals typically have a shiny metallic luster on fresh surfaces, and they conduct heat and electricity well. Many are magnetic or paramagnetic. Most are rigid solids.
- The transition metals include the majority of the important structural metals and the elements with which they are alloyed.
- The nonmetals of Groups 13–18 are characterized by being gases, liquids, or crumbly solids at room temperature. They typically are poor conductors of heat and electricity.
- The metalloids that occur between the metals and nonmetals exhibit intermediate properties of luster, rigidity, conductivity, color, and other features. They are most important in the semiconductor industry and other high-technology applications.
- The halogens are the most chemically reactive elements because they require only one electron to attain a stable noble-gas configuration.
- The noble gases are essentially inert. They have completely full outer electron shells so they are very stable.

Coming to Terms

Term	Page
periodic table	105
periodic law	107
transuranium elements	107
group	111
family	111
period	111
series	111
International Union of Pure and Applied Chemistry (IUPAC)	111
North American Convention Periodic Table	111
representative group	111
metal	111
nonmetal	112
metalloid	112
lanthanide series	112
actinide series	112
atomic radius	113
first ionization energy	115
electron affinity	115
electronegativity	116
descriptive chemistry	118
hydrocarbons	119
alkali metals	120
alkaline-earth metals	120
big bang	121
transition metals	123
inner transition metals	124
paramagnetism	124
post-transition metals	125
metalloids	125
semiconductors	125
oxide	130
sulfide	130
halogen	131
noble gas	132

Review Questions

Concept Review

1. Identify the scientist who
   a. played the lead role in developing the structure of the modern periodic table.
   b. formulated the concept of triads.
   c. proposed that elemental properties varied in octaves.
   d. predicted the existence of several missing elements at the time of his work.
   e. ordered the periodic table by atomic number.
   f. devised the commonly used electronegativity scale.

2. What is the purpose of the periodic table?

3. Why were several elements in odd places in Mendeleev's table? How was the problem corrected?

4. Give two names for horizontal rows in the periodic table.

5. Give two names for vertical columns in the periodic table.

6. Which groups are considered the representative groups?

7. Sketch a small periodic table, like the one below, and on it shade areas for metals, nonmetals, and metalloids. You do not need to draw in all of the cells or label any elements.

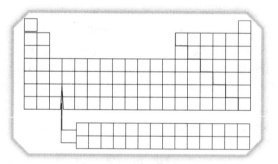

8. Use the periodic table to complete the following table.

Name	Symbol	Z	Period	Group
cadmium				
		56		
	Sn			
			4	1

9. Why are sodium cations smaller than sodium atoms? Why are chlorine anions larger than chlorine atoms?

10. Of the eighteen elements in the fourth period, which
   a. has the largest atomic radius?
   b. has the smallest first ionization energy?
   c. has the largest electron affinity?
   d. has the lowest electronegativity?
   e. has the highest electronegativity?
   f. are electrical semiconductors?
   g. is one of the least reactive elements?

11. Which of the stable alkaline-earth metals (Be, Mg, Ca, Sr, and Ba) fit the following descriptions? (Use Figures 5-9 through 5-14.)
    a. has the largest atomic radius
    b. has the largest ionization energy
    c. has the largest ionic radius
    d. has the smallest electronegativity

12. What is most atypical about the alkali metals compared to other metals?

13. Why were pure metals like gold, silver, and copper known in Old Testament times but metals like sodium, aluminum, and potassium not discovered until relatively recently?

14. Why was chlorine isolated before fluorine? Does this apply to other elements? If so, to which ones?

15. Identify each of the following elements as an actinide, an alkali metal, an alkaline-earth metal, a halogen, a lanthanide, a metalloid, a noble gas, a transition metal, or a post-transition metal. Use the most specific term that applies.
    a. antimony
    b. tungsten
    c. argon
    d. iron
    e. bromine
    f. cesium
    g. cerium
    h. uranium
    i. calcium
    j. lithium

16. Given the elements Al, Au, Br, Ca, Cs, F, H, He, Hg, I, K, Mg, Na, O, P, S, and Si, which satisfy the following conditions? (*Note*: More than one element may be used to answer a question.)
    a. gaseous at room temperature
    b. liquid at room temperature
    c. soft metal
    d. a constituent of table salt
    e. found in salt replacements
    f. responsible for making water hard
    g. relatively inactive solid metal
    h. found in glass and many minerals
    i. a constituent of bauxite
    j. glows in the dark
    k. called brimstone in the Bible
    l. most electronegative element
    m. required for proper function of thyroid gland
    n. used in balloons
    o. a gas chemically similar to alkali metals

17. Choose three of the periodic families and explain how an element from each of those families could be used to show biblical care and concern for other humans.

True or False

18. The first evidence of the noble gases was discovered by Rayleigh and Ramsay in 1895.

19. Metals are known for their thermal and electrical conductivity, and nonmetals (nonconductors) are particularly good insulators.

20. Since energy has to be added to an atom to remove an electron, ionization energies are always positive.
21. Electronegativity is the main property that determines how strongly an isolated neutral atom holds on to its valence electrons.

Serving God as a Metallurgist

Job Description
A metallurgist investigates the properties and use of metals and alloys. This research helps engineers, designers, and builders to determine which material or fabrication process would best meet strength, cost, safety, and flexibility requirements.

A metallurgist may specialize in extractive, physical, or mechanical metallurgy. Extractive metallurgists focus on extracting metals from ores and refining them efficiently. Physical metallurgists study the structure and physical properties of various metals under a range of conditions to use them better. Mechanical metallurgists create and improve processes for metalworking, such as casting, rolling, forging, and drawing.

Possible Workplaces
Metallurgists spend much of their time in a laboratory. They may be working directly with metals and alloys, or with computer simulations. Their research may be based in a university or private industry. Many work in teams with other engineers and business professionals.

Some field work is required of physical or mechanical metallurgists who investigate suspected failures of metals, for instance, a bridge collapse or aircraft crash.

Education
A college degree (BS) is a prerequisite, and many metallurgists possess an advanced degree (MS or PhD) in metallurgy, materials science, or materials engineering. A metallurgist must have a good grasp of chemistry, applied physics, and complex math. All metallurgists must be able to communicate clearly and to think analytically in problem-solving situations. Mechanical metallurgists working in industry must have thorough knowledge of product design and production processes so that they can make recommendations for efficient and safe use of metals and alloys.

Dominion Opportunities
A metallurgist may devise more efficient processing procedures for ore extraction or for metal recycling, thus lessening the impact on the environment. A concern for others can be expressed through the creation and improvement of safety procedures in the dangerous field of metalworking. Metallurgists who deal with the strength and longevity of metals can work with engineers to create better vehicles, aircraft, bridges, buildings, and even artificial body parts that preserve and extend life.

CHEMICAL BONDS

DOMINION SCIENCE PROBLEM

Frequent Visits to Space

The Apollo space program put men on the moon and won the space race for the United States. But Apollo had only a few missions into space. After Apollo, NASA envisioned frequent trips into space. That frequency could be made possible by a reusable, cost-efficient spacecraft. But the huge, expendable Saturn V rockets used for the Apollo missions required enormous volumes of refrigerated fuels, and the Apollo spacecraft lacked sufficient crew and cargo space. How could NASA regularly, efficiently, and safely launch more manned missions?

6

6A Basics of Bonding 140
6B Types of Bonds 143
6C Properties of Compounds 153

Facet
Diamonds 154

6A Basics of Bonding

6.1 Chemical Bonds and the Octet Rule

The second law of thermodynamics implies that natural systems tend toward a state of minimum energy. We can see this in everyday events. For example, balls roll downhill, leaves hang limp after the wind has calmed, and hot lava spontaneously cools.

Similarly, an atom naturally minimizes how much energy it has. In Chapter 4, we learned that its electrons can exist only at certain quanta of energy, and they seek the lowest energy level. Under certain conditions, an atom becomes more compact because the average distance of the electrons from the nucleus decreases. With a smaller average distance, the electrons' potential energy relative to the nucleus is lowered. We also saw that when subshells and the entire outer s and p sublevels are full, atoms are smallest and thus the most stable. Noble gases are good examples of atoms with stable electron configurations.

Because of these inherent properties, atoms naturally link with other atoms to form **chemical bonds**. Those bonds rearrange electrons to maximize stability. Bonding usually releases energy as the atoms' electron configurations attain lower energy states. This energy change may be released as heat, light, or some other form of energy.

An atom's ability to gain or lose electrons determines its chemical reactivity. If atoms can lose, gain, or share enough electrons to attain a noble-gas electron configuration by bonding, they will be more stable. For most elements on the periodic table, a full valence shell contains eight electrons. Thus, by attaining a full valence shell, atoms maximize stability. This principle is called the **octet rule**.

The octet rule does not apply to transition and inner transition metals and to the first four or five elements in the periodic table. In the case of hydrogen, lithium, and beryllium, the nearest noble-gas configuration has only two electrons. For transition and inner transition metals, not only valence electrons but also high-energy electrons from deeper d and f sublevels are removed or shared during bonding. Thus, a full valence octet in the metal cation may not affect the reactivity of these metals.

Many elements give up electrons to attain a full valence shell. Alkali and alkaline-earth metals tend to lose the one or two outer s sublevel electrons when they bond with other elements. Losing those electrons exposes the underlying noble-gas electron configuration from the previous period.

Elements rearrange the smallest number of electrons possible in bonding processes. Atoms with fewer than four valence electrons tend to lose electrons because they have relatively low ionization energies. Atoms with more than four tend to gain electrons because they have relatively high electron affinities.

6A Section Objectives

After finishing this section, you should be able to

- explain why atoms bond, using thermodynamic principles of minimum energy and stability.
- characterize bonds as covalent, ionic, or metallic on the basis of their constituent elements and their interactions with electrons.
- describe the role that electron affinity plays in bonding processes.
- explain the relationship between electron location and polarity in polar covalent molecules.
- predict the type of bond that will form between two atoms, given their location on the periodic table.

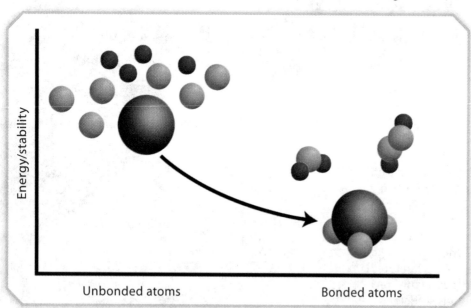

6-1 A simplified representation of the consequences of bonding. Often the electron structures of bonded atoms are more stable and at a lower energy state than those of unbonded atoms.

octet: oct- (L. *octo*—eight) + -et (suffix meaning "a group of")

EXAMPLE PROBLEM 6-1

Acquiring an Octet of Valence Electrons

State two ways in which (a) sodium atoms and (b) oxygen atoms could attain eight valence electrons. Which process is more likely for each element? Why?

Solution
a. Sodium, a Group 1 element with one valence electron, could attain eight valence electrons by gaining seven electrons. Or, it could lose the one valence electron in the third energy level, leaving the second energy level with its eight electrons as the valence shell. It will likely lose the one electron, since it is loosely held. It would be nearly impossible to force seven extra electrons into the atom because it has a low affinity for additional electrons.
b. Oxygen, a Group 16 element with six valence electrons, could gain two electrons to fill its valence shell, or it could lose the six valence electrons, uncovering the two electrons in the first energy level. It is more likely to gain two electrons because it has a high electron affinity and a high first ionization energy, which would make removing electrons difficult.

6.2 Types of Chemical Bonds

Bond type is affected by how atoms attain a stable electron configuration. If two atoms both have strong electron affinities, neither completely removes electrons from the other atom. However, they can share electrons to complete an octet. Bonds formed by sharing electrons are called **covalent bonds**. These most often occur between nonmetals.

When atoms with very different electron affinities combine, the atom with greater affinity takes electrons from the other atom. This electron transfer forms a cation and an anion. Those ions are held together through attraction of opposite charges in an **ionic bond**. Typically, ionic bonds involve a metal bonded to a nonmetal.

Atoms with weak electron affinities usually have only a few loosely held valence electrons. These atoms may bond by sharing their easily lost electrons among many atoms. These electrons are mobile and no longer are associated with any specific nucleus. Metal atoms typically interact this way to form **metallic bonds** when in the solid state.

In this chapter, we will examine each of these bond types and the compounds that contain them.

Chemical Bonds and Your Future

Why is it important to learn all you can about chemical bonds now, even if you aren't particularly interested?

6.3 Polarity and Bond Character

Covalent and ionic bonds are affected by a property called **polarity**. Polarity is the tendency of an object to form two localized regions of opposite character. For example, the earth has a north magnetic pole and a south magnetic pole; it is magnetically polarized. A battery has two electrical poles where positive and negative charges collect; it is electrically polarized. Similarly, a bond can be described as polar if electrons are unequally shared between two atoms.

In the simplest case, let's look at diatomic hydrogen and diatomic fluorine. Each hydrogen has only a single valence electron. If they are brought close together, each will attempt to grab the other's electron to fill its valence energy level and achieve a more stable electron configuration. Since neither can take the electron from the other, they share the electrons equally. The electrons spend most of their time between the nuclei of the hydrogen atoms. No poles form because the hydrogen atoms are identical. A diatomic hydrogen molecule is held together by a completely nonpolar bond, where the positive nuclei are attracted to the negative central region containing the bonding pair of electrons.

Similarly, each fluorine atom has three pairs of electrons and a single unpaired electron in its valence shell. They have identical, very strong electron affinities. Each needs just one more electron to have a full octet. Electrons are shared equally, and, just as with hydrogen, the bond is nonpolar.

Orbital notation can be used to show which electrons are shared. Notice that the shared electrons have opposite spins.

$$F: [He]\ \underset{2s}{\uparrow\downarrow}\ \underset{2p}{\uparrow\downarrow\ \uparrow\downarrow\ \uparrow\downarrow\ \uparrow}$$

$$F: [He]\ \underset{2s}{\uparrow\downarrow}\ \underset{2p}{\uparrow\downarrow\ \uparrow\downarrow\ \uparrow\downarrow\ \downarrow}$$

shared pair of electrons in an F_2 molecule

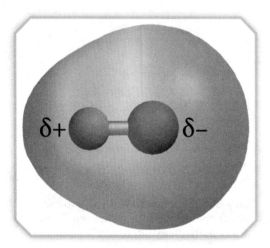

6-2 Chemists use the lowercase Greek letter delta (δ) to indicate the presence of a partial charge around a molecule.

> Electronegativity is the measure of an atom's ability to attract electrons to itself in a chemical bond.

> The uppercase delta (Δ) is used to represent change in many scientific expressions.

Problem-Solving Strategy 6-1
There is no definite boundary between covalent and ionic bonds. If the bond polarity is small and the bonding elements are not both metals, then the bond character is more covalent than ionic.

Now, what happens if a hydrogen atom and a fluorine atom bond? The hydrogen has a single valence electron and the fluorine atom needs only a single electron to complete its valence octet. The fluorine atom has a greater electron affinity than the hydrogen atom. When they bond, the electron is pulled much closer to the fluorine and spends a greater amount of time around the fluorine nucleus than around the hydrogen nucleus. On average, the fluorine end of the molecule will be negative and the hydrogen end will be positive. (These partial charges are indicated by the δ− and the δ+ in Figure 6-2.) Thus, the hydrogen-fluorine bond is electrically polar.

Linus Pauling noticed that covalent bonds between different elements (e.g., hydrogen and fluorine) require more energy to break than the bonds between atoms of either pure element alone (e.g., hydrogen and hydrogen or fluorine and fluorine). Pauling explained these stronger bonds as an effect of polarity. Polar bonds are stronger because the opposite charges of the polarized atoms hold the atoms together. This effect was quantified by assigning electronegativity (EN) values to each element, depending on their ability to attract electrons to themselves in a chemical bond. (See Chapter 5.)

Bonds range from being nonpolar to being highly polar. The polarity of such a bond depends on the difference of the atoms' electronegativities (ΔEN).

What happens if we attempt to bond two atoms with very different electronegativities? Let's combine rubidium, an active metal (EN = 0.8), with fluorine, an active nonmetal (EN = 4.0). For a rubidium and fluorine bond, ΔEN = 3.2. The shared electrons spend nearly all their time around the fluorine atom and almost no time around the rubidium atom. The electron transfer is not complete because the electron still spends a short time around the rubidium nucleus. But the fluorine has essentially captured the valence electron from the rubidium atom, resulting in a positively charged rubidium cation and a negatively charged fluorine anion.

Completely covalent bonds and highly ionic bonds exist at the opposite ends of a spectrum of bond polarities. Small-to-medium bond polarities are characteristic of covalent bonds, while very high bond polarities exist in ionic bonds. We will refer frequently to this concept as we discuss each kind of bond in more detail.

It is easy to think of bonds as falling into one of these categories—covalent, ionic, or metallic—as if they each have very distinct boundaries. But classifying bonds is not that straightforward. It is possible to have a completely covalent bond between two atoms of the same nonmetal. But because bonds between different elements are not 100% covalent, they become more ionic as differences in electronegativities become greater. Many covalent bonds are better described as polar covalent. It is impossible to have a completely ionic bond resulting from electron transfer because that would imply that the donor atom had no affinity for the donated electrons. A more accurate way to describe bonds is to designate them as predominantly covalent or predominantly ionic.

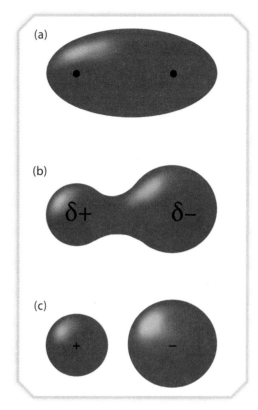

6-3 A comparison of the charges found in (a) a covalent diatomic molecule, (b) a polar covalent bond, and (c) an ionic bond where no electron sharing takes place

6A Section Review

1. Generally, why do chemical bonds form between atoms?
2. State the octet rule. To what elements does it apply?
3. What are the three main bond types that can form between atoms?
4. What chiefly determines the polarity of a bond?
5. Use the periodic table to determine the type of bond that would form between (a) cesium and sulfur, (b) chlorine and bromine, (c) magnesium and fluorine, and (d) silver and mercury.
6. (True or False) A bond between two atoms can be 100% covalent, but it can never be 100% ionic.

6B Types of Bonds
6.4 Covalent Bonding

Many familiar compounds consist of covalently bonded elements. Those elements have similar electronegativities, lacking only a few valence electrons for an octet. They are usually nonmetals and are often close to each other on the periodic table. When they bond, they maintain influence over their valence electrons. To acquire a stable noble-gas electron configuration, atoms share electrons to fill vacancies in their valence orbitals.

Elemental chlorine (Cl_2) is an example of a completely covalent molecule. An isolated, neutral chlorine atom has seven valence electrons. Each chlorine atom needs one more electron to attain an octet. Since both atoms in the bond have the same electronegativity, neither can completely pull the required electron from the other to itself. Instead, both atoms share their unpaired electrons, filling their $3p$ orbitals and thus completing their valence shells. The two shared electrons that make up a pair are called a *bonding pair* in contrast to the other valence electron pairs that do not participate in bonding. These are called lone pairs or nonbonding pairs of electrons.

6B Section Objectives
After finishing this section, you should be able to
- describe how atoms form covalent bonds.
- name the elements that exist as diatomic molecules.
- draw Lewis structures of covalent compounds and polyatomic ions when given their chemical formulas.
- discuss the general arrangement of bonded atoms for each type of bond.
- differentiate between a formula unit and a molecule.
- summarize the current model of metallic bonding in your own words.

Halogen Bonds and Loving Your Neighbor

Scientists studying halogen bonds at Oregon State University found that they have a structure unique from other bonds. They predicted that a better understanding of these bonds could help develop antibiotics that contain halogens. How is developing medicines an example of using chemistry to love your neighbor?

Electrons in bonding pairs have the highest probability of being found between the nuclei of the bonded atoms. There they form a dense region of negative charge. This region attracts the positive nuclei, holding the atoms together. This attraction is called the **electrostatic force**, which is the force exerted between electrical charges. Opposite charges attract, and like charges repel, according to the law of charges.

In the case of our two chlorine atoms, the positive nucleus of each atom is attracted to the negative electrons of the other atom, reducing its potential energy and increasing stability. As the atoms

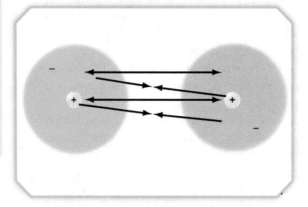

6-4 The single-headed arrows show the attraction between opposite charges, and the double-headed arrows show the repulsion between like charges.

draw closer, however, they reach a point at which the repulsion between their positively charged nuclei balances the attraction for the electron cloud.

6.5 Diatomic Elements

The atoms of hydrogen, nitrogen, oxygen, and the halogens (fluorine, chlorine, bromine, and iodine) are not stable by themselves in the presence of other elements. Their strong electron affinities force them to quickly bond with other atoms in their vicinity. For these elements to exist in a pure form, they occur as diatomic molecules. Their formulas are H_2, N_2, O_2, F_2, Cl_2, Br_2, and I_2. Memorize this list for future reference.

6.6 Lewis Structures

The American chemist Gilbert Lewis developed a useful notation that illustrates how atoms form molecules. **Lewis structures** are two-dimensional diagrams that show the bonds between different atoms. Covalent bonds are symbolically represented in several ways. Covalent bonds involve one or more pairs of electrons shared between two atoms. Each electron pair contributes to the valence-electron structure of both atoms involved in the bond. When illustrating the formation of a covalent bond using electron dot notation, the unbonded element symbols are usually written with the unpaired electrons positioned between the bonding atoms. After bonding, the pair of dots lies between the bonded elements. When using Lewis structures, a dash is substituted for the bonding pair to simplify the diagram. Each dash thus represents two electrons. The nonbonding electron pairs are retained. See the following example of covalent bond notation for the Group 17 element, chlorine.

$$:\!\ddot{C}l\!\cdot\; +\; \cdot\ddot{C}l\!: \;\rightarrow\; :\!\ddot{C}l\!:\!\ddot{C}l\!: \;\text{or}\; :\!\ddot{C}l\!-\!\ddot{C}l\!:$$

6-5 These are the elements that form diatomic molecules. Predict the number of covalent bonds that form in each molecule based on the number of valence electrons in each element.

6-6 Gilbert Lewis (1875–1946) was an American chemist who performed detailed research to describe valence electrons and isotopes. He was the first to explain the covalent bond. He also made significant contributions to the acid-base theory and chemical thermodynamics.

As we noted in the previous subsection, hydrogen can form a diatomic molecule. However, rather than sharing electrons to complete a valence octet, hydrogen atoms share their single electrons to complete their 1s sublevel, which is their valence shell. In this manner, they act much like chlorine or another Group 17 element when they bond together.

$$H \cdot + \cdot H \rightarrow H : H \text{ or } H - H$$

Covalent compounds are also formed when two different nonmetals combine. Probably the most familiar covalent compound on the earth's surface is water, a compound of hydrogen and oxygen. A lone oxygen atom has two unpaired electrons in its valence p subshell. It needs two electrons to complete its valence octet. Hydrogen atoms need only one electron to complete their valence shells. Both the oxygen and the hydrogen atoms can attain complete valence shells by sharing electrons. The central oxygen atom forms covalent bonds with two hydrogen atoms.

$$H \cdot + H \cdot + \cdot \ddot{O} : \rightarrow H : \ddot{O} : \text{ or } H - \ddot{O} :$$
$$\qquad\qquad\qquad\quad H \qquad\quad H$$

Sometimes sharing one electron pair is not enough to fill an atom's valence octet. For example, oxygen must share two electron pairs to attain a noble-gas electron configuration. It can accomplish this by bonding with two other atoms, as in water. It can also share two unpaired electrons with another oxygen to form double covalent bonds.

$$: \ddot{O} \cdot + \cdot \ddot{O} : \rightarrow : \ddot{O} :: \ddot{O} :$$
$$\text{or } : \ddot{O} = \ddot{O} :$$

Triple covalent bonds form when atoms share three pairs of electrons. Diatomic nitrogen molecules in the atmosphere consist of two nitrogen atoms bonded by a triple covalent bond.

$$: N ::: N : \text{ or } : N \equiv N :$$

Many organic compounds contain carbon atoms that form a triple bond with other atoms. One example is acetylene, used for welding torches.

$$H : C ::: C : H \text{ or } H - C \equiv C - H$$

The following guidelines are helpful for creating Lewis structures of covalent molecules, especially complex ones.

1. Lewis structures illustrate only molecular compounds consisting of covalent bonds.
2. Electron sharing occurs when one nonmetal atom bonds with another (e.g., hydrogen and fluorine) or when two identical nonmetal atoms bond (e.g., fluorine and fluorine).
3. Each nonmetal generally forms covalent bonds to achieve eight electrons in its valence shell, including bonding and nonbonding electrons. Hydrogen is a major exception, since its valence shell holds only two electrons.
4. Electrons are normally shared in pairs since two electrons of opposite spin occupy empty orbitals.

> **Problem-Solving Strategy 6-2**
> When drawing Lewis structures, use a dash between atoms to represent each pair of bonding electrons. Doing this will simplify your diagrams and clearly differentiate between bonding and nonbonding electron pairs.

5. Polyatomic molecules, which contain three or more atoms, have one or more central atoms bonded to surrounding atoms. Since hydrogen can share only one pair of electrons, it can never be the central atom in a polyatomic molecule. Atoms with valence structures similar to Group 14 atoms (the carbon family) are more likely to be the central atom in a polyatomic molecule.

6. The number of valence electrons for nonmetals (except hydrogen) can be predicted by the formula

    valence e⁻ = group number − 10.

    The number of covalent bonds in which nonmetals can participate is predicted by the formula

    covalent bonds = 8 − valence e⁻.

There are a few exceptions to these rules that we will observe.

Steps for Drawing Lewis Structures

As an example, we will construct the Lewis structure for formaldehyde (H_2CO), a chemical used as a preservative. You will see that Lewis structures reveal something about the shape of the actual molecule. The shape affects its chemical properties.

1. Write the electron dot symbol for each element in the compound.

    Oxygen: ·Ö: Carbon: ·Ċ: Hydrogen: H·

2. Determine how many valence electrons are available.

    H: 2 × 1 e⁻ = 2 e⁻
    C: 1 × 4 e⁻ = 4 e⁻
    O: 1 × 6 e⁻ = 6 e⁻
    Total: 12 e⁻

3. Place the element with the largest number of unpaired electrons in the center of the structure. These unpaired electrons are essentially bonding sites. This element will likely be the central atom and the least electronegative. If necessary, reposition individual dots so that there is at least 1 dot on each of the four sides of the atom's symbol.

    ·Ċ·   (dot repositioned)

> **Problem-Solving Strategy 6-3**
>
> When drawing Lewis structures of simple compounds, the most important step is to correctly identify the central atom. Look for the element that has the greatest number of unpaired electrons or bonding sites around its symbol after splitting pairs to fill empty sides. That element will likely be the central atom.

4. Place the remaining element symbols for the compound around the central atom. Initially, position them adjacent to the unpaired electrons of the central atom.

5. Create as many single bonds as possible by pairing up adjacent electrons.

    Formaldehyde (H_2CO):   H:Ċ:Ö:
                                 H

The carbon and oxygen atoms still have an unpaired electron, and each needs an additional electron to complete its valence octet.

6. Create a second bond between the carbon and oxygen to give each eight valence electrons.

$$H:\overset{..}{C}::\overset{..}{O}: \quad \text{or} \quad \overset{H}{\underset{H}{\diagup}}C = \overset{..}{\underset{..}{O}}:$$

Notice that oxygen has two lone or nonbonding pairs of electrons.

7. Check your structure by counting up the total number of electrons represented in the molecule and confirm that it equals the total available valence electrons from Step 2.

EXAMPLE PROBLEM 6-2

Drawing Lewis Structures

Draw the Lewis structures for the following compounds:
a. trichloromethane ($CHCl_3$)
b. sulfur dioxide (SO_2)

Solution

a. Write the electron dot symbols for carbon, hydrogen, and chlorine.

$$\cdot\overset{..}{C}: \text{ or } \cdot\overset{\cdot}{\underset{\cdot}{C}}\cdot \quad \cdot\overset{..}{\underset{..}{Cl}}: \quad H\cdot$$

Determine how many valence electrons are available.
C: $1 \times 4\ e^- = 4\ e^-$
H: $1 \times 1\ e^- = 1\ e^-$
Cl: $3 \times 7\ e^- = 21\ e^-$
Total: $26\ e^-$

Carbon will be the central atom since it has the greatest number of unpaired electrons. Position the three chlorine atoms and one hydrogen adjacent to unpaired electrons.

$$H\cdot \ \cdot\overset{\overset{\displaystyle :\overset{..}{Cl}:}{\cdot}}{\underset{\underset{\displaystyle :\overset{..}{Cl}:}{\cdot}}{C}}\cdot \ \cdot\overset{..}{\underset{..}{Cl}}: \rightarrow H:\overset{\overset{\displaystyle :\overset{..}{Cl}:}{}}{\underset{\underset{\displaystyle :\overset{..}{Cl}:}{}}{C}}:\overset{..}{\underset{..}{Cl}}: \quad \text{or} \quad H-\overset{\overset{\displaystyle :\overset{..}{Cl}:}{|}}{\underset{\underset{\displaystyle :\overset{..}{Cl}:}{|}}{C}}-\overset{..}{\underset{..}{Cl}}:$$

Verify that each atom has an octet of electrons and the hydrogen has a pair. The hydrogen may be placed on any side of the carbon atom.
Verify that 26 valence electrons are used.

b. Write the electron dot symbols for sulfur and oxygen.

$$\cdot\overset{\cdot}{\underset{..}{S}}: \quad \cdot\overset{..}{\underset{.}{O}}:$$

Determine how many valence electrons are available.
S: $1 \times 6\ e^- = 6\ e^-$
O: $2 \times 6\ e^- = 12\ e^-$
Total: $18\ e^-$

Sulfur will be the central atom, because there is one sulfur atom and two oxygen atoms. Position the two oxygen atoms on opposite sides of the sulfur atom. (The reason for this will

(continued)

become apparent.) Create the single bonds between sulfur and the oxygen atoms.

$$:\ddot{O}\cdot \quad \cdot\ddot{S}\cdot \quad \cdot\ddot{O}: \rightarrow :\ddot{O}:\ddot{S}:\ddot{O}:$$

Note that the sulfur atom has eight electrons surrounding it, but neither oxygen atom has eight. Each oxygen is short one electron to make a full octet. There is no way to give each atom eight electrons by forming a single bond. By forming a double bond with one of the oxygen atoms, however, all atoms can acquire full valence octets. It makes no difference which oxygen forms the double bond.

$$:\ddot{O}::\ddot{S}:\ddot{O}: \quad \text{or} \quad :\ddot{O}=\ddot{S}-\ddot{O}:$$

Verify that 18 valence electrons are used.
Whenever the central atom has two sets of double bonds, the Lewis structure is always arranged in a line that reflects the actual molecule's shape. We will investigate molecular shapes more thoroughly in Chapter 7.

6.7 Polyatomic Ions

If certain elements with high electron affinities, such as oxygen, are part of a polyatomic molecule, one or more of the missing valence electrons can sometimes be filled by nearby extra electrons instead of sharing electrons with other atoms. In these cases, a covalently bonded anion forms, called a **polyatomic ion**. Examining the Lewis structure of a polyatomic ion will show that every atom in the molecule has an octet. The charge of the ion shows how many extra electrons were used to achieve an octet.

$$\text{Atomic ion:} \quad :\ddot{Cl}: + 1e^- \rightarrow \left[:\ddot{Cl}:\right]^- \quad \text{or} \quad Cl^-$$

$$\text{Polyatomic ion:} \quad :\ddot{O}:H + 1e^- \rightarrow \left[:\ddot{O}:H\right]^- \quad \text{or} \quad OH^-$$

Just as the chlorine ion (Cl$^-$) has an extra electron, the hydroxide ion (OH$^-$) has one more electron in addition to those supplied by the oxygen and hydrogen atoms.

The Lewis structures of polyatomic ions can be drawn just like covalent compounds. The only difference is that any unpaired valence electrons remaining after bonding are paired up with extra electrons that give the anion its charge. Square brackets are placed around Lewis structures of the ions, and the charge is written as a superscript outside the brackets.

Example Problem 6-3
Drawing Polyatomic Ion Lewis Structures

Write the Lewis structure of the ammonium ion (NH$_4^+$).
Solution
Write the electron dot symbols for hydrogen and nitrogen.

$$H\cdot \quad \cdot\dot{N}:$$

Count the number of available electrons. The single nitrogen atom contributes five, and each of the four hydrogen atoms gives one. The total of nine electrons must be reduced by one because the ion has a +1 charge, meaning it has lost one electron. Assemble the four hydrogen atoms around the central nitrogen and form as many bonds as possible by sharing the eight available electrons with the central atom.

$$\left[\begin{array}{c} H \\ | \\ H-N-H \\ | \\ H \end{array} \right]^{+}$$

Note that the Lewis structure for ions is placed in brackets and the net charge of the ion is written as a superscript.

Problem-Solving Strategy 6-4
When drawing Lewis structures of polyatomic anions, any additional electrons added must equal the charge of the ion. Supply this number of dots to fill the vacancies.

Most polyatomic ions are anions, formed by acquiring additional electrons to make up their octet. In relatively rare cases, however, some neutral covalent molecules can acquire a positive charge, forming polyatomic cations. These ions form by attracting hydrogen cations to relatively negative regions on the surface of an overall neutral molecule. Some important examples of these cations include the hydronium ion (H_3O^+) and the ammonium ion (NH_4^+).

Polyatomic ions often act as single charged particles in chemical reactions and solutions. Their stable electron structures often allow them to survive chemical reactions to form compounds without splitting up. In Chapter 8 you will learn more about polyatomic ions and the compounds they form.

6.8 Ionic Bonding

When sodium, a soft, shiny, reactive metal, is placed in an atmosphere of poisonous chlorine gas, a violent chemical reaction takes place. The result is sodium chloride, an edible white crystalline compound commonly called table salt. The explosive release of heat and light indicates that the resulting compound contains much less energy than the sum of the energy in the original elements. What happened?

Sodium atoms in Group 1 have a single valence electron, just one electron away from an octet. Their first ionization energy is very small. Chlorine, in Group 17, has a very high electron affinity, needing one electron to complete its stable valence octet. When these two elements react, chlorine atoms essentially steal the loosely held electrons from the sodium atoms. Both the sodium cations and chlorine anions now have stable valence octets, and the excess energy is released as heat and light. The difference of these elements' electronegativities is large, as is typical of ionic bonds.

$$\Delta EN_{NaCl} = |EN_{Cl} - EN_{Na}|$$

$$\Delta EN_{NaCl} = |3.2 - 0.9|$$

$$\Delta EN_{NaCl} = 2.3$$

6.9 The Structure of Ionic Compounds

Chemical compounds formed from metals and nonmetals are called ionic compounds. A sample of an ionic compound, such as a crystal of sodium chloride, consists of many bonded cations and anions held together by a strong electrostatic attraction. Ionic compounds are not made of distinct molecules like covalent compounds. Even though ionic compounds contain charged particles, the sum of all the ionic charges in a sample equals zero, so they are electrically neutral. The ratio of cations to anions needed to achieve this neutrality determines the chemical formula for the compound, called a **formula unit**. For example, let us examine the electron dot symbols for table salt.

Atoms before bonding: Na· $\xrightarrow{1e^-}$ ·Cl:

Ions after electron transfer: $[Na]^+ \; [:Cl:]^-$

Ratio: 1 Na$^+$: 1 Cl$^-$ or NaCl
Resulting compound:

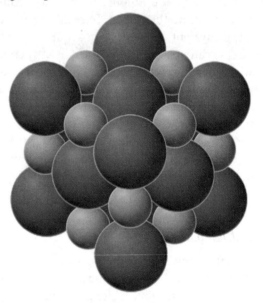

$[Xx]^{n+}$

The notation above is used to represent the electron dot notation of a metal cation depleted of its valence electrons because it is participating in ionic bonding. This cation has acquired the valence electron configuration of the preceding noble gas by giving up its valence electrons.

From the example above, we can see that the formula unit for table salt is NaCl. Although there are no molecules in an ionic compound, it is often convenient to work with its formula unit. We do not call a formula unit a molecule because it does not exist separately like a group of covalently bonded atoms. For example, no single particle of sodium chloride exists except perhaps in concentrated salt solutions.

Let us look at a more complicated example. In the formation of calcium fluoride, calcium atoms lose electrons to strongly electronegative fluorine atoms (ΔEN = 3.0). Calcium atoms need to lose two electrons to attain an octet, but fluorine atoms need only one electron. Therefore, every calcium atom in the compound gives its electrons to two fluorine atoms. As with sodium chloride, this combination can involve many ions, but the ratio is always two fluorine anions to every calcium cation.

Atoms before bonding: Ca: $\xrightarrow{2e^-}$:F:
:F:

Ions after electron transfer: $[Ca]^{2+} \; \begin{matrix}[:F:]^- \\ [:F:]^-\end{matrix}$

Ratio: 1 Ca^{2+} : 2 F^- or CaF_2
Resulting compound:

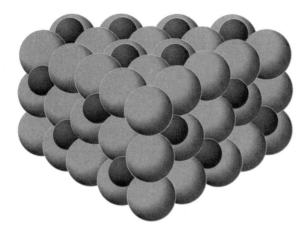

Ions in ionic compounds assemble themselves into a closely packed arrangement. An orderly, three-dimensional pattern develops that depends on the ratio and relative size of the anions and cations. The ions position themselves as close as possible to oppositely charged particles and as distant as possible from similarly charged particles. This orderly arrangement of ions is called a **crystal lattice**. We will discuss crystal lattices more in Chapter 11.

6.10 Metallic Bonding

The metallic bond forms entirely different kinds of materials. If they were limited to the octet rule, metal elements would have a problem forming bonds among themselves. Several properties of metals prevent them from easily losing or gaining electrons to acquire a noble-gas configuration. As a group, most metal atoms need six or more electrons to fill their valence shells. Also, the transition and inner transition metals have gaping holes in their interior d and f sublevels, which must be filled before the valence shell. Since they have low electron affinities, they cannot take electrons from other metal atoms. What, then, holds the atoms in a piece of metal together?

A different type of bond must form. The theory that explains how metal atoms bond together also explains why metals have such unique properties. This model is called the **electron-sea theory** (or the free-electron theory). According to this model, metals are an extensive crystal lattice of metal cations surrounded and submerged in a "sea" of mobile electrons. Since metals have a low first ionization energy, their electrons are shared among adjacent metal atoms and lose their association with their parent atoms. Because of their mobility, these electrons are shared among all the atoms and are said to be delocalized. The negative charge of the **delocalized electrons** acts as a "glue" to hold the positively charged metal ions together.

Metallic bonds are possible because metals generally have low electronegativities, and the difference between their electronegativites is small. Metallic bonds do not occur between a relatively few

6-7 Crystals of table salt reveal the underlying cubic arrangement of their crystal lattice.

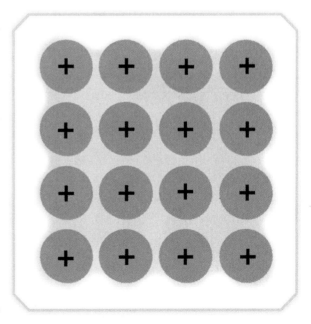

6-8 The metallic bond exists only between numerous metal cations bound by an "electron sea."

atoms, though that is possible for covalent or even ionic compounds. The metallic bond can exist only for an arrangement of metal atoms in a crystal lattice.

6-1 Bonds and Their Properties

Elements that bond	Valence electrons	Type of bond	Bond polarity
nonmetal/nonmetal	tightly shared	covalent	low-to-medium
metal/nonmetal	transferred	ionic	high
metal/metal	widely shared	metallic	n/a

EXAMPLE PROBLEM 6-4
Predicting Bond Character

Predict the predominant type of bond formed between atoms of the following elements:
a. nitrogen and oxygen
b. silver and copper
c. cesium and fluorine

Solution
a. Nitrogen and oxygen are both nonmetals with similar electronegativities. Their bond will be covalent.
b. Silver and copper are both metals, so the bond character is metallic.
c. Cesium is a metal, and fluorine is a highly electronegative nonmetal. With the vast difference between their electronegativites, the bond character is ionic.

6B Section Review

1. Which kind of element forms covalent bonds? Give two reasons for this.
2. What forces hold covalently bonded atoms together?
3. With the exception of hydrogen, what number of valence electrons do all atoms seek to have after forming bonds?
4. Which seven elements occur naturally as diatomic, covalent molecules?
5. State two ways that covalent polyatomic particles can become ions.
⊙ 6. Draw the Lewis structures for the following:
   a. PO_4^{3-}
   b. $COBr_2$
   c. SeO_4^{2-} (Hint: Selenium shares 12 e⁻.)
   d. $SiHCl_3$
7. Why do we not use the term *molecule* to refer to the product of an ionic bond?

8. What term describes the arrangement of particles in an ionic compound?

9. Predict the formula units of ionic compounds that form from the following atoms based on the number of valence electrons of each:
   a. Mg and Br
   b. Al and Cl
   c. Rb and I
   d. Sr and F
   e. Ca and O

10. What kinds of elements normally participate in metallic bonding? State two factors that make metallic bonding possible.

11. Give one similarity and one difference between metals and solid ionic compounds.

6C Properties of Compounds

6.11 Properties of Covalent Compounds

Covalent compounds generally consist of distinct molecules. Various intermolecular forces, discussed in later chapters, hold those molecules together. At room temperature, covalent compounds are gases, liquids, or solids with relatively low melting points. Because the forces between the molecules are relatively weak, it takes little energy to separate them. The solids generally lack the density, hardness, and rigidity of metals and ionic substances. Covalent compounds may exist in a wide array of colors and lusters, and they generally are poor conductors of heat and electricity.

Some notable exceptions to these general characteristics are materials known as **network covalent substances**. These atoms are covalently bonded into a continuous three-dimensional network. The best-known example is diamond, the hardest-known natural substance. A diamond consists of a three-dimensional crystalline array of carbon atoms, each bonded to four adjacent carbon atoms. Similarly, network covalent compounds called silicates form a wide variety of minerals in the earth's crust. Quartz is a common network covalent compound of silicon and oxygen. Network covalent materials often occur as relatively hard, brittle crystals with high melting points, glassy lusters, and unusual electrical properties.

6.12 Properties of Ionic Compounds

Ionic bonds give the compounds they form distinct properties. Since ions are held in place by strong electrostatic forces, the resulting compounds are dense, brittle, and hard solids. The melting points of many of these compounds are 800 °C or higher, illustrating just how strong their bonds are. The brittleness is also related to the forces between ions. If one row or layer of these ions is shifted even slightly, the misalignment creates repulsive forces, causing them to fracture. Because of

6C Section Objectives

After finishing this section, you should be able to
- describe the general properties of compounds according to their predominant bond type.
- explain how the mobility of electrons affects how compounds conduct electricity and heat.
- explain how the intermolecular forces between compound particles affect their melting point.

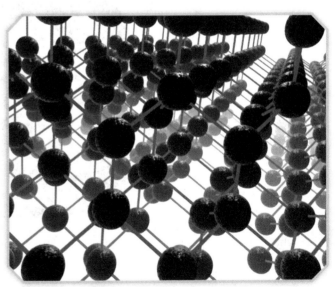

6-9 Diamonds owe their hardness to the rigid network structure of their carbon atoms. The crystalline structure is similar to that of ionic compounds.

their orderly structures, crystals of ionic compounds can usually be split, or cleaved, along a flat surface.

Solid ionic compounds are very poor conductors of electricity and heat. Conductivity depends on mobile electrical charges, such as the delocalized electrons in metals. In ionic compounds, all valence electrons are tightly bound, so these compounds make effective insulators. However, if the compounds are soluble in water or are molten, charged ions themselves are free to move, and the dissolved or molten forms can easily conduct electricity.

6.13 Properties of Metals and Metal Alloys

The electron-sea theory explains many characteristics of metals and their **alloys**. Delocalized electrons are free to move and carry electrical current and thermal energy. When a metal object is heated, its atoms vibrate faster. This motion is transferred to the delocalized electrons.

> An **alloy** is a mixture of the atoms of a metal with another element where the mixture has metallic properties.

DIAMONDS
Facets of Chemistry

In January 1933, a poor prospector and his native helper were searching the diamond field of Pretoria, South Africa. The helper bent down and picked up an earth-encrusted lump. He handed it to the old prospector, Jacobus Jonker. Out of habit, Jonker wiped away the mud from the rather ordinary-looking rock. Then he stared at his hand in disbelief. He was holding a rare "blue-white" diamond about the size of a hen's egg. Needless to say, Jonker spent a sleepless night guarding the diamond. He could not relax until he had it safely deposited in the vaults of the Diamond Corporation. Jonker sold his diamond for $315,000—what seemed like a fantastic amount to him. The new owners, however, knew that $315,000 was a small price to pay. With proper cutting and polishing, the original stone could yield several gems whose combined value would be several times more than that of the rough diamond.

The following year, many leading European diamond cutters submitted plans for cutting the huge diamond. But the lone American expert, Lazare Kaplan, declared that the stone would be ruined if it were cut to the plans that had been submitted. He was one against many, yet the owners finally decided to trust him with their 726-carat treasure. Kaplan planned, measured, and scrutinized for a year and then announced that he was ready. With the help of his son, Leo, he cut a groove on a line of cleavage. Now came the nerve-racking moment: the famous diamond might be shattered into useless fragments, or it might split as planned. Leo held a steel rule in the groove and Kaplan gave it a sharp tap. The Jonker diamond fell apart as planned. The first cleavage yielded a 35-carat chunk. After two more cleavages, the rest of the division was completed by sawing. All of this work increased the value and beauty of the rough diamond.

To determine the value of a gem, diamond cutters have to consider the four Cs: carat, clarity, color, and cut. Of these, the carat, or weight, usually affects the value most. A diamond's weight is measured in carats—a unit of weight equal to 200 mg. Unfortunately, cutting a

Their momentum is carried from atom to atom, increasing atomic vibrations throughout the material. In this way, thermal energy moves quickly through a metallic object.

The luster or shine of metals is explained by the combination of the principles of quantum theory, spectroscopy, and free electron theory. According to quantum theory, in any given metal the delocalized valence electrons move within a number of different orbitals. Metals in the *d* block on the periodic table have empty outer *p* sublevel orbitals and many empty orbitals in the *d* sublevel. All of this unused space allows electrons to move freely within a band of different energy sublevels of an individual atom or to nearby atoms. Therefore, delocalized electrons are able to absorb and instantly re-emit nearly any wavelength of light shining on the metal as they jump between allowed valence energy levels and their ground state. When this factor is combined with the tendency of metallic crystalline structures to contain flat reflective surfaces, the result is the familiar shiny, metallic, and sometimes mirrorlike luster of metals.

diamond usually reduces its weight by one-half. Diamond owners are compensated for the weight loss because cutting increases clarity by removing flaws. A diamond cutter looks for flaws by immersing the diamond in a liquid that bends light rays just as much as diamonds do. This fluid makes the position of any flaws easy to note, and they are avoided when the diamond is cut. The color is also a factor in the value of a gem. A completely colorless stone, referred to as "blue-white," is the most valuable. Other colors are valuable only if the color is definite and attractive. Finally, the cut of the diamond influences the value. The shape of the stone should conform to certain proportions, and the facets must be symmetrical, the same size, and well polished.

The cutting of a diamond involves two steps: dividing and faceting. Although dividing a diamond is only the first step, it is the most important. That is why Kaplan studied the Jonker diamond for a year before attempting to divide it. Two methods can be used to divide diamonds: cleaving and sawing. A diamond can be cleaved in only four directions. This limitation is due to the geometric arrangement of the covalently bonded carbon atoms that form the diamond. To cleave a diamond, cutters form a small scratch in the crystal with a sharp diamond point. They then insert a steel blade into the groove and give it a sharp blow with a mallet. While a diamond will not cleave with ordinary wear, it will cleave cleanly with correct preparation and a hard blow. A diamond can be sawed in only nine directions. The diamond is sawed with a thin disk of phosphor bronze that revolves 5000 to 6000 times each minute. Initially, a mixture of diamond powder and olive oil serves as the abrasive on the disk. In time, however, diamond dust from the cut diamond replaces the original dust. The covalent bonds in the crystal are so strong that sawing a one-carat diamond may take as long as eight hours.

The final step in cutting the diamond is faceting. The object of faceting is to grind smooth surfaces that will allow the light to enter through the top and be internally reflected as many times as possible. The facets also break up light into its component colors and give the diamond its characteristic fire. Faceting for the common round, brilliant gem begins with grinding the stone on a coarse silicon carbide wheel until it is the general size and shape desired. When the stone is satisfactorily shaped, it is closely examined for surface imperfections and chipping. The intricate pavilion facets are then cut and polished. Again the gem is inspected for chipping. Next, the crown facets are cut and polished. After the diamond is cleaned, it is ready for sale.

Cutting the Jonker diamond produced twelve gems for sale. The largest weighed 143 carats and retained the title of the Jonker diamond. All were of the finest blue-white color, and their total value was $2,000,000—quite the profit over the original price Mr. Jonker received for his unexpected find.

Delocalized electrons are also responsible for the malleability and ductility of most metals. Metal atoms are closely packed into vast three-dimensional crystalline structures held together by the "sea" of delocalized electrons. If a force such as a hammer blow displaces some of the atoms, they tend to move along planes within the metal firmly held in place by the "electron sea." (See Figure 6-10.) Metals, unlike ionic compounds, do not cleave as adjacent planar surfaces move across each other. The shape of a metal object can be changed by hammering or stretching without significantly affecting its strength.

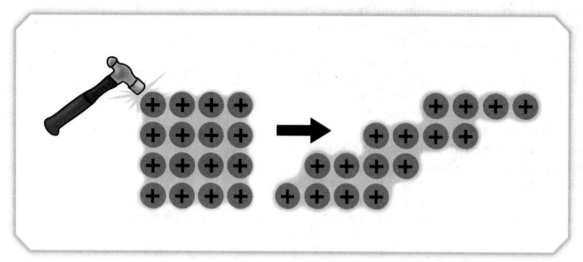

6-10 A force can cause metal ions to move through the sea of delocalized electrons but still remain intact. This contributes to the malleability and ductility of metals.

6.14 Using Chemical Bonds to Solve Problems

How can chemical bonds and the properties of the compounds they form solve NASA's problem? NASA's solution to launch more manned space flights was a reusable spacecraft: the space shuttle. You have probably seen at least one shuttle launch, either in person, on the Internet, or on TV. Do you know how the shuttle system works?

Each shuttle system consists of three parts: the planelike orbiter, which houses astronauts and equipment, two solid-fuel rocket boosters (SRBs), and a disposable external tank containing fuel for the main engines. The external tank is jettisoned just before the shuttle enters space and incinerates in the upper atmosphere. The SRBs, which are reusable, provide about 83% of the liftoff thrust to send the orbiter into the upper atmosphere. After their fuel is spent a little over a minute after launch, they are jettisoned and parachute into the ocean to be retrieved.

Each of the two solid rocket boosters consists of a 45 m (149 ft) long, 3.7 m (12 ft) wide, 1.27 cm (0.5 in.) thick steel canister, containing an igniter, nozzle, and solid fuel. The solid fuel, which has the texture of a pencil eraser, is stable, does not need refrigeration, and is safer than previous fuels. It is lighter and more compact, avoiding some of the problems of the Saturn V fuels. However, once the SRBs are ignited, they burn until their fuel supply is exhausted. Previously, NASA had been hesitant to use solid fuel in manned missions because its burn cannot be regulated. In order to control the rate of burn, the concentration of the propellant varies throughout the booster, adjusting the amount of thrust it provides to minimize impact on the orbiter and astronauts.

This propellant consists of ammonium perchlorate (NH_4ClO_4) and powdered aluminium (Al). It also contains a catalyst, which accelerates the reaction, and binders and fillers to hold the mixture together. The ammonium perchlorate and aluminum react to produce aluminum oxide, aluminum chloride, water vapor, and nitrogen gas. As the bonds in the fuel are broken and new bonds are created to form the products, a great surplus of energy is released. Consequently, this reaction is exothermic, raising the temperature in the SRBs to about 3200 °C (5800 °F). This high temperature rapidly expands the water vapor and nitrogen, causing them to exit through the nozzle and accelerate the rocket.

6C Section Review

1. Why are covalent compounds usually gases, liquids, or solids with low melting points?
2. How does the structure of ionic compounds affect their ability to conduct electricity?
3. How does the electron configuration of metals affect their appearance?

DS 4. Ammonium perchlorate consists of two polyatomic ions: ammonium (NH_4^+) and perchlorate (ClO_4^-). What kind of bonding exists between the atoms in each of these polyatomic ions?

DS 5. What kind of bond forms between the atoms of aluminum in the SRBs? Why do you think that it is involved in such an explosive reaction? (*Hint*: Think of aluminum's electron configuration.)

6-11 An extremely exothermic reaction releases water vapor and nitrogen to propel the space shuttle.

Coming to Terms

chemical bond	140
octet rule	140
covalent bond	141
ionic bond	141
metallic bond	141
polarity	141
electrostatic force	144
Lewis structure	144
polyatomic ion	148
formula unit	150
crystal lattice	151
electron-sea theory	151
delocalized electron	151
network covalent substance	153
alloy	154

Chapter Review

Chapter Summary

- Atoms form chemical bonds in order to attain the low-energy, stable electron configurations of the noble gases. Atoms form bonds by transferring, accepting, or sharing electrons. This principle is called the octet rule.

- There are three chemical bond types defined by the relationship of electrons to the bonded atoms. Covalent bonds share electrons between two atoms. Ionic bonds transfer electrons from one atom to another. Metallic bonds share electrons among many atoms.

- Differences in electronegativities determine bond polarity and character.

- Covalent molecules can be conveniently represented by Lewis structures.

- Covalently bonded polyatomic ions may form when atoms capture stray electrons or hydrogen ions.

- Covalent compounds tend to be gases, liquids, soft solids, or brittle and crumbly solids (although there are notable exceptions). They tend to be poor conductors of heat and electricity.

- Ionic compounds consist of crystal lattices. The ratio of cations to anions needed to cancel each other's charge determines the empirical formula for the compound.

- Ionic compounds generally form solid crystalline substances with high melting points. In their solid state they are good insulators.

- Metals consist of many metal cations immersed in a "sea" of mobile, delocalized electrons.

Review Questions

Concept Review

1. Why do atoms form chemical bonds?

2. Why does the octet rule not always refer to a stable arrangement of eight valence electrons?

3. Considering electron affinity alone, state the most probable way (losing, gaining, sharing) in which the following atoms could attain a valence octet. Also state the number of electrons involved.
    a. K
    b. Ca
    c. Ga
    d. Ge
    e. As
    f. Se
    g. Br
    h. Kr
    i. C

4. Which of the three bond types is distinctly different from the other two? Explain why this is so.

5. Why does hydrogen act like a Group 17 element when forming covalent bonds?

6. Why are the diatomic elements not found in nature as single atoms?

7. Identify the types of atoms (metal or nonmetal) in the following compounds and then tell whether the compounds are predominantly ionic, covalent, or metallic.
   a. sodium chloride (NaCl)
   b. bronze (Sn-Cu-Zn alloy)
   c. magnesium bromide ($MgBr_2$)
   d. carbon dioxide (CO_2)
   e. stainless steel (Fe-Ni-Cr alloy)

DS 8. What forms of energy are produced when the propellant in the space shuttle SRBs ignites? Where does this energy come from?

DS 9. If so much energy is released when aluminum and ammonium perchlorate react, which do you think is more stable—these two substances or the aluminum oxide and aluminum chloride they produce?

DS 10. What do the crashes and deaths involved in the losses of the shuttles *Challenger* in 1986 and *Columbia* in 2003 highlight about space travel? What do these deaths reveal about fulfilling God's command to have dominion over the earth in a fallen world?

True or False

11. Many bonded atoms attain a stable electron configuration without having a valence octet.

12. There is no purely ionic bond.

13. There is no purely covalent bond.

14. It is not possible to identify a metallic bond between two or a few metal atoms. A metallic bond exists only in the presence of a great many metal atoms.

15. A valence energy band containing innumerable energy levels accounts for the mirrorlike luster associated with metals.

Application

⊙ 16. The following covalent compounds contain only single covalent bonds. Draw their Lewis structures.
   a. H_2
   b. HCl
   c. CH_4
   d. CF_2Cl_2
   e. H_2S

⊙ 17. The following covalent compounds or ions each contain at least one double or triple bond. Draw their Lewis structures.
   a. H_2CO
   b. CS_2
   c. CN^-
   d. CO_2
   e. C_2H_4
   f. C_2H_2

⊙ 18. Draw the Lewis structures of the polyatomic ions below.
   a. OH^-
   b. PO_4^{3-}
   c. ClO_4^-
   d. SO_3^{2-}

Bond Theories and Molecular Geometry

Chapter 7

7A Bond Theories	161
7B Molecular Geometry	166
Facet	
The Nose Knows	176

Dominion Science Problem

Saving the Soldiers

In war, one army targets another, seeking to immobilize and destroy it. Yet there is also an almost invisible army that seeks to heal—the massive network of medical experts who treat battlefield casualties. Soldiers wounded in distant lands can receive top treatment in US hospitals within thirty-six hours of being wounded in combat. But the most critical moments are just after injury. Blood loss is the leading cause of death on the battlefield. Corpsmen and field medics need a high-tech, portable, effective technique to stop bleeding quickly, preventing life-threatening blood loss. How can we improve critical care medical techniques to further lower the death rate among soldiers?

7A Bond Theories

7.1 A Limitation of Lewis Structures

In Chapter 6, you learned to use Lewis structures to represent molecules whose atoms are held together by covalent bonds. Lewis structures clearly show the bonds between atoms, but they are so simple that they leave out valuable information about the compounds they represent. For instance, a Lewis structure tells nothing about the three-dimensional shape of a molecule, which can determine many of its physical and chemical properties. Is the shape of the water molecule

$$H-\ddot{\underset{..}{O}}-H \quad \text{or} \quad H-\underset{|}{\overset{..}{\underset{H}{O}}}:$$

or is it something else? Because of the limitations of Lewis structures, there is no way to tell. For a more complete description of chemical bonds and the molecular shapes that they produce—especially their three-dimensional structures—we must use the atomic quantum theory.

> **7A Section Objectives**
> After finishing this section, you should be able to
> - discuss the key features and limitations of the valence bond theory, molecular orbital theory, and bond hybridization.
> - describe the formation of sigma and pi bonds.
> - state exceptions to valence bond theory, such as molecular resonance and exceptions to the octet rule.

7.2 Orbitals and Valence Bond Theory

As you learned in Section 4B, electrons occupy various sublevels of the principal energy levels. These sublevels, named s, p, d, and f, contain a varying number of orbitals. Those orbitals are modeled as spatial regions where electrons have a high probability of being found. Each orbital can hold a maximum of two electrons. The s sublevel has one orbital, the p sublevel has three orbitals, the d sublevel has five orbitals, and the f sublevel has seven orbitals. The s and p orbitals have relatively simple shapes, while the d and f orbitals are much more complex.

Covalent bonds involve two atoms sharing pairs of valence electrons with opposite spins. These shared electrons occupy partially filled valence orbital regions that overlap between bonded atoms. When the two bonding orbitals are superimposed, the overlapping space containing both electrons becomes available to both nuclei. Effectively, both atoms acquire another valence electron that fills the vacancy in that particular orbital. This concept is called the **valence bond theory**. This theory is just one of several models that describe the formation of covalent bonds. It is important to realize that no single theory of bonding accounts for everything that chemists observe about covalent bonds.

> The valence bond theory is also called the localized electron theory.

7.3 Sigma and Pi Bonds

The hydrogen molecule (H_2) is the simplest diatomic molecule. According to valence bond theory, the spherical $1s$ orbital of each hydrogen atom overlaps and merges with the other to share a pair of electrons with opposite spins. A region of high electron density forms on a line, called the *bond axis*, that connects two nuclei. When orbitals overlap on the bond axis, a **sigma (σ) bond** is formed. This bond formation is an "end-to-end" type of overlap. Figure 7-1 illustrates a sigma bond.

> The lowercase Greek letter σ (sigma) is equivalent to the Roman s, as in s orbital.

Valence *p* orbitals also form sigma bonds. Halogens, such as fluorine (F₂), form a single sigma bond when each atom's unfilled, dumbbell-shaped valence *p* orbitals overlap end to end.

In addition, a sigma bond may form if an *s* and a *p* orbital overlap end to end, as in hydrogen fluoride (HF). Sigma bonds are the strongest type of covalent bond because the region of highest electron probability lies on the bond axis. This location maximizes the effectiveness of the electrostatic attraction between the electrons and the nuclei.

Double and triple bonds form when more than one set of orbitals from the *p* sublevel overlap. One of the bonds is almost always a sigma bond formed by an end-to-end overlap of orbitals. The second bond is different. The two lobes of the orbital (in Figure 7-3, the *p* orbital) extend toward the corresponding lobes of the other atom's orbital. The lobes overlap above and below the bond axis to form two regions of high electron density parallel to the bond axis. Even though two separate regions overlap, they are both part of *one* bonding orbital containing two electrons. Such a side-to-side orbital overlapping is called a **pi (π) bond**. Molecular oxygen,

$$:\ddot{O} = \ddot{O}:$$

contains a sigma bond and a single pi bond.

For triple bonds, as in molecular nitrogen, one of the bonds is a sigma bond formed by two lobes of the *p* orbital overlapping end to end. The two remaining bonds are pi bonds. Pairs of *p* orbitals overlap above and below the bond axis and to either side of the bond axis to form these pi bonds.

Pi bonds are weaker than sigma bonds because the electrons are spread out over a greater volume of space around the bond axis. Also, the electrons in pi bonds are less attracted to the nuclei than in sigma bonds because the region of highest density is not along the bonding axis, and so the electrons pull on the nuclei at an angle. Since pi bonds almost always occur with a sigma bond, however, the combination of a sigma and pi bond together is stronger than either bond by itself. As would be expected, triple bonds are stronger than either double or single bonds between atoms.

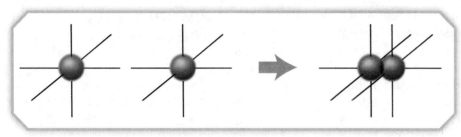

7-1 A single sigma bond forms when *s* orbitals overlap end to end.

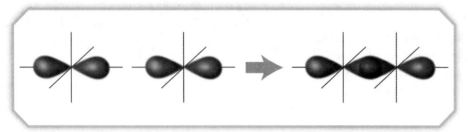

7-2 A single sigma bond also forms when *p* orbitals overlap end to end.

The lowercase Greek letter π (pi) is equivalent to the Roman *p*, as in *p* orbital.

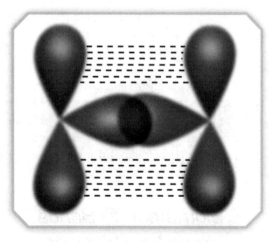

7-3 The second bond in a double bond forms when two orbitals overlap side by side in a pi bond.

7-4 The third bond in a triple bond forms when another set of *p* orbitals forms a pi bond.

7.4 Molecular Resonance

Multiple bonds introduce another limitation of Lewis structures. Consider the sulfur dioxide molecule (SO$_2$). It can be drawn as follows.

$$\ddot{\text{O}} = \text{S}:$$
$$\quad\ \ |$$
$$\quad :\ddot{\text{O}}:$$

However, there is no reason why the double bond could not exist between the sulfur and the other oxygen as follows.

$$:\ddot{\text{O}} - \text{S}:$$
$$\qquad \|$$
$$\qquad \ddot{\text{O}}$$

Because the double bond is stronger than the single bond, there should be a difference in chemical properties between the two ends of the molecule. However, experimental evidence reveals that the molecule acts as if it is completely symmetrical. The sulfur-oxygen bonds must be identical. Chemists show that these two Lewis structures are equivalent by inserting a double-headed arrow between them.

$$\ddot{\text{O}} = \text{S}: \longleftrightarrow :\ddot{\text{O}} - \text{S}:$$
$$\quad\ \ |\qquad\qquad\qquad \|$$
$$\quad :\ddot{\text{O}}:\qquad\qquad\qquad \ddot{\text{O}}$$

For some molecules, no single Lewis structure can completely describe the distribution of electrons. Such molecules exhibit **resonance**. A molecule displaying resonance does *not* oscillate between two (or more) possible electron configurations. Rather, the bonds take on an intermediate character because electrons are mobile and their wave functions can occupy different regions of a molecule at the same time. You can think of these bonds as being the average of the single and double bond possibilities. The sulfur dioxide molecule acts as though there are one and a half bonds between the central sulfur and each oxygen.

One familiar example of a molecule exhibiting resonance is ozone (O$_3$).

$$:\ddot{\text{O}} - \ddot{\text{O}} = \ddot{\text{O}}: \longleftrightarrow :\ddot{\text{O}} = \ddot{\text{O}} - \ddot{\text{O}}:$$

Another is the nitrate ion (NO$_3^-$).

$$\left[:\ddot{\text{O}} - \text{N} = \ddot{\text{O}}: \atop \ \ \ \ \ | \atop \ \ \ \ :\ddot{\text{O}}: \right]^- \longleftrightarrow \left[:\ddot{\text{O}} = \text{N} - \ddot{\text{O}}: \atop \ \ \ \ \ | \atop \ \ \ \ :\ddot{\text{O}}: \right]^- \longleftrightarrow \left[:\ddot{\text{O}} - \text{N} - \ddot{\text{O}}: \atop \ \ \ \ \ \| \atop \ \ \ \ \ \ddot{\text{O}} \right]^-$$

In each case, there are multiple "correct" Lewis structures, but the bonds are experimentally found to be identical.

7-5 Ozone is a triatomic form of oxygen and a major component of smog.

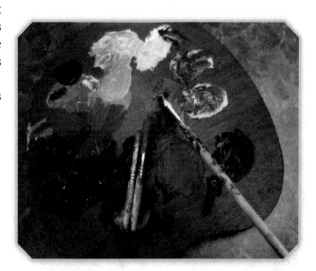

7-6 Resonance bonds are like a mixture of two primary colors. The new mixture does not oscillate between the two original colors but forms a new color.

7.5 Octet Exceptions

Up to this point, the goal in drawing Lewis structures has been to always have eight valence electrons for every atom (except hydrogen). However, some molecules form without an octet. There are three ways this can happen.

First, some molecules have an odd number of valence electrons to share. In these cases, one pair is short one electron. Nitrogen

> The single electron on the nitrogen in nitrogen dioxide is called a *free radical*. Free radicals can build up in our bodies as a byproduct of metabolism and are thought to contribute to disease and the effects of aging. Foods rich in antioxidants rid the body of free radicals. So eat your vegetables!

dioxide (NO_2), a gaseous byproduct of combustion engines, is one such example. Nitrogen contributes five valence electrons, while the two oxygen atoms bring a total of twelve. There is no way to make seventeen electrons form octets, so one atom ends up with seven electrons. The Lewis structure is written as follows.

$$\ddot{\ddot{O}} = \underset{\underset{\ddot{\ddot{O}}:}{|}}{N} \cdot$$

Second, other elements, such as boron, are content with fewer than four pairs of shared electrons. Boron tends to form molecules in which it is surrounded by only six valence electrons. For example, boron has just three valence electrons, and fluorine already has seven of the eight electrons it needs for stability. Boron forms three single bonds with the highly electronegative fluorine atoms.

$$:\ddot{\ddot{F}} - \underset{}{\overset{\overset{:\ddot{\ddot{F}}:}{|}}{B}} - \ddot{\ddot{F}}:$$

In a third group of molecules, more than eight valence electrons are shared with a central atom. Sulfur hexafluoride (SF_6) is a common example. In order to use all forty-eight valence electrons, each of the fluorine atoms forms a single bond (one shared pair) with the central atom, sulfur. Sulfur thus shares six pairs of electrons. Such situations are called expanded octets and occur only in periods three and higher. The valence bond theory explains these exceptions to the octet rule by saying that the extra electrons are shared with orbitals in the *d* sublevel. For example, the sulfur in sulfur hexafluoride fits eight of the shared electrons into its 3*s* and 3*p* sublevels, and the final four electrons into the first two 3*d* orbitals.

7-7 Most berries are good sources of antioxidants, which reduce the free radicals in your body.

7.6 Molecular Orbital Theory

According to valence bond theory, diatomic oxygen should be colorless and nonmagnetic. When a group of scientists liquefied diatomic oxygen, they were surprised to find that it was pale blue and demonstrated magnetic properties. Neither of these characteristics was predicted by the current molecular models. It was time for a new theory.

The valence bond theory begins to explain why molecules have certain shapes. However, this model does not explain all observations. For example, the valence bond theory prohibits stray electrons that could account for unusual chemical and physical properties of certain compounds. The concept of resonance structures was devised to explain why some molecules do not fit the localized electron view of the valence bond theory. As the scientists working with liquid oxygen discovered, valence bond theory must not correctly model the bonds between oxygen atoms.

A more recent and complex model of bonding explains some of those observations. The **molecular orbital theory** suggests that the orbitals of a molecule's atoms are replaced by totally new orbitals when a molecule forms. Each molecule has a unique set of orbitals, equal in number to the sum of the atomic orbitals in the original atoms. Some

7-8 Liquid oxygen can be held by a magnetic field.

orbitals encircle two, three, four, or even more atoms; often they encircle the entire molecule. The molecular orbital theory ranks the resulting orbitals in order of increasing energy. Electrons from the original atoms fill low-energy molecular orbitals before they fill high-energy orbitals. The arrangement of electrons in these orbitals tells chemists whether bonds will form; whether they will be single, double, or triple bonds; and whether the bonds will contain unpaired electrons. The presence of unpaired electrons in molecular orbitals has been associated with paramagnetism, the attraction of the molecule to a magnetic field. This explanation of paramagnetism lets the molecular orbital theory predict the magnetic qualities of diatomic oxygen.

7-9 The pair on the left are interfering with each other; the pair on the right are reinforcing each other's efforts.

In molecular orbital theory, there are both *bonding* and *antibonding* orbitals. When a molecule forms, the electrons in the atomic orbitals can either reinforce or interfere with each other because they behave like waves. When electrons reinforce each other, bonding orbitals form. Bonding orbitals are located between the nuclei. The electrons in these orbitals stabilize the molecule because atoms can share them. They are lower in energy than the atomic orbitals from which they form, so they contribute to bonding. This is how energy is stored in bonds.

Antibonding orbitals form when atomic orbitals combine in an unfavorable manner. If the electron waves destructively interfere with each other, an orbital forms on the outside of the molecule, far from the two nuclei. Antibonding orbitals are higher in energy than the atomic orbital from which they form. The electrons in an antibonding orbital spend little time between the two nuclei, so these antibonding orbitals destabilize the molecule. Molecular orbital theory can thus be used to predict if a bond will form between two atoms. If equal numbers of electrons inhabit bonding orbitals and antibonding orbitals, or if there are more electrons in antibonding orbitals, then no bond will form.

7A Section Review

1. What is necessary for a covalent bond to form according to the valence bond theory?
2. Compare and contrast sigma and pi bonds.
3. Define *molecular resonance*.
4. Why don't bonds form between all atoms according to molecular orbital theory?
5. Draw the Lewis structures for these molecules or ions.
    a. CO_3^{2-}
    b. XeF_2
    c. BeH_2
6. (True or False) According to the molecular orbital theory, bonding orbitals are higher in energy than the atomic orbitals from which they form.

7B Molecular Geometry

7.7 VSEPR and Molecular Shape

7B Section Objectives

After finishing this section, you should be able to

- predict and name a molecule's general shape when given its Lewis structure.
- draw electron configurations representing orbital hybridization.
- predict the existence and direction of a dipole moment in a molecule when given its Lewis structure.

VSEPR (VES per). The term *valence shell electron pair repulsion* is abbreviated VSEPR but pronounced like the word *vesper*.

One of the basic principles of electricity is that opposite charges attract and like charges repel. Charge repulsion determines the shape of molecules held together by covalent bonds. The repulsion of bonding and unbonded or lone electrons in a molecule limits the possible arrangements of covalent bonds around the central atom. Chemists developed the **valence shell electron pair repulsion (VSEPR) theory** to account for these observations.

VSEPR theory focuses on the locations of highest electron density surrounding the central atom in a molecule. The Lewis structure is used to determine the number of regions of electron density. An area of electron density can be

- a nonbonding or lone pair of electrons,
- a single bond,
- a double bond, or
- a triple bond.

In the VSEPR model, these areas of negative electrical charge repel each other until they are as far apart as is geometrically possible. According to valence bond theory, only a few general molecular shapes are possible. Let us examine these in turn.

Carbon tetrachloride (CCl_4), which was once used as a dry-cleaning solvent, has four areas of electron concentrations around the carbon atom—four single bonds.

7-10 When four balloons are tied together, they naturally assume the shape of a tetrahedron.

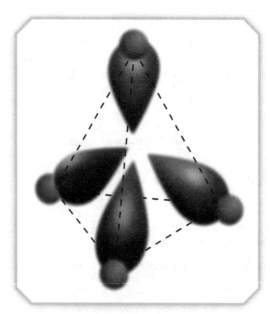

7-11 A tetrahedral molecule contains four bonds to the central atom. The angles between the bonds are 109.5°.

How are the chlorine atoms arranged around the central carbon? The single bonds and the chlorine atoms repel each other. Geometrically, the chlorine atoms are farthest from each other when they are positioned at the vertices of a tetrahedron, which is a polyhedron with four faces, each of which is a regular triangle (Figure 7-11). The carbon atom is located at the center of the tetrahedron. The shape of such a molecule is said to be **tetrahedral**. In general, any molecule in which the central atom is surrounded by four areas of electron density in its Lewis structure assumes an approximately tetrahedral arrangement.

Formaldehyde (CH_2O) has three regions of electron density surrounding the central carbon atom—one double bond and two single bonds.

$$\overset{\overset{\text{H}}{|}}{\text{H} - \text{C} = \overset{..}{\text{O}}\!:}$$

The Lewis structure indicates three areas of electron density that surround the central atom. All involve bonding pairs, so the resulting molecule will be Y-shaped, with all four atoms lying in a single plane (Figure 7-12). This arrangement is called **trigonal planar**.

Carbon dioxide (CO_2) has two regions of electron concentrations around the central carbon—two sets of double bonds.

$$:\!\overset{..}{\text{O}} = \text{C} = \overset{..}{\text{O}}\!:$$

A molecule with two concentrations of electrons around the central atom has a **linear** shape, regardless of the nature of the concentrations (Figure 7-13). The bonds are 180° apart and point in opposite directions. Diatomic (two-atom) molecules of compounds and elements are linear as well.

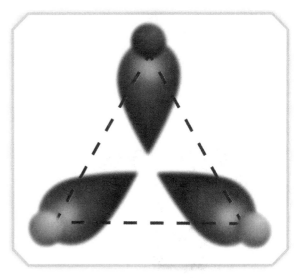

7-12 Molecules having three concentrations of bonds or electron pairs around the central atom are Y-shaped, or trigonal planar. The angles between bonds are approximately 120°.

7.8 Orbital Hybridization

Among the inadequacies of the valence bond theory is its explanation of how carbon bonds. In its ground state, carbon has two partially filled p orbitals and one full s orbital in its valence shell.

$$[\text{He}]2s^2 2p^2$$

According to valence bond theory, it would seem probable that carbon could fill this p orbital by forming two covalent bonds. That

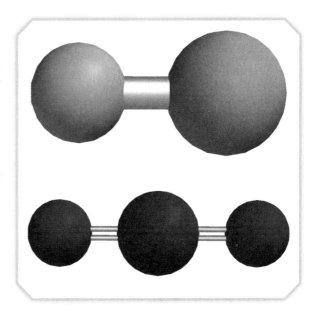

7-13 Molecules with two concentrations of bonds or electrons around the central atom are linear. Three-atom linear molecules have bonds that are 180° apart. Two-atom molecules are always linear. The examples shown here are HF (top) and CO_2 (bottom).

Photo courtesy of Rebreather Labs

7-14 Rebreather units, like this Pelagian DCCCR unit, allow divers to go deeper and stay down longer because they chemically recycle the exhaled carbon dioxide.

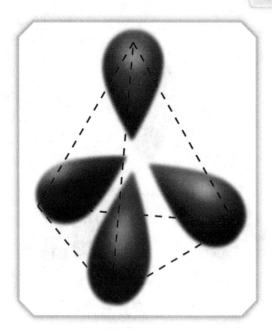

7-15 The four sp^3 hybrid orbitals in carbon. Note that these orbitals exist only when carbon has formed covalent bonds with other atoms.

> Remember that the hybridization model of orbitals is just a refinement of a theory that scientists use to describe bonding. No theory explains everything we observe about bonding. It does not need to since scientists don't claim to understand every detail about how the universe works. However, orbital hybridization and VSEPR theory do help account for the shapes of many molecules. You can think of scientific theories as tools that help scientists work with scientific data.

> The existence of an unused orbital in boron trifluoride makes it chemically active, especially when combined with molecules having a central atom with a lone pair of electrons. These electrons will jump to fill the unused orbital in boron, bonding the two molecules together.

Problem-Solving Strategy 7-1
The number of hybrid orbitals formed is always equal to the number of s and p orbitals containing unpaired electrons after promotion.

explanation violates the octet rule, because it leaves an entire p orbital completely empty. In addition, carbon routinely forms four covalent bonds. Sometimes the four bonds are equivalent, as in methane (CH_4). A modification of bonding theory was necessary to account for these observations.

One explanation for the behavior of carbon suggests that one of the paired electrons in the $2s$ orbital is moved or "promoted" to the empty p orbital. Then carbon would have four unpaired electrons.

$$[\text{He}] \quad \underline{\uparrow\downarrow}_{2s} \quad \underline{\uparrow}\,\underline{\uparrow}\,\underline{}_{2p} \quad \xrightarrow{\text{promote}} \quad [\text{He}] \quad \underline{\uparrow}_{2s} \quad \underline{\uparrow}\,\underline{\uparrow}\,\underline{\uparrow}_{2p}$$

However, the different energies of the s and the p orbitals could not form identical bonds with the four hydrogen atoms.

The most recent theory is that the single s and three p orbitals combine to form four orbitals within a single valence energy level. The process by which new kinds of orbitals with equal energies are formed from a combination of orbitals with different energies is called orbital **hybridization**. Orbital hybridization is common between nonmetals and even in metals involved in covalent bonding. In the case of carbon, one s and three p orbitals form the new orbital. The four new orbitals are known as sp^3 hybrid orbitals. The new carbon orbital notation is as follows.

$$[\text{He}] \quad \underline{\uparrow}_{2s} \quad \underline{\uparrow}\,\underline{\uparrow}\,\underline{\uparrow}_{2p} \quad \xrightarrow{\text{hybridize}} \quad [\text{He}] \quad \underline{\uparrow}\,\underline{\uparrow}\,\underline{\uparrow}\,\underline{\uparrow}_{sp^3}$$

Why do atoms form hybrid orbitals in the first place? Promoting electrons to higher energy levels requires extra energy. However, the main lobes of the individual hybrid orbitals are much larger than the lobes of s or p orbitals. They can point to and overlap the orbitals of other atoms more effectively, forming stronger bonds. The increase in stability from forming bonds more than offsets the energy required to create hybrid orbitals.

Other hybrids of s and p orbitals are possible. For example, in the covalent molecule boron trifluoride (BF_3), the central boron atom is bonded to three fluorine atoms, where each can share only one electron. Boron has three valence electrons—a pair in the $2s$ orbital and a single p orbital electron. In order to form three identical bonds with the three fluoride atoms, it must promote one of its s electrons to an empty p orbital, and then hybridize its s and p orbitals to produce three equivalent hybrid orbitals. Since two p orbitals were involved, they are called sp^2 hybrid orbitals. The empty third p orbital that is "left over" (unhybridized) is not involved in bonding in this molecule.

$$\begin{array}{c} :\!\ddot{\text{F}}\!: \\ | \\ :\!\ddot{\text{F}}\!-\text{B}-\ddot{\text{F}}\!: \end{array}$$

Notice that boron shares only six electrons by forming three sigma bonds with fluorine. This compound is a classic example of how a more stable electron configuration is attained through hybridization rather than by following the octet rule. This is one of the octet rule exceptions mentioned in Subsection 7.5. The new boron orbital notation is as follows.

$$\begin{array}{c} \text{2s} \quad \text{2p} \\ [\text{He}] \; \uparrow\downarrow \; \uparrow \; \underline{} \; \underline{} \end{array} \xrightarrow{\text{promote}} \begin{array}{c} \text{2s} \quad \text{2p} \\ [\text{He}] \; \uparrow \; \uparrow \; \uparrow \; \underline{} \end{array} \xrightarrow{\text{hybridize}} \begin{array}{c} sp^2 \quad \text{2p} \\ [\text{He}] \; \uparrow \; \uparrow \; \uparrow \; \underline{} \end{array}$$

Consider the following molecules, all of which contain four sp^3 hybrid orbitals. These orbitals seek the vertices of a regular tetrahedron because of mutual repulsion. (See Subsection 7.7.) Note how the tetrahedral arrangement determines the locations of bonded atoms and of lone pairs of electrons.

CH_4. All four sp^3 orbitals of the central carbon in methane are involved in bonding. A hydrogen atom is located at the end of each bond, and each carbon-hydrogen bond is identical. The methane molecule is thus arranged so that a hydrogen atom is at each vertex of a tetrahedron. Methane and similar molecules are said to have a tetrahedral shape.

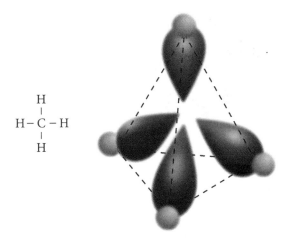

NH_3. The ammonia molecule has only three bonding pairs of electrons around the central atom. The three hydrogen atoms occupy three of the four tetrahedral vertices defined by the sp^3 orbitals. The fourth vertex is occupied by the lone pair of electrons. Since we determine the shape of a molecule by looking at the positions of its atoms, this molecule is **pyramidal**—a triangular pyramid.

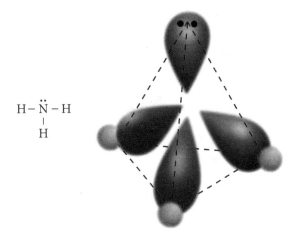

> **Problem-Solving Strategy 7-2**
> We determine the shape of a molecule by looking at the arrangement of atoms around the central atom, not by looking at the unbonded pairs.

H₂O. Water molecules consist of two hydrogen atoms bonded to a central oxygen. The hydrogen atoms occupy two of the tetrahedral vertices defined by the sp^3 orbitals. Note from the Lewis structure that there are two unbonded electron pairs. These occupy the other two vertices, although recent research indicates that these electron pairs are more evenly distributed around the highly electronegative oxygen. If one considers only the arrangement of the atoms, a line drawn from the center of one hydrogen through the oxygen to the other hydrogen forms an angle. Thus, this kind of molecule is called a **bent** molecule. Typically, the angle is 109.5°—the same as in a regular tetrahedron. In water, the angle between the bonds is 104.5° since the nonbonding electrons repel more than the bonded pairs do.

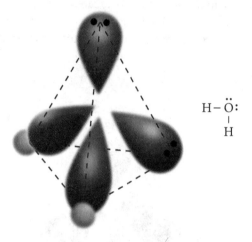

HF. While there is technically no central atom in a hydrogen fluoride molecule, the four sp^3 orbitals belong to the fluorine and define the tetrahedral arrangement of electron concentrations around the molecule. Thus, fluorine is considered the central atom. The single bond between the hydrogen and fluorine atoms produces a simple linear shape. The three lone pairs occupying the other hybrid orbitals are held close to the highly electronegative fluorine.

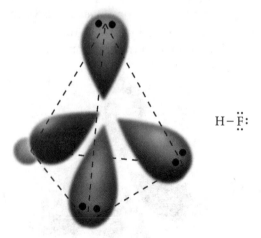

The same basic principles hold true when determining the shapes of sp^2 and sp hybrid compounds. Table 7-1 provides examples for you to study.

7-1 Representative Molecular Shapes

Example	Number of electron regions	Number of bond regions	Lewis structure	Spatial arrangement	Geometry
CH_2Br_2 dibromomethane	4 sp³ hybrid	4	H–C(–H)(–:Br:)(–:Br:)	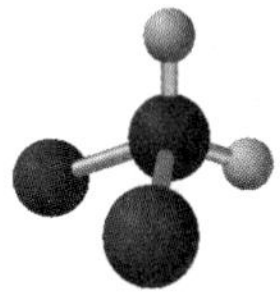	tetrahedral
PH_3 phosphorus hydride		3	H–P̈–H, H	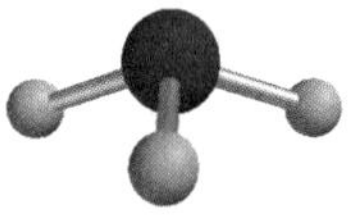	pyramidal
H_2S hydrogen sulfide		2	H–S̈:, H		bent (109.5°)
HCl hydrogen chloride		1	H–C̈l:	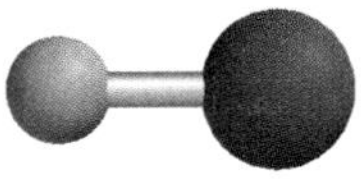	linear
BI_3 boron iodide	3 sp² hybrid	3	:Ï–B–Ï:, :Ï:	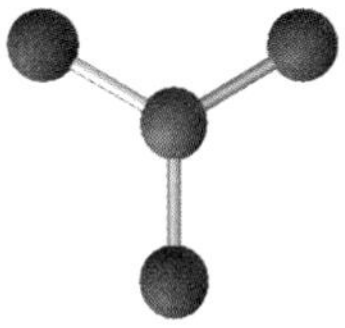	trigonal planar
GeF_2 germanium fluoride		2	:F̈–Ge: ‖ F̈: (resonance)	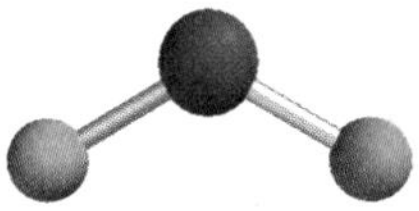	bent (120°)
BeF_2 beryllium fluoride	2 sp hybrid	2	F–Be–F	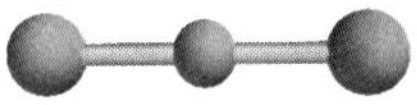	linear
CO carbon monoxide		1	:C≡O:		linear

EXAMPLE PROBLEM 7-1

Predicting Molecular Shapes

Predict the shapes of the following molecules:

a. :Ö–Ö=Ö:

b. H–C≡N:

c. :F̈–N̈–F̈:
   |
   :F̈:

(continued)

Solution

a. There are three areas of electron density around the central oxygen. However, since only two of those regions involve bonds with other atoms (in a resonance structure), triatomic oxygen's shape is bent.
b. There are two regions of electrons around the central carbon in hydrogen cyanide. The three atoms lie in a straight line. This molecule's shape is linear.
c. Nitrogen trifluoride has four regions of electron density around the central nitrogen. Since only three atoms are bonded to the nitrogen, the molecule has a pyramidal shape.

7.9 Polar Covalent Bonds

Diatomic molecules have no polarity. But what about covalent bonds between atoms with different electronegativities? How do electronegativity differences affect how electrons are shared?

Let us examine hydrogen chloride (HCl), a familiar diatomic gaseous molecule that has a moderately large electronegativity difference ($\Delta EN = 1.0$). The pair of electrons in the single sigma bond occupies mainly the region between the atoms along the bond axis. However, chlorine's much larger nuclear charge tends to attract the electrons more effectively than hydrogen's. The densest region of electron probability in the bond shifts toward the chlorine. The charges of these regions of the molecule are not integer values (e.g., 1+, 1−, etc.) because there is no complete transfer of electrons from one atom to the other. When the electrons are shared, though unequally, the bond is called a **polar covalent bond**.

Partial charges on molecules are shown by the lowercase δ (delta), implying a charge differential. The partial charges on the hydrogen chloride molecule are symbolized below.

$$\overset{\delta^+}{H} : \overset{\delta^-}{\ddot{\underset{..}{Cl}}} :$$

The molecule is not completely covalent like a diatomic molecule, nor is it completely ionic.

Both complex molecules and simple molecules can have polar bonds if there is an electronegativity difference in participating atoms. Since bond polarities have important effects on the chemical and physical properties of a compound, chemists developed a special symbol used with Lewis notation that shows the polarity of each bond in a molecule. This symbol is an arrow with a plus sign attached to the arrow's tail (see below). The arrow points toward the more electronegative atom in the bond with the partial negative charge. The plus sign indicates the less electronegative atom in the bond with the partial positive charge. You can use this notation to quickly determine where the partial positive and negative regions are on a molecule.

$$\overset{+\longrightarrow}{H} : \ddot{\underset{..}{Cl}} :$$

All bonds formed between atoms of different elements are polar to some extent. The magnitude of the bond polarity depends on the ΔEN

7-16 The red area shows the zone of highest negative charge.

7-17 Because the charge distribution on diatomic oxygen is balanced, the molecule is nonpolar.

of the two elements. The greater the difference in electronegativity, the more polar the bond will be. For simple molecules consisting of two different elements, polar bonds mean polar molecules. In other words, a molecule formed from a single polar bond will itself be polar, with distinct positive and negative partial charges. Polar molecules like water have important properties that make them very useful.

7.10 Dipole Moment: A Measure of Polarity

If a molecule has polar covalent bonds, does that always mean that the molecule is polar? You might think so, but another factor is involved—the geometric symmetry of the molecule itself. As you learned earlier in this section, molecules can take on a variety of shapes. The polarities of individual bonds may be quite high, but if those bonds are arranged symmetrically, they balance each other. As a result, the molecule overall is nonpolar.

Figure 7-19 shows surface charge diagrams of four molecules: two polar and two nonpolar. Note that in the cases of methane (CH_4) and carbon dioxide (CO_2), the bonds between the outer atoms and the central atom are polar but identical. The molecular symmetry arranges the partial charges so that there are no "ends" or poles that have different partial charges. In the cases of water (H_2O) and ammonia (NH_3), the asymmetry of the molecules, caused by the unbonded electron pairs, ensures that there are distinct regions of negative and positive charges on the molecular surfaces.

When scientists map the strength of charge around a molecule, an image emerges that bears little resemblance to the arrangement of the atoms. The most significant features are the locations of the

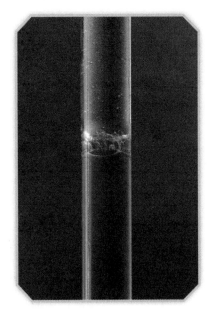

7-18 Water molecules are polar; oil is not. Therefore, the two will not mix.

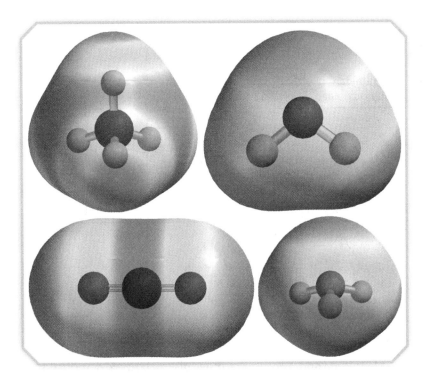

7-19 Surface charge diagrams for methane (top left), carbon dioxide (bottom left), water (top right), and ammonia (bottom right). Red indicates maximum negative charge, and blue indicates maximum positive charge.

positive and negative partial charges. For asymmetrical molecules, one end is predominantly negative and the other positive.

Both the bent and the pyramidal shapes are asymmetrical and so always produce polar molecules if they contain polar covalent bonds. This rule holds true even if all the outer atoms have the same electronegativities. The basic symmetrical molecular geometric shapes are tetrahedral, trigonal planar, and linear. Other less common shapes also are symmetrical. Symmetrical molecules are not polar if the outer atoms are the same, but even symmetrical molecules are polar if the outer atoms are different. Consider two linear molecules: carbon dioxide (CO_2) and carbonyl sulfide (COS). Carbon dioxide is nonpolar because the oxygen atoms balance each other. Carbonyl sulfide, on the other hand, is polar because the outer atoms, oxygen and sulfur, have different electron affinities. Table 7-2 summarizes these guidelines.

7-2 When Molecules Are Polar

Number of atoms bonded to central atom	Four concentrations of electrons	Three concentrations of electrons	Two concentrations of electrons
4	tetrahedral—polar if outer atoms have different EN		
3	pyramidal—always polar	trigonal planar—polar if outer atoms have different EN	
2	bent—always polar	bent—always polar	linear—polar if outer atoms have different EN
1	linear—polar if atoms have different EN	linear—polar if atoms have different EN	linear—polar if atoms have different EN

Magnitude and direction of molecular polarity is described by a quantity known as the **dipole moment (μ)**. It is the vector sum of all bond polarities that exist within the molecule. The Greek letter μ (mu) is used as the symbol for this quantity. The larger the value of μ, the more polar the bond or molecule. As before, a crossed arrow is used to point from the more positive region to the more negative region(s).

7-3 Dipole Moments of Several Substances

H—H	Cl—F	H—F	Cl—C(Cl)(Cl)—Cl	Cl—C(H)(Cl)—Cl	H—N(H)—H
μ = 0.00	μ = 0.89	μ = 1.83	μ = 0.00	μ = 1.04	μ = 1.47

EXAMPLE PROBLEM 7-2

Determining Molecular Polarity

Determine whether the following molecules are polar. If they are, indicate the direction of their dipole moments with a crossed arrow.

a. CO_2 b. H_2S c. ClBr

Solution

The first thing you must do in each case is to draw the molecule's Lewis structure. The drawing can be used to predict the molecule's shape using VSEPR theory. If the molecule is asymmetrical, it must be polar. If the molecule is symmetrical, you can determine the polarity of the molecule by examining the distribution of charge around the central atom and calculating the ΔEN for each bond.

a. CO_2 has the following Lewis structure.

$$:\!\ddot{O}=C=\ddot{O}\!:$$

Because there are two regions of electrons, CO_2 is linear—a symmetrical shape. Since both outer atoms have the same electronegativity, they are "pulling" on the carbon equally, resulting in a nonpolar molecule.

b. H_2S has the following Lewis structure.

$$H-\ddot{\underset{|}{S}}\!:$$
$$H$$

Because it has four electron regions but only two are occupied by atoms, its shape is bent. Since all bent shapes are asymmetrical, H_2S is polar. The dipole moment arrow always points toward the region of a molecule containing lone pairs of electrons.

$$H-\ddot{\underset{|}{S}}\!:$$
$$H$$

By implication, the positive end of this molecule is shared by both hydrogen atoms.

c. ClBr must be linear, since it is a diatomic molecule. The ΔEN between chlorine and bromine is 3.0 − 2.8 = 0.2. Therefore, electrons spend more time closer to the chlorine, resulting in the following dipole moment.

$$\overset{\longleftarrow}{Cl-Br}$$

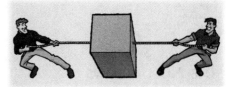

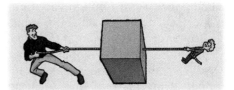

7-20 These three situations are like polarity in compounds.

7.11 Using Charge Attraction to Solve Problems

Compounds in the human body consist of molecules that are usually much more complex than the ones we have discussed, but they are affected by the same factors: orbital hybridization, polarity, and electron interactions. How can we use these properties to quickly stop bleeding in soldiers with serious wounds? How is bleeding stopped in the human body, anyway? You may remember from previous science courses that platelets cause a blood clot to form by creating a mesh of blood cells. With large battlefield wounds, this process must be accelerated as much as possible.

A pharmaceutical company has developed a bandage made of *chitosan*, a compound extracted from the shells of shrimps and other crustaceans. Chitosan is a long, complex sugar molecule that is positively charged. When applied, it attracts the negatively charged

The Nose Knows
Facets of Chemistry

Imagine walking through your front door, smiling as you smell the aroma of sugar cookies baking in the oven. Your nose sends a signal to your brain—someone's been baking! Instantly you interpret this signal on the basis of previous eating experiences. In fact, the memory might even cause your mouth to water if you are hungry.

Now imagine walking in the same door another time—only this time your nose wrinkles in disgust as it tells you that the city sewer system has backed up, flooding your house with raw sewage. There is no chance that you would confuse that stench with the cookie smell!

Compared to vision and hearing, your sense of smell involves relatively simple hardware. The number of chemical receptors in your nose is small compared to other senses. The short distance from the nose to the brain means very few synapses or cell connections are necessary. However, scientists still don't understand all the physics and chemistry that transform an airborne molecule into a smell perception.

Like the sense of taste, smell is a form of *chemoreception*; that is, chemicals stimulate specialized nerve cells to create the sensation. Though taste is thought to be limited to four or five different main flavors, there might be fifty, one hundred, or even more different types of smell receptors. Because two or more smells sometimes blend to create a new one, some humans can clearly distinguish up to a thousand different aromas.

For something to have a smell, it must first release molecules into the air; that is, it must be *volatile*. Com-

plex or heavy molecules are unlikely to be volatile and are consequently less likely to be associated with odors. It doesn't take much, though, to stimulate your nose. The receptors in the roof of your nasal cavity are so sensitive that even a single molecule may be enough to trigger a response.

A lining of mucus covers and protects these sensors. The odor molecule must first pass through the mucus to announce its arrival. Mucus is a thick, water-based substance. Because water molecules have a bent structure with a dipole moment, they are polar. Polar substances tend to mix with other polar substances, but not with non-polar materials. Therefore, the arriving molecule must be at least slightly polar to have a chance of getting through.

Assume that the odor molecule has evaporated, floated up into your nasal sinuses, and navigated through the mucus. At this point, you are still unaware of its presence. To register as a smell, it must stimulate a smell receptor cell, which sends a signal to the brain. Stimulating the receptor is the least understood part of the sense of smell. Like all fields of science, various models are offered to explain the process.

One theory compares smell to a lock and key. Certain molecules are said to fit into a specific type of receptor, turning on a switch that sends a signal to the brain. This theory, published in the 1970s, speculates

surfaces of erythrocytes (red blood cells) to quickly form a clot. It can seal a hole in the heart in sixty seconds. With this and other technologies, about 90% of the soldiers wounded in the wars in Iraq and Afghanistan have survived, up from 76% in the Vietnam War. Although chitosan is the second-most common substance in biomass after cellulose, a four-inch-square bandage costs ninety dollars!

7.12 Seeking the Perfect Bonding Model

The Bohr atomic orbital model, the valence bond model, and the molecular orbital model of molecular bonding each have their places in chemistry. Since none of them accounts for all features of molecular bonds, scientists use the one that best explains a particular situation. It would of course be better to have a single theory that

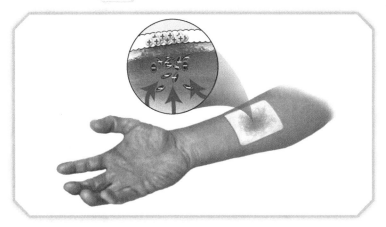

7-21 The positively charged chitosan bandage attracts the negatively charged erythrocytes to accelerate clotting.

that there are only seven different kinds of smell receptors. Many molecules, though, can activate two or more receptors to produce a new smell. Consequently, it is called the *stereochemical theory*.

Another theory explores the possibility that an odor molecule vibrates at a certain frequency. When it comes near an odor receptor, it causes mol-

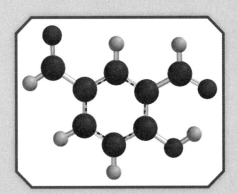

vanillin

ecules within that cell to begin vibrating at the same frequency, thus stimulating a response. This model considers electrons to be waves.

The most widely accepted theory is more complex and concerns the size, geometric shape, and electron arrangement of the odor molecule. Experimental evidence has shown that similar molecules with very slight differences are recognized as distinct odors. In some way, the odor receptor cells are able to read a molecule and send a corresponding electrical impulse to the brain. This signal is initiated by a receptor protein that recognizes the arriving odor molecule. The receptor protein then binds with the molecule at the cell surface and activates other proteins within the cell to send an electrochemical signal to the brain.

A molecule called vanillin creates the pleasing vanilla aroma of sugar cookies. It has a low molecular weight, making it quite volatile. That's why it is easily vaporized in the baking process and becomes airborne.

Hydrogen sulfide gas molecules are the clue that sewage has entered your home. This gas, often described as having a rotten egg smell, is given off whenever sulfur-containing compounds are broken down by bacteria. The offensive odor of sewage is caused by these small but potent molecules. At low levels, these molecules are merely an unpleasant nuisance, but at higher levels they may cause headaches, respiratory complications, and even death.

God could have designed our noses as mere breathing mechanisms. Instead, He created them with the complex ability to detect smells. Some smells provide great pleasure to humans. Consider the aroma of freshly baked cookies or bread, the scent of the fresh air after a storm, or the pleasure of flowers and perfumes. Our sense of smell also enhances our sense of taste, so the chemical reactions in your nose actually help you enjoy a juicy steak or chewy brownie. All of this is a gift from God to enhance our enjoyment of life on the earth (Eccles. 3:13).

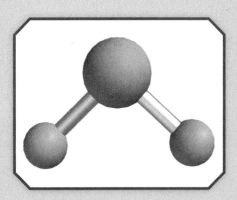

hydrogen sulfide

God also designed our noses to deal with the realities of the Fall. Unpleasant smells are linked to dangers. The repulsive smells of sewage or rotting food warn us of their potential danger. This capability testifies of God's care for His creation. God created the complex chemical processes of smell for our protection and our pleasure.

explains every observation, and scientists continue to work toward such a solution. But we should not be troubled that science does not allow us to completely understand the universe on the subatomic level. God has not commissioned us to know everything about our world in its minutest detail. He has commissioned us to rule over and manage the creation (Gen. 1:26–28). The model making done by scientists is a valuable contribution to fulfilling this commission.

The limitations of scientists in actually discovering the details of how the chemical world works also remind us of the greatness of God. The unseen chemical world that functions within and all around us is still in many ways mysterious. But God designed the way the chemical world works, and He understands every intricacy of molecular bonding.

7B Section Review

1. What basic principle of physics governs VSEPR theory? What property of molecules is significantly affected by that principle?

2. How are sp^3 hybrid orbitals produced? Which group of elements would most likely bond using sp^3 hybrid orbitals?

3. Why are bonds in most covalent compounds polar? How do polar bonds contribute to polar molecules?

DS 4. How are positively charged chitosan and the negatively charged surface of erythrocytes different from the charge distribution in polar molecules?

DS 5. Can you think of some ways that the cost of chitosan bandages can be decreased so that they will be more widely available?

6. (True or False) No single theory accounts for all aspects of bonding in compounds.

⊙ 7. Predict the shapes of the following:

a. $\begin{bmatrix} :\ddot{Cl} - \ddot{O}: \\ | \\ :\ddot{O}: \end{bmatrix}^-$

b. $:\ddot{Br} - \underset{\underset{:\ddot{O}:}{\|}}{C} - \ddot{Br}:$

c. $:\ddot{I} - \underset{\underset{:\ddot{I}:}{|}}{\ddot{N}} - \ddot{I}:$

⊙ 8. Draw the Lewis structure for the following compounds. Determine whether their molecules are polar or nonpolar. If they are polar, indicate the direction of the dipole moment.
   a. CF_4
   b. NI_3
   c. SBr_2

Chapter Review

Chapter Summary

- The valence bond theory states that two elements form a bond when they share two unpaired, oppositely spinning electrons. The orbitals containing the shared electrons overlap.

- The molecular orbital theory states that atomic orbitals are replaced by molecular orbitals when atoms combine to form a molecule. If the contributing atomic orbitals reinforce each other, a bonding orbital is formed. Atomic orbitals that interfere with each other form antibonding orbitals.

- The valence shell electron pair repulsion (VSEPR) theory accounts for the way molecules assume specific arrangements. Regions with a high density of valence electrons repel each other, forcing bonds and lone pairs of electrons to position themselves as far apart as possible around the central atom.

- Bond hybridization theory describes how electrons from s and p orbitals with different energies can form multiple bonds with the same energy. The arrangement of the hybridized orbitals determines the shape of the resulting molecule.

- If a molecule has polar bonds that are symmetrically arranged, it has no dipole moment. If the molecule is asymmetrical, the molecule has a dipole moment.

Coming to Terms

valence bond theory	161
sigma (σ) bond	161
pi (π) bond	162
resonance	163
molecular orbital theory	164
valence shell electron pair repulsion (VSEPR) theory	166
tetrahedral	166
trigonal planar	167
linear	167
hybridization	168
pyramidal	169
bent	170
polar covalent bond	172
dipole moment (μ)	174

Review Questions

Concept Review

1. Why is a sigma bond stronger than a pi bond?
2. How do single and double bonds behave in a molecule that exhibits resonance?
3. How does boron's position in the periodic table hint that it is stable when bonded to only three atoms?
4. What two characteristics of liquid oxygen contradict predictions from the valence bond theory but are explained by the molecular orbital theory?
5. Discuss the difference between the tetrahedral, pyramidal, and trigonal planar molecular geometries.
6. An atom from Group 13 forms hybridized orbitals as it bonds. Describe how this happens, identify the type of hybridization, and state how many hybridized orbitals are formed.
7. Explain how a tetrahedral molecule could have a dipole moment.

True or False

8. All sublevels hold the same total number of electrons.
9. A triple bond consists of a single sigma bond and two pi bonds. (An example of a molecule with a triple bond is atmospheric diatomic nitrogen.)

10. The atmospheric pollutant sulfur dioxide (SO_2), a molecule exhibiting resonance, has single and double bonds that switch back and forth so rapidly that chemists cannot detect their state.

11. A single molecule may have a dipole moment for each polar bond in the molecule.

Application

12. Make simple drawings that show the orientation of the hybridized orbitals around an atom for *sp*, *sp²*, and *sp³* orbitals.

13. Draw the Lewis structures for these compounds.
    a. hydroxide ion (OH^-)
    b. boron trihydride (BH_3)
    c. hydrogen cyanide (HCN)
    d. phosphorus pentachloride (PCl_5)
    e. sulfur hexafluoride (SF_6)
    f. chlorine trifluoride (ClF_3)

14. Predict the molecular shapes for the compounds in Question 13, parts a–c.

15. How many polar "connections" are found between atoms in each of the molecules below? (One "connection" may consist of a single, double, or triple bond.)
    a. H_2
    b. HCl
    c. CO_2
    d. CH_4
    e. CF_2Cl_2
    f. H_2S
    g. H_2CO
    h. C_2H_4
    i. C_2H_2
    j. CS_2
    k. CO

16. Which of the molecules in Question 15 are polar (i.e., have a dipole moment)?

17. Evaluate each of these Lewis structures. If they are polar, copy them and add the crossed arrow notation to show the direction of the charge.

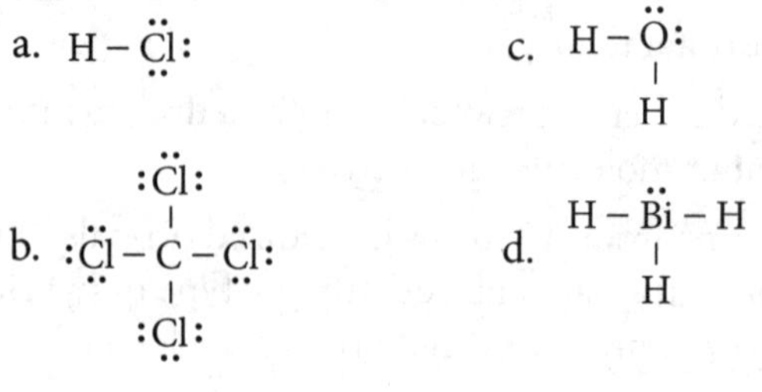

    a. H—C̈l:

    b. :C̈l—C(—C̈l:)(—C̈l:)—C̈l:

    c. H—Ö:
           |
           H

    d. H—Bi—H
           |
           H

18. Do some research on how antioxidants rid the body of free radicals. What foods contain antioxidants?

Chemical Composition and Reactions

CHAPTER 8

DOMINION SCIENCE PROBLEM

Waste Not, Want Not
The United States generates about forty-two million tons of hazardous waste every year. You are partially responsible! You create hazardous wastes when you develop photos, dry-clean your clothes, use pesticides, paint your bedroom, gas up your car, replace batteries, and use electricity! Hazardous wastes can threaten the health of humans and the environment unless they are treated before disposal. Is it possible to decrease this risk and maybe even find a use for some of these wastes?

8A Oxidation Numbers and Formulas	182
8B Nomenclature	187
8C Writing Equations	196
8D Types of Reactions	202
Facet	
Drinking Chemicals	207

8A Oxidation Numbers and Formulas

8.1 Introduction

Chemistry is much more than a study of the elements that comprise matter. It also examines reactions that occur between elements and compounds. To use chemistry effectively, you must be able to write proper formulas and name them. Then you can write balanced equations and classify them by reaction type.

8.2 Oxidation Numbers and Their Rules

Chapter 6 discusses how electrons are lost, gained, or shared when bonds form. These bonds involve specific numbers of electrons, usually from the valence shells of the participating atoms. Most representative elements stabilize by forming bonds to attain a stable valence octet of electrons. Scientists have a "bookkeeping" system to keep track of electrons involved in bonding. This system uses oxidation numbers when writing formulas for chemical compounds. Although electrons are often shared to some extent, oxidation numbers assign each electron to a particular element in the compound.

Oxidation numbers (ON) represent the number of electrons that an atom in a compound must gain or lose to return to its neutral state. A negative number means the bonded atom has gained that many negatively charged electrons, and a positive number means the atom has given up that many electrons. Metals in compounds with low electronegativities have positive oxidation numbers, and nonmetals with high electronegativities have negative oxidation numbers. An oxidation number can also refer to the charge of an ion in an ionic compound.

Oxidation numbers were originally assigned to elements in compounds after experimentally determining the composition of the compounds. After countless thousands of analyses, chemists developed rules that govern how oxidation numbers are assigned to elements. These rules allow scientists to predict how an element typically combines with other elements. As with any system of scientific rules, there are always exceptions.

Rule 1 (the free-element rule). The oxidation number of atoms in their natural form and of atoms in pure elements is zero. Individual atoms such as Fe, Na, Ar, and He have oxidation numbers equal to zero. This rule also applies to covalently bonded elements, such as diatomic molecules. The electrons in the covalent bonds of these elements are equally shared, because all participating atoms have equal electronegativities.

Rule 2 (the ion rule). The oxidation number of a monatomic ion is equal to the charge of the ion. When a Br atom gains an electron to become a Br⁻ ion, it has an oxidation number of –1. This number shows that the neutral atom has gained one electron. Similarly, when a Mg atom loses two electrons to become a Mg^{2+} ion, its oxidation number becomes +2.

Rule 3 (the zero-sum rule). The sum of the oxidation numbers of all the atoms in a compound must be zero. Compounds are not electrically charged.

8A Section Objectives

After finishing this section, you should be able to

- assign oxidation numbers to the elements in compounds or polyatomic ions when given their formulas.
- write formulas for ionic compounds when given the elements that are involved.
- identify the oxidation numbers for atoms that have multiple oxidation states.
- assign oxidation numbers to atoms in polyatomic ions.

The terms *oxidation number* and *oxidation state* mean the same thing.

When ionic bonds form, negative and positive charges are evenly balanced. Ions with a +1 charge and ions with a −1 charge combine in a 1 : 1 ratio. A +2 ion combines with two −1 ions, thus neutralizing the differences in charge.

$$\overset{+1}{Na^+} + \overset{-1}{Cl^-} \longrightarrow \overset{0}{NaCl}$$

$$\overset{+2}{Mg^{2+}} + \overset{-1}{2Cl^-} \longrightarrow \overset{0}{MgCl_2}$$

This rule applies to covalent bonds between nonmetals as well. Shared electrons reside closer to the more electronegative atom in a covalent bond. Therefore, the more electronegative atom is assigned a negative oxidation number based on how many electrons it acquires by sharing. The less electronegative atom is assigned a positive oxidation number for the number of electrons it "donates" by sharing with the more electronegative atom. Generally, the element with the highest electronegativity in a compound determines the oxidation number of the other elements in the compound.

C'MON ... WE'RE IRRESISTIBLE.

Rule 4 (the specific oxidation number rule). Certain elements regularly have the same oxidation numbers in compounds. This makes writing chemical formulas much easier.

A. *Alkali metals* (Group 1) always have a +1 oxidation number in compounds.

B. *Alkaline-earth metals* (Group 2) always have a +2 oxidation number in compounds.

C. *Hydrogen* usually has a +1 oxidation number when bonded to another nonmetal because the compound typically contains a more electronegative element. Hydrogen has a −1 oxidation number when bonded to metals to form compounds called *metallic hydrides*.

D. *Oxygen* always has a −2 oxidation number except when bonded to fluorine. It is the second most electronegative element after fluorine, so this is a reliable assumption. A rare exception to this rule is the peroxide ion, O_2^{2-}, in which oxygen has a −1 oxidation number.

E. *Halogens* (Group 17) have an oxidation number of −1 when bonded to metals. When halogens are bonded to other nonmetals, the element with the higher electronegativity is assigned the negative number. *Fluorine* always has an oxidation number of −1 because it is the most electronegative element.

8-1 Oxygen has a −1 oxidation number in hydrogen peroxide (H_2O_2).

8.3 Finding Oxidation Numbers

The few rules above determine the oxidation numbers for elements in multitudes of compounds. If one rule contradicts another, the rule listed first should be followed. If you can't figure out the oxidation number of an atom from the rules above, you can use an algebraic equation to solve for it. You can write an equation based on the fact that the sum of all the oxidation numbers in a compound must equal zero (Rule 3).

Rule Summary

Rule 1	free atoms = 0
Rule 2	ion charge = oxidation number
Rule 3	compound sum = 0
Rule 4A	Group 1 = +1
Rule 4B	Group 2 = +2
Rule 4C	H = +1 or −1
Rule 4D	O = −2 or −1
Rule 4E	Group 17 = −1
Rule 5	sum of ONs in polyatomic ion = charge

EXAMPLE PROBLEM 8-1
Determining Oxidation Numbers

Determine the oxidation number of each element in the following compounds:

a. Na_2O b. H_2SO_4

Solution

a. The sum of the oxidation numbers of all the atoms in the compound must add up to zero (Rule 3). The sodium atom always has a +1 oxidation number (Rule 4A), and the oxygen atom is normally –2 (Rule 4D). You can check these oxidation numbers using an algebraic equation.

$$2(Na's\ ON) + O's\ ON = 0$$
$$2(+1) + -2 = 0$$

b. Because hydrogen is less electronegative than oxygen and sulfur, it has a +1 oxidation number (Rule 4C). Oxygen has an oxidation number of –2 (Rule 4D). Use an equation to determine the oxidation number of sulfur for this compound.

$$2(H's\ ON) + S's\ ON + 4(O's\ ON) = 0$$
$$2(+1) + S's\ ON + 4(-2) = 0$$
$$S's\ ON = +6$$

Solving this equation, we determine that sulfur's oxidation number must be +6 (Rule 3).

We can use oxidation numbers to write chemical formulas, especially for ionic compounds. Given the elements in an ionic compound, we can determine how many atoms of each element must be present to have a neutral compound.

For binary ionic compounds, a simple way to write the formula is to use the absolute value of the oxidation number of one ion as the subscript for the other element. (This technique is sometimes called the *criss-cross method*.) Since the formulas for ionic compounds must be in the simplest ratio possible, formula subscripts may need to be reduced by dividing each subscript by a common factor. Subscripts of 1 are assumed when no subscript is visible.

> One exception to the simplification step is the peroxide ion, O_2^{2-}. For example, disodium peroxide has the formula Na_2O_2, not NaO, because the peroxide ion exists as a covalently bonded diatomic particle. Formulas for ionic compounds of more than two elements are also exceptions to the criss-cross method.

EXAMPLE PROBLEM 8-2
Writing Ionic Compound Formulas

Write the formula for the ionic compounds of
a. barium and iodine.   b. calcium and oxygen.

Solution

a. Barium is an alkaline-earth metal with an oxidation number of +2 (Rule 4B). An iodine atom (a halogen) always has a –1 oxidation number when combined with a metal (Rule 4E). For the sum of the oxidation numbers to equal zero (Rule 3), two iodine atoms must be included in the formula. The criss-cross method leads to this solution.

$$\overset{+2}{Ba} + \overset{-1}{I} \longrightarrow \overset{+2\ -1}{Ba\diagdown I} = Ba_1I_2 = BaI_2$$

b. Calcium is also an alkaline-earth metal, so it has an oxidation number of +2 (Rule 4B). Oxygen always has a –2 oxidation number when combined with metals (Rule 4D).

$$\overset{+2}{Ca} + \overset{-2}{O} \longrightarrow \overset{+2}{Ca}\overset{-2}{O} = Ca_2O_2 = CaO$$

Though the first formula, Ca_2O_2, has equal negative and positive charges, it is incorrect because it is not in the simplest possible ratio.

8.4 Atoms with Multiple Oxidation States

Some atoms can have more than one oxidation number, depending on the other atoms to which they bond. Transition metals especially change oxidation numbers. These elements have outer energy levels that are very close together. They also have d and f sublevel electrons. Oxidation numbers for the transition metals are determined by how many electrons participate in chemical bonds. For example, iron can form $FeCl_2$ as well as $FeCl_3$. Oxidation numbers of transition metals that form common alloys and ionic compounds must be either memorized or looked up in tables.

Some nonmetals can also have more than one oxidation state. For example, nitrogen can take nearly any oxidation number from +5 to –3. Table 8-1 shows the compounds of nitrogen and oxygen. Each of these compounds is very different. For example, N_2O and NO_2 both make you go to sleep; however, NO_2 would do so quickly and permanently!

8-1 Compounds of Nitrogen and Oxygen

Compound	Name	Oxidation state	Properties
N_2O_5	dinitrogen pentoxide	+5	white or colorless crystalline solid, melting point of 30 °C
NO_2	nitrogen dioxide	+4	toxic brownish gas, used in manufacture of explosives
N_2O_3	dinitrogen trioxide	+3	pale blue liquid that decomposes at 3 °C
NO	nitric oxide	+2	colorless gas, used to make nitric acid
N_2O	nitrous oxide	+1	laughing gas, used as an anesthetic

If nitrogen, sulfur, phosphorus, or carbon bond to highly electronegative elements, they are assigned a positive oxidation number. If they bond to a less electronegative element, they can attract electrons and are assigned a negative oxidation number.

The periodic table in Figure 8-2 shows the common oxidation states of all elements except the inner transition metals. The shaded elements on the chart are the ones used most frequently in this course. Become familiar with this table to help you write chemical formulas correctly.

8.5 Oxidation Numbers and Polyatomic Ions

Polyatomic ions are covalently bonded groups of atoms that carry a charge. Since polyatomic ions are found in many compounds and reactions, they have their own oxidation number rule.

Rule 5. The oxidation numbers of all the atoms in a polyatomic ion add up to its charge.

For example, consider the hydroxide ion, OH^-. The oxygen atom has an oxidation number of –2, and the hydrogen atom +1. The sum

8-2 Oxidation numbers of common elements

H +1,-1																	He
Li +1	Be +2											B +3	C +4,+2,-4	N +5,+4,+3,+2,+1,-3	O -1,-2	F -1	Ne
Na +1	Mg +2											Al +3	Si +4,+3,-4	P +5,+3,-3	S +6,+2,+4,-2	Cl +7,+1,+5,-1	Ar
K +1	Ca +2	Sc +3	Ti +4,+3,+2	V +5,+3,+4,+2	Cr +6,+3,+2	Mn +7,+4,+2,+5	Fe +3,+2	Co +3,+2	Ni +2	Cu +2,+1	Zn +2	Ga +3	Ge +4,-4	As +5,+3,-3	Se +6,+4,-2	Br +5,+3,+1,-1	Kr +4,+2
Rb +1	Sr +2	Y +3	Zr +4	Nb +5,+4	Mo +6,+4,+3	Tc +7,+4,+3	Ru +8,+4,+6,+3	Rh +4,+3,+2	Pd +4,+2	Ag +1	Cd +2	In +3	Sn +4,+2	Sb +5,+3,-3	Te +6,+4,-2	I +7,+1,+5,+3,-1	Xe +6,+4,+2
Cs +1	Ba +2	La +3	Hf +4	Ta +5	W +6,+4	Re +7,+6,+4	Os +8,+4	Ir +4,+3	Pt +4,+2	Au +3,+1	Hg +2,+1	Tl +3,+1	Pb +4,+2	Bi +5,+3	Po +2	At -1	Rn
Fr +1	Ra +2	Ac +3															

of these numbers is –1, which equals the charge on the ion. Polyatomic ions survive most chemical reactions without breaking up. Using the charge of the ion as you would use an oxidation number simplifies many problems.

EXAMPLE PROBLEM 8-3
Oxidation Numbers in Polyatomic Ions

What is the oxidation number of the lead atom in $Pb(OH)_2$?

Solution

Pb's ON + 2(OH⁻'s ON) = 0

Pb's ON + 2(–1) = 0

Pb's ON = +2

The same technique used to find formulas of two-element ionic compounds can be used for polyatomic ionic compounds. The polyatomic ion is treated as a single unit. If more than one polyatomic atom is present in the formula, each needs its own parentheses.

EXAMPLE PROBLEM 8-4
Writing Formulas

What is the formula of the compound that contains ammonium (NH_4^+) ions and phosphate (PO_4^{3-}) ions?

Solution

The oxidation numbers of all the atoms in a compound must add up to zero. Three ammonium ions are required to balance the –3 of the phosphate ion. Using the criss-cross method we get the following.

$$\overset{+1}{NH_4} + \overset{-3}{PO_4} \longrightarrow \overset{+1}{NH_4}\overset{-3}{PO_4} = (NH_4)_3(PO_4)_1 = (NH_4)_3PO_4$$

Algebraic equations can be used to find the oxidation numbers of individual atoms in polyatomic ions. The following equation yields the oxidation state of the S atom in a sulfate ion (SO_4^{2-}). Both Rules 4D and 5 are needed here.

S's ON + 4(O's ON) = −2

S's ON + 4(−2) = −2

S's ON + −8 = −2

S's ON = +6

8A Section Review

1. Write the formula for the ionic compounds of
   a. aluminum and chlorine.
   b. lithium and fluorine.
2. What is the formula of the compound that contains iron (II) (Fe^{2+}) and sulfate (SO_4^{2-}) ions?
3. What is the formula of the compound that contains hydrogen (H^+) and phosphate (PO_4^{3-}) ions?
4. Determine the oxidation number of each element in $KClO_3$, potassium chlorate.
5. Contrast elements that have negative oxidation numbers with those that have positive oxidation numbers.

8B Nomenclature

8.6 Naming Compounds

Do the names *soda ash* and *epsomite* mean anything to you? Chemical compounds have been given many names throughout history. Some names, like soda ash, described how a compound looked or behaved. Other names gave a compound's origin. Epsomite was often found near the English town of Epsom. In America, this compound, $Mg(SO_4)_2$, is called Epsom salts. Although the backgrounds of these names are interesting, they don't give us very much information. You can see why it is difficult to write formulas from the common names of compounds given in Table 8-2.

8B Section Objectives

After finishing this section, you should be able to
- name compounds from their formulas.
- write formulas of compounds from their names.
- know and use the Greek prefixes to name compounds.
- know and use polyatomic ions to name compounds.
- correctly identify binary compounds, polyatomic ionic compounds, metals with multiple oxidation states, hydrates, and acids.

8-2 Common Names of Some Industrial Chemicals

Common name	Chemical name	Formula
oil of vitriol	sulfuric acid	H_2SO_4
caustic lime	calcium hydroxide	$Ca(OH)_2$
lye	sodium hydroxide	NaOH
soda ash	sodium carbonate	Na_2CO_3
milk of magnesia	magnesium hydroxide	$Mg(OH)_2$

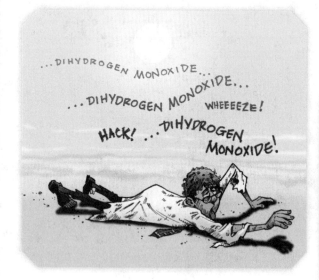

As more and more compounds were discovered and synthesized, chemists realized that they could not continue to rely on memorized names. In the twentieth century, the International Union of Pure and Applied Chemistry (IUPAC) developed a systematic way to name compounds. This system of naming, called a **nomenclature**, follows a standardized set of rules. IUPAC names of compounds are packed with information. They tell which elements are present in the compound, what type the compound is, intermolecular attractions, and the general properties of the compound.

Today the term *soda ash* is used very little. Instead, people usually refer to this compound as sodium carbonate. The name *epsomite* has given way to a more informative name, magnesium sulfate. Some common names like *water* and *ammonia* are still used today but only when the compound is very familiar.

8-3 Epsom salts (magnesium sulfate) have many practical uses. In the bath, they soothe muscle pains and soften skin. In the garden, they are a great plant food, providing essential magnesium.

8.7 Binary Covalent Compounds

The flow chart in Figure 8-4 will help you correctly name compounds. This chart provides an organized method for identifying an unnamed formula.

Let's first consider covalent compounds composed of only nonmetals. Some covalent compounds called acids have specific properties. You may already be familiar with some acids, such as the hydrochloric acid in your stomach or the sulfuric acid used in car batteries. You can readily recognize acids as covalent compounds that begin with hydrogen, although there are exceptions. Acids and covalent compounds that are not acids have different nomenclatures. **Binary** (two-element) **covalent compounds** that are not acids are named using a system of Greek prefixes.

The **Greek prefix system** indicates how many atoms of each element are in a binary covalent compound. Commonly used prefixes are listed in Table 8-3. The prefix *mono-* is used only for the second element in the compound unless it is needed for emphasis or clarity. When it is used, any doubled vowels are omitted. (For instance, CO is named carbon monoxide, not carbon mono-oxide.) The least electronegative element comes first, followed by the more electronegative element. The ending of the last element is changed to *-ide*.

8-3	Greek Prefixes
Prefix	Number
mono-	1
di-	2
tri-	3
tetra-	4
penta-	5
hexa-	6
hepta-	7
octa-	8
nona-	9
deca-	10

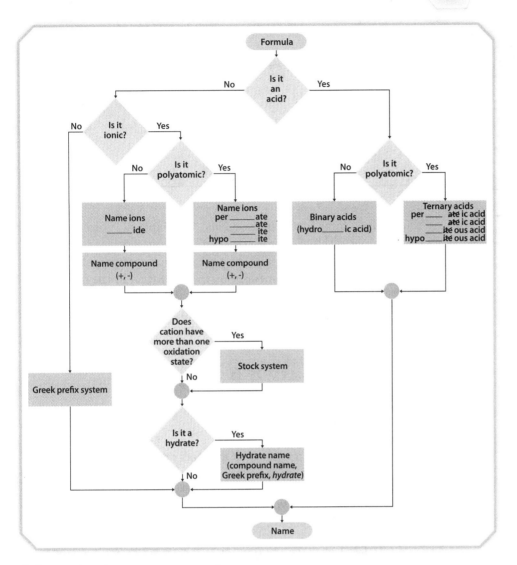

8-4 Flow chart for naming compounds

Example Problem 8-5

Naming Covalent Compounds

Name PCl_3.

Solution

PCl_3 is not an acid, since its formula does not begin with hydrogen. It is not ionic, because all its elements are nonmetals. It should thus be named according to the Greek prefixes. *Phosphorus trichloride* is the accepted name. The prefix *mono-* is generally not used for the first element in the compound.

8.8 Binary Ionic Compounds

Ionic compounds are not normally named with the Greek prefix system. Binary ionic compounds, which are two-element compounds consisting of a metal and nonmetal, are named after the two ions involved. Positive ions use the same name as their parent atoms (sodium atoms form sodium ions). Negative ions have an *-ide* ending

Prefixes are not needed in binary ionic compounds. You can figure out from its oxidation number how many of each element are present in the compound.

8-4 -ide Endings for Some Common Compounds

Compound	Name
NaCl	sodium chloride
MgO	magnesium oxide
Al₂O₃	aluminum oxide
K₂S	potassium sulfide
MgH₂	magnesium hydride

8-5 Names of Nonmetallic Ions

B boride	C carbide	N nitride	O oxide	F fluoride
	Si silicide	P phosphide	S sulfide	Cl chloride
		As arsenide	Se selenide	Br bromide
		Sb antimonide	Te telluride	I iodide
				At astatide

(chlorine atoms form chloride ions). See Table 8-4 for examples of the names of binary ionic compounds. Table 8-5 lists examples of -*ide* endings for nonmetals. For a binary ionic compound, the name of the positive ion (cation) appears first, followed by the name of the negative ion (anion). For metal hydrides, the metal appears first, followed by the hydrogen because the hydrogen is more electronegative than other elements in these compounds.

EXAMPLE PROBLEM 8-6

Naming Binary Ionic Compounds

Name CaI_2.

Solution

Since CaI_2 is a compound of a metal and a nonmetal, it is a binary ionic compound. Its name, *calcium iodide*, is formed by listing the ions. Note that the name must end in -*ide* since it is a binary compound. *Iodine* is changed to *iodide*.

8.9 Polyatomic Ionic Compounds

Tables 8-6 and 8-7 list the formulas and names of common polyatomic ions. You need to memorize these lists. Some generalizations can help make this table easier to learn. The only positive polyatomic ions are the ammonium ion (NH_4^+) and the mercurous ion (Hg_2^{2+}). Anions that contain oxygen and one other element are called **oxyanions**. These ions often have two or more forms with different numbers of the same elements. For example, perchlorate, chlorate, chlorite, and hypochlorite all contain chlorine and oxygen in differing amounts. If there are only two forms of oxyanions, the form with fewer oxygen atoms ends in -*ite* and the form with more oxygen atoms ends in -*ate*. For example, SO_3^{2-} is the sulfite ion and SO_4^{2-} is the sulfate ion. For some polyatomic ions, especially those that contain halogens, there are more than two oxyanions. In such cases, the form with fewer oxygen atoms than the -*ite* ion has the prefix *hypo-*, and the form with more oxygen atoms than the -*ate* ion has the prefix *per-*. You can see how this works for the oxyanions of chlorine and bromine in Table 8-6.

8-6 Common Oxyanions

Charge	Greatest number of oxygens: per___ate	Base number of oxygens: ___ate	Fewer number of oxygens: ___ite	Fewest number of oxygens: hypo___ite
-1	perbromate BrO_4^-	bromate BrO_3^-	bromite BrO_2^-	hypobromite BrO^-
-1	perchlorate ClO_4^-	chlorate ClO_3^-	chlorite ClO_2^-	hypochlorite ClO^-
-1		hydrogen sulfate (bisulfate) HSO_4^-	hydrogen sulfite (bisulfite) HSO_3^-	
-1	periodate IO_4^-	iodate IO_3^-	iodite IO_2^-	hypoiodite IO^-
-1		nitrate NO_3^-	nitrite NO_2^-	
-1	permanganate MnO_4^-			
-2		carbonate CO_3^{2-}		
-2		oxalate $C_2O_4^{2-}$		
-2		chromate CrO_4^{2-}		
-2		dichromate $Cr_2O_7^{2-}$		
-2		silicate SiO_3^{2-}		
-2		sulfate SO_4^{2-}	sulfite SO_3^{2-}	
-2		thiosulfate $S_2O_3^{2-}$		
-3		arsenate AsO_4^{3-}	arsenite AsO_3^{3-}	
-3		borate BO_3^{3-}		
-3		phosphate PO_4^{3-}	phosphite PO_3^{3-}	

8-7 Other Polyatomic Ions

Name	Formula
acetate	$C_2H_3O_2^-$
amide	NH_2^-
ammonium	NH_4^+
azide	N_3^-
cyanate	OCN^-
cyanide	CN^-
hydrogen carbonate (bicarbonate)	HCO_3^-
hydrogen sulfide	HS^-
hydroxide	OH^-
mercurous	Hg_2^{2+}
peroxide	O_2^{2-}
tetraborate	$B_4O_7^{2-}$
thiocyanate	SCN^-

Once you know the names of polyatomic ions, you can name the **polyatomic ionic compounds** that contain them. Simply name the cation and then the anion, just as with binary ionic compounds. Table 8-8 gives some examples of polyatomic ionic compounds and their names.

> If more than one polyatomic ion is required to make a compound neutral, parentheses are placed around the ion and the subscript is placed outside the parentheses.

8-8 Compounds with Polyatomic Ions

Compound	Name
Na_2SO_4	sodium sulfate
KOH	potassium hydroxide
$Ba(ClO_3)_2$	barium chlorate
$Ca(ClO)_2$	calcium hypochlorite
$Mg(NO_2)_2$	magnesium nitrite
NH_4Cl	ammonium chloride

Some anions in Tables 8-6 and 8-7 contain hydrogen. Consider these polyatomic ions as one single ion. That is, when they combine with metal ions, simply follow the same procedure as for other polyatomic ionic compounds—name the cation and the anion. For example, $NaHCO_3$ is called sodium hydrogen carbonate (or sodium bicarbonate) and $Mg(HS)_2$ is called magnesium hydrogen sulfide.

EXAMPLE PROBLEM 8-7
Naming Polyatomic Ionic Compounds

Name NH_4BrO_3.

Solution
The compound NH_4BrO_3 is not an acid, and it is ionic, so we simply name the ions. The name *ammonium bromate* is formed from the names of the two polyatomic ions included in the formula.

EXAMPLE PROBLEM 8-8
Writing Formulas for Polyatomic Ionic Compounds

What is the formula for potassium dichromate?

Solution
The dichromate ion ($Cr_2O_7^{2-}$) has an oxidation number of -2. Two potassium atoms with an oxidation number of $+1$ can form a neutral compound. The formula is $K_2Cr_2O_7$.

8.10 Ionic Compounds and Multiple Oxidation States

If the metal in an ionic compound can have more than one oxidation number, a Roman numeral is placed after the element's name to show its oxidation number. This convention is called the **Stock system** or sometimes the **Roman numeral system**.

> The *Stock* system is named for the German inorganic chemist Alfred Stock, who proposed it in 1919.

Example Problem 8-9

Stock System Nomenclature

Name $Hg(BrO_3)_2$.

Solution

Following the flow chart, you see that $Hg(BrO_3)_2$ is not an acid, that it is ionic, and that it contains a polyatomic ion. Mercury is a metal that can have more than one oxidation number. The last part of the name will be bromate. Since two bromate ions with a –1 oxidation number are present, the mercury atom must have a +2 oxidation number. The Stock system name, therefore, is mercury (II) bromate.

Table 8-9 gives examples of ionic compounds containing metals with more than one oxidation number. You will need to learn which metals occur in more than one oxidation state so that you will know when to use Roman numerals in chemical names.

8-9 Stock System Compound Names	
Compound	Stock system name
Hg_2I_2	mercury (I) iodide
HgI_2	mercury (II) iodide
$CuBr$	copper (I) bromide
$CuBr_2$	copper (II) bromide
$FeCl_2$	iron (II) chloride
$FeCl_3$	iron (III) chloride
CoC_2O_4	cobalt (II) oxalate
$Co_2(C_2O_4)_3$	cobalt (III) oxalate
SnO	tin (II) oxide
SnO_2	tin (IV) oxide
$PbSO_4$	lead (II) sulfate
$Pb(SO_4)_2$	lead (IV) sulfate

8-5 Different cations of the same metal can combine with the same anion to form very different compounds. On the left is lead (II) oxide; on the right is lead (IV) oxide.

Example Problem 8-10

Stock System Formulas

What is the formula for lead (II) phosphate?

Solution

The Roman numeral *II* identifies the +2 oxidation state of lead. Phosphate ions have a –3 charge. A combination of three lead atoms for every two phosphate ions results in an electrically neutral compound. You can find this ratio using the criss-cross method introduced in Subsection 8.3. The formula is $Pb_3(PO_4)_2$.

8.11 Hydrates

Hydrates are compounds that hold a characteristic amount of water in their crystalline structures. This water is called the *water of hydration*.

Water molecules combine with certain compounds in specific ratios due to their crystalline structures. Formulas of these compounds indicate the presence and number of water molecules by a centered dot followed by the number of water molecules ($Na_2CO_3 \cdot 7H_2O$).

To name hydrates, the word *hydrate* preceded by a Greek prefix is added to the end of an ionic compound's name. In order to identify the dehydrated form of a compound as distinguished from its hydrates, the term *anhydrous* is used. An **anhydrous** compound has no water in its crystalline structure.

Table 8-10 lists a compound in its anhydrous form and several of its hydrate forms.

8-10	*Hydrate Compound Names*
Formula	Name (compound's name, Greek prefix + *hydrate*)
Na_2CO_3	sodium carbonate (anhydrous)
$Na_2CO_3 \cdot H_2O$	sodium carbonate monohydrate
$Na_2CO_3 \cdot 7H_2O$	sodium carbonate heptahydrate
$Na_2CO_3 \cdot 10H_2O$	sodium carbonate decahydrate

8.12 Binary Acids

Acids are covalent compounds usually beginning with hydrogen. Acids consisting of hydrogen and one other nonmetal are called **binary acids**. When they are liquids, these covalent compounds are named differently than other covalent compounds. For example, as a gas, HCl is called hydrogen chloride. When HCl is dissolved in water, however, it forms an acid called hydrochloric acid. In general, the names of binary acids include the prefix *hydro-* (referring to the hydrogen), the root name for the nonmetal with an *-ic* ending, and the word *acid*. Table 8-11 gives some examples of these name changes.

8-6 Plaster of Paris is a well-known hydrate. Its formula, $CaSO_4 \cdot \frac{1}{2}H_2O$, is systematically named calcium sulfate hemihydrate (hemi for $\frac{1}{2}$).

For sulfur, the entire name is used as the root rather than the shortened *sulf-* that was used in the anion name.

8-11	*Binary Acid Names*	
Formula	Common name (gases)	Acid name (hydro___ic acid)
HCl	hydrogen chloride	hydro*chlor*ic acid
HBr	hydrogen bromide	hydro*brom*ic acid
H_2S	hydrogen sulfide	hydro*sulfur*ic acid

8.13 Ternary Acids

Ternary acids contain three elements: hydrogen, oxygen, and another nonmetal. The oxygen and other nonmetal are often combined in a polyatomic ion. The names of ternary acids are derived from the anions in the acids. If the anion's name ends in *-ate*, the ending changes to *-ic* and the word *acid* is added. If the anion's name ends in *-ite*, the ending changes to *-ous* and the word *acid* is added. Table 8-12 lists several examples.

8-12 Ternary Acid Names

Anion in ternary acid		Ternary acid	
Formula	Name	Formula	Name
ClO_4^-	perchlorate	$HClO_4$	perchloric acid
ClO_3^-	chlorate	$HClO_3$	chloric acid
ClO_2^-	chlorite	$HClO_2$	chlorous acid
ClO^-	hypochlorite	$HClO$	hypochlorous acid
NO_3^-	nitrate	HNO_3	nitric acid
NO_2^-	nitrite	HNO_2	nitrous acid
PO_4^{3-}	phosphate	H_3PO_4	phosphoric acid
SO_4^{2-}	sulfate	H_2SO_4	sulfuric acid
SO_3^{2-}	sulfite	H_2SO_3	sulfurous acid

Once again you will note that the entire name for sulfur is used as the root; also, the root for phosphorus is *phosphor* instead of the shortened *phosph-*, which is used in the oxyanion name.

EXAMPLE PROBLEM 8-11
Ternary Acid Nomenclature

Name $HBrO_4$.

Solution
The formula $HBrO_4$ starts with hydrogen and is therefore an acid. Since the compound contains a polyatomic ion, its name is derived from the name of the ion. The BrO_4^- ion is the perbromate ion. The *-ate* ending changes to *-ic*. The correct name is *perbromic acid*.

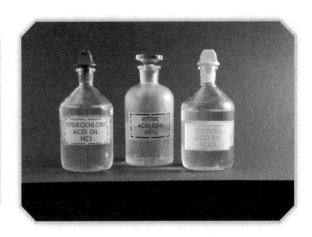

8-7 Although laboratory acids are prepared with different strengths, all should be handled with care since they can burn skin or damage clothing.

8B Section Review

1. Why are common names not always useful?
2. Name the following compounds:
   a. P_2S
   b. $BaCl_2$
   c. $CaSO_4$
   d. FeC_2O_4
3. Name the following polyatomic ions:
   a. NH_4^+
   b. CN^-
   c. $C_2H_3O_2^-$
   d. NO_3^-
   e. MnO_4^-
4. Give a formula for the compound copper (II) sulfate pentahydrate.
5. When do acids follow the same naming conventions as other covalent compounds?
6. Write the correct binary acid (formula and name) that forms from the ion F^-.
7. Write the correct ternary acid (formula and name) that forms from the ion BO_3^{3-}.
8. (True or False) Covalent compounds do not form hydrates.

8C Writing Equations

8.14 Signs of Reactions

Dissolving an antacid tablet produces carbon dioxide bubbles. Invisible nitric oxide gas from car exhaust turns into reddish brown nitrogen dioxide when it reacts with atmospheric oxygen. Burning natural gas releases energy to heat a home. In each of these cases, one set of chemicals produces another set of chemicals, changing their properties in the process. There are visible signs of chemical changes that hint at the unseen molecular changes taking place. Some of the most obvious signs of a chemical change include the formation of a gas, a change in color, or the release of heat or light.

Knowing that a reaction occurred is just the starting point. What substances reacted? What substances were produced? How much of each substance was used or produced? Chemists pack the answers to these questions into an expression called a **chemical equation**.

8.15 What Equations Do

Unlike mathematical equations, chemical equations do not contain an equal sign that equates two quantities. Instead, they represent processes called chemical reactions. To accurately describe reactions, they must do several things.

Chemical equations must identify all the substances involved in a reaction. For example, some water softeners remove calcium salts such as calcium hydrogen carbonate from hard water by adding calcium hydroxide. The two compounds react to form water and calcium carbonate, which settles out of the solution. This information can be put into a word equation. A **word equation** communicates all the substances involved in a chemical reaction but contains no information about the quantities of the materials involved. Below is the word equation for removing calcium salts from hard water.

calcium hydrogen carbonate + calcium hydroxide ⟶ water + calcium carbonate

Substances present before a reaction occurs appear on the left-hand side of a chemical equation and are called **reactants**. Substances that are produced by a reaction appear on the right-hand side and are called **products**. In the word equation above, calcium hydrogen carbonate and calcium hydroxide are reactants, and water and calcium carbonate are products. An arrow separates the reactants from the products and shows the direction of the reaction.

Chemical equations must also show the composition of substances involved in a reaction. Since formulas communicate more information than names, they are used in equations. See how the word equation from above changes.

$$Ca(HCO_3)_2 + Ca(OH)_2 \longrightarrow H_2O + CaCO_3$$

Chemical equations must also account for all the atoms involved in a reaction. According to the *law of mass conservation*, matter cannot be created or destroyed in chemical reactions. The atoms that go into a reaction must come out of the reaction. Different compounds will be formed, but the total number and type of atoms will remain constant.

8C Section Objectives

After finishing this section, you should be able to

- use coefficients and subscripts properly in chemical equations.
- use and interpret the symbols that denote the physical states of substances or special reaction conditions.
- write balanced formula equations for chemical reactions when given word equations.

8-8 The carbon dioxide bubbles show that the tablets are reacting chemically with the water.

The arrow shows that a reaction has occurred and can be read as "reacts to produce" or just "yields."

Write formulas in chemical reactions using the rules in Section 8A.

As it now stands, the chemical equation does not follow the law of mass conservation. See the difference between the number of each atom on the reactant and product sides of the equation.

$$2 \text{ Ca atoms} \longrightarrow 1 \text{ Ca atom}$$

$$4 \text{ H atoms} \longrightarrow 2 \text{ H atoms}$$

$$2 \text{ C atoms} \longrightarrow 1 \text{ C atom}$$

$$8 \text{ O atoms} \longrightarrow 4 \text{ O atoms}$$

If you total these atoms up, sixteen atoms in the reactants produce eight atoms in the products. Can sixteen atoms really turn into eight atoms? Absolutely not! To be correct, the equation must show equal numbers of atoms before and after the reaction.

Chemical equations that follow the law of mass conservation are called **balanced chemical equations**. Balancing an equation involves adjusting the number of atoms, ions, molecules, or formula units by placing numbers called coefficients in front of the formulas. *Coefficients* tell how many atoms, ions, molecules, or formula units are present. When you count atoms, multiply a coefficient by every element in the formula it precedes. No coefficient implies that there is just one atom or molecule.

See how the equation for removing calcium salts from hard water is balanced. Now the number of atoms on each side of the equation is the same.

$$Ca(HCO_3)_2 + Ca(OH)_2 \longrightarrow 2H_2O + 2CaCO_3$$

You will learn how to balance chemical equations by adjusting coefficients in an equation until mass is conserved. This method is called balancing by inspection.

> Reactions that cause atoms to change identity by causing them to lose or gain protons are called *nuclear reactions* and are written in nuclear equations. You will study these in Chapter 20.

8.16 Balancing Equations by Inspection

As mentioned previously, chemical equations must be balanced so that equal numbers of each atom appear on both sides of the equation. Here are general guidelines to help you balance equations by inspection. Consider this word equation for the reaction that occurs when nitrogen monoxide mixes with oxygen in the atmosphere.

nitrogen monoxide + oxygen ⟶ nitrogen dioxide

1. Write formulas for all reactants and products. Make sure that all the formulas are correct. Once they are, *do not change them!*

$$NO + O_2 \longrightarrow NO_2$$

2. Check to see if the equation is already balanced.

$$1 \text{ N atom} \longrightarrow 1 \text{ N atom}$$

$$3 \text{ O atoms} \longrightarrow 2 \text{ O atoms}$$

3. Adjust coefficients until there are equal numbers of atoms on both sides of the arrow.

> Never change a compound's subscripts in order to balance an equation.

> **Problem-Solving Strategy 8-1**
> When balancing equations by inspection, start with the most complicated molecules and save simple molecules until last.

In the example above, more oxygen atoms are needed in the product. The only way to get more oxygen atoms in the product is to add another NO_2 molecule. Changing the coefficient to 2 means you now have two of every atom in the molecule.

$$NO + O_2 \longrightarrow 2NO_2$$

Now there are two nitrogen atoms in the product but only one in the reactants. Putting a 2 in front of the NO will produce equal numbers of every atom on both sides.

$$2NO + O_2 \longrightarrow 2NO_2$$

4. Check to be sure that coefficients are whole numbers in the simplest ratio possible.

If the above equation were $NO + \frac{1}{2}O_2 \longrightarrow NO_2$, it would be balanced, but not all the coefficients would be whole numbers. To rid a chemical equation of fractions, multiply all the coefficients by a common denominator so that they become whole numbers.

$$2NO + 2(\tfrac{1}{2}O_2) \longrightarrow 2NO_2$$

$$2NO + O_2 \longrightarrow 2NO_2$$

If the above equation were $4NO + 2O_2 \longrightarrow 4NO_2$, it would again be balanced, but the coefficients would not be in the simplest possible ratio. Dividing each coefficient by the greatest common factor yields the lowest possible ratio.

$$\tfrac{4}{2}NO + \tfrac{2}{2}O_2 \longrightarrow \tfrac{4}{2}NO_2$$

$$2NO + O_2 \longrightarrow 2NO_2$$

8.17 Special Symbols in Equations

Special symbols pack additional information into equations. Double arrows between reactants and products show that a reaction happens in the reverse direction as well as the forward direction. Such reactions are called **reversible reactions**.

$$3Fe + 4H_2O \rightleftharpoons Fe_3O_4 + 4H_2$$

Symbols in chemical equations can also indicate the physical states of reactants and products. Gases are indicated by (g). If the gas is produced, an upward arrow, ↑, is sometimes used. Liquids are represented by (l). If a substance is dissolved in water, (aq), meaning "aqueous," appears after the formula. Since acids always appear as dissolved in water, we should put (aq) after every acid. For example, H_2SO_4 is hydrogen sulfate, but H_2SO_4 (aq) is sulfuric acid. Solids are indicated by (s). If the solid falls out of a solution, it is called a **precipitate**. The process of precipitation is sometimes noted with a downward arrow, ↓. See examples of these symbols in the following reactions.

$$3Fe\ (s) + 4H_2O\ (g) \rightleftharpoons Fe_3O_4\ (s) + 4H_2\ (g)$$

$$3Fe\ (s) + 4H_2O\ (g) \rightleftharpoons Fe_3O_4 \downarrow + 4H_2 \uparrow$$

8-9 A centrifuge is a machine that spins test tubes rapidly to cause any precipitates to fall out of the solution.

$$AgNO_3\ (aq) + NaCl\ (aq) \longrightarrow AgCl\ (s) + NaNO_3\ (aq)$$

$$AgNO_3\ (aq) + NaCl\ (aq) \longrightarrow AgCl\downarrow + NaNO_3\ (aq)$$

Symbols above and below the reaction arrow communicate special reaction conditions. A delta (Δ) above the arrow means that the reactants are heated. Other descriptions of pressure, light, or specific temperatures can also be placed above the arrow. **Catalysts** are substances that change the rate of the reaction but do not undergo permanent changes themselves. Their symbols can also be placed above the arrow.

$$2KClO_3\ (s) \xrightleftharpoons{Fe_2O_3,\ \Delta} 2KCl\ (s) + 3O_2\ (g)$$

A summary of the symbols used in chemical equations appears in Table 8-13.

> To read an equation with symbols above the reaction arrow, you say "in the presence of..." and name the compound or condition.

8-13 Symbols Used in Chemical Equations	
Symbol	Use
+	between reactants and between products
$\longrightarrow$	means "yields" or "produces"; separates reactants from products
=	same as a single arrow
$\rightleftharpoons$	used in place of a single arrow for reversible reactions
(g)	indicates a gaseous reactant or product
$\uparrow$	sometimes used to indicate a gaseous product
(s)	indicates a solid reactant or product
$\downarrow$	sometimes used to indicate a solid product
(l)	indicates a liquid reactant or product
(aq)	indicates that the reactant or product is in an aqueous solution (dissolved in water)
$\xrightarrow{\Delta}$	indicates that heat must be supplied to reactants before a reaction occurs
$\xrightarrow{MnO_2}$	indicates that a catalyst must be supplied to reactants before a reaction occurs

Example Problem 8-12
Write a Balanced Equation

Solid iron and gaseous chlorine react to form an iron (III) chloride powder. Write and balance the chemical equation for this reaction using all the information given.

Solution

First, write the formulas. Iron atoms can react as individual atoms, but gaseous chlorine atoms exist as diatomic molecules. The formula of iron (III) chloride can be found with the aid of oxidation numbers.

$$Fe\,(s) + Cl_2\,(g) \longrightarrow FeCl_3\,(s)$$

Second, count the atoms.

$$1\ Fe\ atom \longrightarrow 1\ Fe\ atom$$
$$2\ Cl\ atoms \longrightarrow 3\ Cl\ atoms$$

Third, adjust the coefficients. Start with the element having the largest subscript in the most complex formula—the Cl in the $FeCl_3$. In order to balance an element with subscripts of 2 and 3, you will need to use the least common multiple, 6.

$$Fe\,(s) + 3Cl_2\,(g) \longrightarrow 2FeCl_3\,(s)$$

The chlorine atoms are balanced, but now the iron atoms are not. Another iron atom is needed.

$$2Fe + 3Cl_2\,(g) \longrightarrow 2FeCl_3$$

Fourth, check the equation. The number of atoms is balanced, and all coefficients are whole numbers in the simplest ratio possible.

8-10 The discoloration of pennies is caused by chemical reactions involving the copper they contain.

Example Problem 8-13
Write a Balanced Equation

Ethane gas (C_2H_6) burns in oxygen gas, producing gaseous carbon dioxide and liquid water. Write a balanced equation for this reaction using all the information given.

Solution

Remember that oxygen is a diatomic molecule.

$$C_2H_6\,(g) + O_2\,(g) \longrightarrow CO_2\,(g) + H_2O\,(l)$$

Start with the most complex formula (C_2H_6) and leave any free elements (O_2) till last. Balancing carbon and hydrogen atoms results in an equation that is still unbalanced.

$$C_2H_6\,(g) + O_2\,(g) \longrightarrow 2CO_2\,(g) + 3H_2O\,(l)$$

Now seven oxygen atoms are on the product side. Since oxygen molecules exist as two-atom units, we divide the number of atoms needed (7) by the number found in each unit (2) to give $\frac{7}{2}$ as a coefficient.

$$C_2H_6\,(g) + \tfrac{7}{2}O_2\,(g) \longrightarrow 2CO_2\,(g) + 3H_2O\,(l)$$

Since not all coefficients are whole numbers, multiplying the entire equation by 2 produces a balanced, simple equation.

$$2C_2H_6\,(g) + 7O_2\,(g) \longrightarrow 4CO_2\,(g) + 6H_2O\,(l)$$

Example Problem 8-14
Balance an Equation

Aqueous aluminum nitrate and aqueous sodium carbonate react to form a precipitate of aluminum carbonate and aqueous sodium nitrite. Write a balanced equation for this reaction.

Solution

$$Al(NO_3)_3 \;(aq) + Na_2CO_3 \;(aq) \longrightarrow Al_2(CO_3)_3 \downarrow + NaNO_3 \;(aq)$$

$$1 \; Al^{3+} \text{ ion} \longrightarrow 2 \; Al^{3+} \text{ ions}$$

$$3 \; NO_3^- \text{ ions} \longrightarrow 1 \; NO_3^- \text{ ion}$$

$$2 \; Na^+ \text{ ions} \longrightarrow 1 \; Na^+ \text{ ion}$$

$$1 \; CO_3^{2-} \text{ ion} \longrightarrow 3 \; CO_3^{2-} \text{ ions}$$

After putting in coefficients and reducing as necessary, the equation becomes

$$2Al(NO_3)_3 \;(aq) + 3Na_2CO_3 \;(aq) \longrightarrow Al_2(CO_3)_3 \downarrow + 6NaNO_3 \;(aq).$$

> Be sure to use parentheses if there is more than one polyatomic ion in a formula. When you are counting atoms in these compounds, multiply all subscripts by the number outside the parentheses.

8.18 Limitations of Balanced Equations

Although balanced equations give much information about reactions, they do have several limitations. The fact that an equation can be written does not mean that the reaction can occur. The equation below is balanced, but the reaction it represents will not occur.

$$Ag + NaCl \longrightarrow Na + AgCl$$

Chemical equations also do not tell whether a reaction goes to completion. Some reactions leave a mixture of reactants and products. They also cannot show how a reaction occurs. Some reactions involve more than one step. The equation does not show the steps of a reaction or the order in which those steps take place. For example, the balanced reaction

$$4C \;(s) + 6H_2 \;(g) + O_2 \;(g) \longrightarrow 2C_2H_5OH \;(l)$$

is the sum of three more basic reactions,

$$4C \;(s) + 4O_2 \;(g) \longrightarrow 4CO_2 \;(g),$$

$$6H_2 \;(g) + 3O_2 \;(g) \longrightarrow 6H_2O \;(l), \text{ and}$$

$$4CO_2 \;(g) + 6H_2O \;(l) \longrightarrow 2C_2H_5OH \;(l) + 6O_2 \;(g).$$

8.19 Using Chemical Reactions to Solve Problems

How can chemical reactions solve the problem of millions of tons of unusable, toxic material that Americans generate? How hazardous wastes are treated depends on why they are considered hazardous. A substance is a hazardous waste if it can catch fire, react, corrode other materials, or cause health problems like cancer.

One method of waste treatment that recovers some use from wastes is incineration, or burning. In this process of waste treatment, hazardous wastes are burned in a furnace to produce electricity.

> **Why Care About the Environment?**
> 1. God created the world, and He was pleased with His work (Gen. 1:31). Even after the Fall, the creation still testifies to God's glory (Ps. 19:1–4). To harm God's world disregards God's love for His creation and detracts from the display of His glory.
> 2. God gave humans the responsibility to act as stewards over His creation (Gen. 1:26–28). To fail to care for the creation is to disobey His command.
> 3. A deteriorating environment also harms humans. Since the second greatest commandment is to love our neighbor (Matt. 22:39), the Christian should not be unconcerned about pollution.

Sometimes, however, the burning of these wastes can produce toxic gases that must be contained and treated. One hazardous waste called 1,1-difluoroethane requires special consideration. What are some of the benefits we get from using this chemical? Does it have any negative side effects? Note that when 1,1-difluoroethane incinerates, it reacts in the following manner, producing poisonous gases as a byproduct.

$$C_2H_4F_2 + O_2 \xrightarrow{\Delta} HF + CO + H_2O \text{ (not balanced)}$$

What are the byproducts created when this chemical incinerates? Do you think there may be better ways that chemical reactions can be used to help the environment?

8C Section Review

1. Balance the following equations:
    a. $P_4 (s) + S_8 (s) \longrightarrow P_4S_3 (s)$
    b. $KClO_3 (s) \xrightarrow{\Delta} KCl (s) + O_2 (g)$
    c. $AgNO_3 (aq) + Cu (s) \longrightarrow Cu(NO_3)_2 (aq) + Ag (s)$
    d. $H_3PO_4 (aq) + Ba(OH)_2 (aq) \longrightarrow Ba_3(PO_4)_2 (s) + H_2O (l)$
    e. $NaHCO_3 (s) \xrightarrow{\Delta} Na_2CO_3 (s) + CO_2 (g) + H_2O (g)$

2. Write balanced chemical equations for the following word equations:
    a. Aqueous calcium hydroxide reacts with gaseous sulfur trioxide, producing a precipitate of calcium sulfate and liquid water.
    b. Solid aluminum metal reacts with aqueous sulfuric acid, producing hydrogen gas and aqueous aluminum sulfate.
    c. Acetylene gas (C_2H_2) burns in oxygen, producing gaseous carbon dioxide and steam.
    d. Table sugar (sucrose, $C_{12}H_{22}O_{11}$) burns in gaseous oxygen, producing gaseous carbon dioxide and gaseous water.
    e. Aqueous aluminum sulfate reacts with aqueous barium hydroxide, producing precipitates of aluminum hydroxide and barium sulfate.

DS 3. Balance the reaction in Subsection 8.19 that shows the combustion of 1,1-difluoroethane.

8D Types of Reactions

Most reactions can be classified by how a chemical change takes place. Grouping similar reactions leads to generalizations that can help us better understand reactions. The major categories of chemical reactions are synthesis, decomposition, single replacement, and double replacement reactions.

8.20 Synthesis Reactions

Synthesis reactions combine two or more substances into a single product. These reactions follow the general pattern

$$A + B \longrightarrow AB.$$

8D Section Objectives

After finishing this section, you should be able to

- define and give the general form for synthesis, decomposition, single replacement, and double replacement reactions.
- classify a reaction as a synthesis, decomposition, single replacement, or double replacement reaction.
- use an activity series to predict whether a single replacement reaction will occur.

8-11 Calcium chloride (CaCl$_2$) is a common salt that is used to lower the freezing point of water to prevent icing.

1. Metals and nonmetals other than oxygen form compounds called salts.
$$Ca\ (s) + Cl_2\ (g) \longrightarrow CaCl_2\ (s)$$

2. Metals combine with oxygen to form metallic oxides.
$$2Mg\ (s) + O_2\ (g) \longrightarrow 2MgO\ (s)$$

3. Nonmetals react with oxygen to form oxides.
$$P_4\ (s) + 5O_2\ (g) \longrightarrow 2P_2O_5\ (s)$$

4. Water and metal oxides form metal hydroxides.
$$H_2O\ (l) + CaO\ (s) \longrightarrow Ca(OH)_2\ (s)$$

5. Water and nonmetal oxides combine to form oxyacids.
$$H_2O\ (l) + SO_3\ (g) \longrightarrow H_2SO_4\ (aq)$$

8.21 Decomposition Reactions

The opposite of a synthesis reaction is a **decomposition reaction**. While synthesis reactions combine substances, decomposition reactions break substances down into two or more simpler substances. Breaking compounds apart usually requires an input of energy. Decomposition reactions have the general form AB $\longrightarrow$ A + B. In this case, AB is a compound while A and B may be elements or compounds.

1. Oxygen can be driven out of some compounds.
   a. Metal oxides are usually very stable; however, some of the compounds of less active metals are decomposed by heating them to high temperatures.
$$2HgO\ (s) \xrightarrow{\Delta} 2Hg\ (l) + O_2\ (g)$$
   b. Metal chlorates, when heated, produce oxygen.
$$2KClO_3\ (s) \xrightarrow{\Delta,\ MnO_2} 2KCl\ (s) + 3O_2\ (g)$$

8-12 Orange ammonium dichromate is decomposed by gentle heating (top). Green chromium (III) oxide, water vapor, and nitrogen gas result when ammonium dichromate decomposes (bottom).

    c. Water is decomposed by an electric current.

$$2H_2O\ (l) \xrightarrow{elec.} 2H_2\ (g) + O_2\ (g)$$

2. Metal hydroxides release gaseous water when heated.

$$Mg(OH)_2\ (s) \xrightarrow{\Delta} MgO\ (s) + H_2O\ (g)$$

3. Metal carbonates release carbon dioxide when heated.

$$CaCO_3\ (s) \xrightarrow{\Delta} CaO\ (s) + CO_2\ (g)$$

4. Some acids decompose into nonmetal oxides and water.

$$H_2CO_3\ (aq) \xrightarrow{\Delta} CO_2\ (g) + H_2O\ (l)$$

> A carbonated drink goes flat when this reaction occurs. The carbonic acid that gives it a tangy taste decomposes at room temperature.

5. Hydrates release their water molecules when heated sufficiently.

$$BaCl_2 \cdot 2H_2O\ (s) \xrightarrow{\Delta} BaCl_2\ (s) + 2H_2O\ (l)$$

8.22 Single Replacement Reactions

In **single replacement reactions**, a reactive element replaces a less reactive element in a compound. These kinds of reactions are also called displacement or substitution reactions. They occur frequently in solutions and involve less energy than synthesis or decomposition reactions. Single replacement reactions generally follow the form

$$A + BZ \longrightarrow B + AZ.$$

1. Atoms of active metals can replace less active ions from solutions and compounds.

   When a piece of solid zinc is placed in a copper (II) chloride solution, the zinc replaces the copper ions in the solution and forces them to precipitate out as solid copper.

   $$Zn\ (s) + CuCl_2\ (aq) \longrightarrow Cu\ (s) + ZnCl_2\ (aq)$$

   Some metals form bubbles of hydrogen gas when placed in acids. In these reactions, metal atoms replace hydrogen ions in acid molecules. The hydrogen bubbles off as hydrogen gas, and a salt forms from the acid molecule.

   $$Mg\ (s) + 2HCl\ (aq) \longrightarrow MgCl_2\ (s) + H_2\ (g)$$

   Very reactive metals react with even water to produce hydrogen gas. In the equation below, writing water as HOH clearly shows how this is a replacement reaction.

   $$2Na\ (s) + 2HOH\ (l) \longrightarrow 2NaOH\ (aq) + H_2\ (g)$$

2. Reactive halogens can replace less reactive halogens that are in solution.

   $$Cl_2\ (g) + 2NaBr\ (aq) \longrightarrow 2NaCl\ (aq) + Br_2\ (g)$$

   When chlorine gas (a reactive element) is bubbled through a sodium bromide solution, chlorine replaces bromine in the solution. The bromine atoms that are replaced escape as bromine gas.

What makes single replacement reactions occur? The cause depends on the activity of the elements that are involved. An active element can be defined as one that has a strong tendency to lose or gain electrons to form bonds. If unbonded atoms are more active than the bonded atoms, a reaction will probably take place. The more active elements force the less active elements from the bond. The most active metal is lithium; the least active is gold. Other metals can be arranged according to their activities in an **activity series**, as shown in Figure 8-14.

Elements at the top of the list are the most active. Note that hydrogen is included even though it is not a metal. You can predict whether a reaction is probable from this series. For instance, will gold react when it is placed in a sodium chloride solution? Gold is at the bottom of the activity series. Its tendency to lose electrons

8-13 Can you identify the products of this reaction between magnesium and hydrochloric acid?

8-14 Activity series of some metals

Li	React with cold water, liberating hydrogen		
K			
Ba		React with steam, liberating hydrogen	
Sr			
Ca			
Na			
Mg			
Al			React with acids, liberating hydrogen
Mn			
Zn			
Cr			
Fe			
Cd			
Co			
Ni			
Sn			
Pb			
H			
Cu			React with oxygen, giving oxides
Ag			
Hg			
Au			Form oxides indirectly

and form ionic bonds is nowhere near that of sodium. Since no reaction will occur, a person can safely wear a gold ring when swimming in the ocean without losing a fortune.

$$Au\ (s) + NaCl\ (aq) \longrightarrow \text{no reaction}$$

With the aid of an activity series, it is also possible to predict which metals will react with acids. If a metal is more active than hydrogen, it can take the place of hydrogen. Because zinc is above hydrogen in the activity series, it will react with an acid such as dilute sulfuric acid.

$$Zn\ (s) + H_2SO_4\ (aq) \longrightarrow ZnSO_4\ (aq) + H_2\ (g)$$

The halogens have their own activity series (Figure 8-15), which follows the same order in which they are listed in the periodic table. A halogen can replace another halogen that is below it in the series. Iodine, at the bottom of the series, cannot replace any of the other halogens because of its low activity.

$$Cl_2\ (g) + MgBr_2\ (aq) \longrightarrow MgCl_2\ (aq) + Br_2\ (g)$$

8-15 Activity series of halogens

8.23 Double Replacement Reactions

In **double replacement reactions**, two compounds switch partners. Equations of double replacement reactions have the general form

$$AY + BZ \longrightarrow AZ + BY.$$

Most double replacement reactions occur in an aqueous mixture of two ionic compounds. A precipitate often indicates that a double replacement reaction has occurred. For example, when lead (II) nitrate and potassium chromate solutions are mixed, a double replacement reaction occurs. Solid lead (II) chromate, a brilliant yellow compound that has been used as a pigment in paint, falls out of the solution.

$$Pb(NO_3)_2\ (aq) + K_2CrO_4\ (aq) \longrightarrow PbCrO_4\ (s) + 2KNO_3\ (aq)$$

An **ionic equation** represents all the particles present before and after the reaction. Ionic equations can be written only for reactions taking place in a solution. It is the aqueous environment that allows the ionization to occur. Soluble ionic compounds (those indicated by *aq*) are broken into separate ions, but insoluble ionic compounds (indicated by *s*) are not.

8-16 When lead (II) nitrate (colorless) is mixed with potassium chromate (clear orange), lead (II) chromate precipitates, leaving potassium nitrate in solution.

Ionic equation: $Pb^{2+}\ (aq) + 2NO_3^-\ (aq) + 2K^+\ (aq) + CrO_4^{2-}\ (aq) \longrightarrow PbCrO_4\ (s) + 2K^+\ (aq) + 2NO_3^-\ (aq)$

Ionic equations can include particles that do not participate in the reaction. These **spectator ions** appear in the reactants and in the products. In the example above, NO_3^- and K^+ are the spectator ions. They do not react with other ions. They are cancelled from both sides of the ionic equation, yielding the net ionic equation. **Net ionic equations** show only the ions that actually react, omitting spectator ions.

$$Pb^{2+}\ (aq) + 2\cancel{NO_3^-\ (aq)} + 2\cancel{K^+\ (aq)} + CrO_4^{2-}\ (aq) \longrightarrow PbCrO_4\ (s) + 2\cancel{K^+\ (aq)} + 2\cancel{NO_3^-\ (aq)}$$

Net ionic equation: $Pb^{2+}\ (aq) + CrO_4^{2-}\ (aq) \longrightarrow PbCrO_4\ (s)$

Drinking Chemicals

Facets of Chemistry

On a hot, steamy day, nothing satisfies like a nice tall glass of chemicals over ice. I'm not talking about chemicals from a pharmacist or chemist. You can get the chemicals I'm thinking of right from your faucet. When it comes right down to it, even pure, distilled water is nothing more than chemicals. Most people drink water that has been purified through several chemical reactions.

Unless you use only bottled water, the water you drink probably comes either from your own well that pulls water from deep underground or from a municipal water treatment plant. This treated water is sometimes called city water. That water may come from a lake, river, well, or underground river called an aquifer. In any of these cases, the water is carefully analyzed and treated to make it safe for industrial and domestic use.

Murky water drawn from a river or lake contains sediments, minerals, dissolved gases, bacteria, microorganisms, and more. Some of these are safe to consume, but many are not. At a water treatment plant, chemicals are added to the water to remove harmful chemicals, kill microorganisms, and make it safe to drink.

After the floating debris and largest particles are removed by screens or filters, most water still contains many medium-sized particles and bacteria. A simple synthesis reaction allows a process called *flocculation* to take place. Calcium oxide reacts with water in this reaction: $CaO + H_2O \longrightarrow Ca(OH)_2$. Calcium hydroxide has twice the volume of the calcium oxide reactant and forms masses called flocs. These flocs gather or coagulate the smaller particles and bacteria, taking them out of suspension. This material is then allowed to settle out (sedimentation) or is filtered. Aluminum sulfate is sometimes used for the same purpose.

A later step called *sterilization* attempts to destroy the viruses, bacteria, and similar organisms that might have slipped through the flocculation process. Chlorine gas reacts with water to form hypochlorous acid as well as hydrogen and chlorine ions. The balanced equation of this reaction is $Cl_2\,(g) + H_2O\,(l) \longrightarrow HClO\,(aq) + H^+\,(aq) + Cl^-\,(aq)$. Hypochlorous acid then ionizes in water to yield hydronium and hypochlorite ions: $HClO\,(aq) + H_2O\,(l) \longrightarrow H_3O^+\,(aq) + OCl^-\,(aq)$. It is the hypochlorite ions that destroy the microorganisms.

Even after the water gets to your home, it may still need to be adjusted for your needs. In many regions, especially where limestone is predominant, excessive amounts of Ca^{2+} or Mg^{2+} ions are in solution. Groundwater dissolves calcium and magnesium, releasing them as ions. These ions are what make water *hard*. The calcium and magnesium cations in hard water precipitate out, forming *scale*, or gunk, in your pipes. This gunk can damage your pipes and water heater. These ions make it difficult for soap to lather and leave behind a sticky soap scum.

Some homeowners address these problems by using a *water softener* unit. Water softeners take advantage of sodium's ability to replace calcium and magnesium ions. This device adds sodium chloride to the water. In a series of replacement reactions, the sodium changes to its ionic form. The chlorine ions combine with the calcium and magnesium ions to precipitate out as calcium chloride and magnesium chloride. These solid precipitates are then collected and can be flushed out with the waste water. Sodium ions remain in the water but do not cause the same problems as the other cations they replace.

About a billion people lack a clean water supply. Clean water is one of the biggest needs worldwide. If the Lord calls you to become a scientist, one day you may be able to develop an inexpensive and portable way to use chemical reactions to make water potable, or drinkable, for people around the world who don't have clean water. Lack of clean drinking water is no small concern. Jesus says that after His return He will commend the righteous with the words, "I was thirsty, and ye gave me drink." When the righteous ask when they did this, Jesus replies, "Inasmuch as ye have done it unto one of the least of these my brethren, ye have done it unto me" (Matt. 25:35, 40).

Many neutralization reactions between acids and bases can be classified as double replacement reactions.

$$HCl\ (aq) + KOH\ (aq) \longrightarrow HOH\ (l) + KCl\ (aq)$$

Written as an ionic equation, it would be as follows.

$$H^+\ (aq) + Cl^-\ (aq) + K^+\ (aq) + OH^-\ (aq) \longrightarrow HOH\ (l) + K^+\ (aq) + Cl^-\ (aq)$$

Net ionic equation: $H^+\ (aq) + OH^-\ (aq) \longrightarrow HOH\ (l)$.

Cl^- and K^+ are the spectator ions. All neutralization reactions between acids and bases have this same net ionic equation.

Double replacement reactions usually reduce the number of ions in solution. Solid precipitates, such as lead (II) chromate, and the formation of the largely non-ionizable water molecules in acid-base reactions (which will be discussed in Chapter 16) reduce the number of ions.

8D Section Review

1. Classify each of the following reactions. Note that the equations are not balanced.
    a. $P_4\ (s) + S_8\ (s) \longrightarrow P_4S_3\ (s)$
    b. $KClO_3\ (s) \xrightarrow{\Delta} KCl\ (s) + O_2\ (g)$
    c. $AgNO_3\ (aq) + Cu\ (s) \longrightarrow Cu(NO_3)_2\ (aq) + Ag\ (s)$
    d. $H_3PO_4\ (aq) + Ba(OH)_2\ (aq) \longrightarrow Ba_3(PO_4)_2\ (s) + H_2O\ (l)$
    e. $NaHCO_3\ (s) \xrightarrow{\Delta} Na_2CO_3\ (s) + CO_2\ (g) + H_2O\ (g)$

2. Using the activity series in Figures 8-14 and 8-15, predict whether the following reactions are possible.
    a. $Ba\ (s) + 2HOH\ (l) \longrightarrow Ba(OH)_2\ (aq) + H_2\ (g)$
    b. $Ba\ (s) + SnCl_2\ (aq) \longrightarrow BaCl_2\ (aq) + Sn\ (s)$
    c. $I_2\ (aq) + MgCl_2\ (aq) \longrightarrow Cl_2\ (aq) + MgI_2\ (aq)$

3. For each of the following equations, identify the spectator ions and write the net ionic equation.
    a. Balanced equation:
    $3HCl\ (aq) + Al(OH)_3\ (s) \longrightarrow 3H_2O\ (l) + AlCl_3\ (aq)$
    Complete ionic equation:
    $3H^+ + 3Cl^- + Al(OH)_3\ (s) \longrightarrow 3H_2O\ (l) + Al^{3+} + 3Cl^-$
    b. Balanced equation:
    $Pb(NO_3)_2\ (aq) + Na_2CrO_4\ (aq) \longrightarrow PbCrO_4\ (s) + 2NaNO_3\ (aq)$
    Complete ionic equation:
    $Pb^{2+} + 2NO_3^- + 2Na^+ + CrO_4^{2-} \longrightarrow PbCrO_4\ (s) + 2Na^+ + 2NO_3^-$

Using Chemical Reactions

Chemical reactions are used to create things that affect us daily. For instance, plastics, gasoline, and even the asphalt you travel on come from refined petroleum. Chemists refine petroleum by isolating various hydrocarbons. The isolation of the different hydrocarbons necessary for these products requires chemical reactions. It took scientists who understood chemical bonds to develop the processes to make these common materials. Chemical reactions are also used in creating medicines, steel, and rubber. If you are called to be a scientist, you may develop additional products using chemical reactions.

Chapter Review

Chapter Summary

- Oxidation numbers help keep track of valence electrons so you can write formulas correctly. A negative number indicates that an element gains electrons, and a positive number indicates that an element loses electrons.
- Many elements have more than one possible oxidation number.
- The sum of the oxidation numbers in a neutral molecule is zero, and ions and polyatomic ions have sums equal to their net charge.
- Greek prefixes are used to indicate the number of atoms in a covalent molecule according to chemical nomenclature. They are also used to indicate the number of waters of hydration in a hydrate.
- Ionic compounds do not use Greek prefixes, but an *-ide* ending is added to the anion that comes second in the formula and in the name.
- Polyatomic ions participate in compounds as single charged units. Those that contain oxygen and one other compound are called oxyanions.
- In metals that have more than one possible oxidation number, the Stock system uses a Roman numeral after the element's name to indicate the correct oxidation state.
- A hydrate is an ionic compound that has water molecules incorporated into its structure in a fixed ratio to the formula units.
- Binary and ternary acids have formulas beginning with hydrogen. They use their own system of nomenclature.
- A balanced chemical equation shows the formulas and symbols for all substances involved in a chemical reaction and accounts for all atoms in both the reactants and the products.
- Coefficients, subscripts, letters, and other symbols provide more details on the quantities and conditions of a reaction.
- The four major categories of chemical reactions are synthesis, decomposition, single replacement, and double replacement.
- Christian scientists ought to use chemistry to benefit God's creation since God created the world to display His glory, gave humans the responsibility to be stewards, and commanded us to love one another.

Coming to Terms

Term	Page
oxidation number	182
polyatomic ion	185
nomenclature	188
binary covalent compound	188
Greek prefix system	188
oxyanion	190
polyatomic ionic compound	192
Stock system	192
Roman numeral system	192
hydrate	193
anhydrous	194
binary acid	194
ternary acid	194
chemical equation	196
word equation	196
reactant	196
product	196
balanced chemical equation	197
reversible reaction	198
precipitate	198
catalyst	199
synthesis reaction	202
decomposition reaction	203
single replacement reaction	204
activity series	205
double replacement reaction	206
ionic equation	206
spectator ion	206
net ionic equation	206

Review Questions

Concept Review

1. Explain why
   a. fluorine's oxidation number is always negative.
   b. the oxidation numbers of alkali metals are always positive.
   c. elements such as P, N, and S have positive oxidation numbers in some compounds but have negative oxidation numbers in others.

2. When the following pairs of atoms bond, which atom has a positive oxidation number?
   a. H, O
   b. Na, S
   c. N, S
   d. Na, H

3. Give the compound name as well as the oxidation number of each atom in the following covalent compounds:
   a. N_2O_3
   b. I_2O_5
   c. P_4O_6
   d. S_2Cl_2
   e. PCl_3
   f. Cl_2O_7

4. Give formulas for the following covalent compounds:
   a. carbon disulfide
   b. sulfur trioxide
   c. boron trichloride
   d. phosphorus pentabromide
   e. dinitrogen pentasulfide
   f. dibromine monoxide

5. Give the compound name as well as the oxidation number of each atom in the following binary ionic compounds:
   a. LiCl
   b. Mg_3N_2
   c. CaO
   d. NaI
   e. Al_2S_3
   f. CuCl

6. Give formulas for the following ionic compounds:
   a. zinc chloride
   b. calcium phosphide
   c. potassium chloride
   d. barium chloride
   e. strontium oxide
   f. calcium chloride

7. Give the compound name as well as the oxidation number of each atom in the following polyatomic ionic compounds:
   a. $AgNO_3$
   b. NH_4NO_3
   c. $NaNO_2$
   d. $Zn_3(PO_4)_2$
   e. $(NH_4)_2S$
   f. $Zn(C_2H_3O_2)_2$

8. Give formulas for the following polyatomic ionic compounds:
   a. ammonium bromate
   b. potassium permanganate
   c. barium phosphate
   d. aluminum acetate
   e. calcium carbonate
   f. barium chromate

9. Give the compound name (using the Stock system) as well as the oxidation number of each atom in the following compounds:
   a. $PbCl_2$
   b. HgS
   c. CoS
   d. $Fe(OH)_3$
   e. $Pb(CrO_4)$
   f. $Sn(C_2H_3O_2)_2$

10. Give formulas for the following compounds:
    a. iron (III) oxide
    b. copper (I) hydroxide
    c. lead (IV) chromate
    d. lead (II) arsenate
    e. tin (IV) chloride

11. Name the following hydrates:
    a. $CaSO_4 \cdot 2H_2O$
    b. $MgSO_4 \cdot 7H_2O$
    c. $Na_2SO_4 \cdot 10H_2O$
    d. $NiSO_4 \cdot 6H_2O$
    e. $Na_2S_2O_3 \cdot 5H_2O$
    f. $FeSO_4 \cdot 7H_2O$

12. Give formulas for the following hydrates:
    a. iron (III) bromide hexahydrate
    b. barium chloride dihydrate
    c. lead (II) acetate decahydrate
    d. cobalt (II) chloride hexahydrate
    e. sodium tetraborate decahydrate
    f. magnesium carbonate pentahydrate

13. Name the following compounds. Give both the chemical name and the acid name.
    a. HF
    b. H_2Te

14. Give formulas for the following compounds:
    a. hydroselenic acid
    b. hydroiodic acid

15. Identify the anion and give the ternary acid name for the following compounds:
    a. H_3AsO_4
    b. $HC_2H_3O_2$
    c. H_2CO_3
    d. $HMnO_4$
    e. $H_2C_2O_4$

16. Give formulas for the following compounds:
    a. periodic acid
    b. chromic acid
    c. phosphoric acid
    d. cyanic acid
    e. bromic acid

DS 17. Name the toxic chemicals produced when 1,1-difluoroethane is burned. (See Subsection 8.19.) If this chemical produces toxic chemicals when it is incinerated, why is it treated this way?

18. List several evidences that tell whether a chemical reaction has occurred.

19. What are three essential functions of a chemical equation?

20. Explain what each highlighted symbol means.

$$2HgO\;(s) \xrightarrow{\Delta} 2Hg\;(l) + O_2\;(g)$$

21. Consider this reaction:

$$HCl\;(aq) + H_2O\;(l) \rightleftarrows Cl^-\;(aq) + H_3O^+\;(aq)$$

   a. Why are two opposite arrows shown?
   b. Why is (aq) written after several of the substances?

DS 22. Give three biblical reasons why you ought to be concerned about caring for the environment.

DS 23. What are some practical things that you can do to minimize the production of hazardous waste?

True or False

24. An element's oxidation number is the same as its group number on the periodic table.
25. Hydrogen may have a +1 or a –1 oxidation number, depending on what element(s) it combines with.
26. The sum of all the oxidation numbers in a compound should equal zero.
27. Latin prefixes indicate the number of atoms of each element that are found in a binary covalent compound.
28. In the name of a binary ionic compound, the cation comes before the anion.
29. In the Stock system, a Roman numeral indicates the number of atoms that come from the metal element.
30. When balancing a chemical equation, you may change subscripts, but never coefficients.
31. A precipitate is a solid product that falls out of a solution as part of a chemical reaction.
32. A synthesis reaction typically has one reactant and two or more products.

Application

33. Balance the following equations if they are not already balanced. If only a word equation is given, write out the correct formulas and then balance the equation.
    a. $BaO_2\,(s) \xrightarrow{\Delta} BaO\,(s) + O_2\,(g)$
    b. $Li\,(s) + H_2O \xrightarrow{\Delta} LiOH\,(s) + H_2\,(g)$
    c. Hydrogen peroxide can decompose to water and gaseous oxygen when exposed to bright sunlight.
    d. Sodium hydride can form when hydrogen gas is bubbled through molten sodium.
    e. $H_2SO_4\,(l) \xrightarrow{\Delta} H_2O\,(g) + SO_2\,(g) + O_2\,(g)$
    f. Ammonia is produced commercially from nitrogen and hydrogen gases.
    g. Ammonium chloride is produced when ammonia and hydrogen chloride vapors mix.
       $NH_3\,(g) + HCl\,(g) \longrightarrow NH_4Cl\,(s)$
    h. Carbon black (pure carbon) is used in rubber tires and black ink. It is produced by decomposing methane. Hydrogen gas is also a product of this reaction.
    i. High-purity silicon is used to produce microcomputer chips. One process for producing chip-grade silicon entails three steps. The first step is to obtain impure silicon from molten sand (SiO_2).
       $SiO_2\,(l) + C\,(s) \longrightarrow Si\,(l) + 2CO\,(g)$
    j. The second step in the production of pure silicon is to produce silicon tetrachloride from the impure silicon.
       $Si\,(s) + Cl_2\,(g) \xrightarrow{\Delta} SiCl_4\,(l)$

k. The last step in silicon production is to pass hot silicon tetrachloride vapor and hydrogen gas through a tube. Pure silicon condenses.
   $SiCl_4\ (g) + H_2\ (g) \xrightarrow{\Delta} Si\ (s) + HCl\ (g)$

l. Ingested barium sulfate causes the intestinal tract to be emphasized in x-ray pictures. It is produced when barium reacts with sulfuric acid. Hydrogen gas is also produced.

m. Milk of magnesia is an aqueous suspension of magnesium hydroxide. When ingested, it reduces the amount of hydrochloric acid in the stomach by neutralizing this acid. This reaction produces magnesium chloride and water.
   $Mg(OH)_2\ (aq) + HCl\ (aq) \longrightarrow MgCl_2\ (aq) + H_2O\ (l)$

⊙ 34. Tell whether each of the reactions in Question 33 is a synthesis, decomposition, single replacement, or double replacement reaction.

⊙ 35. Predict whether the following single replacement reactions will occur in standard conditions. Base your answers on the activity series given in the text.
   a. $BaSO_4 + Ca \longrightarrow CaSO_4 + Ba$
   b. $BaCl_2 + Br_2 \longrightarrow BaBr_2 + Cl_2$
   c. $Ni(OH)_2 + Mg \longrightarrow Mg(OH)_2 + Ni$
   d. $2FeCl_3 + 3Mg \longrightarrow 3MgCl_2 + 2Fe$
   e. $Al_2(SO_4)_3 + 2Fe \longrightarrow Fe_2(SO_4)_3 + 2Al$

⊙ 36. Complete ionic equations for double replacement reactions are given below. For each equation, identify the spectator ions and write the net ionic equation.
   a. Balanced equation:
      $AgNO_3\ (aq) + HCl\ (aq) \longrightarrow AgCl\ (s) + HNO_3\ (aq)$
      Complete ionic equation:
      $Ag^+ + NO_3^- + H^+ + Cl^- \longrightarrow AgCl\ (s) + H^+ + NO_3^-$
   b. Balanced equation:
      $CaCl_2\ (aq) + Na_2CO_3\ (aq) \longrightarrow CaCO_3\ (s) + 2NaCl\ (aq)$
      Complete ionic equation:
      $Ca^{2+} + 2Cl^- + 2Na^+ + CO_3^{2-} \longrightarrow CaCO_3\ (s) + 2Na^+ + 2Cl^-$
   c. Balanced equation:
      $H_2SO_4\ (aq) + 2KOH\ (aq) \longrightarrow K_2SO_4\ (aq) + 2H_2O\ (l)$
      Complete ionic equation:
      $2H^+ + SO_4^{2-} + 2K^+ + 2OH^- \longrightarrow 2K^+ + SO_4^{2-} + 2H_2O\ (l)$
   d. Balanced equation:
      $Mg(OH)_2\ (s) + 2HCl\ (aq) \longrightarrow MgCl_2\ (aq) + 2H_2O\ (l)$
      Complete ionic equation:
      $Mg(OH)_2\ (s) + 2H^+ + 2Cl^- \longrightarrow Mg^{2+} + 2Cl^- + 2H_2O\ (l)$
   e. Balanced equation:
      $Ba(NO_2)_2\ (aq) + Na_2SO_4\ (aq) \longrightarrow BaSO_4\ (s) + 2NaNO_2\ (aq)$
      Complete ionic equation:
      $Ba^{2+} + 2NO_2^- + 2Na^+ + SO_4^{2-} \longrightarrow BaSO_4\ (s) + 2Na^+ + 2NO_2^-$

Serving God as a Science Illustrator

Job Description
Science illustrators are artists who work in the various fields of science. They make detailed drawings for textbooks and other publications used by medical personnel, industry employees, and students. They may diagram equipment pieces and setups used in experiments or help illustrate or design manufacturing and industrial processes and equipment. Science illustrators may illustrate the steps surgeons take during operations or even help to fabricate artificial body parts, such as ears or eyes, for patients who need them.

Possible Workplaces
Many science illustrators are employed by commercial art studios that create illustrations for textbooks and other publications. Some work for hospitals, universities, and research institutions, where they sometimes also teach. Others work for museums; pharmaceutical companies; commercial art studios; advertising, marketing and public-relations firms; and scientific product manufacturers. There are also many freelance and work-from-home opportunities.

Education
A science illustrator needs a bachelor's degree, and maybe a master's degree. While at school, he should have taken courses in art and in the biological sciences. The illustrator must know how to use the computer as a design tool. He should be familiar with the printing process and have knowledge of how to prepare computer documents for commercial printing. Other requirements include the ability to adapt to changing technology and production methods, good communication and people skills, knowledge and ongoing interest of current design trends and practices, and multitasking and organizational skills.

Dominion Opportunities
Lifelike science illustrations are important to physicians, teachers, workers, builders, researchers, and students because they show details of structures, reactions, and equipment that are difficult to see in the body, classroom, or workplace. Science illustrators have a wonderful opportunity to positively and accurately depict detailed aspects of God's creation.

Chemical Calculations

CHAPTER 9

DOMINION SCIENCE PROBLEM

Producing Sulfuric Acid

Industry uses all kinds of important chemicals, called *commodity chemicals*, to make products we use. Other than water, American industries use sulfuric acid (H_2SO_4) more than any other chemical. In fact, the amount of sulfuric acid a country produces is closely related to its industrial strength. Sulfuric acid is used to produce fertilizers, batteries, steel, paper, gasoline, medicines, dyes, and nylon. But sulfuric acid is extremely corrosive and releases a great amount of heat when dissolved in water. How can industry mass-produce pure forms of this extremely useful chemical?

9A The Mole
9B Stoichiometry 216
Facet 227
Clearing the Air 238

Chapter Nine

9A Section Objectives

After finishing this section, you should be able to

- state Avogadro's number.
- describe the significance of Avogadro's number.
- calculate the mass, the number of particles, or the number of moles present in a given chemical substance when one of the quantities is given.
- distinguish between structural, molecular, and empirical formulas.
- calculate the percent composition of a substance when given masses.
- calculate an empirical formula from the percent composition of a substance.

The official definition of Avogadro's number is the number of carbon atoms in exactly 12 g of ^{12}C.

Although there are 6.022×10^{23} units in one mole, it is not a counted or defined number as are the units dozen and gross. This is important when considering significant digits in calculations. In this book we will use four significant digits for mole calculations.

Notice that the particles that make up a mole of helium are atoms. Similarly, the particles that make up a mole of water are molecules, the particles that make up a mole of salt are formula units, and the particles that make up a mole of hydrogen plasma are cations.

9A The Mole

9.1 Avogadro's Number and the Mole

How many molecules are in a test tube of water or a nugget of gold, do you suppose? Sextillions! Yet these are the sizes of samples that chemists often work with. It is nearly impossible to work with single molecules or atoms in the laboratory. So how do we know how many particles are in a sample?

Because atoms are so small and are nearly impossible to count, scientists devised a unit called the mole to make counting atoms more practical. A **mole** (mol) is the amount of substance contained in 6.022×10^{23} particles. The number 6.022×10^{23} is called **Avogadro's number** (N_A) in honor of the Italian physicist Amedeo Avogadro (1776–1856). In 1811, Avogadro determined the volume of one mole of a gas.

1 mol of He atoms = 6.022×10^{23} He atoms

1 mol of H_2O molecules = 6.022×10^{23} H_2O molecules

1 mol of NaCl formula units = 6.022×10^{23} NaCl formula units

Because the mole contains such a huge number of particles, it is used to describe only very small objects on the atomic scale. No one has ever heard of a mole of bricks or a mole of golf balls because no one has ever manufactured 602.2 sextillion of them! If one mole of marbles were spread out over the earth's surface, the earth would be covered 1.4 miles deep!

Example Problem 9-1
Using Avogadro's Number

How many atoms are in a 4.5 mol sample of helium?

Solution
Since one mole of any substance contains 6.022×10^{23} objects, this relationship can be used as a conversion factor to change 4.5 mol into the number of atoms.

$$\frac{4.5 \text{ mol} \times 6.022 \times 10^{23} \text{ atoms}}{1 \text{ mol}} = 2.7 \times 10^{24} \text{ atoms}$$

9.2 Molar Mass

If you were to compare the mass of a dozen grapes to the mass of a dozen apples, you would find that the apples have considerably more mass. The number of objects is the same, but each apple has a much greater mass. Similarly, a mole of carbon has a different mass than a mole of copper atoms. Can you guess which one contains more mass?

Scientists use the mole to relate the atomic mass listed on the periodic table to the mass in grams. The mass of a mole of substance can be found by expressing the object's mass in atomic mass units and converting it to grams. The mass of one mole of any pure substance is called its **molar mass**. These particles can be ions, atoms, molecules, or formula units.

For example, a hydrogen atom has a relative mass of about 1 atomic mass unit (u). Scientists have experimentally proven that 6.022×10^{23}

9-1 One mole each of various substances. Each sample contains about 6.022×10^{23} atoms but has a different mass.

For one mole of a pure substance, $1u = 1g$.

You may have noticed on the periodic table that carbon's atomic mass is not 12 but 12.01. In fact, with the exception of some of the transuranium elements that exist only under controlled laboratory conditions, none of the elements have whole number masses. Most elements exist in more than one isotopic form, and the value on the table is a weighted average.

hydrogen atoms have a mass of about 1 g. Avogadro's number was assigned its value so that the atomic mass of an element in atomic mass units equals its molar mass in grams per mole. A carbon-12 atom has a mass of 12 u, and so 6.022×10^{23} carbon-12 atoms have a mass of 12 g. Thus, the periodic table can be used to find not only the masses of individual atoms, molecules, or formula units, but also the mass of one mole of these particles. How can one mole of hydrogen atoms have a different mass than one mole of carbon atoms? Remember that a mole, like a dozen, always contains a specific number of objects. The mass of each object determines the molar mass.

EXAMPLE PROBLEM 9-2

Mass and Moles

Calculate the mass of 0.5000 mol of helium atoms.

Solution

On average, a helium atom has a mass of 4.003 u. This number, expressed in g, is the mass of one mole of helium atoms, 4.003 grams per mole.

$$\frac{0.5000 \text{ mol He}}{} \bigg| \frac{4.003 \text{ g He}}{1 \text{ mol He}} = 2.002 \text{ g He}$$

The arrows between mass, moles, and number of particles in Figure 9-2 show possible routes to take when solving a problem. Along the connecting lines are formats for the conversion units.

What is the mass of 5.00 mol of potassium atoms? The given value is expressed in moles (5.00 mol K), and the value to be calculated is the mass in grams. The flow chart shows that the conversion relating moles to molar mass is the only step that needs to be taken. Solving this problem on the basis of units alone, you know from the periodic table that 1 mol K = 39.10 g K. This relationship contains both the given unit (mol) and

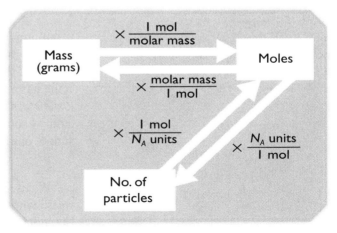

9-2 Mass-to-moles-to-particles conversion

> **Christians and Math**
> This chapter asks for you to think mathematically. Math is necessary for scientists to make use of various chemicals. The ability to think mathematically is part of the image of God in man and is one of the abilities that God gave humans so they could maximize the usefulness of the world God created for them (Gen. 1:26–28).

> Moles are always the "bridge" between the mass of a substance and the number of particles of that substance.

the desired unit (g), forming the only conversion factor needed for this problem.

$$\frac{5.00 \text{ mol K}}{} \cdot \frac{39.10 \text{ g K}}{1 \text{ mol K}} = 196 \text{ g K}$$

How many copper atoms are in a 4.00 g sample of pure copper wire? A look at Figure 9-2 shows that there is no step that will directly convert from the mass in grams to the number of units in that sample. The number of moles must be calculated as an intermediate step. From the periodic table, 1 mol of copper atoms is contained in 63.55 g, providing the following conversion factor.

$$\frac{4.00 \text{ g Cu}}{} \cdot \frac{1 \text{ mol Cu}}{63.55 \text{ g Cu}} = 0.0629 \text{ mol Cu}$$

Once the number of moles has been found, the last step on the flow chart can be taken.

$$\frac{0.0629 \text{ mol Cu}}{} \cdot \frac{6.022 \times 10^{23} \text{ atoms Cu}}{1 \text{ mol Cu}} = 3.79 \times 10^{22} \text{ atoms Cu}$$

> **EXAMPLE PROBLEM 9-3**
> *Mass and Atoms*
> How many atoms are in 33.3 mg (3.33×10^{-2} g) of gold? (This amount is about the size of the period at the end of this sentence.)
> **Solution**
> Figure 9-2 shows that first grams should be changed to moles, and then the moles should be changed to atoms.
>
> $$\frac{3.33 \times 10^{-2} \text{ g Au}}{} \cdot \frac{1 \text{ mol Au}}{197.0 \text{ g Au}} \cdot \frac{6.022 \times 10^{23} \text{ atoms Au}}{1 \text{ mol Au}}$$
>
> $$= 1.02 \times 10^{20} \text{ atoms Au}$$

9.3 Compounds and the Mole

Compounds contain two or more bonded atoms that behave as one unit. The masses of the molecules can be found simply by adding the masses of the atoms they contain. For example, a water (H_2O) molecule has a mass of 2(1.008 u) + 1(16.00 u) = 18.02 u, where 1.008 u is the mass of a hydrogen atom and 16.00 u is the mass of an oxygen atom. A sodium chloride formula unit (NaCl) has a mass equal to the sum of one atom of sodium and one atom of chlorine.

$$1(22.99 \text{ u}) + 1(35.46 \text{ u}) = 58.45 \text{ u}$$

> One mole of water contains two moles of hydrogen atoms and one mole of oxygen atoms.

We use the molar mass of compounds and formula units just as we do with atoms. In any of these situations, we can easily express the molar mass in *grams per mole* (g/mol). As shown in Table 9-1, the value of relative mass and the value of the molar mass are the same.

> For any sample of a pure substance, 1 u = 1 g/mol.

9-1 Masses of Compounds

Object	Relative mass	Molar mass
H_2 molecule	2.016 u	2.016 g/mol
H_2O molecule	18.02 u	18.02 g/mol
NaCl formula unit	58.45 u	58.45 g/mol

EXAMPLE PROBLEM 9-4
Calculating Molar Mass

Find the molar mass of $Al_2(SO_4)_3$.

Solution

Each formula unit contains two aluminum, three sulfur, and twelve oxygen atoms. A mole of $Al_2(SO_4)_3$ consists of 2 mol of aluminum atoms, 3 mol of sulfur atoms, and 12 mol of oxygen atoms.

$$\frac{2 \text{ mol Al} \mid 26.98 \text{ g Al}}{1 \text{ mol Al}} = 53.96 \text{ g Al}$$

$$\frac{3 \text{ mol S} \mid 32.07 \text{ g S}}{1 \text{ mol S}} = 96.21 \text{ g S}$$

$$\frac{12 \text{ mol O} \mid 16.00 \text{ g O}}{1 \text{ mol O}} = 192.0 \text{ g O}$$

$$1 \text{ mol } Al_2(SO_4)_3 = 342.2 \text{ g } Al_2(SO_4)_3$$

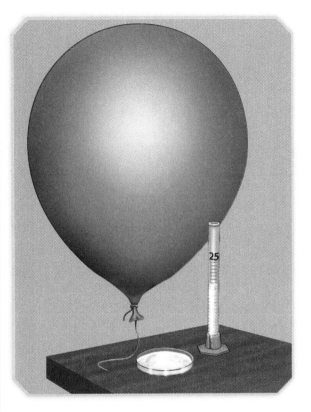

9-3 One mole of each of the substances from Table 9-1 is shown here.

9.4 Types of Formulas

There are several ways to describe the composition of a chemical substance. **Structural formulas** show the types of atoms involved, the exact composition of each molecule, and the arrangement of chemical bonds. These formulas are informative, but they can also be difficult to draw and often take up large amounts of space. The structural formula for water is simply H–O–H, but other structural formulas, such as acetic acid's (CH_3COOH), can become quite complex.

$$\begin{array}{c} H \\ | \\ H-C-C \\ | \\ H \end{array} \begin{array}{c} O \cdots H-O \\ \diagup \quad \diagdown \\ \diagdown \quad \diagup \\ O-H \cdots O \end{array} \begin{array}{c} H \\ | \\ C-C-H \\ | \\ H \end{array}$$

Molecular formulas show the types and numbers of atoms involved as they appear in the molecule. These formulas are more convenient than structural formulas, but they do not show the shapes of the molecules, the locations of the bonds, or the types of bonds present.

Empirical formulas tell what elements are present and give the simplest whole-number ratio of atoms in the compound. Empirical formulas have already been used to describe the composition of ionic compounds. When used for a molecular compound, the empirical formula might represent the actual molecular composition if the molecular formula contains a simple ratio—as in H_2O. On the other hand, the empirical formula of a molecular compound may not represent the molecule. Ethene's empirical formula is CH_2 because there

9-4 "Fine, boys. Tomorrow, we'll work on acetic acid."

are twice as many hydrogen atoms as there are carbon atoms. In this case the molecular formula is not the simplest ratio, so the empirical and molecular formulas are different. Notice also that the empirical formula of diatomic chlorine shows just one atom.

9-2 Molecular and Empirical Formulas

Compound	Molecular formula	Empirical formula
water	H_2O	H_2O
ethene	C_2H_4	CH_2
chlorine	Cl_2	Cl

All of the compounds listed in Table 9-3 have the same empirical formula—CH_2O—but have very different molecular formulas. Notice that the empirical formula and the molecular formula for formaldehyde are the same.

9-3 Compounds with the Empirical Formula CH_2O

Molecular formula	Name
CH_2O	formaldehyde
$C_2H_4O_2$ (CH_3COOH)	acetic acid
$C_3H_6O_3$	L-lactic acid
$C_4H_8O_4$	D-threose
$C_5H_{10}O_5$	D-ribose
$C_6H_{12}O_6$	D-fructose
$C_7H_{14}O_7$	D-mannoheptulose (from avocados)

9.5 Percent Composition

Percent composition describes the mass composition of a compound by showing what percentage of its total mass comes from each element. The percent composition deals with the masses of the atoms as well as the number of atoms in a molecule.

Percent compositions often look totally different from empirical formulas. For example, although water has only one oxygen atom, this one atom contains the majority of the compound's mass. The two hydrogen atoms do not contribute much mass because they are so small. Likewise, ethene has two hydrogen atoms for every carbon atom, but more mass comes from carbon because it is much more massive.

A laboratory analysis of a substance is usually expressed as a percent composition. It is essential to understand that "percent" literally means "per hundred." A general setup to calculate any percent is part/whole × 100%. Suppose that a 60.00 g sample of water were

9-5 These circle graphs show the compositions of water, ethene, and chlorine.

decomposed into its elements and that 53.28 g of oxygen and 6.72 g of hydrogen gas was produced. The percent composition of water could be calculated from these results as follows.

Hydrogen: There are 6.72 g of hydrogen in 60.00 g of water.

$$\frac{\text{part}}{\text{whole}} \times 100\% = \frac{6.72 \text{ g H}}{60.00 \text{ g H}_2\text{O}} \times 100\%$$

$$= 0.112 \times 100\% = 11.2\%$$

Oxygen: There are 53.28 g of oxygen in 60.00 g of water.

$$\frac{\text{part}}{\text{whole}} \times 100\% = \frac{53.28 \text{ g O}}{60.00 \text{ g H}_2\text{O}} \times 100\%$$

$$= 0.8880 \times 100\% = 88.80\%$$

Percent composition for a specific compound does not change based on the sample size because of the law of definite composition for compounds. The **law of definite composition** says that *the ratio of elements in a compound is constant for every particle in that compound.*

EXAMPLE PROBLEM 9-5
Percent Composition

A laboratory analysis of a 30.00 g sample of $Al_2(SO_4)_3$ showed that it contained 4.731 g of aluminum, 8.436 g of sulfur, and 16.833 g of oxygen. What is the percent composition of this compound?

Solution

Al: $\dfrac{4.731 \text{ g Al}}{30.00 \text{ g Al}_2(SO_4)_3} \times 100\% = 0.1577 \times 100\% = 15.77\%$ Al

S: $\dfrac{8.436 \text{ g S}}{30.00 \text{ g Al}_2(SO_4)_3} \times 100\% = 0.2812 \times 100\% = 28.12\%$ S

O: $\dfrac{16.833 \text{ g O}}{30.00 \text{ g Al}_2(SO_4)_3} \times 100\% = 0.5611 \times 100\% = 56.11\%$ O

Problem-Solving Strategy 9-1

One way to check yourself when calculating percent composition is to make sure that all the percentages for the atoms involved add up to 100%.

You can also calculate the percent composition of a compound directly from a formula.

Example Problem 9-6

Percent Composition

Find the percent composition of $Al_2(SO_4)_3$.

Solution

First, assume you have a one-mole sample of the compound. In this case, each mole of aluminum sulfate would contain the following:

2 mol of Al (26.98 g/mol) = 53.96 g
3 mol of S (32.07 g/mol) = 96.21 g
12 mol of O (16.00 g/mol) = 192.0 g
Total 342.2 g

Al: $\dfrac{53.96 \text{ g Al}}{342.2 \text{ g Al}_2(SO_4)_3} \times 100\% = 0.1577 \times 100\% = 15.77\%$ Al

S: $\dfrac{96.21 \text{ g S}}{342.2 \text{ g Al}_2(SO_4)_3} \times 100\% = 0.2812 \times 100\% = 28.12\%$ S

O: $\dfrac{192.0 \text{ g O}}{342.2 \text{ g Al}_2(SO_4)_3} \times 100\% = 0.5611 \times 100\% = 56.11\%$ O

Notice that the percentages are the same using either method. This equivalence is due to the law of definite composition.

Example Problem 9-7

Percent Composition

How many grams of oxygen would a 65.00 g sample of $Al_2(SO_4)_3$ contain?

Solution

Because of the law of definite composition, any other sample of $Al_2(SO_4)_3$ will have the same percent composition that the 30.00 g sample or hypothetical 1 mol sample did in the above problems. The problem could be worked by unit analysis as in the preceding sample problem. It could also be done as a simple multiplication problem.

$$\dfrac{65.00 \text{ g Al}_2(SO_4)_3 \quad | \quad 16.833 \text{ g O}}{30.00 \text{ g Al}_2(SO_4)_3} = 36.47 \text{ g O}$$

$65.00 \text{ g Al}_2(SO_4)_3 \times 0.5611 = 36.47 \text{ g O}$

9.6 Calculations with Empirical Formulas

Empirical formulas contain the information necessary to calculate the percent composition of compounds since they preserve the ratio of atoms in a compound. To find percent composition, the mole ratio in the empirical formula must be converted to a mass ratio through a series of calculations.

The formula H₂O means that there are exactly 2 mol of hydrogen atoms for every 1 mol of oxygen atoms. One mole of water contains 2 mol of hydrogen atoms and 1 mol of oxygen atoms. Expressed in masses, 1 mol of water contains 2.016 g of hydrogen atoms and 16.00 g of oxygen atoms.

$$\frac{2 \text{ mol H}}{} \cdot \frac{1.008 \text{ g H}}{1 \text{ mol H}} = 2.016 \text{ g H}$$

$$\frac{1 \text{ mol O}}{} \cdot \frac{16.00 \text{ g O}}{1 \text{ mol O}} = 16.00 \text{ g O}$$

The total mass of one mole of water is 18.02 g. Now the percent composition of water can be found.

$$\frac{2.016 \text{ g H}}{18.02 \text{ g H}_2\text{O}} \times 100\% = 0.1119 \times 100\% = 11.19\% \text{ H}$$

$$\frac{16.00 \text{ g O}}{18.02 \text{ g H}_2\text{O}} \times 100\% = 0.8879 \times 100\% = 88.79\% \text{ O}$$

Laboratories often do the reverse process, using the percent composition of an unknown compound determined experimentally to calculate its empirical formula. Suppose that a chemist is given 100.0 g of an unknown compound and is told to determine its empirical formula. After a careful analysis in an analytical laboratory, the chemist concludes that 75.00 g (75.00%) of the sample's mass is carbon. The other 25.00 g (25.00%) come from hydrogen. Since empirical formulas are based on mole ratios rather than mass ratios, these percentages need to be converted to moles using molar mass.

Because carbon atoms have much more mass than hydrogen atoms, the empirical formula is nothing like $C_{75}H_{25}$ or C_3H_1. To determine the mole composition, the mass composition of the sample must be used.

9-6 Finding the mass of a substance before and after a reaction can reveal the substance's empirical formula.

$$\frac{75.00 \text{ g C}}{} \cdot \frac{1 \text{ mol C}}{12.01 \text{ g C}} = 6.245 \text{ mol C}$$

$$\frac{25.00 \text{ g H}}{} \cdot \frac{1 \text{ mol H}}{1.008 \text{ g H}} = 24.80 \text{ mol H}$$

According to the above calculation, the empirical formula for this compound could be written as $C_{6.245}H_{24.80}$. Although this formula is numerically accurate, it is not in its final form because empirical formulas must be written as whole number ratios. Dividing both numbers by the smaller number gives the simplest form of the ratio and guarantees that one of the subscripts will be a 1.

$$\text{mol C} : \text{mol H} = \frac{6.245}{6.245} : \frac{24.80}{6.245} = 1 : 3.97 \text{ or } 1 : 4$$

The empirical formula of the compound is C_1H_4, or CH_4.

This flow chart shows the steps to convert from percent composition to an empirical formula.

Percent composition	Mass composition (for a 100 g sample)	Mole composition	Mole ratio	Empirical formula
75.00% C 25.00% H	75.00 g C 25.00 g H	6.245 mol C 24.80 mol H	$\dfrac{1 \text{ mol C}}{4 \text{ mol H}}$	CH_4

Notice that it will often be necessary to round slightly to get to a whole number. Experimental error or rounding during calculation can make the final results slightly off. Sometimes calculations produce ratios such as 1:1.5 or 1:1.33. These should not be rounded off. In general, if the decimal portion of the part of the ratio is greater than 0.1 and less than 0.9, you should not round off. The ratio 1:1.5 is equivalent to the whole-number ratio of 2:3. The ratio 1:1.33 is equivalent to the whole-number ratio 3:4. Insight and practice are necessary to know when to round off and when to make another ratio. One method of obtaining a whole number ratio is to express the ratio in fractions having a common denominator, then multiply through by the denominator to cancel.

> **Problem-Solving Strategy 9-2**
> Though the percent composition of any quantity of a sample will be the same, if we choose the sample size to be exactly 100 g, then the "%" unit can simply be changed to "g," and the calculation is simplified.

> **EXAMPLE PROBLEM 9-8**
> *Calculating Empirical Formulas from Percent Composition*
>
> A laboratory analysis of an unknown gas has determined that the gas is 72.55% oxygen and 27.45% carbon by mass. What is the empirical formula of the compound?
>
> **Solution**
> *Percent composition:*
> 72.55% O
> 27.45% C
> *Mass composition:*
> In a 100 g sample of the gas, there will be 72.55 g of oxygen and 27.45 g of carbon.
> *Mole composition of sample:*
>
> $$72.55 \text{ g O} \times \dfrac{1 \text{ mol O}}{16.00 \text{ g O}} = 4.534 \text{ mol O}$$
>
> $$27.45 \text{ g C} \times \dfrac{1 \text{ mol C}}{12.01 \text{ g C}} = 2.286 \text{ mol C}$$
>
> *Mole ratio:*
>
> mol O : mol C = 4.534 : 2.286 reduced to lowest terms is
>
> $$\text{mol O : mol C} = \dfrac{4.534}{2.286} : \dfrac{2.286}{2.286} = 1.983 : 1.000$$
>
> *Empirical formula:*
> For every 2 mol of oxygen, there is 1 mol of carbon. The empirical formula must be CO_2.

> Notice that in these problems we use the molar mass of each element, even for those that naturally exist as diatomic molecules.

EXAMPLE PROBLEM 9-9

Calculating Empirical Formulas from Mass Composition

A 5.000 g sample of an unknown compound contains 1.844 g of nitrogen and 3.156 g of oxygen. Find the empirical formula.

Solution

The mass composition of the sample is already known, so we can bypass one step on the flow chart.

Mass composition:
1.844 g N
3.156 g O

Mole composition of sample:

$$\frac{1.844 \text{ g N}}{1} \cdot \frac{1 \text{ mol N}}{14.01 \text{ g N}} = 0.1316 \text{ mol N}$$

$$\frac{3.156 \text{ g O}}{1} \cdot \frac{1 \text{ mol O}}{16.00 \text{ g O}} = 0.1973 \text{ mol O}$$

Mole ratio:

mol N : mol O = 0.1316 : 0.1973 reduced to lowest terms is

$$\text{mol N : mol O} = \frac{0.1316}{0.1316} : \frac{0.1973}{0.1316} = 1.000 : 1.499$$

The ratio should be 1 : 1.5 (N : O).

Empirical formula:

The ratio 1 : 1.5 is in its simplest form, but the numbers are not whole numbers. To put the ratio in whole numbers, express the numbers as fractions.

$$1 : 1.5 = \frac{2}{2} : \frac{3}{2}$$

Eliminate the fractions by multiplying through by the common denominator.

$$\left(\frac{2}{1}\right)\frac{2}{2} : \frac{3}{2}\left(\frac{2}{1}\right) = 2 : 3$$

The empirical formula of the compound is N_2O_3.

Sometimes molecular formulas can be determined when the molar mass of the actual compound and its empirical formula is known. The molar mass of the compound is divided by the molar mass of the empirical formula. This whole number is multiplied by the subscripts in the empirical formula to produce the compound's molecular formula.

Example Problem 9-10

Calculating Molecular Formulas from Percent Composition

Caffeine, in coffee and some carbonated beverages, is 5.170% hydrogen, 16.49% oxygen, 28.86% nitrogen, and 49.48% carbon by mass. The molar mass of a caffeine molecule is 194.20 g/mol. Find its molecular formula.

Solution
Percent composition:
5.170% H
16.49% O
28.86% N
49.48% C

Mass composition:
In a 100 g sample of the compound, there will be 5.170 g of hydrogen, 16.49 g of oxygen, 28.86 g of nitrogen, and 49.48 g of carbon.

Mole composition of sample:

$$\frac{5.170 \text{ g H}}{1} \times \frac{1 \text{ mol H}}{1.008 \text{ g H}} = 5.129 \text{ mol H}$$

$$\frac{16.49 \text{ g O}}{1} \times \frac{1 \text{ mol O}}{16.00 \text{ g O}} = 1.031 \text{ mol O}$$

$$\frac{28.86 \text{ g N}}{1} \times \frac{1 \text{ mol O}}{14.01 \text{ g N}} = 2.060 \text{ mol N}$$

$$\frac{49.48 \text{ g C}}{1} \times \frac{1 \text{ mol C}}{12.01 \text{ g C}} = 4.120 \text{ mol C}$$

Mole ratio:
Finding the ratios of these elements is done by dividing all of their mole compositions by the lowest value, 1.031 mol. This gives us whole number ratios.

$$\text{H: } \frac{5.129 \text{ mol}}{1.031 \text{ mol}} = 4.974$$

$$\text{O: } \frac{1.031 \text{ mol}}{1.031 \text{ mol}} = 1.000$$

$$\text{N: } \frac{2.060 \text{ mol}}{1.031 \text{ mol}} = 1.998$$

$$\text{C: } \frac{4.120 \text{ mol}}{1.031 \text{ mol}} = 3.996$$

Molecular formula:
The ratios above give an empirical formula of $C_4H_5N_2O$, which has a molar mass of 97.10 g/mol. The molar mass of caffeine is divided by the molar mass of its empirical formula to give a ratio.

$$\frac{194.20 \text{ g/mol}}{97.10 \text{ g/mol}} = 2.000$$

This whole number is multiplied by the subscripts in the empirical formula, $(C_4H_5N_2O)_2$, to produce the molecular formula $C_8H_{10}N_4O_2$.

9A Section Review

1. What is Avogadro's number, and what is its physical significance?
2. How many atoms are found in 600. g of iron?
3. Give one benefit and one disadvantage of using structural formulas.
4. What is the difference between an empirical formula and a molecular formula?
5. Give the structural, molecular, and empirical formulas of hydrogen peroxide.
6. How many moles of atoms are in 1.00 lb (454 g) of lead?
7. Find the molar mass of vitamin A ($C_{20}H_{30}O$).
8. DDT, an insecticide banned in the United States, has the formula $C_{14}H_9Cl_5$. Find its percent composition.
9. Find the molecular formula for acetone (molar mass 58.0 g/mol), used to make nail polish and other industrial products. Its composition is 62.0% carbon, 10.4% hydrogen, and 27.6% oxygen.

9B Stoichiometry

9.7 Introduction to Stoichiometry

The United States produces about ten million tons of phosphoric acid each year. Most of this is used to make fertilizers. One of the most common methods to produce this acid is to grind phosphate rock to a powder and mix it with sulfuric acid in water. The products include the intended phosphoric acid plus a hydrated form of calcium sulfate. A simplified equation for this reaction is

$$Ca_3(PO_4)_2\ (s) + 3H_2SO_4\ (aq) + 6H_2O\ (l) \rightleftharpoons 2H_3PO_4\ (aq) + 3CaSO_4 \cdot 2H_2O\ (s).$$

Chemical engineers who manage this process must carefully track all the compounds involved. For instance, they need to know how much product is needed to supply the market and how much of each reactant is needed to make that product. They can calculate quantities of reactants and products using the mole and mass relationships between them. **Stoichiometry** is about the mathematical relationships between the amounts of reactants and products in a chemical reaction.

9.8 Mole-to-Mole Conversions

You're getting ready to go to a party, and you decide to bring some crisped rice treats. You pull out your recipe.

CRISPED RICE TREATS (makes 24 2×2-inch squares)

3 tablespoons of margarine

One 10-ounce package of marshmallows (about 40)

6 cups of crisped rice cereal

> **9B Section Objectives**
> After finishing this section, you should be able to
> - solve stoichiometric conversions between moles, masses, and the number of particles for any substance in a chemical reaction when given a balanced chemical equation and one of the quantities.
> - identify the limiting and excess reactants in a chemical reaction when given one of the quantities.
> - calculate the amount of excess reactant in a chemical reaction.
> - calculate the theoretical yield in a chemical reaction when given the quantity of a reactant.
> - calculate the percent yield when given the actual yield from a chemical reaction.

> stoichiometry (STOI kee OM e tree): stoichio- (Gk. *stoicheion*—elements) + -metry (Gk. *metron*—measure)

9-7 If you wanted to scale the recipe up to make more of these treats, you would use a form of stoichiometry.

If you were really hungry, you could scale the recipe up to make more treats. You could also reduce the recipe if you needed to. As long as you preserve the ratios of the ingredients, you can adjust the recipe to suit your needs. For example, the ratio of cereal to marshmallows is 6 cups to 40 marshmallows. Suppose you had only 4.5 cups of cereal and sufficient amounts of the other ingredients. How many marshmallows would you need? You could use the recipe relationships and unit analysis as follows.

$$\frac{4.5 \text{ cups cereal}}{1} \cdot \frac{40 \text{ marshmallows}}{6 \text{ cups cereal}} = 30 \text{ marshmallows}$$

Further, you could determine how many treats you could make (assuming they were the same size) or how much margarine is needed. All of these can be determined from the relationships given in the recipe.

$$\frac{4.5 \text{ cups cereal}}{1} \cdot \frac{24 \text{ treats}}{6 \text{ cups cereal}} = 18 \text{ treats}$$

Similarly, chemical equations show the ratios of substances involved in a reaction. Balanced chemical equations contain the numerical information necessary for detailed calculations. The coefficients in front of each substance show how many atoms, molecules, and moles are involved. For example, the equation

$$P_4 + 5O_2 \longrightarrow 2P_2O_5$$

means that one P_4 molecule and five O_2 molecules form two P_2O_5 molecules. The coefficients can also represent the number of moles of a compound or element. The coefficients in balanced equations are the key to stoichiometric calculations. They give the numerical

God and the Order of the Universe

When chemists work with substances, they find that quantitative relationships between specific elements in chemical reactions are consistent. The measurements that chemists make are based on the orderly nature of these chemical reactions. Substances react in an orderly way because that is the way that God created them to work. As chemists look into the details of how chemicals react differently in various combinations, it should cause them to marvel at the wisdom of God.

relationships between the substances in a reaction, like the ingredients in the recipe. This balanced equation shows that

1 mol P_4 reacts to form 2 mol P_2O_5 (a 1:2 ratio),

1 mol P_4 reacts with 5 mol O_2 (a 1:5 ratio), and

5 mol O_2 react to form 2 mol P_2O_5 (a 5:2 ratio).

Even if the original quantities were cut in half, the ratios would still hold true. In each case, the coefficients from the balanced equation give the ratios between the moles of one substance and the moles of another substance. These are known as **mole ratios**.

Ratios from the coefficients of a balanced chemical equation can be used as conversion factors to convert from moles of one substance to moles of another. Figure 9-8 illustrates how to do a mole-to-mole stoichiometric conversion. Coefficients can be used to calculate the number of moles of one substance when given the number of moles of another substance.

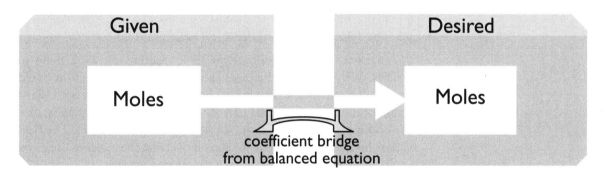

9-8 Mole-to-mole conversion

For example, suppose a chemical engineer wanted to produce 25.0 mol of diphosphorus pentoxide. How many moles of phosphorus would he burn? From the previous example, we can see that 1 mol of phosphorus produces 2 mol of diphosphorus pentoxide. Using that information, he can set up the following equation to solve the problem.

$$\frac{25.0 \text{ mol } P_2O_5}{} \times \frac{1 \text{ mol } P_4}{2 \text{ mol } P_2O_5} = 12.5 \text{ mol } P_4$$

The amount of oxygen required is calculated from either the 12.5 mol of phosphorus or from the 25.0 mol of diphosphorus pentoxide. Both methods use ratios formed from the coefficients in the balanced equation.

$$\frac{12.5 \text{ mol } P_4}{} \times \frac{5 \text{ mol } O_2}{1 \text{ mol } P_4} = 62.5 \text{ mol } O_2$$

$$\frac{25.0 \text{ mol } P_2O_5}{} \times \frac{5 \text{ mol } O_2}{2 \text{ mol } P_2O_5} = 62.5 \text{ mol } O_2$$

Problem-Solving Strategy 9-3
Be sure to get into the habit of using the complete formula and unit labels, not just "g" or "mol." Specify the substance in each case so you can see how the units cancel to get the desired result.

Example Problem 9-11
Mole-to-Mole Conversions

If 25.0 mol of diphosphorus pentoxide reacts with water to form phosphoric acid, how many moles of water are required? The balanced equation of the reaction is

$$P_2O_5 + 3H_2O \longrightarrow 2H_3PO_4.$$

Solution

This problem amounts to changing from moles of one substance (P_2O_5) to moles of another substance (H_2O). Their relationship is found in the balanced equation, where the coefficients show that 3 mol of water are needed for every 1 mol of diphosphorus pentoxide—the "coefficient bridge."

$$\frac{25.0 \text{ mol } P_2O_5}{} \cdot \frac{3 \text{ mol } H_2O}{1 \text{ mol } P_2O_5} = 75.0 \text{ mol } H_2O$$

9.9 Mass-to-Mole Conversions

Conversions between substances in a reaction must be done in moles. Why? Because the balanced equation—the recipe—is based on moles, not mass. The coefficients in chemical reactions show molar ratios, not mass ratios. In order to convert from one substance to another, the amount of the substance must be in moles. The flow chart for mass-to-mole conversions shows that the moles of the given substance must be calculated as an intermediate step.

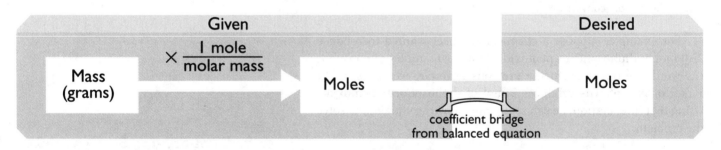

9-9 Mass-to-mole conversion

Calculate the moles of diphosphorus pentoxide that would result from the burning of 1.55 kg (1550 g) of phosphorus from the equation in Subsection 9.8. Before the conversion factor

$$\frac{2 \text{ mol } P_2O_5}{1 \text{ mol } P_4}$$

can be used, the mass of phosphorus in grams must be converted to moles. Remember from Section 9A that the molar mass of a substance is numerically equal to the atomic mass. The molar mass of phosphorus is 30.97 × 4 g/mol, or 123.9 g/mol.

$$\frac{1550 \text{ g } P_4}{} \cdot \frac{1 \text{ mol } P_4}{123.9 \text{ g } P_4} = 12.5 \text{ mol } P_4$$

The 12.5 mol of phosphorus can now be converted to moles of diphosphorus pentoxide using the mole ratio obtained from the balanced equation.

$$\begin{array}{c|c} 12.5\ \cancel{\text{mol P}_4} & 2\ \text{mol P}_2\text{O}_5 \\ \hline & 1\ \cancel{\text{mol P}_4} \end{array} = 25.0\ \text{mol P}_2\text{O}_5$$

You can use unit analysis to make one equation using both the gram-to-mole and the mole-to-mole conversion factors.

$$\begin{array}{c|c|c} 1550\ \cancel{\text{g P}_4} & 1\ \cancel{\text{mol P}_4} & 2\ \text{mol P}_2\text{O}_5 \\ \hline & 123.9\ \cancel{\text{g P}_4} & 1\ \cancel{\text{mol P}_4} \end{array} = 25.0\ \text{mol P}_2\text{O}_5$$

EXAMPLE PROBLEM 9-12

Mass-to-Mole Conversions

How many moles of phosphoric acid can be formed from 3550 g of diphosphorus pentoxide?

$$\text{P}_2\text{O}_5 + 3\text{H}_2\text{O} \longrightarrow 2\text{H}_3\text{PO}_4$$

Solution

Note that the given value is in grams and the desired value is in moles. Since the conversion factors from the balanced equation are based on a mole ratio, the 3550 g of diphosphorus pentoxide must first be converted to moles. The molar mass of P_2O_5 is 141.9 g/mol.

$$\begin{array}{c|c} 3550\ \cancel{\text{g P}_2\text{O}_5} & 1\ \text{mol P}_2\text{O}_5 \\ \hline & 141.9\ \cancel{\text{g P}_2\text{O}_5} \end{array} = 25.0\ \text{mol P}_2\text{O}_5$$

Now the moles of diphosphorus pentoxide can be converted to moles of phosphoric acid according to the ratio of the coefficients in the above reaction.

$$\begin{array}{c|c} 25.0\ \cancel{\text{mol P}_2\text{O}_5} & 2\ \text{mol H}_3\text{PO}_4 \\ \hline & 1\ \cancel{\text{mol P}_2\text{O}_5} \end{array} = 50.0\ \text{mol H}_3\text{PO}_4$$

Or you may combine steps using one long bridge.

$$\begin{array}{c|c|c} 3550\ \cancel{\text{g P}_2\text{O}_5} & 1\ \cancel{\text{mol P}_2\text{O}_5} & 2\ \text{mol H}_3\text{PO}_4 \\ \hline & 141.9\ \cancel{\text{g P}_2\text{O}_5} & 1\ \cancel{\text{mol P}_2\text{O}_5} \end{array} = 50.0\ \text{mol H}_3\text{PO}_4$$

9.10 Mass-to-Mass Conversions

When the mass of one substance in a reaction is known, the mass of a second substance can be calculated. This type of problem is the most practical presented in this chapter because chemists frequently do this in their laboratories. To do so, you must be able to convert between mass and moles, so the preceding problems have prepared you for this task. The moles of the known and unknown substances must be determined before the unknown mass can be calculated.

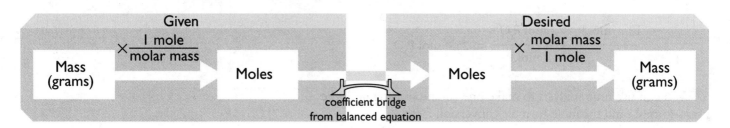

9-10 Mass-to-mass conversion

> **EXAMPLE PROBLEM 9-13**
> *Mass-to-Mass Conversions*
>
> What mass of water will react with 3550 g of diphosphorus pentoxide?
>
> $$P_2O_5 + 3H_2O \longrightarrow 2H_3PO_4$$
>
> **Solution**
> The mass of diphosphorus pentoxide must be expressed as moles before the molar conversion can be done. Once the number of moles of water is known, the mass of water can be calculated.
>
> $$\frac{3550 \text{ g } P_2O_5}{} \times \frac{1 \text{ mol } P_2O_5}{141.9 \text{ g } P_2O_5} \times \frac{3 \text{ mol } H_2O}{1 \text{ mol } P_2O_5} \times \frac{18.016 \text{ g } H_2O}{1 \text{ mol } H_2O} = 1350 \text{ g } H_2O$$

The number of atoms or molecules involved in a reaction can also be calculated from known molar quantities. Suppose that the sample problem above asked, "How many water molecules will react with 3550 g of diphosphorus pentoxide?" The solution to the problem is the same until the step in which the number of moles of water is converted to the mass of water. The conversion factor

$$\frac{6.022 \times 10^{23} \text{ molecules}}{\text{mol}}$$

can be used to convert the number of moles of water to the number of molecules of water.

$$\frac{3550 \text{ g } P_2O_5}{} \times \frac{1 \text{ mol } P_2O_5}{141.9 \text{ g } P_2O_5} \times \frac{3 \text{ mol } H_2O}{1 \text{ mol } P_2O_5} \times \frac{6.022 \times 10^{23} \text{ molecules } H_2O}{1 \text{ mol } H_2O}$$
$$= 4.52 \times 10^{25} \text{ molecules } H_2O$$

With this addition, the flow chart for stoichiometric problems as seen in Figure 9-11 is the completed "stoichiometric roadmap." As on any map, you must first determine where you are and where you want to arrive. The left side of the "map" shows the various starting points of the stoichiometric trip—the given information of the question. All roads lead to the "coefficient bridge," which is determined from the balanced equation. Once across the bridge, you can map your way to the various destinations (moles, mass, or number of particles) using bridge notation.

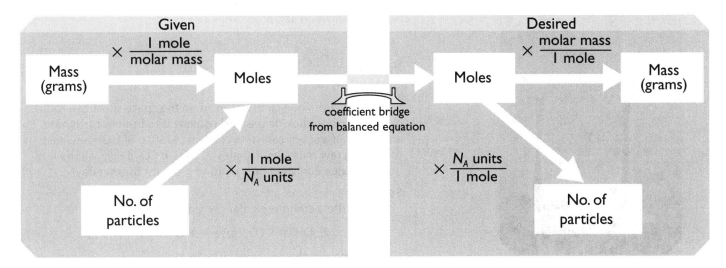

9-11 All conversions

EXAMPLE PROBLEM 9-14
Mass-to-Mass Conversions

How many grams of sodium chloride decompose to yield 27 g of chlorine gas?

Solution

First, write a complete balanced equation.

$$2NaCl \xrightarrow{\Delta} 2Na + Cl_2$$

Next, following Figure 9-11, convert from grams of chlorine gas to moles of chlorine gas, then to moles of salt, and then to grams of salt.

$$\frac{27 \text{ g Cl}_2}{} \times \frac{1 \text{ mol Cl}_2}{71 \text{ g Cl}_2} \times \frac{2 \text{ mol NaCl}}{1 \text{ mol Cl}_2} \times \frac{58 \text{ g NaCl}}{1 \text{ mol NaCl}} = 44 \text{ g NaCl}$$

9.11 Limiting Reactants

Up to this point in our study of chemistry, we have assumed that chemical reactions all start with measured amounts of reactants and end with specific amounts of products. In fact, nature is rarely so tidy. Even in a laboratory where conditions are controlled and reactants are carefully measured, at least one reactant is used up before the others, preventing more products from forming. This reactant is called the **limiting reactant**. Obviously, these conditions will leave at least one, and possibly several, other reactants in their original unchanged states. We call these leftover substances **excess reactants**.

9-12 When oxygen is the limiting reactant in the combustion of propane, the orange color indicates that some of the propane reactant has not completely reacted (left). A properly adjusted flame has ample oxygen for the reaction, and there is no excess propane reactant (right).

Imagine that you are having a cookout for your sports team, and fourteen people say they will eat only cheeseburgers. You have plenty of hamburgers and rolls, but only twelve slices of cheese. In this case, the cheese is the limiting reactant because it limits how many cheeseburgers can be produced. The extra rolls and "uncheesed" hamburgers are the excess reactants.

Example Problem 9-15
Limiting Reactant

Lithium hydroxide canisters used on the space shuttles capture exhaled carbon dioxide gas and convert it to lithium carbonate and water. If a set of canisters contains 5750. g of LiOH and each of the six crew members exhales 21 mol of CO_2 daily, will the lithium hydroxide canisters be a limiting reactant for that day?

Solution

First, write a complete balanced equation.

$$2LiOH\,(s) + CO_2\,(g) \longrightarrow Li_2CO_3\,(s) + H_2O\,(l)$$

Next, convert the grams of LiOH to moles.

$$\frac{5750.\text{ g LiOH}}{} \cdot \frac{1\text{ mol LiOH}}{23.95\text{ g LiOH}} = 240.1\text{ mol LiOH}$$

From the mole ratio in the balanced equation above, you can see that it takes two moles of LiOH to react with every one mole of CO_2 (2 mol LiOH : 1 mol CO_2). Consequently, because the six astronauts would contribute 126 mol of CO_2, you would need 2(126 mol) or 252 mol of LiOH to process all of the CO_2. Clearly, 240.1 mol of LiOH is less than the 252 mol required, so LiOH is the limiting reactant.

Example Problem 9-16
Excess Reactant

In Example Problem 9-15, how many grams of excess reactant are there?

Solution

You already know that the CO_2 is the excess reactant because there was some left over after all the LiOH was consumed in the reaction. Since the mole ratio tells you that 240 mol of LiOH can react with 120 mol CO_2, you know that you have 6 mol of CO_2 that are in excess (126 mol − 120 mol). Now you just convert the excess mol back to g.

$$\frac{6\text{ mol }CO_2}{} \cdot \frac{44.01\text{ g }CO_2}{1\text{ mol }CO_2} = 264.1\text{ g }CO_2$$

You have 264.1 g of excess CO_2 reactant.

9.12 Percent Yield

Even if a scientist carefully measures out the proper number of grams of reactants based on their mole ratio, there is a high probability that he will find less mass in his products than expected. In the problems found in this chapter, you have been calculating the theoretical yield. The **theoretical yield** is the maximum amount of product that could be created from a given amount of reactant.

There are many reasons why experimental yield will probably not match the theoretical yield. In many reactions, there are competing side reactions that take place that are not figured into the calculations. In some cases, the products must be purified first and some material may be lost in the process. Some dissolved products may cling to the inside of glassware, or liquids may evaporate. Limiting reactants may prevent some product from forming. The measured amount of product at the end of a reaction is called the **actual yield**. The actual yield is normally expressed in grams of the product.

Chemists often need to know how efficient a reaction is, especially in industry, where costs must be carefully controlled. For industry, a more important figure than theoretical or actual yield is the percent yield. The **percent yield** is the percentage of the theoretical yield that was actually produced. It is calculated by taking a ratio of the actual yield to the theoretical yield and multiplying by 100%.

EXAMPLE PROBLEM 9-17

Percent Yield

Aluminum hydroxide is found in many antacid tablets because it reacts with hydrochloric acid from the stomach to form aluminum chloride and water.

a. If 15.00 g of aluminum hydroxide is found in a single tablet, determine the theoretical yield of aluminum chloride that would form if you assume there was an excess amount of hydrochloric acid.
b. If 23.00 g aluminum chloride were measured in the products, calculate the percent yield.

Solution

First, write the complete balanced equation:

$$Al(OH)_3 \, (s) + 3HCl \, (aq) \longrightarrow AlCl_3 \, (aq) + 3H_2O \, (l).$$

a. To answer this question, you must first convert from grams to moles of $Al(OH)_3$, then convert moles of $Al(OH)_3$ to moles of $AlCl_3$, and then convert moles of $AlCl_3$ to grams.

$$\frac{15.00 \text{ g } Al(OH)_3}{} \times \frac{1 \text{ mol } Al(OH)_3}{78.00 \text{ g } Al(OH)_3} \times \frac{1 \text{ mol } AlCl_3}{1 \text{ mol } Al(OH)_3} \times \frac{133.3 \text{ g } AlCl_3}{1 \text{ mol } AlCl_3}$$

$$= 25.64 \text{ g } AlCl_3$$

The theoretical yield of this reaction is then 25.64 g $AlCl_3$.

b. The percent yield is calculated by dividing the actual yield by the theoretical yield and multiplying by 100%.

$$\frac{23.00 \text{ g } AlCl_3}{25.64 \text{ g } AlCl_3} \times 100\% = 89.70\%$$

9.13 Using Stoichiometry to Solve Problems

How can stoichiometry solve the problem of a way to mass-produce sulfuric acid? Despite sulfuric acid's reactivity and the risk that it presents to human health, engineers have found a way to produce it from water, sulfur, and oxygen—all common and readily available substances—using a process of chemical reactions called the *contact process*.

First, elemental sulfur from mines is combusted with oxygen.

$$S\ (s) + O_2\ (g) \longrightarrow SO_2\ (g)$$

In the next step, the sulfur dioxide just produced reacts with oxygen again in the presence of a catalyst called vanadium (V) oxide (V_2O_5).

$$2SO_2\ (g) + O_2\ (g) \xrightarrow{V_2O_5} 2SO_3\ (g)$$

The sulfur trioxide produced in this second step could react with water to form sulfuric acid, but this reaction releases so much heat that it is replaced with two additional steps. The sulfur trioxide reacts with sulfuric acid from a different source to form a substance called oleum.

$$H_2SO_4\ (l) + SO_3\ (g) \longrightarrow H_2S_2O_7\ (l)$$

Oleum reacts with water to form sulfuric acid that is 98–99% pure, just what industries need.

$$H_2S_2O_7\ (l) + H_2O\ (l) \longrightarrow 2H_2SO_4\ (l)$$

Though the contact process is an efficient and economical way to mass-produce sufficiently pure sulfuric acid, it has its problems. It releases toxic gases that must be treated, and this treatment is costly. Perhaps one day, if God calls you to be a chemist, you could develop a less costly way to treat these gases. Or perhaps you could develop an alternative to sulfuric acid that does not produce toxic gases.

9B Section Review

1. The reaction that occurs when iron rusts is shown below.

    $$4Fe\ (s) + 3O_2\ (g) \longrightarrow 2Fe_2O_3\ (s)$$

    a. How many moles of rust can be produced from 3.2 moles of iron metal?
    b. How many moles of oxygen will be consumed in this process?
    c. What mass of iron metal is needed to form 100.0 g of rust?
    d. How many molecules of oxygen will be needed to react with 50.0 g iron metal?

2. In regard to the reaction introduced in Question 1:
    a. If 200.0 g of Fe react fully with O_2, what is the theoretical yield of Fe_2O_3?
    b. If the actual yield is 252.7 g, what is the percent yield?

DS ⊙ 3. If about 47 thousand metric tons of sulfuric acid are industrially produced every year in the United States, how many moles are produced? (1 T = 1000 kg)

DS 4. Write the net reaction for the production of sulfuric acid.

DS ⊙ 5. About 60% of the sulfuric acid produced worldwide is used to produce phosphoric acid for fertilizers. That reaction is

$$Ca_5F(PO_4)_3 + 5H_2SO_4 + 10H_2O \longrightarrow 5CaSO_4 \cdot 2H_2O + HF + 3H_3PO_4.$$

If about 16,000 metric tons of phosphoric acid is produced every year in the United States, how many metric tons of sulfuric acid are required?

DS 6. Can you think of a way to reduce emissions from the contact process or to lower the cost of sulfuric acid production?

DS 7. Since sulfuric acid is toxic and polluting, should Christians oppose its creation and use? Why or why not? In answering this question, consider the benefits of the products in which it is used. Consider also the benefits of not using the chemicals.

Clearing the Air
Facets of Chemistry

Chances are pretty good that you haven't worried about being poisoned by your own carbon dioxide today. Under normal circumstances, the carbon dioxide you exhale diffuses into the atmosphere. Even with every living thing on the planet releasing this gas, plants are able to recycle enough of it during the process of photosynthesis to keep it well within safe breathing levels.

In closed spaces like the space shuttle, however, things are different. Even with enough oxygen, high levels of carbon dioxide can cause an astronaut to develop headaches, muscle twitches, disorientation, or convulsions. These disorders occur because elevated carbon dioxide levels in the blood begin to poison the cells. Prolonged exposure to elevated CO_2 levels ultimately causes death.

In the 1960s, NASA engineers developed a simple technology to filter carbon dioxide from the air. The process relies on the tendency of lithium hydroxide to react with carbon dioxide. This same chemical reaction is still used today with slightly different equipment in space shuttle orbiters.

Lithium hydroxide is a solid compound that reacts readily with carbon dioxide to produce water and the salt lithium carbonate. The reaction proceeds as follows:

$$2LiOH\;(s) + CO_2\;(g) \longrightarrow Li_2CO_3\;(s) + H_2O\;(l)$$

Lithium hydroxide is housed in vented canisters that allow air to pass through them. An adequate supply of these one-time-use canisters is stowed in a special container under the middeck floor of the shuttle. After insertion into the Environmental Control and Life Support Systems,

Astronauts replacing the LiOH canisters beneath the middeck of the Space Shuttle *Atlantis*

the canisters trap the excess carbon dioxide from the air and convert it to lithium carbonate. Five sets of fans circulate the air of the orbiter through these canisters. A crew of seven astronauts produces enough carbon dioxide to require the canisters to be

This jerry-rigged assembly, nicknamed the "mailbox," allowed the LiOH canisters from the Command/Service Module to be used in the Lunar Module.

replaced every eleven hours. After the lithium hydroxide canisters are depleted, they are removed from the filtering system and returned to their original storage containers.

Stoichiometric calculations allow engineers to calculate how many canisters are needed based on the number of crew members and the length and activities of the mission.

You may have heard of the explosion and system failure aboard the Apollo 13 mission in 1970. As a result, the astronauts were forced to evacuate the Command/Service Module and return to Earth in the Lunar Module. Had the three astronauts been unable to retrofit the square lithium hydroxide canister from the crippled Command/Service Module for use in the round cylinder of the Lunar Module's air-handling system, they would not have made it back alive.

The lunar module, which served as their unlikely lifeboat, contained just enough lithium hydroxide canisters to process the carbon dioxide of two people for the two days they were to be on the moon. After the explosion, it had to support all three astronauts for the four-day trip back to Earth. Providentially, the men were able to retrieve the lithium hydroxide canisters from the Command Module. Although many NASA spacecraft components are interchangeable, in this rare case the units were quite different. With the help of some cardboard, plastic, spacesuit hoses, and duct tape, they connected them into the system. This real-life interaction of chemistry and engineering saved their lives. Once again, stoichiometry saved the day.

Chapter Review

Chapter Summary

- Chemists have devised a unit called the mole to simplify calculations related to chemical reactions. A mole is defined as the amount of substance in 6.022×10^{23} (Avogadro's number) particles.

- The molar mass of a substance is the mass found in one mole of its particles.

- The atomic masses found on the periodic table can be equated with the unit grams per mole when calculating the molar mass of a pure substance.

- Structural formulas, molecular formulas, and empirical formulas emphasize different characteristics of molecules. These different formulas are useful in different situations.

- An empirical formula gives the simplest whole-number ratio of atoms in a compound. In some cases, it is identical to the molecular formula.

- Percent composition is used to express the ratio of masses in a compound.

- Stoichiometry is about the mathematical relationships between the amounts of reactants and products in a chemical reaction.

- Stoichiometric calculations rely on mole ratios from balanced chemical reactions to show the relationships of their masses.

- In most real chemical reactions, a limiting reactant is the first to be completely consumed. When there is no more limiting reactant, the reaction stops generating further products. Unused reactants are called excess reactants.

- Balanced chemical equations can be used to calculate theoretical yield through stoichiometry. In reality, the actual yield is often less than the theoretical yield. The ratio of the two is called the percent yield.

Coming to Terms

Term	Page
mole	216
Avogadro's number (N_A)	216
molar mass	216
structural formula	219
molecular formula	219
empirical formula	219
percent composition	220
law of definite composition	221
stoichiometry	227
mole ratio	229
limiting reactant	233
excess reactant	233
theoretical yield	235
actual yield	235
percent yield	235

Review Questions

Concept Review

1. Describe the relationship between Avogadro's number and a mole of a substance.

2. When determining a compound's empirical formula, if your calculated mole amounts are not equal to whole numbers, what must you do?

3. What term describes the calculations used to relate the amounts of reactants and products in a chemical reaction?

4. Explain why excess reactants often remain in a chemical reaction.

5. Give at least two reasons why the actual yield of a chemical reaction is typically less than the calculated theoretical yield.

True or False

6. The molar mass of an element is obtained by taking its atomic mass from the periodic table and changing the unit from atomic mass units to grams per liter.
7. A compound's empirical formula is in some cases the same as its molecular formula.
8. The coefficients in a balanced chemical equation show the mole ratio of the substances in the reaction.

Application

9. Give the molar mass for H, Sc, As, I, and U. Give your answers with four significant digits and correct units.
10. How many atoms are in the following?
    a. 12.01 g of C
    b. 16.00 g of O
    c. 1.008 g of H
11. How many moles
    a. of Fe are in 37.0 g of Fe?
    b. of Kr are in 4.58×10^{20} Kr atoms?
    c. of $NaIO_3$ are in 3.25×10^{26} formula units of $NaIO_3$?
12. Calculate the molar mass of the following compounds:
    a. N_2O_3
    b. Mg_3N_2
    c. $AgNO_3$
13. How many
    a. Fe atoms are in 0.256 mol of Fe?
    b. Kr atoms are in 3.87 g of Kr?
    c. $AlCl_3$ formula units are in 6.17 mol of $AlCl_3$?
    d. $NaIO_3$ formula units are in 8.58 g of $NaIO_3$?
14. What is the mass in grams of
    a. 6.58 mol Fe?
    b. 8.58×10^{28} atoms Kr?
    c. 1.05 mol $AlCl_3$?
    d. 3.17×10^{18} formula units $NaIO_3$?
15. Iron (II) sulfate ($FeSO_4$) is a therapeutic agent for iron deficiency anemia. It is administered orally as iron (II) sulfate heptahydrate. If 0.300 g of $FeSO_4 \cdot 7H_2O$ contains 0.0603 g of Fe, 0.0346 g of S, 0.190 g of O, and 0.0151 g of H, what is the percent composition of $FeSO_4 \cdot 7H_2O$?
16. What mass of vitamin E contains 0.500 g of C, given that vitamin E is 80.8% C by mass?
17. Epsom salts that are used as a laxative consist of $MgSO_4 \cdot 7H_2O$. How many grams of Mg are present in 0.0250 mole of $MgSO_4 \cdot 7H_2O$, which is 9.86% Mg?
18. Limestone, which is foundational to cement, consists of calcium carbonate. If 35.80 kg of $CaCO_3$ contain 4.296 kg C, 14.33 kg Ca, and 17.17 kg O, what is the percent composition of $CaCO_3$?

19. Aspirin is the common name for acetyl salicylic acid. If 100 g of aspirin contains 60.00 g C, 4.480 g H, and 35.53 g O, what is its empirical formula?

20. Lidocaine is a widely used local anesthetic. A laboratory analysis of lidocaine reveals that a 5.000 g sample of lidocaine contains 3.588 g C, 0.473 g H, 0.598 g N, and 0.342 g O.
    a. What is the percent composition of lidocaine?
    b. What is the empirical formula of lidocaine?

21. Chlorine was the first of the halogens to be isolated. C. W. Scheele carried out the following reaction in 1774:

    $4NaCl\ (aq) + 2H_2SO_4\ (aq) + MnO_2\ (s) \longrightarrow 2Na_2SO_4\ (aq) + MnCl_2\ (aq) + 2H_2O\ (l) + Cl_2\ (g)$.

    a. If you start with 1.00 g NaCl and an excess of the other reactants, how many grams of Cl_2 will be produced?
    b. If you start with 1.0 g H_2SO_4 and an excess of the other reactants, how many grams of Cl_2 will be produced?
    c. If you start with 1.0 g MnO_2 and an excess of the other reactants, how many grams of Cl_2 will be produced?
    d. How many grams of NaCl must react to produce 1.00 g Cl_2?
    e. How many grams of H_2SO_4 must react to produce 1.00 g Cl_2?
    f. How many grams of MnO_2 must react to produce 1.00 g Cl_2?
    g. How many grams of Na_2SO_4 will be produced along with 1.00 g Cl_2?
    h. How many grams of $MnCl_2$ will be produced along with 1.00 g Cl_2?
    i. How many grams of H_2O will be produced along with 1.00 g Cl_2?
    j. Show that the law of mass conservation is upheld in the production of 1.00 g Cl_2.

22. Hematite (Fe_2O_3) is converted to molten iron in a blast furnace and is then poured into molds. The balanced equation is

    $Fe_2O_3\ (s) + 3CO\ (g) \longrightarrow 2Fe\ (l) + 3CO_2\ (g)$.

    a. How many moles of Fe_2O_3 must react to produce 262 mol of Fe?
    b. How many moles of CO_2 are produced by the reaction of 64.0 mol of CO?
    c. How many moles of Fe_2O_3 must react to produce 760 kg of Fe?
    d. How many grams of CO_2 will be produced when 40.0 mol of CO react?
    e. What mass (in kg) of Fe is produced when 299 kg of Fe_2O_3 react?

23. Bleach is an aqueous solution of sodium hypochlorite. A bleach solution may be prepared when chlorine gas is bubbled through aqueous sodium hydroxide.

$$2NaOH\ (aq) + Cl_2\ (g) \longrightarrow NaCl\ (aq) + NaOCl\ (aq) + H_2O\ (l)$$

a. How many moles of NaOCl will be produced from 5.73 mol of Cl_2?
b. How many moles of NaOH must react to produce 13.7 mol of NaCl?
c. How many grams of H_2O will be produced if 0.750 mol of NaOH react?
d. How many grams of Cl_2 must react to produce 6.70 mol of NaCl?
e. What mass of NaOCl will be produced when 65.5 g of NaOH react?
f. How many grams of NaOH will react with 37.5 g of Cl_2?
g. What is the theoretical yield of NaCl in grams in this reaction if 100 g of NaOH react?
h. If instead of the theoretical yield from Question (g) you measured 68.45 g of NaCl, what is the percent yield?

GASES

DOMINION SCIENCE PROBLEM

Car Casualties
Chances are you or someone you know has been in a car accident. Though some cause only minor inconveniences, others cause major injuries and even death. In 2005, there were over 6.4 million car accidents in the United States, injuring almost 2.9 million people and killing 42,636 people. This figures out to one death every 13 minutes, and that's just in the United States. Is there some way to minimize injuries and especially deaths from car accidents?

CHAPTER 10

10A	Properties of Gases	244
10B	Gas Laws	249
10C	Gases and the Mole	258
	Facet	
	A Breath of Fresh Air	269

10A Properties of Gases

10.1 Kinetic-Molecular Description of Gases

Have you ever let a helium balloon go, watching it rise into the sky and drift with the wind? It seems like it will never stop rising until it reaches space! At some point, however, the balloon does stop rising. Either the helium no longer has a density low enough to make the balloon continue to rise, or the balloon pops when it is stretched beyond its capacity due to the difference of internal and atmospheric pressure. The kinetic-molecular theory explains the behavior of gases, such as in a helium-filled balloon, on the basis of particle motion. This theory makes several predictions.

1. *Gas particles are tiny compared to the great distances between them.* These particles have unique sizes and masses. The average distance an oxygen molecule must travel between collisions is about 20 000 times its own diameter!

2. *Particles move at random, with high velocities, in all directions, and at many different speeds.* Collisions constantly change the speeds and directions of the particles.

3. *Particles do not interact with each other or the walls of their container except during momentary collisions.* This means that any gravitational, electrical, or chemical forces between the molecules can be ignored. Many particle collisions occur every second. A single air molecule at 0 °C undergoes 5 billion collisions every second!

4. *Collisions between gas particles are elastic. That is, they conserve energy.* Two colliding particles separate with the sum of their energies unchanged. Although their directions and individual velocities change, the net kinetic energy is the same.

5. *The average kinetic energy of the gas particles is directly proportional to the temperature of the gas in kelvins.* As gas temperature rises, the particles move faster and experience more collisions.

10.2 The Physical Properties of Gases

The kinetic-molecular theory is a very workable model for explaining the physical properties of gases we observe.

Diffusion and Effusion

When a gas enters a vacuum or mixes with another gas, it spreads out to uniformly fill the entire volume over time. This process is called **diffusion**. Diffusion occurs because gas molecules are in constant motion. **Effusion** is a related process in which gas particles pass through a tiny opening into an evacuated chamber or space. Both of these processes are directly related to the speed and constant motion of gas molecules.

In 1846, Thomas Graham experimented with various gases. In his studies, he allowed gases at the same temperature to effuse into a vacuum—a space containing no matter. He discovered that the

---

10A Section Objectives
After finishing this section, you should be able to
- list the properties of gases according to the kinetic-molecular theory.
- show how the kinetic-molecular theory explains what we observe about gases.
- demonstrate how and why changes in temperature and pressure affect the volume of a gas.
- convert measurements between different units of pressure.

Low Density of Gases
Compared to solids and liquids, gases have very low mass per unit of volume. This is not surprising since gases have so much empty space between their particles. Most gases have densities about $\frac{1}{1000}$ of the densities of their solid or liquid states.

With few exceptions, gases at room temperature are molecules.

rate of effusion was inversely related to the gases' molar masses. Mathematically, **Graham's law of effusion** states that the rate of effusion for a gas is inversely proportional to the square root of its molar mass.

$$\text{rate of effusion} \propto \frac{1}{\sqrt{\text{molar mass}}}$$

This same law allows us to calculate the ratio between effusion rates of two different gases using the following formula.

$$\frac{\text{rate of effusion for gas 1}}{\text{rate of effusion for gas 2}} = \frac{\sqrt{M_2}}{\sqrt{M_1}}$$

Science and Common Grace
This chapter mentions several important scientists, and one is noted to have been a devout Christian. If Christians are to be involved in science for the benefit of their neighbor, what should their reaction be when unbelievers produce scientific benefits for their fellow humans?

Example Problem 10-1
Calculating an Effusion Rate Ratio

Calculate the ratio of effusion rates between nitrogen (N_2) and argon (Ar).

Solution
First, you must calculate the molar masses of each gas. Diatomic nitrogen is 28.02 g/mol and argon is 39.95 g/mol. Second, you must solve for the ratio of effusion rates using Graham's law of effusion.

$$\frac{\text{rate of effusion for } N_2}{\text{rate of effusion for Ar}} = \frac{\sqrt{M_{Ar}}}{\sqrt{M_{N_2}}} = \sqrt{\frac{39.95 \text{ g/mol}}{28.02 \text{ g/mol}}} = 1.194$$

This result indicates that the lighter nitrogen gas will effuse 1.194 times faster than the argon gas.

Permeability
The ability of a gas to mingle with or pass through another porous substance is **permeability**. This process occurs because the constantly moving gas particles move into spaces between other particles. This permeability is why a sealed balloon with no leaks gradually loses volume over a few days. Gas particles, especially helium particles, can permeate the particles in the balloon to escape into the air.

Fluidity
Gas particles are able to glide past one another like particles in liquids. Both liquids and gases can be poured. Of course, gas particles are much farther apart. Both gases and liquids are considered **fluids** because they have the ability to flow and take the shape of their container.

Compressibility and Expansibility
Though liquids do not easily compress, gases have the ability to change their volume to fit their containers. High pressures can squeeze gases into smaller volumes. This property is called **compressibility**. When gases experience lower pressure, they quickly expand to fill the available space. This property is called **expansibility**.

10-1 Helium gas is compressed into a canister for storage and shipping, but it quickly expands when released to inflate balloons.

Gases can expand without limit, and they always fill their containers. Empty spaces between constantly moving gas molecules make these properties possible.

10.3 How Gases Cause Pressure

Gas molecules collide with each other billions of times each second. They can bang into other objects like trees, buildings, people, and the walls of their containers. Although individual collisions are not very forceful, they add up to produce a significant force. **Pressure** is the average force exerted per unit area when molecules collide against a boundary. It is the observable result of billions of molecular collisions.

Pressure is measured in force per unit area. Normal atmospheric pressure at sea level is 14.7 **pounds per square inch (psi)**. When atmospheric pressure is measured with a **barometer**, another unit of pressure can also be used. Mercury barometers allow air pressure to support a column of mercury. As pressure increases, the column of mercury rises. The pressure and the length of the column can be expressed in **millimeters of mercury (mm Hg)**. For example, normal atmospheric pressure at sea level can support a column of mercury 760 mm high. This pressure is expressed as 760 mm Hg. Another common name for millimeters of mercury is the **torr**, named after Evangelista Torricelli, the inventor of the barometer.

Another measure of pressure is the **atmosphere (atm)**. One atmosphere is simply the normal atmospheric pressure at sea level at 45° latitude. A pressure of 2.0 atm is double the normal pressure at sea level, and 0.5 atm is half the normal pressure. The SI unit of pressure is the **pascal (Pa)**, which is named for Blaise Pascal. A pressure of 1.0 atm is equal to $1.013\,25 \times 10^5$ Pa, or 101.3 kPa.

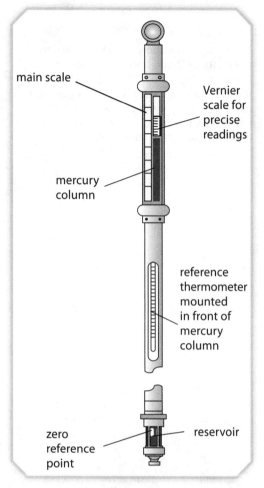

10-2 A mercury barometer. Would the column of mercury in a barometer be shorter or taller on top of Mount Everest? in Death Valley?

A pascal is defined as the pressure of one newton acting on an area of one square meter: 1 Pa = 1 N/m².

Blaise Pascal (pas KAL) (1623–62) was a French mathematician, physicist, and philosopher known for his theory of probability and his work with pressure. He is also famous for his argument for the existence of God, called Pascal's wager.

10-3 An astronaut's spacesuit is pressurized to 0.29 atm. Without pressure, his body fluids would boil so that his tissues would expand, causing death.

$$1 \text{ atm} = 760 \text{ mm Hg} = 760 \text{ torr} = 14.7 \text{ psi} = 101\,325 \text{ Pa} = 101.3 \text{ kPa}$$

These equalities and their reciprocals can be used as conversion factors.

$$\frac{1 \text{ atm}}{760 \text{ torr}} \quad \frac{1 \text{ atm}}{14.7 \text{ psi}} \quad \frac{760 \text{ torr}}{14.7 \text{ psi}} \quad \frac{1 \text{ atm}}{101\,325 \text{ Pa}}$$

Example Problem 10-2

Converting Pressures

Convert 1.20 atm to torr and psi.

Solution

$$\frac{1.20 \text{ atm} \mid 760 \text{ torr}}{1 \text{ atm}} = 912 \text{ torr}$$

$$\frac{1.20 \text{ atm} \mid 14.7 \text{ psi}}{1 \text{ atm}} = 17.6 \text{ psi}$$

The conversions from atmospheres to torr and from atmospheres to pascals are defined and have infinite significant figures. Any calculation using 14.7 psi as a conversion factor will be limited to three significant digits because it is a measured quantity.

10.4 Pressure, Volume, and Temperature

A gas's volume depends not only on how many gas molecules are present but also on its temperature and pressure. A gas at a given temperature contains molecules moving at many different speeds. These molecules exert a pressure on the walls of their container. The average kinetic energy of these molecules determines its temperature. Figure 10-4 shows how gas molecules behave at different temperatures. At 0 °C many hydrogen molecules move at velocities near 1500 m/s. Some move slower; some move faster. The average velocity of 1500 meters per second causes the temperature to be 0 °C. At 500 °C hydrogen molecules move faster, making the average velocity greater. Let's look at three different scenarios to see how a gas's pressure, temperature, and volume are all related.

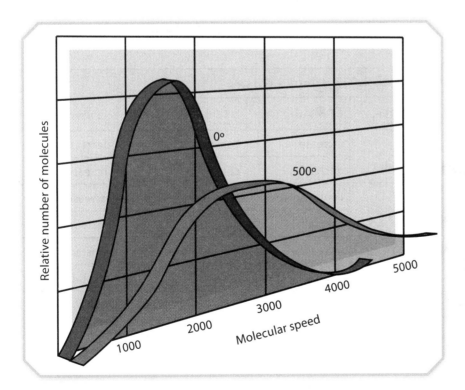

10-4 Notice how the average speed of hydrogen molecules shifts with the temperature increase.

10-5 This device can be used to demonstrate the inverse relationship between volume and pressure.

Case 1: An Inverse Relationship

When temperature is constant and pressure increases, volume decreases. When pressure decreases, volume increases.

Case 2: A Direct Relationship

When pressure is constant and temperature increases, volume also increases. When pressure is constant and temperature decreases, volume also decreases.

Case 3: A Direct Relationship

When volume is constant and temperature increases, pressure also increases. When volume is constant and temperature decreases, pressure also decreases.

Case 1

What happens if pressure increases on a volume of gas? As pressure goes up and temperature is held constant, the volume of a gas decreases. This phenomenon is seen when the plunger of a sealed, air-filled syringe is pushed down. When external forces act upon a gas, they oppose the effect of all the submicroscopic collisions. When external forces exceed the internal pressure of a gas, they squeeze the gas into a smaller space. Volume therefore decreases when external pressure increases, and it increases when the internal pressure increases. Internal pressure increases with an increasing quantity of the gas, such as when you inflate a tire with an air pump. If the external pressure decreases, the volume increases if the gas is in a flexible container.

Case 2

Hot-air balloonists can inflate their empty balloon using the heat from a gas burner. Hot air inflates the balloon because an increase in temperature increases the gas's volume. This example shows that when pressure is constant and temperature increases, volume also increases. On the other hand, if temperature decreases, volume decreases when pressure is constant.

Case 3

At higher temperatures, gas molecules have a higher average kinetic energy. Molecules move faster, collide more often, and strike with more force. The pressure is bound to increase because of the additional, more forceful collisions. If the gas is confined to a fixed volume, pressure will increase as the temperature increases. Thus, it may be necessary to add air to your car tires when the weather becomes much colder.

Table 10-1 summarizes the effects of pressure, temperature, and volume on gases.

10-1 Changes in Gases			
Case	Pressure (P)	Temperature (T)	Volume (V)
1	increase	constant	decrease
1	decrease	constant	increase
2	constant	increase	increase
2	constant	decrease	decrease
3	increase	increase	constant
3	decrease	decrease	constant

10-6 Food cooks more rapidly in a pressure cooker because the fixed volume causes the pressure and temperature to increase together.

10A Section Review

1. Why can you smell a cake baking in the kitchen when you are in your bedroom?
2. How does the kinetic theory explain how you can smell coffee and bacon at the same time?

3. Calculate the ratio of the effusion rates between helium and radon gases.

4. Why is the carbon dioxide gas that carbonates fountain drinks stored in sturdy canisters?

5. How are mm Hg, torr, psi, atm, and Pa related to each other mathematically?

6. Express a pressure of 0.85 atm in torr, psi, and Pa. Remember significant figures.

7. Which of these pressures is the lowest?

   1.1 atm, 16.7 psi, 740 mm Hg, 111 458 Pa

8. What will happen to the volume of a gas in a closed container if you increase the external pressure when the temperature remains constant?

9. Explain why a balloon becomes smaller when placed in a freezer.

10B Gas Laws

10.5 Standard Conditions

So far, we have qualitatively described how gases behave. However, using a series of gas laws based on the qualitative kinetic theory, we can make these descriptions quantitative. These gas laws can be used to calculate volumes, pressures, and temperatures of gases at various conditions.

Because the volume of a gas can change with temperature and pressure, reporting a volume without specifying these conditions is meaningless. Scientists have defined a **standard temperature and pressure (STP)** to be used when measuring gases. Standard temperature is 0 °C or 273 K, and standard pressure is 760 torr, 101 325 Pa, or 1 atm.

10.6 Boyle's Law: Pressures and Volumes

Robert Boyle was the first to measure the effects of pressure on the volume of a gas. He found that pressure was inversely proportional to volume (Case 1 from Table 10-1). He summed up this relationship in a mathematical relationship, called **Boyle's law**: *The volume of a dry gas is inversely related to the pressure if the temperature is held constant.* In equation form, this law is

$$PV = k,$$

where P is pressure, V is volume, and k is a constant.

The product of pressure and volume remains the same for any value of pressure and volume, if the temperature is constant. The equation below is a form of Boyle's law that can be used to solve problems when pressures and volumes of gases change.

$$P_1V_1 = k = P_2V_2$$

10B Section Objectives

After finishing this section, you should be able to

- state Boyle's, Charles's, Gay-Lussac's, and the combined gas laws qualitatively and mathematically.
- apply appropriate gas laws to adjust pressures, volumes, and temperatures to a new set of given conditions.
- describe how the Kelvin temperature scale was formulated from gas behavior.
- state the law of partial pressures.
- use the law of partial pressures to adjust the pressure of a gas collected by water displacement to its partial pressure when given the atmospheric pressure and temperature.

Robert Boyle (1627–91) was an Irish-born English physicist and a devout Christian. He was most noted for his critique of Aristotelian chemistry and alchemy. He also investigated the properties of gases and discovered the importance of air to life and combustion.

Boyle's law: $P_1V_1 = P_2V_2$

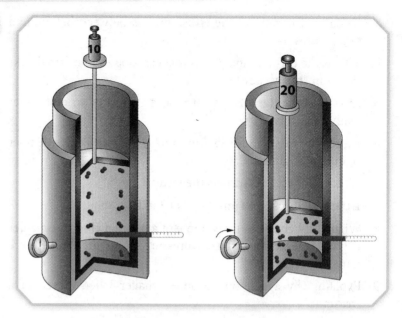

10-7 As pressure increases, volume decreases.

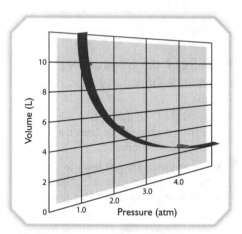

10-8 This graph shows the inverse relationship between pressure and volume in gases.

Problem-Solving Strategy 10-1

Make sure that all pressures, volumes, and temperatures in the gas laws use consistent units. This allows them to cancel out in calculation. For example, if V_1 has different units from V_2, then convert one so units can cancel.

Jacques Charles (1747–1823) was a French physicist who built his own balloon and inflated it using hydrogen. It stayed aloft for forty-five minutes before landing in a field, where it was destroyed by superstitious farmers.

Example Problem 10-3
Using Boyle's Law

A sample of gas occupies 450 mL when it is under a pressure of 1.02 atm. What volume will it occupy if the pressure is increased to 850 torr?

Solution

First, associate all numeric data in the problem with a variable. Notice that P_1 must be converted to torr before substituting it into Boyle's law to solve for V_2.

$V_1 = 450$ mL
$P_1 = 1.02$ atm $= 780.$ torr
$P_2 = 850$ torr

$$P_1V_1 = P_2V_2$$
$$(780. \text{ torr})(450 \text{ mL}) = (850 \text{ torr})(V_2)$$
$$\frac{(780. \cancel{\text{torr}})(450 \text{ mL})}{850 \cancel{\text{torr}}} = V_2$$
$$V_2 = 410 \text{ mL}$$

10.7 Charles's Law: Temperatures and Volumes

Charles's law, formulated by Jacques Charles, describes the relationship between a gas's temperature and volume (Case 2 from Table 10-1). Careful measurements of the volume of a gas at different temperatures and constant pressures could yield the data listed in Table 10-2.

You can see from this table that temperature and volume are directly proportional when pressure is constant. Graphing these values with temperature on the x-axis and volume on the y-axis shows that the volume would approach zero at a temperature of –273 °C. This extrapolation of gas volumes led to the development of the Kelvin

10-2 Relationship Between Temperature and Volume

Temperature (°C)	Volume (mL)
127	1000
−73	500
−173	250

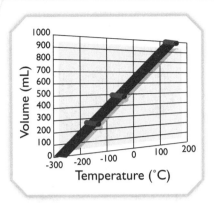

temperature scale. On this scale, the lowest temperature that is theoretically possible (−273 °C) is labeled 0 K. The two scales are related by the formula K = °C + 273. Table 10-3 compares Celsius temperatures with Kelvin temperatures at the same volume.

10-3 Celsius and Kelvin Temperatures Compared

Temperature (°C)	Temperature (K)	Volume (mL)
127	400	1000
−73	200	500
−173	100	250

Notice that the unit for Kelvin temperature is the kelvin, not degrees Kelvin.

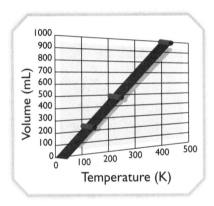

When the Kelvin temperature scale is used, the relationship between temperature and volume can easily be seen. **Charles's law** states that *when the pressure on a sample of a dry gas is held constant,*

Charles's law: $\dfrac{V_1}{T_1} = \dfrac{V_2}{T_2}$

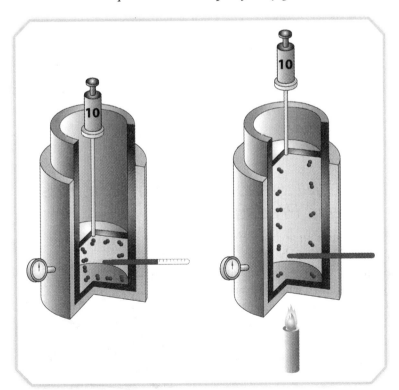

10-9 As temperature increases, volume increases.

the Kelvin temperature and the volume are directly related. Mathematically, this law is stated as follows.

$$\frac{V}{T} = k$$

If the temperature decreases, the volume must also decrease to keep the constant the same. No matter what changes occur, the ratio between volume and temperature in kelvins will be the same for a sample of gas at constant pressure.

$$\frac{V_1}{T_1} = k = \frac{V_2}{T_2}$$

Example Problem 10-4
Using Charles's Law

A sample of gas occupies 430 mL when it is at a temperature of 25 °C. What volume will it occupy when it is at standard temperature?

Solution

First, associate all numeric data in the problem with a variable. From Subsection 10.5, we know that standard temperature is 273 K.

$V_1 = 430$ mL
$T_1 = 25$ °C
$T_2 = 273$ K

Since T_1 is in degrees Celsius, it must be converted to kelvins. Once values have been substituted, the equation can be rearranged to solve for V_2.

$$\frac{V_1}{T_1} = \frac{V_2}{T_2}$$

$$\frac{430 \text{ mL}}{25 \text{ °C} + 273 \text{ K} = 298 \text{ K}} = \frac{V_2}{273 \text{ K}}$$

$$\frac{(430 \text{ mL})(273 \text{ K})}{298 \text{ K}} = V_2$$

$$V_2 = 390 \text{ mL}$$

10-10 A balloon immersed in liquid nitrogen shrinks because the air inside contracts as it cools. As it warms back up to room temperature, it will resume its original volume.

10.8 Gay-Lussac's Law: Temperatures and Pressures

The pressure in a car's tires increases as tires heat up on a lengthy trip. A basketball loses its bounce when left outside on a cold winter day. Both of these changes are examples of **Gay-Lussac's law**: *Pressure is directly proportional to temperature in kelvins for a fixed mass of gas held in a constant volume.* (This law is Case 3

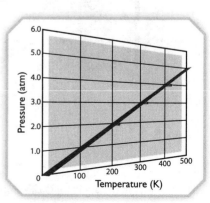

Joseph Louis Gay-Lussac (1778–1850) was a French chemist and physicist. He published the law bearing his name after studying atmospheric gases extensively.

Gay-Lussac's law: $\dfrac{P_1}{T_1} = \dfrac{P_2}{T_2}$

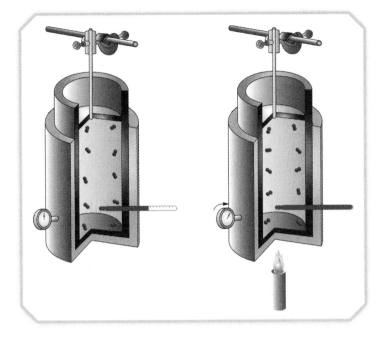

10-11 As temperature increases, pressure increases at a constant volume.

from Table 10-1.) The law is expressed mathematically as follows.

$$\frac{P}{T} = k$$

Friction with the road raises the temperature of the air in the tires. Gas molecules move faster, collide more often, and transfer more force to the inner walls of the tires. Since the volume is fixed, pressure increases. In the case of the basketball, the molecules move more slowly at colder temperatures. The air pressure in the ball decreases accordingly. As long as volume is held constant, Gay-Lussac's law holds true for many different pressures and temperatures.

$$\frac{P_1}{T_1} = k = \frac{P_2}{T_2}$$

10-12 When tires heat up due to friction with the road, the warmer air increases in pressure, even though the volume has not changed.

EXAMPLE PROBLEM 10-5

Using Gay-Lussac's Law

Before Molly began a car trip, she measured the air pressure in her car tires and found that it was 32 psi at 18 °C. After two hours of driving, she found that the pressure had increased to 34 psi. What was the new temperature of the air in her tires?

Solution

In this problem, the pressure is increasing. In order for the pressure to increase, the temperature must also increase. First, associate all numeric data in the problem with a variable.

P_1 = 32 psi
T_1 = 18 °C
P_2 = 34 psi

(continued)

Since T_1 is measured in degrees Celsius, it must be converted to kelvins. Once values have been substituted, the equation can be rearranged to solve for T_2.

$$\frac{P_1}{T_1} = \frac{P_2}{T_2}$$

$$\frac{32 \text{ psi}}{18 \text{ °C} + 273 \text{ K} = 291 \text{ K}} = \frac{34 \text{ psi}}{T_2}$$

$$\frac{(34 \text{ psi})(291 \text{ K})}{32 \text{ psi}} = T_2$$

$$T_2 = 310 \text{ K}$$

10.9 Combined Gas Law

The gas laws we have discussed so far have applied only to situations in which one quantity is held constant. For Boyle's law, temperature is constant. For Charles's law, pressure is constant. For Gay-Lussac's law, volume is constant. But what happens when all three quantities change? These three laws can be combined to form a single equation called the **combined gas law**.

$$\frac{PV}{T} = k$$

$$\frac{P_1 V_1}{T_1} = k = \frac{P_2 V_2}{T_2}$$

The advantage of the combined gas law is that it uses one equation to solve problems in which pressure, volume, and temperature all change.

combined gas law: $\dfrac{P_1 V_1}{T_1} = \dfrac{P_2 V_2}{T_2}$

The three other gas laws can be derived from the combined gas law. For example, if temperature is constant, $T_1 = T_2$. The T variables can be cancelled from the combined gas law to produce Boyle's law. Use this fact to help you remember the gas laws by memorizing only one equation.

EXAMPLE PROBLEM 10-6
Using the Combined Gas Law

A gas has a volume of 3.6 L when it is under a pressure of 1.05 atm and a temperature of −15 °C. What will its volume be at STP?

Solution
$V_1 = 3.6$ L
$P_1 = 1.05$ atm
$T_1 = -15$ °C
$P_2 = 1.00$ atm
$T_2 = 273$ K

As before, T_1 must be converted to Kelvin temperature.

$$\frac{P_1 V_1}{T_1} = \frac{P_2 V_2}{T_2}$$

$$\frac{(1.05 \text{ atm})(3.6 \text{ L})}{(-15 \text{ °C} + 273 \text{ K}) = 258 \text{ K}} = \frac{(1.00 \text{ atm})(V_2)}{273 \text{ K}}$$

$$\frac{(1.05 \text{ atm})(3.6 \text{ L})(273 \text{ K})}{(258 \text{ K})(1.00 \text{ atm})} = V_2$$

$$V_2 = 4.0 \text{ L}$$

10.10 Dalton's Law of Partial Pressures: Mixtures of Gases

It is rare to find a naturally occurring pure gas. Even gases produced in laboratories contain some impurities. These mixtures complicate gas law calculations. The English chemist John Dalton formulated a law to describe how gaseous mixtures behave, called **Dalton's law of partial pressures**. This law states that *the total pressure of a mixture of gases equals the sum of the partial pressures.*

Dalton's law of partial pressures:
$P_{total} = P_1 + P_2 + P_3 + \ldots P_n$

Suppose that 1 L of oxygen at STP is added to 1 L of nitrogen gas at the same temperature and pressure in a 1 L container. When molecules of gases do not chemically react, they behave independently of each other. Thus, oxygen molecules exert a pressure of 760 torr just as they did before they were mixed with the nitrogen gas. The nitrogen gas exerts the same pressure, so the total pressure from the two gases is 1520 torr.

10-13 Atmospheric pressure is the sum of the partial pressures of the gases that make up the atmosphere.

As another example, a sample of dry air contains 78% nitrogen, 21% oxygen, and 1% argon. These percentages are based on mass. To calculate the partial pressures of each gas, stoichiometry is used to convert grams to moles. When we do so, we find that 67.6% of the pressure (514 torr) comes from nitrogen, 31.8% of the pressure (242 torr) comes from oxygen, and 0.6% of the pressure (4 torr) comes from argon. The sum of all the partial pressures equals the total pressure of the mixture.

In the laboratory, chemists often collect a sample of a gas by trapping it at the top of a water-filled container. The gas bubbles up through the water, collects at the top, and forces the water out the bottom. This technique is called collecting over water, or collection by water displacement. The gas is initially pure, but as the gas bubbles through the water, some water evaporates and mixes with the gas being collected. Accurate measurements of the gas cannot be made when it is mixed with water vapor. The extra water molecules exert a pressure called **vapor pressure**. The total pressure, which equals the atmospheric pressure, is made up of the partial pressures of the gas and water vapor. To find the pressure due to the gas, the pressure from the water vapor must be subtracted from the atmospheric pressure. The pressure from the water vapor depends upon its temperature. Table 10-4 lists vapor pressures of water at various temperatures.

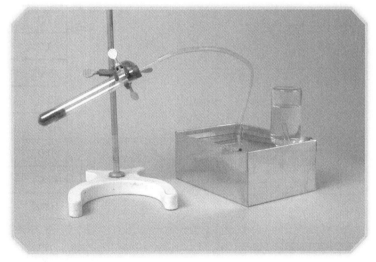

10-14 Gases generated in the lab are often collected by water displacement. These collected gases now contain some water vapor.

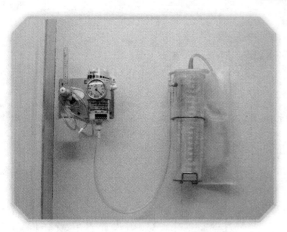

10-15 Ambulance equipment humidifies the oxygen by bubbling it through water. Without this precaution, a patient's mouth, throat, and lungs would become dehydrated.

10-4 Vapor Pressure of Water at Various Temperatures

Temperature (°C)	Vapor pressure (torr)	Vapor pressure (kPa)
0.0	4.585	0.6113
5.0	6.543	0.8723
10.0	9.209	1.228
15.0	12.788	1.705
20.0	17.535	2.338
25.0	23.756	3.167
30.0	31.838	4.245
35.0	42.175	5.623
40.0	55.357	7.381
45.0	71.88	9.583
50.0	92.55	12.34
55.0	118.04	15.74
60.0	149.48	19.93
65.0	187.54	25.00
70.0	233.8	31.17
75.0	289.1	38.54
80.0	355.3	47.37
85.0	433.6	57.81
90.0	525.83	70.11
95.0	633.90	84.51
100.0	760.00	101.3

Example Problem 10-7
Adjusting for Vapor Pressure

A 46 mL sample of O_2 gas is collected over water at 25 °C when the atmospheric pressure is 102 kPa. What volume of pure oxygen is this at STP?

Solution
To use the combined gas law, you must find the volume of pure oxygen. Use Table 10-4 to find the pressure of water vapor at 25 °C.

$$P_{total} = P_{O_2} + P_{H_2O}$$
$$P_{O_2} = P_{total} - P_{H_2O}$$
$$P_{O_2} = 102 \text{ kPa} - 3.167 \text{ kPa}$$
$$P_{O_2} = 98.8 \text{ kPa}$$

Now we have what we need to substitute into the combined gas equation. Notice that Celsius temperatures must be converted.

$V_1 = 46$ mL
$P_1 = 98.8$ kPa
$T_1 = 25$ °C
$P_2 = 101.3$ kPa
$T_2 = 273$ K

$$\frac{P_1 V_1}{T_1} = \frac{P_2 V_2}{T_2}$$

$$\frac{(98.8 \text{ kPa})(46 \text{ mL})}{(25 \text{ °C} + 273 \text{ K}) = 298 \text{ K}} = \frac{(101.3 \text{ kPa})(V_2)}{273 \text{ K}}$$

$$\frac{(98.8 \cancel{\text{kPa}})(46 \text{ mL})(273 \cancel{\text{K}})}{(298 \cancel{\text{K}})(101.3 \cancel{\text{kPa}})} = V_2$$

$$V_2 = 41 \text{ mL}$$

10B Section Review

1. Look at Table 10-2 again. After converting all the temperatures to kelvins, graph these values, with temperature on the *x*-axis and volume on the *y*-axis.
   a. What is the shape of the graph?
   b. What is the slope of the graph? How does this relate to Charles's law, when it is in the slope-intercept form $V = kT$?
   c. If temperature is in kelvins, what is the *y*-intercept of the graph? What significance does this have?

2. What do you think the graph of pressure vs. volume would look like? the graph of temperature vs. pressure?

3. When dealing with gases, why is a standard value for temperature and pressure important?

4. A gas has a volume of 95 mL at a pressure of 930 torr. What volume will the gas occupy if the pressure is increased to 970 torr and the temperature remains constant?

5. A gas has a volume of 111 mL at a temperature of 32 °C. What volume will the gas occupy at standard temperature if pressure is held constant?

6. A particular sample of pantothenic acid, a B vitamin, gives off 72.6 mL of nitrogen gas at 23 °C and 795 torr. What is the volume of the nitrogen at STP?

7. A sample of 40.0 mL of hydrogen is collected by water displacement at a temperature of 20 °C. The barometer reads 751 torr. What is the volume of the hydrogen at STP?

10C Gases and the Mole

When 1 mol of water decomposes into its elements, it produces 16 g of oxygen and 2 g of hydrogen. Yet when the gases are collected separately and measured, there is twice as much hydrogen as oxygen by volume. Why is this so?

10.11 The Law of Combining Volumes: Gases in Reactions

Stoichiometry affects how much of each kind of gas is produced in a chemical reaction. In addition to his studies of pressure and temperature, Gay-Lussac studied the chemical reactions of gases. In particular, he measured and compared the volumes of gases that reacted with each other. One reaction that he studied was between hydrogen and chlorine to form hydrogen chloride gas. He found that if the gases had identical pressures and temperatures, 1 L of hydrogen combined with 1 L of chlorine formed 2 L of hydrogen chloride gas. When he studied the reaction between hydrogen and oxygen to form water, he found different ratios between their combining volumes. When another chemist investigated the reaction between nitrogen and hydrogen to form ammonia, he found yet another set of volume ratios.

$$H_2 + Cl_2 \longrightarrow 2HCl \quad \text{(mole ratio)}$$

$$1\text{ L } H_2 + 1\text{ L } Cl_2 \longrightarrow 2\text{ L } HCl \quad \text{(volume ratio)}$$

$$2H_2 + O_2 \longrightarrow 2H_2O \quad \text{(mole ratio)}$$

$$2\text{ L } H_2 + 1\text{ L } O_2 \longrightarrow 2\text{ L } H_2O \quad \text{(volume ratio)}$$

$$N_2 + 3H_2 \longrightarrow 2NH_3 \quad \text{(mole ratio)}$$

$$1\text{ L } N_2 + 3\text{ L } H_2 \longrightarrow 2\text{ L } NH_3 \quad \text{(volume ratio)}$$

In 1808, Gay-Lussac formulated the **law of combining volumes**: *Under equivalent conditions, the volumes of reacting gases and their*

10C Section Objectives

After finishing this section, you should be able to

- state the law of combining volumes.
- use Avogadro's number to relate the number of particles and the volume of a sample of gas.
- use the molar volume of a gas along with the flow chart to perform stoichiometric conversions.
- determine the molar mass of a gas from its density.
- state the limitations of the kinetic-molecular theory's description of real gases.
- use the ideal gas law to solve for pressure, volume, temperature, or moles of gaseous substance present when given three of the four variables used in the ideal gas law.

10-16 Volcanic eruptions involve chemical reactions as well as changes in the temperature, pressure, and volume of gases.

gaseous products are expressed in ratios of small whole numbers. Although he did not know it at the time, the ratios of these small whole numbers were the ratios of moles of reactants to moles of products. This law, along with the law of definite proportion, eventually led chemists to understand the relationship between atoms, molecules, and compounds.

10.12 The Molar Volume of a Gas

Gay-Lussac's law of combining volumes led Amedeo Avogadro to hypothesize that under equivalent conditions, equal volumes of gases contain the same number of molecules. His experimentation verified this, and further work led to the development of **Avogadro's law**: *The volume of a gas, maintained at a constant temperature and pressure, is directly proportional to the number of moles of the gas.*

Every equivalent volume of gas in a reaction contains the same number of molecules, if they are at the same temperature and pressure. When hydrogen and chlorine react, 1 L of hydrogen contains the same number of molecules as the 1 L of chlorine. One hydrogen molecule reacts with one chlorine molecule to form two hydrogen chloride molecules. Since twice as many hydrogen chloride molecules are produced, 2 L of gas result. In the formation of water from hydrogen and oxygen, the 2 L of hydrogen contain twice as many molecules as the 1 L of oxygen. The 2 L of water vapor contain the same number of molecules as the 2 L of hydrogen and twice the number of molecules as the 1 L of oxygen. Likewise, the volumes of gases involved in the formation of ammonia (NH_3) show the relative number of molecules.

Avogadro's ideas eventually helped establish the mole as a unit of measure. After Avogadro's death, Avogadro's number was determined and named in his honor. Experiments also determined how many molecules were present in a given volume. At STP, a volume of 22.4 liters contains 6.022×10^{23} molecules, or 1 mole, of a gas. For this reason, 22.4 liters is called the **molar volume** of a gas. No matter what the gas, 1 mole at STP occupies approximately 22.4 liters, or in SI units, 0.0224 m^3.

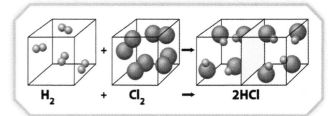

10-17 $H_2 + Cl_2 \longrightarrow 2HCl$

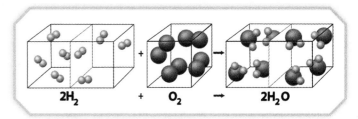

10-18 $2H_2 + O_2 \longrightarrow 2H_2O$

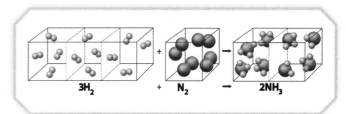

10-19 $3H_2 + N_2 \longrightarrow 2NH_3$

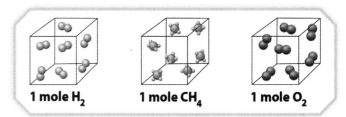

10-20 One mole each of H_2, CH_4, and O_2

1 mole of gas at STP = 22.4 L

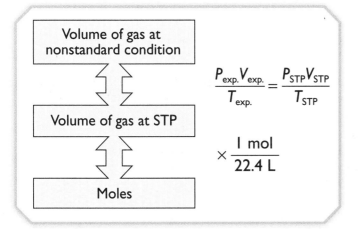

10-21 The conversion of moles to volume

EXAMPLE PROBLEM 10-8
Calculating Gas Volumes at STP
What volume will 4.00 mol of ammonia occupy at STP?

Solution
Since 1 mol of any gas fills 22.4 L at standard conditions, 1 mol NH_3 = 22.4 L NH_3.

$$\frac{4.00 \text{ mol } NH_3}{} \left| \frac{22.4 \text{ L } NH_3}{1 \text{ mol } NH_3} \right. = 89.6 \text{ L } NH_3 \text{ at STP}$$

EXAMPLE PROBLEM 10-9
Calculating Gas Volumes at Nonstandard Conditions
What volume will 2.50 mol of hydrogen gas occupy at 300. K and 400. torr?

Solution

$$\frac{2.50 \text{ mol } H_2}{} \left| \frac{22.4 \text{ L } H_2}{1 \text{ mol } H_2} \right. = 56.0 \text{ L } H_2 \text{ at STP}$$

The combined gas law can now be used to adjust this volume to nonstandard conditions.
V_1 = 56.0 L
P_1 = 760. torr
T_1 = 273 K
P_2 = 400. torr
T_2 = 300. K

$$\frac{P_1 V_1}{T_1} = \frac{P_2 V_2}{T_2}$$

$$\frac{(760. \text{ torr})(56.0 \text{ L})}{273 \text{ K}} = \frac{(400. \text{ torr})(V_2)}{300. \text{ K}}$$

$$\frac{(760. \text{ torr})(56.0 \text{ L})(300. \text{ K})}{(273 \text{ K})(400. \text{ torr})} = V_2$$

$$V_2 = 117 \text{ L}$$

EXAMPLE PROBLEM 10-10
Calculating Moles of a Gas at Nonstandard Conditions
A sample of oxygen gas occupies 1.00 L when its temperature is 190. K and its pressure is 129 kPa. How many moles of oxygen are present?

Solution
The given solution must first be adjusted to standard conditions.
V_1 = 1.00 L
P_1 = 129 kPa
T_1 = 190. K
P_2 = 101.3 kPa
T_2 = 273 K

$$\frac{P_1V_1}{T_1} = \frac{P_2V_2}{T_2}$$

$$\frac{(129 \text{ kPa})(1.00 \text{ L})}{190. \text{ K}} = \frac{(101.3 \text{ kPa})(V_2)}{273 \text{ K}}$$

$$\frac{(129 \cancel{\text{kPa}})(1.00 \text{ L})(273 \cancel{\text{K}})}{(190. \cancel{\text{K}})(101.3 \cancel{\text{kPa}})} = V_2$$

$$V_2 = 1.83 \text{ L at STP}$$

Once the gas's volume at STP has been calculated, the number of moles can be found.

$$\frac{1.83 \cancel{\text{L}}}{} \bigg| \frac{1 \text{ mol O}_2}{22.4 \cancel{\text{L}}} = 0.0817 \text{ mol O}_2$$

10.13 The Densities of Gases

As with all types of matter, the density of a gas is defined as mass per unit volume. Gas density is measured in grams per liter. Some people assume incorrectly that all gases have at least similar densities. Helium (0.1785 g/L) has such a low density it can be used to lift weather balloons. Air at 25 °C and 1 atm has a density of 1.185 g/L. Some gases, such as nitrogen dioxide (1.977 g/L), have densities so high that they immediately sink in air and roll along the ground.

Gases have different densities because their molar masses vary. One mole of any gas occupies 22.4 L at STP. Molar mass determines the density of the gas. Gases with small molecules and low molar masses will have low densities. The density of a gas is a function of the molar mass. Gases with large molecules have greater densities.

$$\text{density at STP} = \frac{\text{g}}{\text{L}} = \frac{\text{g/mol}}{\text{L/mol}} = \frac{\text{molar mass}}{\text{molar volume}}$$

This equation is a powerful tool to calculate the density of gases if the molar mass is known.

10-22 In 1986, a huge amount of heavy carbon dioxide gas was released from deep within Lake Nyos in Cameroon. This cloud of gas flowed down through a river valley, poisoning many cattle and killing more than 1700 people.

10-23 One mole of any gas occupies 22.4 L at STP. Every gas has the same number of particles in one mole; however, the mass of each gas is different.

EXAMPLE PROBLEM 10-11
Calculating Gas Density

What is the density of hydrogen gas?

Solution

Hydrogen gas is a diatomic molecule. Molar mass can be calculated from the periodic table.

$$\text{density of } H_2 = \frac{\text{molar mass}}{\text{molar volume}} = \frac{2.016 \text{ g/mol}}{22.4 \text{ L/mol}} = 0.0900 \text{ g/L}$$

If the density and molar mass of a gas are known, the density equation can be rearranged to determine the molar volume.

$$\text{density} = \frac{\text{molar mass}}{\text{molar volume}}$$

$$\text{molar volume} = \frac{\text{molar mass}}{\text{density}}$$

EXAMPLE PROBLEM 10-12
Calculating Molar Volume of a Gas

If the density of oxygen gas at STP is 1.429 g/L, find the molar volume of the gas.

Solution

Oxygen gas is a diatomic molecule. Molar mass can be calculated from the periodic table.

$$\text{molar volume of } O_2 = \frac{\text{molar mass}}{\text{density}} = \frac{32.00 \text{ g/mol}}{1.429 \text{ g/L}} = 22.39 \text{ L/mol}$$

Analytical chemists can use this equation to find the molar mass of an unknown gas. They can measure its density and then solve for its molar mass.

$$\text{density} = \frac{\text{molar mass}}{\text{molar volume}}$$

$$\text{molar mass} = \text{density} \times \text{molar volume}$$

EXAMPLE PROBLEM 10-13
Calculating Molar Mass of a Gas

One mole of an unknown gas has a density of 2.144 g/L at STP. What is its molar mass?

Solution

$$\begin{aligned}\text{molar mass} &= \text{density} \times \text{molar volume} \\ &= 2.144 \text{ g/L} \times 22.4 \text{ L/mol} \\ &= 48.0 \text{ g/mol}\end{aligned}$$

Further tests could prove that this gas is ozone, which has a molar mass of 48.00 g/mol.

10.14 Stoichiometric Conversions with Gases

Chapter 9 shows how the coefficients in balanced chemical equations give the molar ratios between reactants and products. The number of particles and their masses can be calculated once the number of moles is known. The information that 1 mole of a gas at STP occupies 22.4 liters can be added to the flow chart used for stoichiometric problems.

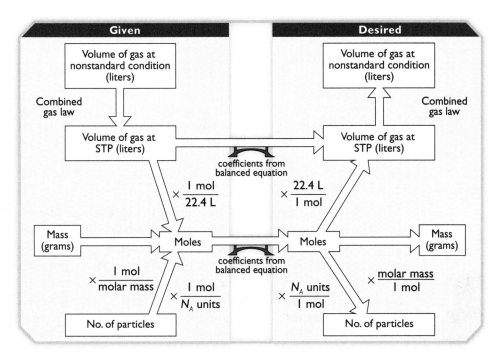

10-24 Flow chart for stoichiometric conversions with gases

EXAMPLE PROBLEM 10-14
Calculating Production Volume of a Gas

When 2.00 mol of calcium react with water to form calcium hydroxide and hydrogen gas, what volume of hydrogen gas will be produced at STP?

Solution
First, write out the reaction and balance it.

$$Ca + 2H_2O \longrightarrow Ca(OH)_2 + H_2$$

Moles of calcium can be used to find moles of hydrogen, which can then be converted to the volume of hydrogen.

$$\frac{2.00 \text{ mol Ca}}{} \times \frac{1 \text{ mol } H_2}{1 \text{ mol Ca}} \times \frac{22.4 \text{ L}}{1 \text{ mol } H_2} = 44.8 \text{ L } H_2 \text{ at STP}$$

Example Problem 10-15

Calculating Mass of Product from the Volume of a Gaseous Reactant

How many grams of water will be produced if 0.500 L of oxygen gas at STP is burned with hydrogen?

Solution
The balanced equation for the reaction is $2H_2 + O_2 \longrightarrow 2H_2O$. The volume of oxygen must be converted to moles of oxygen. The moles of oxygen can be used to find moles of water, which can then be converted to the mass of water.

$$\frac{0.500 \text{ L } O_2}{} \times \frac{1 \text{ mol } O_2}{22.4 \text{ L } O_2} \times \frac{2 \text{ mol } H_2O}{1 \text{ mol } O_2} \times \frac{18.016 \text{ g } H_2O}{1 \text{ mol } H_2O} = 0.804 \text{ g } H_2O$$

These conversions apply only to gases at STP. If gases are not at STP, then their volumes must be adjusted before they are converted to moles.

10.15 Ideal Gases

Did you know that gases do not exactly obey Boyle's law for all pressures and temperatures? For that matter, Charles's law falls short, too. Gas laws do not govern how gases behave; they merely describe. Boyle and Charles provided accurate descriptions of the volume, pressure, and temperature of gases, but their laws are valid only in specific circumstances. As earlier chapters have noted, scientists create models to help them work with the world around them. Through science, man tries to see patterns and to generalize about the universe. The scientist should not expect to be able to explain everything in the world exactly as it is; he ought to recognize his limitations as a human. Scientists do try to continue to refine, improve, and replace existing models by using the skills and abilities that God has given them.

Because the kinetic-molecular theory's description of gases is simple, it cannot exactly describe the behavior of most gases. An **ideal gas** is a gas whose behavior is perfectly predicted by the kinetic-molecular theory. Though no real-world gases are ideal, some come very close at certain temperatures and pressures. The kinetic-molecular theory makes several assumptions about real gases that are not always accurate. Let's reexamine two of the postulates of the kinetic-molecular theory.

1. *Gas particles are tiny compared to the great distances between them.* The kinetic-molecular theory basically assumes that molecules are masses that take up no volume. Since molecules, especially larger ones, take up at least some volume, this postulate of the kinetic-molecular theory does not match the characteristics of real gases. The assumption that gas particles are separated by great distances is valid for gases at high temperatures and low pressures. Great pressures or extremely low temperatures can force gases to approach the point of *liquefaction*. Under these conditions, gas molecules slow down considerably, and the spaces between them shrink. In these conditions, real gases behave quite differently from ideal gases.

Camp Chef

10-25 Propane gas remains in a liquid state while in the tank, but it rapidly vaporizes as pressure is released, allowing the gas to flow through the hose.

2. *Particles do not interact with each other or the walls of their container except during momentary collisions.* Under normal conditions, gas molecules move so fast and are so far apart that intermolecular forces have little chance to act. Yet they always interact to some degree. When a gas is near liquefaction, these forces have an even greater impact on the molecules. Consequently, the gas molecules start sticking together and actual pressures and volumes are smaller than expected.

In general, gases behave most like the kinetic theory predicts when they have low molar mass, are under low pressure, and are at a relatively high temperature.

The molar volume of all gases is 22.4 L, right? Not exactly. The volume 22.4 L is only a useful approximation. When gaseous molecules are large, molar volumes decline slightly because intermolecular forces become significant. Table 10-5 shows the difference in molar volume for some common gases.

10-5 Standard Molar Volumes for Some Common Gases (0 °C)			
Gas	Gas density (g/L)	Molar mass (g/mol)	Standard molar volume (L)
H_2	0.0899	2.016	22.428
He	0.1785	4.003	22.426
Ne	0.9002	20.18	22.425
N_2	1.251	28.02	22.404
O_2	1.429	32.00	22.394
Ar	1.784	39.95	22.393
CO_2	1.977	44.01	22.256
NH_3	0.7710	17.03	22.094
Cl_2	3.214	70.90	22.063

10.16 The Ideal Gas Law

Boyle's law states that volume is inversely proportional to pressure when temperature remains constant. Charles's law states that volume is directly proportional to the Kelvin temperature when pressure is constant. It is also true that the volume is directly proportional to the number of moles of gas (*n*) in the sample.

$$V \propto 1/P$$

$$V \propto T$$

$$V \propto n$$

The ideal gas law is expressed as $PV = nRT$.

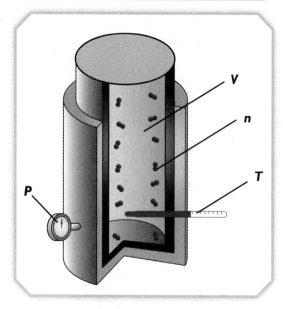

10-26 Pressure, temperature, volume, and number of moles are all considered by the ideal gas law.

The universal gas constant, R, can be expressed as

$$\frac{0.0821 \text{ L} \cdot \text{atm}}{\text{mol} \cdot \text{K}},$$

$$\frac{62.36 \text{ L} \cdot \text{torr}}{\text{mol} \cdot \text{K}}, \text{ or}$$

$$\frac{8.31 \text{ m}^3 \cdot \text{Pa}}{\text{mol} \cdot \text{K}}.$$

Problem-Solving Strategy 10-2
The units for pressure and volume in a problem can direct you to the value of R to use.

When these are combined, they give us the proportion

$$V \propto \frac{nT}{P},$$

or

$$PV \propto nT.$$

In mathematics, you can change a proportionality into an equation by using a constant. With the right numerical constant, this proportion becomes an equation called the **ideal gas law**.

$$PV = nRT$$

R, the **universal gas constant**, relates the units of pressure, volume, temperature, and quantity. Its value and its units depend on the units used for *P*, *V*, *n*, and *T*. The ideal gas law can be rearranged to solve for the value of *R*.

$$R = \frac{PV}{nT}$$

Since 1 mol of gas at 1 atm of pressure and 273 K occupies approximately 22.4 L, these values can be substituted into the equation.

$$R = \frac{(1 \text{ atm})(22.4 \text{ L})}{(1 \text{ mol})(273 \text{ K})}$$

$$R = 0.0821 \text{ L} \cdot \text{atm}/(\text{mol} \cdot \text{K})$$

When pressure is measured in torr instead of atmospheres, *R* has a value of 62.36 L · torr/(mol · K). In SI units, *R*'s value is 8.31 m³ · Pa/(mol · K).

Example Problem 10-16
Calculating Moles of a Gas from the Volume at Nonstandard Conditions

How many moles of a gas are present in a 2.4 L sample at 1.25 atm of pressure and 27 °C?

Solution
Identify the variables and express them in the correct units.
V = 2.4 L
P = 1.25 atm
T = 27 °C
Since the pressure is given in atmospheres and the volume in liters, the value for R that should be used is R = 0.0821 L · atm/(mol · K).

$$PV = nRT$$

$$\frac{PV}{RT} = n$$

$$\frac{(1.25 \text{ atm})(2.4 \text{ L})}{[0.0821 \text{ L} \cdot \text{atm}/(\text{mol} \cdot \text{K})](300. \text{ K})} = n$$

$$n = 0.12 \text{ mol}$$

If chemists measure the mass, pressure, volume, and temperature of an unknown gas, they can find the gas's molar mass. They first use the ideal gas law to find the number of moles present. They then form a ratio of the mass of the sample to the number of moles in the sample (n). This ratio is reduced to produce the molar mass of the gas.

$$\frac{\text{mass of sample}}{\text{number of moles in sample }(n)} = \frac{\text{molar mass}}{1 \text{ mole}}$$

EXAMPLE PROBLEM 10-17

Calculating the Molar Mass of a Gas from its Mass at Nonstandard Conditions

A 5.04×10^{-4} kg sample of gas occupies 4.57×10^{-4} m³ at 9.63×10^4 Pa and 293 K. What is the molar mass of this unknown gas?

Solution

The first step is to use the ideal gas law to determine how many moles are present.

$V = 4.57 \times 10^{-4}$ m³
$P = 9.63 \times 10^4$ Pa
$T = 293$ K

Since pressure is in Pa and volume is in cubic meters, the constant $R = 8.31$ m³·Pa/(mol·K) can be used.

$$PV = nRT$$

$$\frac{PV}{RT} = n$$

$$\frac{(9.63 \times 10^4 \text{ Pa})(4.57 \times 10^{-4} \text{ m}^3)}{\left[8.31 \text{ m}^3 \cdot \text{Pa/(mol} \cdot \text{K)}\right](293 \text{ K})} = n$$

$$n = 0.0181 \text{ mol}$$

Now we can find molar mass. Since molar mass is in units of g/mol, we must convert the given mass to grams.

$$\frac{0.0504 \text{ g}}{0.0181 \text{ mol}} = \frac{27.8 \text{ g}}{\text{mol}}$$

The molar mass of the unknown gas is 27.8 g/mol.

10.17 Using Gases to Solve Problems

Can you figure out some way to use gases, their properties, and the reactions they undergo to minimize injuries and deaths in car accidents? The answer might be right under your nose.

By federal regulation, all cars made since 1998 are equipped with front airbags, and with good reason. Thousands of lives have been saved by airbags. Driver, passenger, and side airbags have played a role in decreasing impact on those who experience car accidents.

How do airbags work? Airbags cushion a body upon impact so that the impact of collision is spread out over a larger area on a person's body. The cushioning must happen fast enough to make a difference. Airbags are designed to activate and inflate faster than the blink of an eye. When a car experiences a collision, a sensor in the car detects rapid deceleration and sends out an electrical impulse.

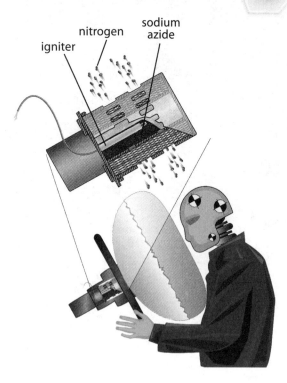

10-27 The rapid production of nitrogen gas inflates airbags to save lives.

The impulse sparks a sample of sodium azide, a chemical also used in solid rockets, causing it to rapidly decompose to form sodium metal and nitrogen gas. In another reaction, the sodium metal reacts with potassium nitrate to form metal oxides and more nitrogen. Nitrogen explodes from these two reactions up to 200 mph (89 m/s) to fill a nylon bag.

$$2NaN_3 \longrightarrow 2Na + 3N_2 (g)$$

$$10Na + 2KNO_3 \longrightarrow K_2O + 5Na_2O + N_2 (g)$$

A following reaction with silicate consumes the hazardous metal oxides. Almost instantaneously, the airbag begins to deflate to provide softer cushioning before impact. Two seconds after the collision, the airbag has completely deflated.

10C Section Review

1. Why is gas collected over water not pure?
2. How are real gases different from ideal gases?
3. When is the behavior of real gases most like ideal ones? When is it most different?
4. If the behavior of real gases cannot be exactly predicted by gas laws, why are they still useful?
5. How many moles of ammonia produce 134.4 L at STP?
6. What is the density of chlorine gas (Cl_2) at STP when the molar mass is 70.91 g/mol?
7. How many mL of propane (C_3H_8) must be burned to form 100.0 mL of CO_2? Use the following equation.

    $$C_3H_8 + 5O_2 \longrightarrow 3CO_2 + 4H_2O$$

8. How many moles of O_2 will occupy 1.00 L at a temperature of –118 °C and a pressure of 49.77 atm? Use the ideal gas law to solve the problem.

DS 9. How much sodium azide is needed to fill a 60.0 L airbag at STP, assuming only the nitrogen from the first equation is needed to fill it?

A Breath of Fresh Air
Facets of Chemistry

This world is filled with gases. The air we breathe is a mixture of many gases, blended in perfect proportion to support life. For example, the two main components, nitrogen and oxygen, compose 99% of the atmosphere. Nitrogen serves mainly to dilute the oxygen to the optimal level for cellular respiration in living things. Because diatomic nitrogen is relatively inert, it does not combine with the oxygen. This is not a fortunate evolutionary accident, but further evidence that God designed a world that works.

The following are some of the primary gases that make up the atmosphere.

Nitrogen (N_2). Nitrogen is a colorless, odorless, tasteless, and relatively inactive gas that makes up 78% (by volume) of the lower atmosphere.

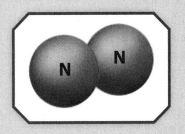

It is captured from the atmosphere by bacteria, which then convert it into a form that is useful for plants. Nitrogen is essential for all living things since it is found in amino acids.

Oxygen (O_2). Oxygen, an active gas in our atmosphere, is colorless, odorless, and tasteless. Oxygen

makes up 21% of the volume of the lower atmosphere. It is slightly denser than air. The sun's ultraviolet radiation can convert diatomic oxygen to *ozone* (O_3). A product of photosynthesis, oxygen is required for the process of cellular respiration, providing energy for all living things.

Argon (Ar). The third-most abundant component of the earth's atmosphere (1%) is a colorless, odorless, tasteless noble gas. Industries make use of argon's chemical inactivity. It is used in light bulbs, due to its inertness, to protect the filaments. When welding aluminum and stainless steel, arc welders use argon to momentarily

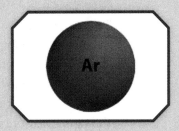

shield the molten metals from the atmosphere. The argon prevents rapid reaction with the oxygen in the air, which would weaken the weld. It is also used in medical laser surgeries.

Helium (He). Although helium, also a noble gas, is the second-most abundant element in the universe, it composes only 0.0005% of the earth's atmosphere. Helium is produced by the decay of radioactive elements in the earth's crust. Natural gas de-

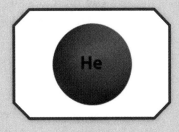

posits serve as the primary source of this odorless, tasteless gas. Helium is chemically inert and has an extremely low density. Consequently, it is useful in providing lift to balloons and dirigibles.

Water vapor (H_2O). Gaseous water molecules are a vital part of the air that we breathe. Without some humidity, throat and lung tissues would dry out. Because temperature is directly related to the amount of water

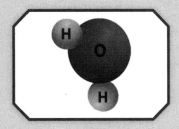

vapor the air can hold, you can see your warm, exhaled breath on a cold day. This happens because the water vapor is condensing back to its liquid state. The role of atmospheric water vapor in maintaining the greenhouse effect to keep the earth warm is yet another evidence of God's design.

Carbon dioxide (CO_2). Carbon dioxide gas is colorless and odorless. It has a slightly sour taste because it combines with saliva to form carbonic acid in the mouth. Plants get the carbon atoms they need for photosynthesis from atmospheric carbon dioxide. Burning fuel, whether in cellular respiration or in a car's engine, produces carbon dioxide.

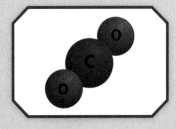

It is one of the gases thought to be responsible for keeping a portion of the earth's thermal energy received from the sun from escaping back into space. One and a half times as heavy as air, pure carbon dioxide gas sinks to a tabletop or floor whenever it is released. It is used to extinguish fires and to make carbonated beverages. Solid carbon dioxide is sold as dry ice.

Coming to Terms

Term	Page
diffusion	244
effusion	244
Graham's law of effusion	245
permeability	245
fluid	245
compressibility	245
expansibility	245
pressure	246
pounds per square inch (psi)	246
barometer	246
millimeters of mercury (mm Hg)	246
torr	246
atmosphere (atm)	246
pascal (Pa)	246
standard temperature and pressure (STP)	249
Boyle's law	249
Charles's law	251
Gay-Lussac's law	252
combined gas law	254
Dalton's law of partial pressures	255
vapor pressure	255
law of combining volumes	258
Avogadro's law	259
molar volume	259
ideal gas	264
ideal gas law	266
universal gas constant (R)	266

Chapter Review

Chapter Summary

- The kinetic-molecular theory describes the nature of gases based on the movement of their particles.
- Key properties of gases include low density, diffusion, effusion, permeability, fluidity, compressibility, and expansibility.
- Pressure results from the collision of molecules against a boundary. It can be expressed using many units.
- Standard temperature and pressure (STP) is defined as 273 K and 1.00 atm and its equivalent in other units of pressure.
- Boyle's law states that when temperature is constant the volume of a dry gas is inversely proportional to the pressure.
- Charles's law states that when pressure is constant the Kelvin temperature of a gas is directly proportional to its volume.
- Gay-Lussac's law states that when volume is constant the pressure of a gas is directly proportional to its Kelvin temperature.
- The combined gas law allows calculations when the pressure, volume, and temperature of a gas all change.
- Dalton's law of partial pressures states that the total pressure of a mixture of gases equals the sum of the partial pressures.
- When gas is collected by water displacement, or "over water," the vapor pressure must be considered in any calculations.
- Gay-Lussac's law of combining volumes states that the volumes of reacting gases and their gaseous products are expressed in ratios of small whole numbers. This principle allows the amounts of gases in moles to be calculated with stoichiometry.
- Avogadro's law states that the volume of a gas, maintained at a constant temperature and pressure, is directly proportional to the number of moles of the gas.
- At STP, 1 mol of any gas occupies approximately 22.4 L.
- The density of gases can be determined by dividing their molar mass by the molar volume.
- The ideal gas law relates the pressure, volume, temperature, and quantity of gases.

Review Questions

Concept Review

1. Use the kinetic theory of gases to explain why
   a. air has a low density.
   b. on a day with no wind, you can smell a dead skunk that is five hundred yards away.
   c. a large volume of air can be pumped into a small bicycle tire.
   d. your egg-salad sandwich can be smelled even through a sealed bag.

2. Use your knowledge of the gas laws to predict what will happen to the
   a. pressure in a can of spray paint when the gas is heated and no gas escapes (Case 3).
   b. volume of a balloon as it is warmed when pressure is constant (Case 2).
   c. volume of an inflated tire tube when it is immersed in cold water and pressure stays the same (Case 2).
   d. pressure in a can of spray paint when some of the paint is released, and the can remains at a constant temperature (Case 1).
3. After pumping air into a bicycle tire, you notice that the nozzle on the pump is quite hot. Why? What gas law applies to this situation?

DS 4. On what properties of gases do airbags rely?

DS 5. Which of the gas laws apply to airbag gases?

DS 6. Most airbags in cars never deploy. Sodium azide used in airbags is actually quite toxic and can react with water to form toxic and explosive hydrazoic acid. How can we prevent these hazardous chemicals from entering the environment, especially from scrapyards?

True or False

7. According to the kinetic-molecular theory, collisions between gas molecules change their directions, but not their speeds.
8. Liquids and gases are both considered fluids.
9. The SI unit of pressure is the atmosphere.
10. Boyle's law explains why tennis balls do not bounce as well when they are cold.
11. The combined gas law relates temperature, pressure, and volume in a single equation.
12. Under equivalent conditions, equal volumes of gases contain the same number of molecules.
13. Ideal gases exist only under carefully controlled laboratory conditions.

Application

⊙ 14. Convert pressure units to fill in the blanks.

Atm	Psi	Torr	kPa
	44.5	2.30×10^3	
500.0			50 650
	32.0		
		14.7	1.960

15. Answer the following questions using Boyle's law:
    a. A gas at 1.00 atm occupies 5.00 L. Under what pressure will the volume be 10.0 L if the temperature remains constant?
    b. A gas occupying 25.0 L has a pressure of 25.0 lb/in.². What volume will it occupy if the pressure changes to 35.0 lb/in.² while the temperature remains constant?
    c. A 3.50 L sample of neon gas has a pressure of 0.950 atm at 20 °C. What would the volume be if the pressure increased to 1.50 atm and the temperature remained constant?
    d. A gas with a pressure of 845 torr occupies 11.0 L. What will the pressure be for a volume of 2.50 L if the temperature remains constant?

16. Answer the following questions using Charles's law:
    a. A sample of nitrogen gas occupies 130. mL at 20 °C. What volume will the gas occupy at 45 °C if the pressure remains constant?
    b. A particular tank of oxygen gas contains 785 L at 21 °C. If the pressure remains constant, what volume will the gas occupy if the temperature is changed to 28 °C?
    c. An experiment calls for 5.50 L of sulfur dioxide (SO_2) at 0 °C and 1.00 atm. What would the volume of this gas be at 30 °C if the pressure remains constant?
    d. At 19 °C, a chemical reaction produces 4.30 mL of oxygen gas. How much oxygen gas would be produced at 45 °C if the pressure remains constant?

17. Answer the following questions using Gay-Lussac's law:
    a. A gas occupying 0.500 L at 1200. torr and 16.5 °C undergoes a temperature change so that the pressure is now 300. torr. If the volume has remained constant, what is the new temperature?
    b. A reaction requires 1.50 mL of ammonia (NH_3) if it occurs at 1.65 atm and 23 °C. If the temperature is changed to 30 °C, what will the new pressure be if the volume remains constant?
    c. A 1.25 mL sample of a colorless, odorless gas had a pressure of 806 torr at 120 °C. If the volume remained constant, what would the gas's pressure be if the temperature dropped to 100 °C?
    d. A 3.20 mL sample of hydrogen sulfide (H_2S) had a pressure of 970 torr at 55 °C. At what temperature would the pressure drop to 900 torr?

18. Answer the following questions using the combined gas law:
    a. If a gas at a pressure of 1.2 atm and at a temperature of 22.0 °C occupies 0.350 L, what pressure will hold the same sample of gas in a volume of 0.050 L if the temperature of the gas increases to 25.0 °C?
    b. A 15.04 L volume of gas at 700 torr of pressure has a temperature of 35.0 °C. What will its temperature be if the pressure increases to 735 torr and the volume increases to 23.8 L?

c. A 40.0 mL sample of hydrogen is collected by water displacement at a temperature of 20 °C. The barometer reads 751 torr. What is the volume of the dry hydrogen at STP?

d. A sample of nitrogen gas at 16 °C and 760 torr has a volume of 2.60 L. What is the volume of the gas at STP?

⊙ 19. Answer the following questions using the ideal gas law:
   a. A 500 mL flask is filled with Kr at STP. How many moles of Kr are present? Calculate the density of Kr at STP.
   b. A weather balloon contains 10.0 L of helium at STP. How many moles of He are present? Calculate the density of He at STP.
   c. A 1.00 L flask is filled with Ar by water displacement at 20.0 °C. The atmospheric pressure is 750.6 torr. After correcting for the water vapor in the flask, determine how many moles of Ar are present. What is the density of Ar at STP?
   d. A tank of oxygen gas contains 89.0 g O_2. If the volume of the tank is 7.5 L, what is the pressure of the O_2 if the temperature is 21 °C?

⊙ 20. A technician produced H_2 gas by reacting zinc (Zn) with sulfuric acid (H_2SO_4). The reaction was 2Zn (s) + H_2SO_4 (aq) ⟶ Zn_2SO_4 (aq) + H_2. He collected the hydrogen in a flask by water displacement. Atmospheric pressure was 737.2 torr, and room temperature was 25.0 °C.
   a. What was the vapor pressure of water in the flask?
   b. What was the pressure of H_2 in the flask?
   c. Suppose that a leak in the gas collection tubes accidentally let a small amount of air into the flask. The unwanted air exerts 25.2 torr of pressure in the flask. How much pressure is the H_2 exerting in the flask?

⊙ 21. An unknown gas has a density of 0.714 g/L at STP. What is its molar mass?

⊙ 22. What is the molar mass of a gas whose density is 5.531 g/L at STP?

⊙ 23. Use the ideal gas law to obtain the answers to the following questions:
   a. How many moles of O_2 will occupy 1.00 L at a temperature of –118 °C and a pressure of 49.77 atm?
   b. What is the temperature of 0.257 mole of O_2 occupying 6.78 L at 0.856 atm?
   c. What is the pressure of 25.6 g of Cl_2 occupying 15.6 L at 28.6 °C?

⊙ 24. When nitroglycerine (227.1 g/mol) explodes, N_2, CO_2, H_2O, and O_2 gases are released initially. Assume that the gases from the explosion cool to standard conditions without reacting further.

$$4C_3H_5N_3O_9 (s) \longrightarrow 6N_2 (g) + 12CO_2 (g) + 10H_2O (g) + O_2 (g)$$

   a. If 16.7 moles of nitroglycerine react, how many liters of N_2 are produced?
   b. What is the total volume of gas (at STP) produced when 1.000 kg of nitroglycerine reacts?

25. A reaction between NH_3 and O_2 is the first step in the preparation of nitric acid (HNO_3) on a commercial scale. The reactants are mixed at STP. The products are produced at 1000 °C (1273 K) and at atmospheric pressure.

$$4NH_3\ (g) + 5O_2\ (g) \xrightarrow{catalyst} 4NO\ (g) + 6H_2O\ (l)$$

a. What volume of NO is produced in the reaction vessel by the reaction of 0.500 mol O_2?
b. What mass of H_2O is produced by the reaction of 15.0 L of NH_3 at STP?
c. How many liters of O_2 must react to produce 33.5 L of NO?

Serving God as an Anesthetist

Job Description
An anesthetist administers appropriate anesthetics according to an anesthesiologist's orders. His purpose is to make sure that a patient does not feel pain during medical and dental procedures. The anesthetist prepares prescribed solutions and administers local, intravenous, spinal, or other anesthetics, according to specified methods and procedures. He keeps the physician informed of the patient's condition during anesthesia.

Possible Workplaces
Anesthetists work at hospitals, outpatient surgery centers, dental offices, and research facilities.

Education Needed
Nurse anesthetists must complete a four-year baccalaureate degree in nursing or a science-related subject. They must also be licensed as registered nurses. The AANA (American Association of Nurse Anesthetists) requires at least one year of full-time nursing experience. However, this field is very competitive, and more experience would result in a better choice of graduate schools or work opportunities. Registered nurses then need to enroll in an accredited program of anesthesia education for an additional two to three years. Most CRNAs (certified registered nurse anesthetists) graduate with a master's degree in either nursing or anesthesia. After completing their formal education, CRNAs must pass a mandatory national certification examination, earn continuing education credits, and meet any requirements of the state in which they are working.

Dominion Opportunities
People going under or coming out of anesthesia are often nervous, if not downright scared! Who would be better than a strong Christian to encourage and exhort other believers to trust the Lord to protect them through the procedure, to pray with those desiring that comfort, to kindly explain what the patient will experience while the anesthesia is being administered, or to demonstrate Christlike concern and care after the procedure? As you model Christian love, even if you aren't able to verbally share the gospel with patients, you would have a great opportunity to "let your light so shine before men, that they may see your good works, and glorify your Father which is in heaven" (Matt. 5:16).

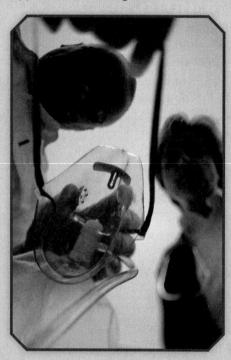

SOLIDS AND LIQUIDS

CHAPTER 11

DOMINION SCIENCE PROBLEM

Lean Screens
How many hours a day do you spend staring at a noisy, eye-straining, bulky cathode-ray tube? Many computer monitors and televisions use heavy cathode-ray tubes to display images. But displays that are portable and easier to look at would make life much better! Is it possible to find a better way to display information?

11A	Intermolecular Forces	276
11B	Solids	279
11C	Liquids	286
	Facet	
	Cryogenics: It's Really Cold in Here!	293

11A Section Objectives

After finishing this section, you should be able to

- compare and contrast dipole-dipole forces, hydrogen bonds, and dispersion forces.
- predict the type(s) of intermolecular force(s) that might occur between a given pair of molecules.
- relate physical properties such as solubility, electrical conductivity, and melting and boiling points to the types of bonding or intermolecular forces that are present.

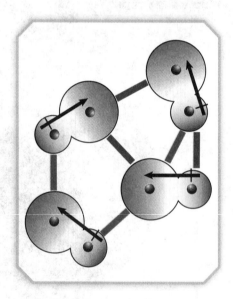

11-1 Dipole-dipole forces are caused by the attraction between the opposite charges on polar molecules.

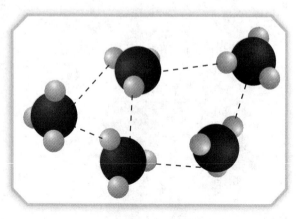

11-2 Hydrogen bonds are special dipole-dipole forces involving hydrogen and a highly electronegative element.

11A Intermolecular Forces

11.1 Defining Intermolecular Forces

Molecules interact with each other—some more than others. The electrostatic attractions between molecules are called **intermolecular forces**. These forces are much weaker than the bonds that hold molecules together. On a scale from 0 to 100, the strength of bonds within a molecule would measure 100 while the strength of forces between molecules would be only 0.001 to 15. Nonetheless, these intermolecular forces influence the physical properties of many substances. They explain why at room temperature hydrogen is a gas, water is a liquid, and sugar is a solid.

Intermolecular forces are classified into three groups, each with a different strength. Each force is the result of attraction between regions of charge on some molecules and regions of opposite charge on other molecules.

11.2 Dipole-Dipole Forces

Consider polar molecules, which have regions of unevenly distributed electrical charge. The positive areas of one molecule attract the negative areas of other molecules. The stronger the polarity is, the stronger the attraction is between the two molecules. These polar molecules align themselves so that the positive end of one molecule is near the negative end of another molecule. This attraction is called a **dipole-dipole force**. The dipole-dipole forces of polar molecules are similar to the electrostatic attraction of ions in crystals of ionic compounds, but they are much weaker. They are strong enough, however, to have a significant influence on the physical properties of many compounds.

11.3 Hydrogen Bonds

Hydrogen bonds are a special type of dipole-dipole interaction in which one of the participating atoms is hydrogen and the other element is either fluorine, oxygen, or nitrogen. Whenever hydrogen is bonded to one of these very electronegative elements, the shared electrons shift away from the hydrogen atom because of the greater electronegativity of the other element. The hydrogen nucleus, essentially a proton, is left exposed. Any negatively charged regions of other molecules containing nitrogen, oxygen, or fluorine interact with the exposed proton. Since hydrogen is so small, interacting molecules can get extremely close to it. The combination of high polarity and proximity produces the strongest of the intermolecular forces.

Hydrogen Bonds in Water

Hydrogen bonds form between the hydrogen atom of one water molecule and the oxygen atom of another water molecule. These bonds, like other intermolecular forces, are not as strong as the covalent bonds in water molecules. While the energy of covalent bonds in water is about 286 000 J/mol, the energy of hydrogen bonds in water is only about 20 000 J/mol. The average hydrogen-bond length between water molecules is 0.177 nm, compared to 0.099 nm for the O-H covalent bond.

Despite the apparent weakness of these bonds, they have a great effect on the physical properties of water. Because there is a significant difference in the electronegativities of hydrogen and oxygen, water molecules are quite polar. The hydrogen bonds cause water to be a liquid at room temperature. Table 11-1 compares water to other similar hydrogen-containing, covalent compounds. Although water has an atomic mass that is much lower than the other compounds, hydrogen bonds raise its boiling and melting points.

11-1	Hydrogen Compounds Compared		
Compound	Molecular mass	Melting point (°C)	Boiling point (°C)
H_2Te	129.6	−49.0	−2.0
H_2Se	80.98	−65.7	−41.3
H_2S	34.07	−85.5	−59.6
H_2O	18.02	0.0	100.0

> DNA, the "messenger of life," is composed of two strands of atoms held together by hydrogen bonds in a helical arrangement. Keratin, the major protein in hair, folds into a helix when a hydrogen atom bonded to nitrogen in one part of the molecule forms a hydrogen bond with an oxygen atom in a different part of the same molecule.

> The stronger a bond is, the shorter the distance is between the bonded atoms.

Water is essential for life. Is it merely coincidental that atmospheric conditions on this planet allow water to exist in the liquid state? Hardly. The Lord providentially engineered the sun's thermal output, the dimensions of the solar system, and the earth's cloud cover to allow water to be a liquid. This water vapor contributes to the greenhouse effect that moderates our temperatures and promotes life. Other planets and moons do not have appreciable amounts of water. Those that do have water have surface temperatures and pressures that will not allow it to exist as a liquid. God says this creation testifies to the fact that He is the Creator God and that He is to be worshiped as the sovereign of the world. "For thus saith the Lord that created the heavens; God himself that formed the earth and made it; he hath established it, he created it not [empty], he formed it to be inhabited: I am the Lord; and there is none else" (Isa. 45:18).

11.4 Dispersion (London) Forces

Even nonpolar substances such as wax and gasoline can exist as solids and liquids at room temperature. Why is this? Dipole-dipole interactions and hydrogen bonds are not the only intermolecular forces. There is a third type of intermolecular force called the **dispersion force**. Nonpolar molecules do not have permanent regions of electrical charge, yet they still attract each other. On average, electrons in a nonpolar bond spend an equal amount of time around each atom. Yet at any point, these electrons can momentarily concentrate at one end of the molecule. These concentrations occur when the negative electrons of one molecule repel the electrons of a nearby molecule. The concentrations form temporary dipoles. Although the imbalance of charges occurs for just an instant, this is long enough to attract an oppositely charged region of a neighboring molecule. Thus, dispersion forces result from the random, unequal dispersion of electrons around the atoms in a molecule. Dispersion forces occur between all molecules, but they are the only intermolecular force acting on nonpolar molecules.

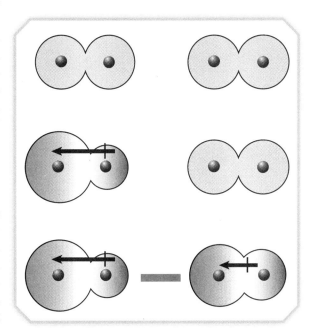

11-3 Dispersion forces act on both polar and nonpolar molecules.

> **Dispersion forces** are sometimes called London forces after Fritz London, the German-American physicist who first described them.

Larger molecules have more electrons and are more likely to participate in dispersion forces at any given time. This tendency affects their physical properties. For example, consider the hydrocarbons. Methane, which has one carbon, is a gas at room temperature. Octane, with a chain of eight carbons, is a free-flowing liquid at room temperature. Octadecane, an eighteen-carbon molecule, is a solid at room temperature. All of these nonpolar compounds exhibit dispersion forces. Their differences in state depend on the number of carbon atoms in each molecule.

Hydrocarbons are nonpolar molecules consisting of only carbon and hydrogen atoms.

Intermolecular Forces Summary
- Dipole-dipole forces exist only between polar molecules and are stronger than dispersion forces.
- Hydrogen bonds are dipole-dipole forces that occur only between polar molecules that have the necessary atoms: H along with N, O, or F.
- Dispersion forces work between all kinds of molecules, polar and nonpolar, but they are the primary forces affecting nonpolar molecules.

Example Problem 11-1
Intermolecular Forces

List the types of intermolecular forces that affect the following compounds:

a. $:\!O=C=O\!:$

b. $H:\!\ddot{F}\!:$

c. $:\!\ddot{B}r:\!\ddot{C}l\!:$

Solution
a. CO_2 is nonpolar, so only dispersion forces act.
b. HF is polar, and it has the necessary elements for hydrogen bonds to form. Thus hydrogen bonds, dipole-dipole forces, and dispersion forces are present.
c. BrCl molecules are polar, so dipole-dipole and dispersion forces are present.

Example Problem 11-2
Bond Comparison

Based on the bonds within each molecule and the forces between each molecule, predict which compound has a higher boiling point.
a. KF or BrF
b. Cl_2 or ICl

Solution
a. KF is an ionic compound that has extremely strong ionic bonds between its units. Its boiling point (1505 °C) is higher than that of the covalently bonded BrF (20 °C), which has only weaker dipole-dipole and dispersion forces between its units.
b. Both Cl_2 and ICl are covalently bonded. ICl molecules are slightly polar, while Cl_2 molecules are not. Since ICl molecules have dipole-dipole interactions as well as dispersion forces, its boiling point (97.4 °C) is higher than that of Cl_2 (−34.6 °C).

Just as ionic, covalent, and metallic bonds greatly affect the physical properties of the compounds that they form, intermolecular forces also affect these physical properties. Table 11-2 summarizes intermolecular forces, the compounds they affect, and their influence on the physical properties of compounds.

11-2 Bond Forces Comparison

Structural units	Forces between units	Properties
ions	ionic bonds	very high melting/boiling points; usually soluble in polar solvents; electrical conductors when molten or dissolved
cations, mobile electrons (metals)	metallic bonds	high melting/boiling points; insoluble; electrical conductors in all states
polar molecules	dispersion, dipole-dipole, possibly hydrogen bonds	melting/boiling points slightly higher than those of nonpolar molecules; usually soluble in polar solvents; nonconductors of electricity
nonpolar molecules	dispersion forces	low melting/boiling points; molecules soluble only in nonpolar solvents; nonconductors of heat and electricity

11A Section Review

1. How are the three types of intermolecular forces similar? How are they different?
2. What type of intermolecular forces would you expect to find between molecules of H_2O? of BrCl? of hexane (C_6H_{14})?
3. (True or False) Wax is nonpolar, yet it exists as a solid at room temperature. Its solidity is best explained by the presence of dispersion forces.

11B Solids

11.5 A Kinetic Description of Solids

The strength of intermolecular forces in a solid affects the movements of solid particles. According to the kinetic-molecular theory of matter, particles in solids are always moving and vibrating. At low temperatures, these particles barely vibrate. As temperatures approach a solid's melting point, the particles have more kinetic energy and the vibrations get faster and bigger. Strong intermolecular forces stifle these vibrations, while weak forces allow the particles to move more freely. For example, table salt exists as a solid up to the scorching temperature of 801 °C because attractions between the Na⁺ ions and the Cl⁻ ions are extremely strong. Nonpolar diatomic bromine molecules, with only weak dispersion forces, can move more freely. Consequently, bromine melts at −7 °C, allowing it to exist as a liquid at room temperature.

Intermolecular forces affect other properties besides a substance's boiling and melting points. Because solid particles have little motion, they are not fluids, have low rates of diffusion, and

11B Section Objectives
After finishing this section, you should be able to
- list and explain the properties of solids using the kinetic-molecular theory.
- relate the structure of the water molecule to its unusual density change when it becomes a solid.
- compare and contrast crystalline and amorphous solids.
- explain the significance of the slopes and plateaus on a warming curve.
- describe what happens on the particle level during melting, freezing, sublimation, and deposition.
- list three factors that influence the formation of a crystal lattice.
- list factors that affect lattice energies.

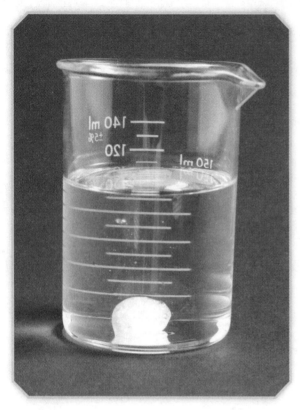

11-4 Solid (frozen) benzene sinks in liquid benzene because it is denser as a solid than as a liquid.

> **Divine Design and Water's Freezing Point**
>
> Because ice is less dense than water, ponds freeze on the surface and not from the bottom up. How does this phenomenon speak to us of God's design and benevolence to His creatures?

are not permeable. If a silver coin and a copper coin were clamped together for several years, a few atoms would diffuse between the metals, but the rate would be imperceptible compared to diffusion in a liquid.

11.6 Density of Solids

Because particles in solids are held close together, they usually have high densities. Since solids are so dense and have fixed shapes and definite volumes, they resist compression. Atoms, molecules, or ions would have to be deformed for solid matter to compress significantly. A substance is typically about 10% denser as a solid than as a liquid. Consequently, when liquid and solid forms of the same substance are together, the solid form almost always sinks to the bottom.

One notable exception to this trend is water. Down to a temperature of 4 °C, water becomes increasingly dense, but as it cools below this temperature, hydrogen bonds and dipole-dipole forces cause water molecules to spread apart from each other, assembling into an open hexagonal lattice. The increased space between the molecules causes ice to be less dense than water. This lower density is why ice cubes float in a glass of water.

11.7 Crystalline and Amorphous Solids

Solids can occur in many shapes and forms. Some solids have naturally orderly shapes, forming regular three-dimensional patterns with distinct edges and sharp angles. When they are shattered, smaller shapes form with similar edges and angles. These solids are called **crystalline solids**. Ionic and metallic solids are usually crystalline. Other solids have no distinct shape or underlying pattern. When they split or shatter, irregular fragments result. Such solids are called **amorphous solids**. Covalent compounds are sometimes amorphous.

The differences between crystalline and amorphous solids result from the particular structures of the solids. The particles in crystalline solids—such as salt, sugar, and monoclinic sulfur—are arranged in well-defined, orderly, three-dimensional patterns.

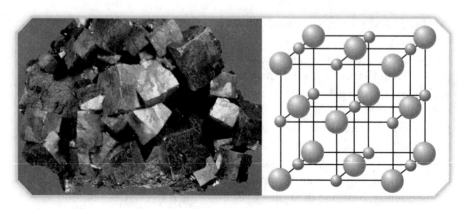

11-5 The internal structure of galena (PbS) is reflected in the external shape of its crystals.

Atoms, ions, or molecules are stacked row upon row, column by column. The patterns, when repeated many times, result in crystals with regular shapes.

Particles in amorphous solids such as rubber, some plastics, asphalt, paraffin, and amorphous sulfur are not arranged in any particular pattern. Their random, disordered structure results in globular microscopic shapes. Some amorphous solids are called *supercooled liquids* or glasses. If a liquid can be cooled fast enough, its particles may not have time to form a preferred crystalline pattern. These molecules are locked in random positions.

11.8 Melting and Freezing

Melting and freezing are changes between the solid and liquid states of a substance. The temperature at which this process takes place is called the **melting point**. Melting is the transition from a solid to a liquid; freezing is the reverse. For pure substances, the freezing point is the same as the melting point. Freezing points do not always occur at cold temperatures. For example, iron freezes at 1538 °C! When a lump of lead is placed in a ladle and heated, the atoms begin to vibrate more vigorously. Temperature, a measure of the vibrations of lead particles, rises. When the melting point of the lead is reached, the temperature stabilizes. At this point, all atoms are at the brink of liquefaction. Additional heat overcomes the attractive forces and melts the solid. Not until all the lead is melted will the temperature resume its upward climb.

11-6 The random arrangement of particles in glass, which is a supercooled liquid, causes glass fragments to have irregular shapes.

11-7 The warming curve for lead shows how temperature changes as heat is added.

The melting points of crystalline and amorphous solids are affected by the arrangements of their particles. Crystalline substances such as ice or lead have distinct melting points. Warming-curve graphs for crystalline substances show clear plateaus that correspond to sharp melting points. An entire sample melts at a clearly defined temperature because all particles are held by nearly identical forces. Amorphous solids, on the other hand, do not have sharp melting points. Their particles are in random positions at different distances from each other. Since the forces vary with distance, not all particles are held together with identical forces. As the temperature rises, only some intermolecular forces are overcome at each specific temperature. Amorphous solids gradually soften as more intermolecular forces are broken. How would a warming curve of an amorphous solid differ from the warming curve for lead? (See Figure 11-7.)

Analytical chemists use melting points to determine the identity and purity of compounds. Many pure substances are crystalline,

11-8 An optical lens in this melting point apparatus allows chemists to observe the point at which crystals begin to melt.

11-9 Iodine crystals sublime to form a purple vapor that deposits as crystals again on the cover of the beaker when chilled by ice.

with their melting points listed in reference tables. Chemists can analyze the purity of a compound by observing its melting-point range. A narrow range indicates a pure sample, while a wide range indicates that impurities are present.

11.9 Sublimation and Deposition

Many caterers use dry ice (frozen carbon dioxide) to refrigerate their foods. When dry ice warms, it changes directly into a gas instead of melting into a liquid as water ice does. This state change bypasses the liquid phase and alleviates the problems of messy puddles. **Sublimation** is the direct change in state from the solid to the gaseous state. Sublimation occurs because individual molecules leave the surface of a solid and become gaseous. For example, the smell of naphthalene or para-dichlorobenzene mothballs comes from the individual molecules that have sublimed. Sublimation causes snow to slowly disappear, even when the temperature remains below freezing. Substances sublime readily if they have many molecules that can easily leave their surfaces.

11-10 The hoarfrost on these plants formed from the deposition of water vapor.

Deposition occurs when a substance changes directly from a gas to a solid—the reverse of sublimation. When frost forms on your windshield or lawn, deposition has taken place. Deposition of water vapor high above the earth forms snowflakes.

11.10 Crystalline Structures

The particles in crystals are arranged in orderly, repeating patterns. These patterns vary, as shown by the wide variety of beautiful shapes found in natural crystals. The beauty and predictability of naturally occurring crystals is yet another of God's fingerprints on our world. Several factors influence how the particles are arranged. Like electrical charges repel to maximize the distance between particles. At the same time, opposite charges seek to minimize the distance between particles. These forces interact to produce the most stable crystal arrangement possible.

A crystal's three-dimensional pattern, or **crystal lattice**, depends on the number and kinds of particles involved, their relative sizes, and the charges of the ions. Scientists have found and classified seven basic classes of crystals. They are cubic, tetragonal, rhombohedral, triclinic, monoclinic, hexagonal, and orthorhombic. (See Figure 11-11.) Scientists mentally divide the natural structures of crystals into building blocks called **unit cells** that contain the fundamental patterns of the lattices. Unit cells usually consist of a specific number of ions. These portions of crystals can be compared to the basic pattern

> A **unit cell** of a crystal lattice is the section of a crystal lattice that contains one formula unit of the compound or one repeating segment of the overall lattice structure.

in a piece of wallpaper. The wallpaper pattern can be repeated indefinitely to cover large walls. Unit cells, when repeated many times in three dimensions, make up crystals. (See Figure 11-12.)

Some of the seven basic classes of crystals can be slightly modified by adding particles to the faces or interiors. A body-centered crystal not only has particles at each of the corners, but also has one in the center (body) of the crystal. A face-centered crystal has particles on the corners and on the plane made by one or more of the sides, but no particle in the center of the crystal. Some, but not all, of the seven basic crystal types have one or both of these modifications. These variations expand the seven basic classes into fourteen lattices.

Because crystals are three-dimensional arrangements of unit cells, they often assume the same shape as their unit cells. Salt crystals, for example, are tiny cubes. When they are formed their units build upon themselves to form larger cubes with approximately the same number of units along each edge. A sodium chloride unit cell, which contains four Na^+ ions and four Cl^- ions, has a length of 0.56 nm (1 nm = 10^{-9} m). A salt crystal 0.5 mm along each edge has nearly one million (10^6) unit cells along each edge and approximately 10^{18} unit cells in its total structure.

However, many crystals found in nature do not have the same shape as their constituent unit cells. Because of varying temperature, pressure, and other environmental factors, unit cells may stack in such a way that a different external structure forms. The "steps" on the surfaces of the crystals are only a fraction of a nanometer

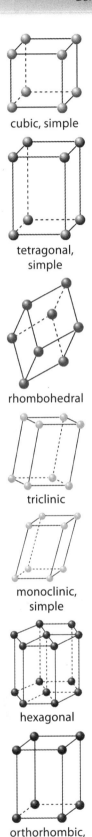

11-11 The seven basic classes of crystals

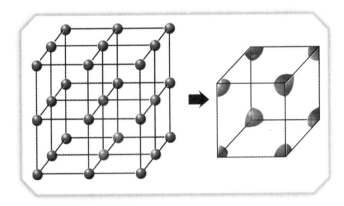

11-12 Unit cells are the basic building blocks for any type of crystal.

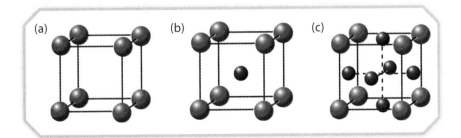

11-13 Simple cubic cells have particles only on the corners (a). A body-centered cubic cell has an extra particle in the center (b). A face-centered cubic cell has a particle at the center of each face but not in the center of the cell itself (c).

11-14 Fluorite has cubic crystals (left). Aragonite's crystals are orthorhombic (center). Rhombohedral crystals are seen in calcite (right).

wide, so they are not visible to the eye. The crystal faces appear to be smooth even though they are ragged on the atomic scale. Despite the fact that all the known minerals have one of the basic crystal structures, mineral crystals can exhibit a great variety of external forms.

11.11 Polymorphs and Allotropes

Some substances can form more than one type of crystal lattice. Such elements and compounds are said to be **polymorphous**. When Ca^{2+} and CO_3^{2-} ions crystallize at a low temperature, they fall into a rhombohedral lattice. Mineralogists call this form of calcium carbonate calcite. When these same ions crystallize at a high temperature, they orient themselves in an orthorhombic lattice and form a substance called aragonite. (See Figure 11-14.)

Pure elements that are polymorphous are called **allotropic elements**. The different forms of allotropic elements are called **allotropes**. Sulfur, phosphorus, and arsenic are a few allotropic elements. When sulfur is in the solid state, it exists as ring-shaped S_8 molecules. These rings can be arranged in either rhombic or monoclinic lattices. Rapid cooling of red liquid sulfur at 300 °C produces yet another allotrope, yellow amorphous sulfur. This globular form hardens or crystallizes after a couple of days.

11.12 Binding Forces in Crystals

Crystals are shaped by electrical forces. Oppositely charged regions attract, and similarly charged regions repel. The balance between attractions and repulsions determines how tightly the particles in the crystal are bound. If attractions just barely overcome the repulsions, the crystal will have weak bonds. If attractions are much greater than repulsions, the crystal will have strong binding forces.

The **lattice energy** of a crystal is the energy released when gaseous particles form crystals. It is equal in magnitude but opposite in sign to the energy necessary to pull a crystal apart. The lattice energy shows the difference in energy between gas particles and those trapped in a crystal.

Binding forces must be overcome whenever a crystal is melted or dissolved. When a crystal melts, thermal energy is used to overcome binding forces. When a crystal dissolves, attractions between the particles of the crystal and the molecules of the solvent are able to break apart the crystal structure. For example, when table salt

---

polymorphous (POL ee MORE fus): poly- (Gk. *polu*—much, many) + -morph- (Gk. *morphe*—form, shape) + -ous

allotrope (AL uh TROHP): allo- (Gk. *allos*—other) + -trope (Gk. *tropos*—turn)

11-15 Sulfur can exist in many different crystalline forms.

Lattice energy is expressed as a negative number because heat is released during crystallization, leaving less energy in the products than in the reactants.

dissolves in water, its crystals absorb thermal energy from the water in order to overcome the lattice energy and break apart the ions. If a large amount of salt is added to a glass of water, you can feel the glass cool as the salt dissolves, using energy from its environment to break the crystal lattice.

What makes a crystal strong? In other words, what determines the lattice energy? The magnitude of the electrical charges in a crystal affects its stability. For example, sodium chloride, with a lattice energy of −787 kJ/mol, involves a sodium ion with a charge of 1^+ and a chloride ion of 1^-. But magnesium oxide, with a lattice energy of −3791 kJ/mol, involves a magnesium ion with a charge of 2^+ and an oxide ion of 2^-. Highly charged particles, like those in ionic compounds, interact strongly to produce strong crystals. The size of the particles also affects the binding forces. Small particles can be more tightly bound than large particles. The geometric structure of the crystal also affects the binding forces because both the structure and the size of the particles involved affect the distance between ions.

> The heat necessary to dissolve one substance in another is called the *heat* or *enthalpy of solution*. We will discuss this further in Chapter 12.

11.13 Types of Crystalline Solids

Crystalline substances can also be classified by the types of particles they contain and how they are held together. This method of classification relies on an understanding of the types of bonds we discussed in Chapter 6.

Atomic Crystals

These solids form only when noble gases (Group 18 elements) freeze. The particles in these solids are individual atoms since noble gases do not typically form molecules. As you might expect, they are soft and are poor conductors of electricity.

Covalent Molecular Crystals

Molecular solids are made of covalently bonded atoms held together by any combination of dipole-dipole forces, hydrogen bonds, and dispersion forces. The intermolecular forces that attract the molecules to each other are weaker than the forces binding atoms together within each molecule. Thus, they have relatively low melting points. Their low melting points explain why molecular compounds are often gases or liquids at room temperature.

Covalent Network Crystals

Some nonmetallic elements such as carbon and silicon are able to bond covalently to form large networks of atoms. The particles in these solids are individual atoms bound together by covalent bonds. These crystals function more like giant molecules. The strength in these three-dimensional arrangements contributes to the generally high melting points of covalent network solids.

Ionic Crystals

Ionic solids are made of a repeating network of ions defined by a unit cell. The ions may be monatomic or polyatomic. Remember from our earlier study that ionic compounds typically form between a metal and a nonmetal. The strong bonds between the alternating positive and negative ions in these crystals cause them to be strong and have high melting points. Many also fracture easily.

11-16 Diamonds are covalent network solids made of carbon atoms bonded in a network that gives them extreme hardness.

Metallic Crystals

Positive metal ions are surrounded by a sea of valence electrons in metallic solids. Because metals vary greatly in the strength of the metallic bonds that form between the metal ions and mobile electrons, metals have a wide range of physical properties. For example, mercury is a liquid at room temperature, while many others remain solid past 1000 °C. Most metals are lustrous and conduct electricity well because of their free electrons.

11B Section Review

1. Explain the difference between crystalline and amorphous solids.
2. Draw warming curves that would represent a crystalline solid and an amorphous solid. Is there a difference between the two curves? If so, why?
3. How is a unit cell like a formula unit?
4. Explain why the lattice energy and the energy of crystallization of a substance differ only in sign but are equal in magnitude.
5. How are covalent network solids similar to ionic solids?
6. Explain why covalent network crystals have relatively high melting points while covalent molecular crystals, as a rule, melt at much lower temperatures.

11-3 Crystalline Solids

Type	Unit particles	Examples
atomic	atoms	Group 18 elements
covalent molecular	molecules	H_2O, CO_2, $C_{12}H_{22}O_{11}$ (table sugar)
covalent network	covalently bonded atoms	diamond (carbon atoms)
ionic	ions	NaCl, K_2SO_4
metallic	metal ions surrounded by mobile electrons	all metals

11C Liquids

If liquids were not so common, we would consider them amazing. Why does water form spherical drops? Why are bubbles round? Why does rubbing alcohol make your skin feel cold? What makes the surface of a liquid in a test tube curve? How can some insects skate across the surface of a pond? Although many properties of liquids are well known, they are not necessarily well understood.

11.14 A Kinetic Description of Liquids

Particles in liquids are held together by intermolecular forces that balance out their kinetic energy. These particles move, but not with the reckless motion of gas particles. They have less energy, and they experience stronger intermolecular forces than gases do because their particles are closer together. Yet the attractive forces holding the molecules together do not totally dominate. There is enough freedom for liquid particles to roll and slide over each other. These particles are not fixed in any one position. Liquids, like gases, can flow and be poured, so they are both fluids.

Liquids have densities that are much greater than gases since there is little empty space between the molecules of liquids. Because of this closeness, liquids cannot significantly expand or compress like gases. People rely on this latter property every time they apply brakes in a car. The pressure that is applied on the liquid near the brake pedal is transferred through the brake line to the brake pads or discs. Since the molecules are already close together, liquids can transfer the incredible pressures of even the most powerful hydraulic

---

11C Section Objectives

After finishing this section, you should be able to

- use the kinetic theory to explain the properties of liquids.
- explain surface tension, viscosity, capillary action, diffusibility, and permeability based on intermolecular attractions.
- identify factors that affect the rate of evaporation.
- trace temperature changes and heat flow during the process of evaporation.
- explain how pressure affects boiling.
- identify the boiling points of different liquids at various pressures from a vapor pressure graph.
- use a phase diagram to predict a substance's state of matter under various temperature and pressure conditions.

11-17 How long do you think it took for the food coloring to diffuse through the water?

systems without being compressed much. The molecules themselves would have to be crushed before a liquid could be appreciably compressed.

11.15 Effects of Intermolecular Attractions

The properties of liquids are greatly affected by intermolecular forces. The attraction between the particles of the liquid is called **cohesion**. The attraction between the particles of the liquid and the particles of other materials is called **adhesion**. Cohesion and adhesion together cause the amazing properties of liquids we observe.

Cohesion

Why can a water strider skate across the surface of a pond? **Surface tension** is the elastic "skin" that forms the surface of liquids. This skin is produced by the imbalance of intermolecular forces on the surface of a liquid. A particle within a volume of liquid has neighbors on all sides. Each neighbor attracts these interior particles equally in all directions. However, a surface particle does not have other molecules on one side. All forces exerted on surface particles are directed toward the interior of the liquid, and the unbalanced forces bind surface molecules together. If the force exerted on the surface of a liquid is less than its surface tension, the object will suspend on the surface of the liquid. This phenomenon allows water striders and other insects to move across the surface of the water without sinking.

Surface tension affects the shapes of liquids. Raindrops are spherical because surface tension pulls water into the shape with the least surface area per unit volume. This action minimizes the unbalanced forces. Drops of mercury, which has more surface tension than water, are even more perfectly spherical than water drops.

Surface tension affects more than just the shape of raindrops. Have you ever tried to use water to get oil or grease off your hands with no soap? Detergents are surface-active agents, or **surfactants**. They break down the normal surface tension of water by interfering with hydrogen bonds and therefore allow grease, oil, and other nonpolar liquids to dissolve. A soap molecule is a long molecule with a polar end and a nonpolar end. The polar end attracts water through dipole-dipole

11-18 The relative incompressibility of hydraulic fluid allows pistons like this to exert great pressure to operate heavy equipment.

11-19 Surface tension causes water droplets to form spheres.

Oil drillers use surfactants to recover 70% of the oil remaining in a "dry" well.

11-20 Strong intermolecular forces make some liquids, such as syrup, viscous.

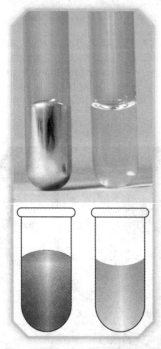

11-21 The surface tension of mercury surpasses its attraction for the glass wall of the test tube (left). The strong attraction of water molecules for the glass causes the meniscus of the water to curve upward (right).

forces, and the nonpolar end attracts dirt, grease, and oil, allowing the water and the other substances to mix.

The attraction of liquid particles for one another affects their ability to flow. Thick and gummy liquids consist of particles that have a high cohesion. Liquids that flow easily consist of particles that have a low cohesion. The degree of cohesion and the shape of fluid particles affect a property called viscosity. **Viscosity** is a liquid's ability to resist flowing. The stronger the attractions between particles, the more viscous the fluid will be. Molasses is viscous, so it can be expected to have some strong attractions between its particles. Thinner, less viscous fluids like water and gasoline pour well and spread out quickly because they have weak intermolecular attractions. Temperature affects the strength of intermolecular forces. At cold temperatures intermolecular forces are more effective on the slow-moving particles. That is why cold syrup flows so slowly.

Adhesion

The same intermolecular forces that cause surface tension affect how a liquid interacts with the particles in its container. Have you ever observed the concave surface of a liquid in a container with a narrow neck? This concave surface, called a **meniscus**, results from intermolecular attractions within the liquid and between the liquid and the container. When attractions between the container and the molecules of the liquid (adhesion) exceed the attractions between the liquid molecules (cohesion), the liquid can climb the walls of the container. In narrow glass tubes called capillary tubes, this effect becomes greatly exaggerated. Water rises up narrow capillary tubes easily in what is called **capillary action**. Mercury, on the other hand, has little attraction for glass but strong internal cohesive forces. These cohesive forces are so strong that the surface molecules are pulled away from the glass. Thus, mercury does not exhibit capillary action.

Other fluid properties that liquids exhibit are *diffusibility* and *permeability*. A drop of food coloring spreads, or diffuses, throughout a glass of water. Spilled milk seeps into, or permeates, a paper towel. Both cases are evidence that liquids consist of moving particles in unfixed positions. However, liquids diffuse and permeate other liquids only when they are similar to one another. For instance, water wets a cotton cloth, but it does not wet wax candles. Water molecules are polar, so they readily adhere to sugar, cotton, skin, and any other materials that have polar regions on their surfaces. Nonpolar surfaces, such as a freshly waxed car, cause water to bead up and roll away. On the other hand, nonpolar liquids wet nonpolar surfaces but not polar surfaces. In each situation, the intermolecular attractions determine whether the liquids are "wet."

11.16 Vaporization

Liquid particles do not all move at the same speed. Many move at similar speeds, but because particles move randomly their speeds cannot be completely uniform. Molecules with above-average speeds sometimes break away from the liquid phase. This process, whether it occurs above or below the boiling point, is called **vaporization**. Vaporization can occur at any temperature, but it happens much more readily at higher temperatures. **Condensation** is the reverse of vaporization—it is the formation of a liquid from its gaseous state.

Evaporation

When vaporization occurs in a non-boiling liquid, it is known as **evaporation**. Evaporation is a cooling process. For example, you might shiver before drying off after a shower or a swim. Liquid particles with the most kinetic energy leave the liquid state and enter the gaseous state. Remaining particles have a lower average kinetic energy. The intermolecular attractions within a liquid also affect the rate of evaporation. Strong attractions restrain particles from evaporating, while weak attractions allow quick evaporation.

Particles that evaporate can easily reenter liquids. Suppose that a liquid evaporates in a partially filled, closed jar. At first, the vapor particles enter the space above the liquid. During their random movements, these particles can bounce back to the surface of the liquid. As more particles evaporate, the air space becomes crowded, making it more likely that some will return to the liquid. The pressure inside the container is determined by the number of molecules that evaporate. The more molecules that are evaporated, the greater the pressure, and vice versa. The pressure that is exerted by the evaporated molecules over a liquid is called the **vapor pressure**.

When enough particles have evaporated and the vapor pressure is high enough, the number of particles that go back into the liquid equals the number of particles that evaporate. At this point, the level of the liquid will remain the same. The volume of the liquid stays constant because the opposite process, condensation, nullifies the effect of evaporation. This situation, called a **dynamic equilibrium**, occurs when the two processes of condensation and evaporation balance each other so that no net effect can be observed.

Boiling

Boiling is a rapid change between the liquid and the gaseous states caused by the application of heat. Vapor escapes from a liquid's surface, as in evaporation. Vapor bubbles also form inside the liquid and rise to the surface.

Boiling occurs when the vapor pressure of a liquid equals the atmospheric pressure. Normally, atmospheric pressure is greater than vapor pressure at room temperature. The pressure prevents liquids from boiling. Table 11-4 on page 290 shows that the vapor pressure of ethanol at 25 °C is only 68 torr. The normal atmospheric pressure of 760 torr can easily suppress this weak tendency to boil. But when the alcohol is heated to 78.4 °C, ethanol's vapor pressure equals normal atmospheric pressure. If heating continues, boiling occurs.

11.17 Vapor Pressure and Boiling Point

Temperature is not the only factor that affects the point at which a liquid boils. Climbers at the summit of Mount Everest would find that water boils at 70 °C, not at 100 °C. At an elevation of 8850 meters (29 035 ft), the atmospheric pressure is only about 225 torr. Boiling occurs more readily because the vapor pressure of the liquid exceeds the lower atmospheric pressure at a lower temperature. When the vapor pressure exceeds 225 torr, the water will boil. Similarly, if

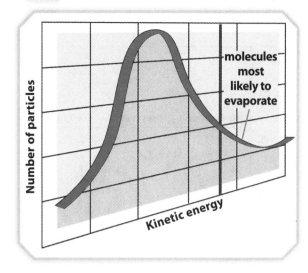

11-22 Frequency distribution of water molecules

11-23 This sealed container of bromine contains liquid (at the bottom) and gaseous bromine vapor that are in dynamic equilibrium.

11-24 Hospital autoclaves use a pressurized environment and superheated steam to sterilize materials.

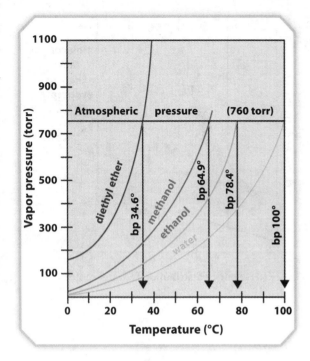

11-25 Vapor pressures of liquids increase steadily with temperature.

atmospheric pressure is higher, a higher temperature is required for a liquid to boil, as is the case in hospital autoclaves.

Because boiling points change with pressure, the **boiling point** of a liquid is defined as the temperature at which the vapor pressure equals the applied pressure. The **normal boiling point** is the temperature at which the vapor pressure equals 760 torr.

When molecules evaporate, they enter the gaseous state. They then move and collide just like any other gaseous molecules. As temperatures rise, the molecules move faster and evaporate more easily. The intermolecular attractions within a liquid also affect the rate of evaporation. Strong attractions restrain particles from evaporating, while weak attractions allow quick evaporation.

As expected, temperature and intermolecular forces affect evaporation and vapor pressure similarly. As the temperature is increased, vapor pressure increases. The vapor pressure is less for a substance with strong intermolecular forces than for a substance with weaker intermolecular forces.

We can use the calculated vapor pressures of various substances to compare the rate at which they will evaporate. Suppose that some mercury, some water, and some diethyl ether are spilled in a room at 25 °C. From Table 11-4, we can see that the mercury will exert 0.001 85 torr of vapor pressure; the water, 23.8 torr; and the ether, 541 torr. These numbers show that the ether will evaporate quickly, the water will take longer, and the mercury will take decades to evaporate. The vapor pressure of a substance at a specific atmospheric pressure can also be used to predict when a substance will boil.

11-4	Vapor Pressures (torr)			
Substance	0 °C	25 °C	50 °C	100 °C
mercury	0.000 185	0.001 85	0.0127	0.273
water	4.58	23.8	92.5	760
ethanol	12.2	68	234	1663
methanol	23.7	122	404	—
diethyl ether	185	541	1216	4859

Example Problem 11-3

Calculating the Boiling Point

What is the boiling point of water that is subjected to an atmospheric pressure of 500 torr?

Solution

Using Figure 11-25, draw a horizontal line from 500 torr on the vertical axis to the curve that represents water's vapor pressure curve. From that intersection, drop a vertical line to the axis representing the temperature. It will intersect at approximately 89 °C. At this temperature, the vapor pressure of water will match the given atmospheric pressure, and the water can boil.

11.18 Distilling Liquids

The processes of vaporization and condensation can be used to separate mixtures. This technique is called **distillation**. Through this method salt water can be purified, and combinations of liquids can be separated into pure samples called *fractions*.

> The separation of a mixture of liquids with different boiling points by distillation is often called *fractional distillation*.

> ### Distillation: Good or Bad
> Distillation is used to make alcoholic drinks stronger. But distillation is also used to purify water and to refine oil. God created man with abilities to manipulate the physical world. How can you determine whether you're using the abilities God gave you to please Him or not?

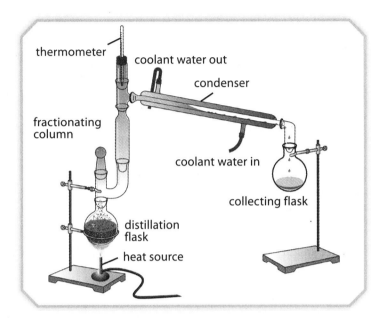

11-26 A simple distillation apparatus

The mixture to be separated is placed into a distillation flask and heated. The temperature rises steadily until it reaches the boiling point of the liquid that boils at the lowest temperature. The vapor of this liquid enters a condenser and flows over its water-cooled glass walls. The vapor soon condenses and drips into a collecting flask. Once all of the first liquid has been vaporized, the temperature can rise until the boiling point of another substance is reached. The process is repeated, and the other liquid is collected in another flask. This technique works well as long as the substances in the mixture have distinctly different boiling points and do not strongly interact. If distillation is precisely controlled, substances with boiling points only a few degrees apart can be separated.

11.19 Phase Diagrams

A **phase diagram** is a graphical way to summarize the temperature and pressure conditions in which a substance exists as a solid, liquid, or gas. Figure 11-27 is a phase diagram of water. All of the points along the liquid-solid curve (AB) are melting points, all of the points along the liquid-gas curve (AC) are boiling points, and all of the points along the solid-gas curve (AD) are sublimation points.

Notice that the line of boiling points and the line of melting points intersect at point A. At this specific pressure and temperature (0.01 °C, 4.6 torr), water can exist in all three states. This specific set of conditions is called the **triple point** of water.

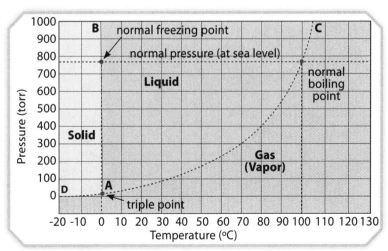

11-27 A phase diagram for water

Example Problem 11-4

Determining the State of Matter

Using Figure 11-27, determine the state of water at the following conditions:
a. 60 °C, 400 torr
b. −10 °C, 1000 torr
c. 80 °C, 200 torr

Solution
a. The point that corresponds to 60 °C and 400 torr is between the line of melting points and the line of boiling points. This region represents the conditions at which water is in the liquid state.
b. The point corresponding to −10 °C and 1000 torr is to the left of the line of melting points. Water is a solid under these conditions.
c. The point corresponding to 80 °C and 200 torr lies beyond the line of boiling points. Water will exist as a gas under these conditions.

11.20 Critical Temperatures and Pressures

Oxygen, hydrogen, nitrogen, and helium, though gases at room temperature, can be liquids under the proper conditions. Low temperatures and high pressures can condense and even solidify any gas. Low temperatures slow the molecules down, and the high pressures pack them together.

Scientists have found that high pressures alone cannot liquefy gases. Their temperatures must be lowered past a certain point. This value, called the **critical temperature** (T_c), is the highest temperature at which a gas can be liquefied. Each gas has its own characteristic critical temperature.

11-28 Oxygen, helium, nitrogen, argon, and carbon dioxide are a few of the gases that can be liquefied and transported in special cryogenic tankers like this one.

11-5 Critical Temperatures and Pressures		
Substance	T_c (K)	T_p (atm)
hydrogen (H_2)	33.1	12.8
nitrogen (N_2)	126	33.5
oxygen (O_2)	155	50.1
carbon dioxide (CO_2)	304	72.9
ammonia (NH_3)	406	112
chlorine (Cl_2)	417	76.1
water (H_2O)	647	218

Hydrogen gas at 35 K cannot be liquefied, even at tremendous pressures. The molecules are moving too quickly. If the temperature is lowered to 33 K, it can be squeezed into a liquid if enough pressure is applied. Gases that have critical temperatures above room temperature

Cryogenics: It's Really Cold in Here!

Facets of Chemistry

In the bizarre world of cryogenics, rubber balls shatter on impact and bananas are so strong that they can be used to pound nails into wood. *Cryogenics* is the science of the supercold. The word *cryogenics* comes from the Greek word *kruos*, meaning "frost." Cryogenics does not deal with the "moderately" cold temperatures of a freezer, but with the intensely cold world of –150 °C and below. Applications of this fascinating area of study are found in such wide-ranging fields as space exploration, food preservation, and medicine.

Supercooled gases are used extensively in the space industry. Liquid hydrogen and oxygen are used

together as a powerful fuel. Life support systems and space refrigeration systems also make ample use of cryogenic techniques.

Another application of cryogenics has been the rapid freezing of food. Some meats, pastries, and even vegetables are frozen rapidly by being passed through cryogenic tunnels. As the food passes through frigid gases, the surface freezes rapidly. The cold temperature then gradually spreads to the core for an even freeze.

Liquefied gases have another interesting connection to food. The eggs, sperm, and even embryos of livestock can be kept frozen in liquid nitrogen for years before being thawed for use and implanted into female animals. Some prize bulls that have been dead for years are still fathering calves through this technology.

Cryosurgery, the use of a freezing probe in place of a surgeon's scalpel, has been used with good success in removing warts, tonsils, and cataracts. Cryosurgery offers the advantages of less pain and freedom from hemorrhaging. Many operations that previously required hospitalization have been replaced by procedures that can be performed in a doctor's office.

One questionable area of experimentation is cryonics. *Cryonics* is the process of freezing and storing a human body in hope that the individual can be brought back to life at some future date. Some men hope that people with incurable diseases could be stored in a deep freeze until a cure is found. This extravagant dream has serious ethical and moral implications. Freezing people before they die involves either the sin of murder or the sin of suicide. If the freezing is done after death, there is no real hope of reviving the person, since, as far as science has been able to determine, the biological processes of death are irreversible. Leading authorities also seriously doubt that an entire human body could ever be frozen intact. When more than a few cells are frozen at a time, it is impossible to maintain a fast, uniform rate of freezing. Consequently, ice crystals form that destroy cell membranes. Despite these facts, people still seek immortality apart from Christ, and unscrupulous "freezatorium" operators continue to offer families false hopes in exchange for huge sums of money.

Death is God's penalty for sin. Isn't it amazing that people will spend huge sums of money in an attempt to cheat death? Man's striving after im-

mortality is only one more example of the sin committed in Eden. Instead of submitting to God, men demand to be like God by ruling themselves. The consequences for demanding godlikeness are always disastrous—always death. The good news is that God in love has provided a solution: those who turn to Jesus in faith will "not perish, but have everlasting life" (John 3:16).

can be liquefied at room temperature by pressure alone. Other gases require a combination of refrigeration and compression.

The pressure that is required to liquefy a gas at its critical temperature is called the **critical pressure** (P_c). Hydrogen gas at its critical temperature can be liquefied under a pressure of 12.8 atmospheres. If the gas is colder than its critical temperature, less pressure is required to liquefy it.

11.21 Using Liquids and Solids to Solve Problems

We have discussed the three most common states of matter—solid, liquid, and gas—in depth. But we know that not all materials fit into these three states. *Liquid crystals* are substances that have some of the properties of solids *and* some of the properties of liquids. These substances are crystals that have some of the properties of crystalline solids. But they are also mobile, making them like liquids. Is there a way to use liquid crystals to solve our dominion science problem?

Liquid crystal displays (LCDs) use liquid crystals to display images using a fraction of the electrical power of cathode-ray tubes. This makes them perfect for battery power. Liquid crystal displays show up in calculators, watches, laptop computers, digital clocks, TVs, and CD players. They display images with better image clarity, little noise, and reduced eyestrain.

How do liquid crystal displays work? Many liquid crystals in an LCD are naturally in a twisted shape that unwinds when they are exposed to an electrical current. This initial state is called the *twisted nematic phase*. When they unwind, liquid crystals change how filtered or polarized light passes through them. The strength of the current affects how the light is transmitted. In an LCD screen, a layer of liquid crystals in the nematic phase is placed on top of an electrode and between two sheets of polarized glass, which filter light in directions perpendicular to each other. Light that passes through the layers is blocked or altered when a current passes through the layer of liquid crystals. The light source can be behind the liquid crystals to backlight them, or a mirror behind this layer can reflect light from an external source. Adding a color filter to this LCD "sandwich" produces different shades of colors.

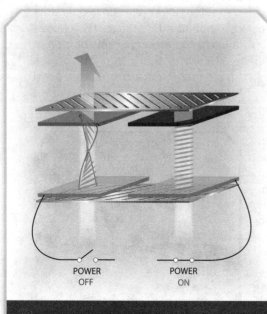

11-29 A highly magnified cross section (top) of a typical LCD (bottom)

11C Section Review

1. What would happen to a paper clip suspended on the surface of water if some detergent were added?

2. Why are the published values of the boiling points of substances always taken at 1 atm?

DS 3. Why can a substance not exist as a liquid above its critical temperature? Your answer should be based on the kinetic-molecular theory.

DS 4. How do the forces that hold liquid crystals together affect their properties?

Chapter Review
Chapter Summary

- Intermolecular forces are related to the presence of temporary or permanently charged regions. These forces influence the physical properties of substances.
- Dipole-dipole forces, hydrogen bonds, and dispersion forces are the three main types of intermolecular attraction.
- The properties of solids and liquids can be described and explained by the kinetic-molecular theory.
- Amorphous solids have no preferred arrangement of particles, whereas crystalline solids form regular three-dimensional patterns. These crystal lattices can be described using the concept of unit cells.
- Warming curves illustrate how heat converts solids to liquids.
- Melting is the change from the solid to the liquid state, and freezing is the reverse process.
- Sublimation is the direct change from the solid to the gaseous state, and deposition is the reverse process.
- The lattice energy of a crystal is reflected in its melting points and solubility.
- Surface tension, capillary action, and viscosity are all dependent on intermolecular forces.
- Vaporization is the change from the liquid to the gaseous state, and condensation is the reverse process.
- There are two types of vaporization. Evaporation is vaporization below the boiling point of a substance, and boiling is vaporization at the boiling point of a substance.
- Boiling takes place when the vapor pressure of a liquid equals the atmospheric pressure.
- Distillation is a technique used to separate mixtures by selectively vaporizing and condensing the components or fractions.
- Phase diagrams show the temperature and pressure conditions under which a substance is solid, liquid, or gaseous.

Coming to Terms

Term	Page
intermolecular force	276
dipole-dipole force	276
hydrogen bond	276
dispersion force	277
crystalline solid	280
amorphous solid	280
melting point	281
sublimation	282
deposition	282
crystal lattice	282
unit cell	282
polymorphous	284
allotropic element	284
allotrope	284
lattice energy	284
cohesion	287
adhesion	287
surface tension	287
surfactant	287
viscosity	288
meniscus	288
capillary action	288
vaporization	288
condensation	289
evaporation	289
vapor pressure	289
dynamic equilibrium	289
boiling	289
boiling point	290
normal boiling point	290
distillation	291
phase diagram	291
triple point	291
critical temperature (T_c)	292
critical pressure (P_c)	294

Review Questions

Concept Review

1. Predict the type(s) of intermolecular force(s) that may act between the molecules in these substances.
    a. CO_2
    b. NH_3
    c. HCl
    d. C_3H_8

2. Based on your understanding of intermolecular forces, which substance in each of the pairs should have the higher boiling point? Give your reasoning.
    a. NF_3, NH_3
    b. $NaCl$, HCl
    c. CF_4, CHF_3
    d. Cl_2, C_2H_5Cl

3. Use the kinetic theory to explain each observation given.
   a. Wax melts near the flame of a burning candle.
   b. Liquid water may be converted into ice cubes in a freezer.
   c. Ginger ale flows to match the shape of the glass.
   d. Water gradually evaporates from a swimming pool.
   e. Water vapor condenses inside house windows on cold days.
   f. Snow gradually disappears, even when the temperature remains below freezing.
   g. Solids and liquids cannot be compressed as much as gases.

4. What is the major difference between a crystal and an amorphous solid?

5. Assume that the particles in table salt (NaCl) vibrate just as forcefully as the particles in lead (Pb). Explain why table salt remains a solid at 500 °C while lead exists as a liquid.

6. When heat is being removed from a liquid, why does the temperature of the liquid at its freezing point remain constant until all of the liquid freezes?

7. After a jar of liquid has been sealed, the level of the liquid decreases slightly because of evaporation. After a slight decrease, the level of the liquid ceases to change. Why?

8. Name three key factors that determine the structure of a crystal.

9. Name three factors that determine the strength of a crystal.

10. Predict which member in each of the following pairs of crystals has the stronger binding forces. Explain the reasons for your prediction.
    a. NaCl, I_2
    b. KBr, NaBr
    c. CaI_2, KI

11. In terms of attractive forces and kinetic energies of particles, explain what happens during the following phase changes. Example: melting—The particles gain enough kinetic energy to overcome the attractive forces that hold them in fixed positions.
    a. boiling
    b. evaporation
    c. freezing
    d. sublimation

12. A white powder contains tiny, cube-shaped grains and melts at a temperature between 141.6 °C and 142.2 °C. Is this solid more likely to be a crystalline solid or an amorphous solid?

13. Fill in the following chart, which summarizes the properties of solids, liquids, and gases.

State	Compressible	Fluid	Density
solid			relatively high
liquid		yes	
gas	yes		

14. What causes surface tension?
15. Water rolls off a duck's back but thoroughly wets a head of human hair. What do these observations reveal about the chemical nature of these two surfaces?
16. Why do raindrops not assume triangular or cubic shapes?
17. How does perspiration cool the skin on a person's body?
18. Why does the surface of water in a glass test tube curve upward at the edges?
19. Automobile engines are designed to be lubricated with motor oils of specific viscosities. These are rated from the lowest viscosity of 5 W up to the highest viscosity of 70 W. What can you conclude about the intermolecular attractions in an oil rated 10 W?
20. What is the essential difference between boiling and evaporation?
21. Water in a truck's radiator can get hotter than 100 °C when the radiator is sealed tightly. How is it possible for water to exist as a liquid at temperatures above its normal boiling point?
22. Crude oil is purified or refined through a process called fractional distillation. In this process the different components of oil (lubricating oil, grease, gasoline, natural gas, etc.) are separated by distillation. What do you conclude about the physical properties of the different components?

DS 23. If liquid crystals have an orderly shape like solids, how do they act like liquids?

DS 24. Do you think liquid crystals would have a low melting point? Why or why not?

DS ⊙ 25. In addition to the obvious weight reduction, what are some other advantages of LCD monitors over the traditional CRT monitors?

True or False

26. The bonds between atoms within a molecule are stronger than the forces that exist between molecules.
27. A hydrogen bond is a special kind of dispersion force that can occur only in a molecule containing hydrogen.
28. In most cases, a substance has a greater density in its solid than in its liquid state.
29. Scientists have identified and described seven basic classes of crystals, which expand to fourteen different lattice structures.
30. The meniscus on the surface of water in a test tube is the result of water's viscosity.

Application

⊙ 31. Refer to Figure 11-27, the phase diagram of water, to determine the state of water at the following conditions:
   a. 80 °C, 300 torr
   b. 100 °C, 900 torr
   c. −10 °C, 500 torr

⊙ 32. Refer to Figure 11-25 to determine
   a. the boiling point of ethanol when it is at normal atmospheric pressure.
   b. the boiling point of methanol at 720 torr.
   c. the atmospheric pressure at which diethyl ether boils at 20 °C.

⊙ 33. A scientist claims to have a cooling apparatus kept at −100 °C by liquid nitrogen. Is this possible? Why or why not?

34. Write the three common states of matter—solid, liquid, and gas—in a triangle with space for long arrows between each pair of terms. Draw an arrow in each direction from each term, and write the term for the phase change that goes in that direction. For instance, the arrow from solid to liquid would be identified as *melting*.

SOLUTIONS

CHAPTER 12

DOMINION SCIENCE PROBLEM
World Thirst and Disease
Many consider polluted drinking water to be the world's greatest health issue. Over two billion people, nearly a third of the world's population, do not have a source of clean water. Thus, they suffer from dehydration, parasites, cholera, dysentery, and other epidemics. In fact, impure water causes about 90% of the diseases in these countries. Children are hit especially hard. About six thousand children die every day from diseases related to contaminated drinking water. To make matters worse, the countries that need the most help are remote or secluded by violence and internal strife. How can this world crisis be remedied?

12A The Dissolving Process 300
12B Measures of Concentration 308
12C Colligative Properties 313
12D Colloids 319

Facet
That's One Sweet Crystal! 309

12A The Dissolving Process

12.1 Defining Solutions

We use solutions and their properties when we add antifreeze to a radiator, salt an icy sidewalk, wear sterling silver, disinfect a cut, or clean windows. We benefit from many products made from solutions. Almost every chemical reaction takes place in solution. People breathe a solution called air, and most of the human body is a water solution. The study of chemistry is saturated with solutions.

Shampoos, soft drinks, and perfumes are mixtures of many ingredients. Their ingredients do not clump together, separate from each other, or fall to the bottom of a container. Their components cannot be filtered out because they are very small, typically less than 1 nm. These products are not heterogeneous mixtures since their compositions are uniform. They are not compounds because no reaction has bound them together.

The substances listed above are **solutions**, which are defined as homogeneous mixtures of variable composition in a single phase. The most abundant substance in a solution is called the **solvent**. In shampoo and soft drinks the solvent is water; in perfume, the solvent is often some type of alcohol. Dissolved substances are called **solutes**. Solutes dissolved in a solvent make a solution.

12.2 Types of Solutions

Liquid Solutions

Most of the solutions we use are liquid solutions because the solvent is a liquid. Liquid solvents can dissolve solids, other liquids, and even gases. When sugar is dissolved in tea, a liquid-solid solution is created. Seawater is a liquid-solid solution of many salts in water.

Liquids can dissolve other liquids as well. Rubbing alcohol, a mixture of 30% water and 70% isopropyl alcohol, is a common liquid-liquid solution. That is, both the solute and solvent are liquids. When two liquids mix to form a homogeneous mixture, they are said to be **miscible**. When liquids separate after mixing, they are said to be **immiscible**. For example, oil and vinegar in Italian salad dressing are immiscible liquids. No matter how hard they are shaken, oil and vinegar will separate into two distinct layers.

Liquid solvents can also dissolve gases. This fact is essential for fish. Fish "breathe" with thin, membranous gills that remove dissolved oxygen from water. Since fish are cold-blooded, the amount of oxygen they remove from the water is determined by their body temperature and the rate at which they move and eat. During the summer months, when there is great competition for dissolved oxygen, large numbers of fish may suffocate in some streams.

Solid Solutions

Solids can also act as solvents. Liquids, gases, and even other solids can dissolve in solid solvents. The most common solid-solid solutions are metal alloys. Brass, a copper and zinc alloy, blends the two metals into a uniform mixture. Gold and mercury can form a solid-liquid solution that is sometimes undesirable—liquid mercury permanently

---

12A Section Objectives

After finishing this section, you should be able to

- describe the parts of a solution.
- categorize examples of solutions into their types.
- explain the dissolving process.
- explain why certain solutes do not dissolve in certain solvents.
- compare the solubilities of different substances.
- describe how a solute can dissolve faster.
- describe how more of a solute can dissolve in a solvent.

The word *solution* is related to the word *dissolve*. Both words come from the Latin word *solvere*, meaning "to loosen."

Properties of Solutions

1. Solutes cannot be filtered out of a solution.
2. Solutes do not settle out of a solution.
3. Solutions are uniform.
4. Solutions are not chemically combined.
5. A solution can vary in its concentration.

miscible (MISS uh bul): misc- (L. *miscere*—to mix) + -ible

immiscible (im MISS uh bul): im- (L. *im*—not) + -misc- (L. *miscere*—to mix) + -ible

12-1 The air that supports life is a solution of gases (left). These protozoans and algae live in a liquid solvent containing dissolved solids and gases (right).

12-2 Fourteen-karat gold is 14/24 (58.3%) gold. It is a solution of two solids, gold and silver. Which is the solvent?

discolors gold jewelry. A solution containing mercury is called an *amalgam*. Hydrogen gas can dissolve in palladium metal and produce a solid-gas solution. Scientists have used this process to purify hydrogen by allowing it to dissolve into the metal and then forcing it out with heat.

Gaseous Solutions

Gases can also act as solvents. Although some scientists may argue the point, there is only one type of permanent gas solution at ordinary pressures. Gaseous solvents dissolving gaseous solutes are the only permanent gas solutions. Gases cannot dissolve liquid or solid particles because gas particles are so much smaller than liquid or solid particles. These liquid and solid particles cannot be sufficiently broken up by gas particles to produce particles small enough to fit the definition of a solution. An example of a gaseous solution is the air we breathe. It is a solution of oxygen, carbon dioxide, and trace amounts of other gases dissolved in approximately 78% nitrogen.

> Not all alloys are solutions. If their particle sizes exceed the size limits for solutions, they are called *heterogeneous alloys*. These alloys are prepared using different methods.

12-1	Solutions		
	Solvents		
Solutes	**gas**	**liquid**	**solid**
gas	gas-in-gas (O_2 in N_2, air)	gas-in-liquid (soft drink)	gas-in-solid (H_2 in Pd)
liquid	does not exist	liquid-in-liquid (vinegar)	liquid-in-solid (amalgam)
solid	does not exist	solid-in-liquid (salt water)	solid-in-solid (alloys)

12.3 Dissolving Process

So how does a solute dissolve into a solvent? For example, how does sugar dissolve in coffee? The dissolving mechanisms for each type of solution are different. For a substance to be dissolved in a liquid solution, attractive intermolecular forces must be overcome within the solute and solvent.

> ### The Bible and Solutions
> Much of this chapter is about how solutions work. The Bible tells us that God created the world, and understanding how solutions work causes us to marvel at His creative skill. This chapter also describes how scientists measure and manipulate solutions. It provides examples of how using our knowledge about solutions affects everyday life: from antifreeze in our car, to disinfectant on a cut, to salt on an icy sidewalk. Using our knowledge fulfills God's Word that man is to subdue the creation. The dominion science problem in this chapter reminds us that sin has affected everything in the world—even solutions. The effects of sin will not be completely reversed until the return of Christ (Rom. 8:22–23), but in the meantime God has given scientists the ability to alleviate some of the physical consequences of the Fall. This duty is part of loving our neighbor.

Dissolving for Liquid Solutions

Solvation is the dissolving process in solid-in-liquid solutions. In solvation, the molecules of the solvent surround and separate the ions or molecules of the solute. In a sugar-water solution, the solvent (water) separates the solute particles (sugar molecules) from each other. Prior to this, energy is needed to overcome the attractions between solvent (water) molecules. The positive regions of the water molecules attach to the negative regions of the sugar molecules. The water molecules then pull sugar molecules away from their neighbors. Negative regions of water molecules attract positive regions of sugar molecules in the same manner. Hydrogen bonds form and dispersion forces take effect. After solvation, diffusion carries the dissolved particles throughout the solution. In the end, an even concentration of solute particles will remain in solution.

Several other terms describe specific types of solvation. Solvation in water is called **hydration**. If the solute is an ionic compound, solvent molecules break up the crystal into ions in the process of **dissociation**. The more general term, *solvation*, describes the dissolving of either ionic or covalent compounds. In some cases, covalent compounds form ions when they dissolve in solution, known as **ionization**.

Thermal energy is necessary to overcome the intermolecular forces between the solvent particles and between the solute particles. Therefore, the first steps of solvation are endothermic, absorbing heat from the surroundings. After the separation, solute and solvent particles mix and begin to attract each other. This mixing produces a more stable arrangement of particles, releasing energy. This part of the solution process is exothermic. The net energy change during the dissolving process is measured as the **enthalpy (heat) of solution** (ΔH_{soln}). This value, specific to each solvent, is expressed as the number of kilojoules per mole (kJ/mol). The concept of enthalpy will be discussed more in Chapter 13 on thermodynamics.

> $+\Delta H_{soln}$ is endothermic.
> $-\Delta H_{soln}$ is exothermic.

> Solutions form intermolecular bonds when they mix.

Dissolving for Solid and Gaseous Solutions

In solid and gaseous solutions, solute and solvent particles mix chiefly by the process of diffusion. In gaseous solutions, molecules mix by random, constant motion. In a solid, the intermolecular forces that hold crystals in place must be overcome by the addition of energy for diffusion to take place.

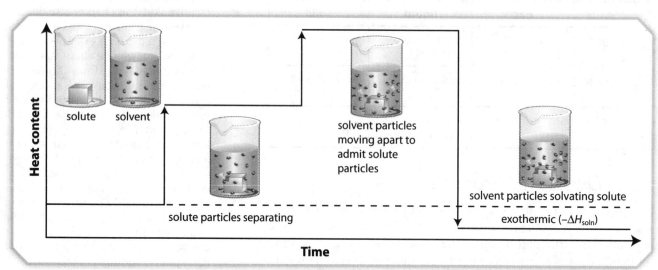

12-3 The changes in heat content during the dissolving process are shown on this graph.

12.4 Solvent Selectivity

Why do acetic acid and water mix, but not oil and water? What determines which substances can mix or dissolve in others? The governing principle in solutions is "like dissolves like." Polar solvents dissolve polar solutes, and nonpolar solvents dissolve nonpolar solutes.

Ionic compounds and polar covalent molecules have regions of electrical charge. These regions of charge can interact with neighboring ions or with other polar molecules. Water (H_2O), acetic acid ($HC_2H_3O_2$), ammonia (NH_3), and hydrogen chloride (HCl) are all polar substances. They mix with water because of dipole-dipole forces. Electrostatic forces can also attract ions and polar molecules, as in a solution of salt water. Positive sodium ions are strongly attracted to the negative portions of water molecules, while negatively charged chloride ions are attracted to the positive portions. When a solute is able to dissolve in a solvent, it is said to be **soluble** in that solvent. **Solubility** is the maximum amount of a solute that can dissolve in a specific solvent under specific conditions, such as temperature and pressure.

Nonpolar substances such as hexane, pentane, and petroleum ether cannot mix with polar substances. They lack dipoles and so cannot form hydrogen bonds. Consequently, they are "squeezed out" of polar solvents, forming distinct layers in the container. If you were to try to mix water and hexane, they would be immiscible and remain in two layers.

> ***Principles of Miscibility***
> - Polar/polar mixtures of liquids and nonpolar/nonpolar liquid solutions are miscible.
> - Polar/nonpolar mixtures of liquids are immiscible.

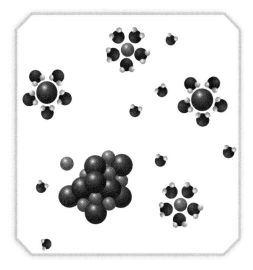

12-4 As sodium chloride dissociates, negative portions of water molecules attach to the Na+ ions. Positive portions of water molecules interact with Cl- ions.

Example Problem 12-1
Comparisons of Solubility

Considering the electrical natures of methanol, hexane, and aluminum chloride, which of these substances can dissolve in water?

methanol hexane aluminum chloride

H–C(H)(H)–OH   H–C(H)(H)–C(H)(H)–C(H)(H)–C(H)(H)–C(H)(H)–C(H)(H)–H $[Al^{3+}][Cl^-]_3$

Solution
Polar water molecules can dissolve polar methanol molecules and ionic aluminum chloride, but they cannot dissolve nonpolar hexane molecules.

"Like dissolves like" is only a general rule; there are several exceptions. Table salt dissolves in water, but calcium carbonate ($CaCO_3$) does not. Both compounds are ionic, so why aren't both soluble in water, which is polar? The strength of ionic attractions within compounds accounts for differences in solubility. Dissociation must overcome the lattice energy of the crystalline structures. It is possible for ions to be bound together so tightly that a specific solvent cannot break them apart. The solute is then described as being **insoluble** in that solvent. Table 12-2 describes the solubility of many ions and compounds.

> Substances such as calcium carbonate (chalk), though considered insoluble in water, do have a limited solubility. Almost no substances have zero solubility.

Soluble salts are those in which at least 1 g of solute will dissolve in 100 g H$_2$O. They are insoluble if less than 0.1 g of solute will dissolve in 100 g H$_2$O. Those that fall between these levels are classified as *slightly soluble*.

12-2 Solubility Rules for Salts in Water

Soluble salts	Insoluble salts
most nitrates (NO$_3^-$)	common sulfides (S^{2-}) *except* for Na$_2$S, K$_2$S, (NH$_4$)$_2$S, MgS, CaS, SrS, and BaS
most Group 1 salts (Li$^+$, Na$^+$, K$^+$, Cs$^+$, Rb$^+$) as well as ammonium salts (NH$_4^+$)	phosphates (PO$_4^{3-}$) and carbonates (CO$_3^{2-}$) *except* for salts containing Na$^+$, K$^+$, or NH$_4^+$ ions
salts of the chloride (Cl$^-$), bromide (Br$^-$), and iodide (I$^-$) ions *except* when any of these combine with Ag$^+$, Pb^{2+}, Cu$^+$, or Hg$_2^{2+}$	hydroxides (OH$^-$) *except* for NaOH and KOH (which are soluble) and Ba(OH)$_2$ and Ca(OH)$_2$ (which are slightly soluble)
most sulfates (SO$_4^{2-}$) *except* for Ag$_2$SO$_4$, BaSO$_4$, PbSO$_4$, Hg$_2$SO$_4$, SrSO$_4$, and CaSO$_4$	oxides (O^{2-}) *except* for Na$_2$O, K$_2$O, SrO, and BaO (which are soluble) and CaO (which is slightly soluble)

EXAMPLE PROBLEM 12-2

Salt Solubility

Using Table 12-2, determine whether the following salts will be soluble in water.

a. KCl
b. PbSO$_4$
c. BeO
d. MgS

Solution

a. soluble
b. insoluble
c. insoluble
d. soluble

12.5 Solution Equilibria

Consider a glass half full of water and half full of salt crystals. During the dissociation process, a solution has not yet reached its maximum solubility. Since more solute can still be dissolved, the solution is said to be **unsaturated**. The thermal energy required for the dissociation of salt is greater than the heat given off in hydration. Heat is therefore absorbed in the solution process, and the solution becomes slightly cooler.

$$H_2O\ (l) + NaCl\ (s) + heat \longrightarrow Na^+\ (aq) + Cl^-\ (aq)$$

Eventually, however, no more salt appears to dissolve. The temperature and the amount of salt at the bottom of the glass remain constant. When a solution contains the maximum amount of solute at a given temperature, it is **saturated**. But this condition does not stop the solution process. The water molecules are still moving, dissociating ions. The amount of salt at the bottom of the glass remains constant because a reverse process also occurs.

$$Na^+\ (aq) + Cl^-\ (aq) \longrightarrow NaCl\ (s) + heat + H_2O\ (l)$$

When both processes occur at the same rate, no noticeable changes occur. This condition is called a **dynamic equilibrium**. We will discuss equilibrium reactions more in Chapter 15.

$$H_2O\ (l) + NaCl\ (s) + heat \longrightarrow Na^+\ (aq) + Cl^-\ (aq)$$

Supersaturation occurs when a solution contains more dissolved solute than it does at equilibrium. For example, the solubility of sodium thiosulfate ($Na_2S_2O_3$) in water is about 50 g/100 mL at room temperature. However, if you raise the temperature of the solution to 100 °C, the solubility increases to 231 g/100 mL; thus, more sodium thiosulfate can be dissolved. If the solution is then slowly and carefully cooled back to room temperature, the excess sodium thiosulfate will not crystallize out. Instead, 231 g of sodium thiosulfate remain in 100 mL of water—more than four times the expected amount! Supersaturated solutions are not in equilibrium but are instead unstable. A scratch on the inner surface of the container, the addition of a small crystal of sodium thiosulfate, or even a sudden disturbance can cause the excess solute to rapidly crystallize, returning the solution to its saturated level.

12-5 Crystallization begins almost immediately when a seed crystal of sodium thiosulfate is added to a supersaturated solution (left). Within seconds, crystals grow throughout the solution (center and right).

Equilibria can also exist for liquid-gas solutions. Carbonated drinks are actually solutions of carbon dioxide. When a bottle of carbonated beverage is opened, carbon dioxide gas escapes as a "fizz." When the cap is replaced, the escaping gas is again confined in the bottle. As the pressure in the bottle builds up, the rate at which carbon dioxide escapes from the solution decreases. Eventually, the rate at which carbon dioxide reenters the solution will match the rate at which it leaves the solution. Because the two processes oppose each other, the drink does not lose all of its carbonation.

12.6 Factors That Affect the Rate of Solution

The rate at which a solute dissolves depends on its solubility and the frequency of collisions with solvent particles. The more collisions there are, the faster the process is. Temperature, stirring, and the amount of surface area of the solute exposed to the solvent all affect the frequency of these collisions.

The temperature of a solution affects the number of molecular collisions. Consider a glass of warm, unsweetened tea. Which should be put into the tea first—sugar or ice? If the sugar is to dissolve quickly, it should be put in first. The active molecules in warm tea will dissolve sugar faster than the more sluggish molecules in iced tea.

12-6 An unopened bottle of soda contains no gas bubbles because the pressure keeps all of the CO_2 dissolved in the solution (top). When that pressure is reduced, the CO_2 can rapidly form bubbles and escape from the liquid (bottom).

Stirring a solvent increases the number of collisions. A moving spoon more rapidly brings the solvent molecules into contact with the sugar at the bottom of the glass.

The number of collisions also depends on the surface area exposed to solvent action. Which dissolves more quickly in tea, granulated sugar or sugar cubes? The granulated sugar does. In a sugar cube, many sugar molecules are shielded from the solvent. In granulated sugar, many more sugar molecules are exposed. Increased surface area speeds up the rate of solution.

12.7 Factors That Affect Solubility

Temperature

Will the solubility of a substance increase or decrease as temperature increases? In general, higher temperatures help solids and liquids to dissolve by increasing the number of particle collisions. Increased temperature also increases solubility because most solids and liquids require heat to dissolve. Suppose that a saturated solution of potassium chlorate ($KClO_3$) is at equilibrium.

$$KClO_3 + H_2O + \text{heat} \longrightarrow \text{solution}$$

Adding heat is equivalent to adding more of a reactant. This heat fuels the forward reaction. With the addition of more heat, more potassium chlorate will dissolve until a new equilibrium is established at the new temperature. Figure 12-7 shows how much $KClO_3$ can be dissolved in 100 g of water at various temperatures.

However, when gases dissolve, the opposite usually occurs—heat is released.

$$CO_2 + H_2O \longrightarrow \text{solution} + \text{heat}$$

Solubility and the Wisdom of God

What are the factors that make solubility work the way it does? What are ways that solubility touches your life every day? If you think carefully about the answers to these questions, you'll realize that solubility is both complex and inextricably tied to the way the world works. Now consider how much forethought it would take for a human to devise such a system.

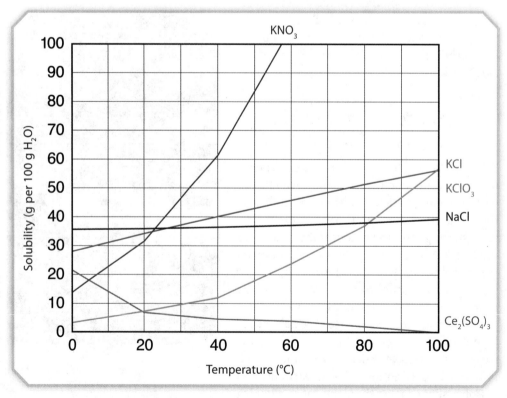

12-7 Notice that solubilities of solids usually increase with higher temperatures, but not all at the same rate.

We could, therefore, think of heat as a product of the process. Adding heat is equivalent to increasing the amount of products. This heat fuels the reverse reaction and causes carbon dioxide to escape from the solution until a new equilibrium is established. Figure 12-8 shows how the solubility of carbon dioxide changes with temperature.

12-3 Solubilities of Some Compounds at Different Temperatures				
Formula	Solubility (g/100 g H$_2$O)			
	0 °C	20 °C	60 °C	100 °C
Ba(OH)$_2$	1.67	3.89	20.94	–
Ce$_2$(SO$_4$)$_3$	21.4	9.84	3.87	–
CO$_2$	1.713	0.878	0.359	–
KCl	28.0	34.2	45.8	56.3
KClO$_3$	3.3	7.3	23.8	56.8
KNO$_3$	13.9	31.6	106	245
NaCl	35.7	35.9	37.1	39.2

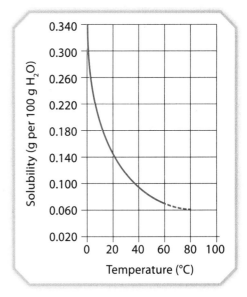

12-8 Carbon dioxide, like most gases, becomes less soluble with higher temperatures.

Example Problem 12-3
Solubility and Temperature Change

Use Table 12-3 to answer the following questions about the solubility of various compounds in water at different temperatures.
a. Of the compounds listed, which compound is the most soluble at 0 °C?
b. Of the compounds listed, which compound is the most soluble at 60 °C?
c. Of the compounds listed, which compounds show decreasing solubility as the temperature increases?

Solution
a. NaCl
b. KNO$_3$
c. Ce$_2$(SO$_4$)$_3$ and CO$_2$

Effects of Pressure Change
Pressure also drastically affects the solubility of gases in liquids. **Henry's law** states that the solubility of gases is directly proportional to the partial pressures of the gases above the solutions. It is expressed as

$$\frac{S_1}{P_1} = \frac{S_2}{P_2},$$

where S is the solubility of two different gases and P is their partial pressure above the gas-liquid solution. This law is illustrated vividly when a bottle of carbonated soft drink is opened. When the 4.5 atm of pressure above the beverage is removed, the solubility of carbon dioxide in the beverage decreases and the excess gas escapes. The escape of a gas from a liquid-gas solution is called **effervescence**.

12A Section Review

1. Why do you think water is often called the universal solvent? What makes it such a good solvent?
2. Explain why the crude oil in an oil spill can be skimmed from the top of the water and collected.
3. How are dissociation and ionization different?
4. Explain why the dissolving process for one substance may be endothermic while for a different substance it may be exothermic.
5. Describe how you would
   a. create a saturated sugar-water solution for making candy.
   b. create a supersaturated sugar-water solution for making candy.
6. The chief engineer at a soft drink bottling plant wants to increase the amount of carbonation in the sodas. What can he do to increase the amount of gas dissolved in the sodas?
7. For each of the following substances, decide whether it is a solution. If it is, list first the solvent and then the primary solute.
   a. sterling silver
   b. chocolate milk
   c. root beer
   d. egg whites
   e. rubbing alcohol
   f. diver's air
8. Suppose that all of the water on our planet had to be replaced by another liquid. List the physical and chemical properties that the new compound would need and suggest a possible replacement. Consider especially the topic of this chapter in your answer.

12B Measures of Concentration

12.8 Defining Concentration

What is the best way to describe how much solute is in a solution? Terms such as *diluted* and *concentrated* refer to relatively small and large amounts of solute in a solvent. These qualitative descriptions are not sufficient to precisely describe the amount of solute in a solution. Quantitative measurements of solutes are much more useful in chemistry. **Concentration** is a measure of the number of solute particles in a certain volume or mass of the solvent, or of the solution itself. Any of these measures of concentration follows the general form of

$$\text{concentration} = \frac{\text{amount of solute}}{\text{amount of solution (or solvent)}}.$$

12B Section Objectives

After finishing this section, you should be able to
- define *concentration*.
- perform calculations using different expressions of concentration.
- use concentrations of solutions to find the mass or moles of solute.

That's One Sweet Crystal!
Facets of Chemistry

What do you get when you take a liquid extract of the plant *Saccharum officinarum*; adjust the pH with calcium oxide; clarify the solution with sulfur dioxide, phosphoric acid, and calcium hydroxide; filter it through activated carbon; and then evaporate off the water? If you control the steps carefully, you have $C_{12}H_{22}O_{11}$—ordinary table sugar. Called sucrose by scientists, sugar is a staple of our diet. It is a preservative for fruits, a flavor enhancer in many foods, and the primary ingredient of candies.

The tiny, sandlike grains of sugar help determine the texture and hardness of some of our sweetest pleasures. To make candy, you must first make a sugar solution. All candies, from jawbreakers to the creamiest soft fudges, begin with supersaturated sucrose solutions.

Because sucrose is a polar covalent molecule, it readily dissolves in water. The attractive forces between the polar water molecules and the sugar molecules separate the sugar molecules from the sugar crystals. Eventually, the water dissolves all of the sucrose it can hold at a given temperature. At this point, the sucrose solution is saturated.

Because most liquid solutions can dissolve greater amounts of solute at higher temperatures, we can create a very concentrated sucrose solution by heating the water. As the solution continues to boil, water molecules are released in the form of gaseous water vapor. The escape of water further concentrates the solution.

Due to the phenomenon of boiling-point elevation, the temperature climbs far above the 100 °C temperature at which pure water boils. The relationship between sugar concentration and boiling point has been carefully studied and graphed. As a result, a confectioner (candy maker) can closely calculate the sugar percentage of a solution by its boiling point. (See graph.)

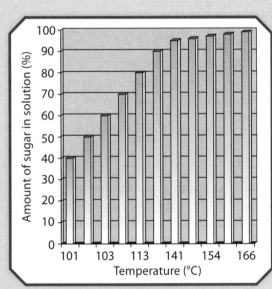

Confectioners carefully control the sugar content of the solution because the texture of each candy type is related to the percentage of sugar it contains. Fudges typically are made from 80% solutions, while a 90% solution works best for caramels. Hard candies like peanut brittle require solutions that are about 97% sucrose.

If a saturated sucrose solution is cooled, it will become supersaturated. At this point, the art of candymaking becomes evident. When the

temperature falls to the range of 21–27 °C, the fun begins. Many things affect how the solution will eventually crystallize. A solid sugar crystal or even a stray dust particle can become a center of crystallization. The rate of cooling is crucial, and even accidentally moving the syrup affects the formation and growth of crystals. It is the number, size, and arrangement of these crystals that give each kind of candy its unique texture.

The most common ways to express concentration are percent by mass, molarity, and molality. Each has its own advantages and unique uses. Although the amounts may be expressed in grams or moles, and in some cases the solute is divided by the amount of solvent rather than the amount of total solution, the same basic steps in all of these calculations are followed.

12.9 Percent by Mass

One common method of expressing concentrations compares the mass of the solute to the mass of the solution. This ratio is known as **percent by mass**. The solute content or salinity of ocean water is usually measured this way. Typically, ocean water has 3.5 g of dissolved salts in every 100 g of solution.

$$\% \text{ by mass} = \frac{\text{mass of solute}}{\text{mass of solution}} \times 100\%$$

$$= \frac{3.5 \text{ g}}{100 \text{ g}} \times 100\% = 3.5\%$$

Although this measure of concentration is used in ecology and some other common concentrations, it is seldom used in chemical applications.

12-9 This compact salinity probe measures the electrical conductivity of the water to determine the number of dissolved ions.

Photo courtesy of Vernier Software & Technology

$$\% \text{ by mass} = \frac{\text{mass of solute}}{\text{mass of solution}}$$

12.10 Percent by Volume

Another common method of expressing concentrations relates the volume of the solute to the volume of the solution. **Percent by volume** is used on the ingredient labels of many household products, especially liquid-liquid solutions. Vinegar is approximately 5% acetic acid by volume. That means that acetic acid contributes 5 mL for every 100 mL of solution.

$$\% \text{ by volume} = \frac{\text{volume of solute}}{\text{volume of solution}} \times 100\% = \frac{5 \text{ mL}}{100 \text{ mL}} = 5\%$$

Although this measure of concentration is used for some foods and medicines, it is not often used in scientific research.

$$\% \text{ by volume} = \frac{\text{volume of solute}}{\text{volume of solution}}$$

12-10 Rubbing alcohol and vinegar are two common products whose concentrations are expressed in percent by volume.

12.11 Molarity

The most common expression of concentration in chemistry uses the moles of solute compared to the volume of the solution. The **molarity** of a solution is the number of moles of solute per liter of solution. For example, a solution containing six moles of HCl per liter of solution would be written "6 M HCl solution" and read as "six molar hydrochloric acid solution."

To prepare a solution with a specific molarity, chemists add the solute to a small amount of solvent, and then slowly dilute the solution up to its required volume. Special pieces of glassware called *volumetric flasks* are carefully calibrated to yield exact amounts of solutions.

$$\text{molarity } (M) = \frac{\text{mol solute}}{\text{L solution}}$$

Example Problem 12-4
Molarity Calculation

What is the molarity of a solution that contains 3.40 mol of solute in 245 mL of solution?

Solution
A concentration is found by dividing the amount of solute by the amount of solution. In the case of molarity, we need the solute to be in moles and the volume to be in liters.

$$\frac{3.40 \text{ mol solute}}{245 \text{ mL sol'n}} \bigg| \frac{1000 \text{ mL}}{1 \text{ L}} = 13.9 \, M$$

If you were given the amount of solute in grams instead of moles, you would first have to use the molar mass of the solute to convert it to moles.

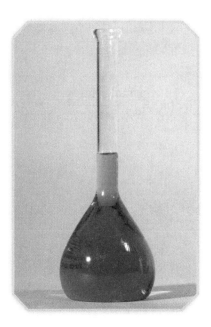

12-11 Volumetric flasks help chemists prepare solutions accurately.

Example Problem 12-5
Mole Calculation from Molarity

You have 0.6 L of a 0.5 M HCl solution. How many moles of HCl does this solution contain?

Solution
The molarity value gives you the number of moles that are in one liter of solution. You can use this ratio as a conversion factor.

$$\frac{0.6 \text{ L sol'n}}{} \bigg| \frac{0.5 \text{ mol HCl}}{1 \text{ L sol'n}} = 0.3 \text{ mol HCl}$$

Example Problem 12-6
Milliliters for Molarity Calculation

A chemical reaction requires 0.180 mol of silver nitrate ($AgNO_3$). How many milliliters of a 0.800 M solution should be added to the reaction vessel to provide this amount?

(continued)

12-12 Laboratory solutions are often measured in molarity.

Molar solutions are convenient to use in chemical reactions. They make it easy for a chemist to measure out a precise number of atoms, molecules, or ions.

Solution
Since M means moles of solute per liter of solution, we can rewrite the concentration as 0.800 mol AgNO$_3$/1 L of solution and use its inverse as a conversion factor.

$$\frac{0.180 \text{ mol AgNO}_3}{} \times \frac{1 \text{ L sol'n}}{0.800 \text{ mol AgNO}_3} \times \frac{1000 \text{ mL}}{1 \text{ L}} = 225 \text{ mL sol'n}$$

12.12 Molality

$$\text{molality } (m) = \frac{\text{mol solute}}{\text{kg solvent}}$$

The **molality** of a solution is defined as the number of moles of solute per kilogram of solvent. It is different from the molarity of the solution, and it is used in different instances. For example, a solution containing six moles of HCl per kilogram of solvent (water) would be written "6 m HCl solution" and read as "six molal hydrochloric acid solution."

Example Problem 12-7
Molality Calculation

What is the molality of a solution composed of 30.0 g of sodium nitrate (NaNO$_3$) and 400. g of water?

Solution
The problem can be solved by setting up an equation using bridge notation with the information given. Since the definition for molality is moles of solute per kilogram of solvent, the grams of water must be converted to kilograms of water. Use the periodic table to calculate the molar mass of NaNO$_3$. The equation is as follows.

$$\frac{30.0 \text{ g NaNO}_3}{400. \text{ g H}_2\text{O}} \times \frac{1000 \text{ g H}_2\text{O}}{1 \text{ kg H}_2\text{O}} \times \frac{1 \text{ mol NaNO}_3}{85.00 \text{ g NaNO}_3} = 0.882 \text{ } m$$

12B Section Review

1. How is solubility related to concentration?
2. List two different units used to express quantities of the solute in the different measures of concentration.
3. What volume units are used in calculating percent by mass?
4. Which is the most common measure of concentration in scientific studies? the least common?
⊙ 5. What mass of K$_2$SO$_4$ would you measure out to prepare 550 mL of a 0.76 M solution?
⊙ 6. A radiator is filled with a mixture of 3.25 kg ethylene glycol (C$_2$H$_6$O$_2$) in 7.75 kg of water. Calculate the molality of this solution.

12C Colligative Properties

12.13 Defining Colligative Properties

The presence of solutes causes solutions to behave differently than their pure solvents. Freezing points, boiling points, vapor pressures, and osmotic pressures all change. These changes do not depend on the types of solutes present. The new properties, called **colligative properties**, depend on the number of particles in solution.

12.14 Vapor Pressure Depression

At the surface of a solution, solute particles fill positions that are normally occupied by solvent molecules. If the solute is less likely to become a gas, or is less *volatile*, than the solvent, then the vapor pressure will be lowered. Since fewer of the more volatile solvent molecules are exposed to the surface, fewer have the chance to evaporate, making the vapor pressure of the solution lower than the vapor pressure of the pure solvent.

If a mole of sugar were added to a liter of water, the vapor pressure would be lower than pure water. If a mole of sodium chloride were added to a liter of water, the vapor pressure would be even lower. Sodium chloride units dissociate into Na^+ and Cl^- ions. These extra particles lower the vapor pressure. If aluminum chloride ($AlCl_3$) were dissolved in a liter of water, four moles of ions for every one mole of aluminum chloride would be released, further decreasing the vapor pressure.

> **12C Section Objectives**
> After finishing this section, you should be able to
> - explain how a solute affects the colligative properties of a solution.
> - describe what happens to the vapor pressure of a solution as its concentration increases.
> - describe what happens to the boiling point of a solution as its concentration increases.
> - calculate boiling-point elevation.
> - describe what happens to the freezing point of a solution as its concentration increases.
> - calculate freezing-point depression.
> - explain the process of osmosis.
> - explain how a solute affects the osmotic pressure of a solution.

12.15 Boiling-Point Elevation

When solute particles lower vapor pressure, they prevent the solvent from boiling at its expected boiling point. Higher-than-normal temperatures are necessary to raise vapor pressures up to atmospheric pressures. The effect that solute particles have on the boiling points of substances is called **boiling-point elevation**.

A 1 molal solution of sugar in water boils at 100.512 °C—an increase of 0.512 °C. The boiling-point elevation is represented by the symbol ΔT_b. The change in boiling temperature is the difference between the new boiling point and the original boiling point.

$$\Delta T_b = b_{new} - b_{original}$$

It has been found that 1 mole of particles—molecules, atoms, or ions—in 1 kilogram of water elevates the boiling point of water 0.512 °C. This value is called the **molal boiling-point constant** (K_b) for water. Other solvents have their own unique molal boiling-point constants. (See Table 12-4.) Note that concentration is always measured in molality for these constants.

The more concentrated a solution is, the greater the boiling-point elevation will be. The degree of boiling-point elevation depends on

12-13 In a pure solvent (left), there is greater vapor pressure than in a solution (right).

> Δ is the capital Greek letter *delta*. It stands for the words *change in*.

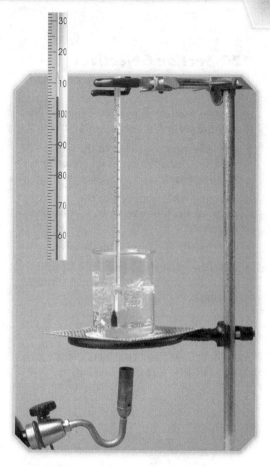

12-14 The concentrated sugar solution boils at a temperature well above 100 °C.

the precise concentration and the molal boiling-point constant of the solvent. This can be calculated by the equation

$$\Delta T_b = K_b m,$$

where ΔT_b is the change in boiling point, K_b is the molal boiling-point constant for the solvent, and m is the concentration in molals.

12-4 Boiling-Point Elevations of Solvents

Solvent	Normal boiling point (°C)	Molal boiling-point constant (K_b) (°C/m)
acetic acid	117.9	3.07
acetone	56.2	1.71
benzene	80.15	2.53
carbon tetrachloride	76.50	5.03
ethanol	78.26	1.22
ether	34.42	2.02
phenol	181.8	3.56
water	100.0	0.512

Example Problem 12-8

Boiling-Point Elevation

What is the boiling point of acetone when 0.500 mol of naphthalene ($C_{10}H_8$) is added to 1.00 kg of acetone? The molal boiling-point constant of acetone is 1.71 °C/m.

Solution

Remember that molality is defined as moles of solute per kilogram of solvent. Thus, a solution of 0.500 mol naphthalene per kilogram of acetone constitutes a 0.500 molal solution.

$\Delta T_b = K_b m$
$\Delta T_b = (1.71 \text{ °C/}m)(0.500\ m)$
$\Delta T_b = 0.855$ °C

To find the new boiling point, add the temperature change to the normal boiling point of acetone. (See Table 12-4.)

$$56.2 \text{ °C} + 0.855 \text{ °C} = 57.1 \text{ °C}$$

Recall that ionic solutes produce more particles in solution than covalent compounds do. When a covalent compound dissolves, it separates into molecules, but ionic compounds dissociate to form two or more ions per formula unit. Thus, the number of particles or moles must be multiplied by the number of released ions per formula unit when you calculate the impact of molality on boiling points. This result is referred to as the *particle molality* of a substance.

EXAMPLE PROBLEM 12-9
Boiling-Point Elevation

If 425 g of magnesium chloride ($MgCl_2$) are added to 675 g of hot water, what is the boiling point of the solution?

Solution

The first step is to determine the molality of the solution.

$$\frac{425 \text{ g MgCl}_2}{675 \text{ g H}_2\text{O}} \left| \frac{1000 \text{ g}}{1 \text{ kg}} \right| \frac{1 \text{ mol MgCl}_2}{95.21 \text{ g MgCl}_2} = \frac{6.61 \text{ mol MgCl}_2}{1 \text{ kg H}_2\text{O}} = 6.61 \text{ m}$$

Because each mole of magnesium chloride dissociates to form one mole of Mg^{2+} ions and two moles of Cl^- ions, the particle molality of the solution is 3×6.61, or 19.8 m. See Table 12-4 for the molal boiling-point constant of water.

$$\Delta T_b = K_b m$$
$$\Delta T_b = (0.512 \text{ °C}/m)(19.8 \text{ } m)$$
$$\Delta T_b = 10.1 \text{ °C}$$

The elevation in boiling point (10.1 °C) is then added to the normal boiling point (100.0 °C) to obtain the new boiling point of 110.1 °C.

EXAMPLE PROBLEM 12-10
Molality from Boiling-Point Elevation

What is the molality of a solution of NaCl in water if the solution has a $\Delta T_b = 0.400$ °C?

Solution

The first step is to rearrange the boiling-point elevation formula to solve for m.

$$\Delta T_b = K_b m$$
$$m = \frac{\Delta T_b}{K_b}$$

Remember that since NaCl is ionic, it dissociates to form two moles for each mole of salt. The temperature change is caused by two moles, but the molal boiling-point constant (K_b) is based on one mole per one kilogram of solvent. Consequently, you will use just half of the ΔT_b for your calculations since each mole caused half of the change.

$$\frac{0.200 \text{ °C}}{0.512 \text{ °C}/m} = 0.391 \text{ } m \text{ NaCl sol}$$

12.16 Freezing-Point Depression

Solutions freeze at lower temperatures than their pure solvents. A 1 molal solution of sugar in water freezes at −1.86 °C, or 1.86 °C lower than the normal freezing point of water. The change in freezing point is called the **freezing-point depression** and is represented by the symbol ΔT_f. This quantity is calculated similarly to the boiling-point elevation.

$$\Delta T_f = f_{new} - f_{original}$$

12-15 The addition of salt to the melting ice surrounding the ice-cream container depresses the temperature of the liquid below 0 °C. This should decrease the time it takes for the cream to freeze.

As with molal boiling-point constants, each solvent has its own characteristic **molal freezing-point constant (K_f)**, which is multiplied by the molality to find the magnitude of the freezing-point depression.

12-5 Freezing-Point Depressions of Solvents

Solvent	Normal freezing point (°C)	Molal freezing-point constant (K_f) (°C/m)
acetic acid	16.6	3.90
benzene	5.48	5.12
ether	−116.3	1.79
phenol	40.9	7.40
water	0.00	1.86

EXAMPLE PROBLEM 12-11
Freezing-Point Depression for Covalent Compounds

How much will the freezing point of phenol change if 4.00 mol of naphthalene are dissolved in 2.00 kg of phenol? Naphthalene is a covalent compound, thus contributing one mole of particles per mole of solute.

Solution
The first step is to determine the molality of the solution.

$$\frac{4.00 \text{ mol sol}}{2.00 \text{ kg solvent}} = 2.00 \, m$$

The molal freezing-point constant of phenol is 7.40 °C/m.

$\Delta T_f = K_f m$
$\Delta T_f = (7.40 \text{ °C}/m)(2.00 \, m)$
$\Delta T_f = 14.8 \text{ °C}$

The freezing point will be 14.8 °C lower than the freezing point of pure phenol.

Antifreeze is added to engine coolant systems for this very reason. Even though it is called "antifreeze," it affects both the freezing point and the boiling point of the water in the system. A 25% (by volume) solution of ethylene glycol (antifreeze) will not only depress the freezing point by about 12 °C but also elevate the boiling point. If antifreeze were not added, the water in the cooling system would freeze in the winter, causing hoses and pipes to burst. In the summer months, a higher boiling point keeps the water from vaporizing completely and keeps the engine from overheating.

In regions that must deal with snow and ice on roads, sodium chloride and calcium chloride are often used to lower the freezing point of water and melt frozen precipitation. Calcium chloride ($CaCl_2$) lowers the freezing point of water more than sodium chloride (NaCl) because it dissociates into more ions than sodium chloride does. A mole of calcium chloride dissociates into three (rather than two) moles of ions. The effective concentration of the solution is tripled, rather than doubled, so there is a greater depression in the freezing point. This depression causes the ice to melt at lower temperatures. Solid calcium chloride also has the added benefit of actively absorbing water from the air to dissolve itself, thus increasing the rate at which it dissolves and lowers the freezing point.

12-16 Antifreeze has a higher boiling point and a lower freezing point than water.

You might think that calcium chloride would be the natural choice for deicing all roads, but there is a catch. Sodium chloride in the raw form of rock salt is much cheaper and is often effective enough to get the job done. In situations with heavy precipitation or extremely low temperatures, calcium chloride may be used alone or in combination with sodium chloride.

Example Problem 12-12

Freezing-Point Depression for Ionic Compounds

Compare the freezing points of a 3.00 m calcium chloride solution and a 3.00 m sodium chloride solution.

Solution

The concentration of the calcium chloride solution will be tripled (3 × 3.00 m = 9 mol/kg), whereas the concentration of the sodium chloride solution will only be doubled (2 × 3.00 m = 6 mol/kg).

For calcium chloride:

$\Delta T_f = K_f m$
$\Delta T_f = (1.86\ °C/m)(9.00\ m)$
$\Delta T_f = 16.7\ °C$ for $CaCl_2$

The freezing point of the calcium chloride solution is 16.7 °C lower than the normal freezing point of water, making the new freezing point −16.7 °C.

For sodium chloride:

$\Delta T_f = K_f m$
$\Delta T_f = (1.86\ °C/m)(6.00\ m)$
$\Delta T_f = 11.2\ °C$ for NaCl

The sodium chloride solution freezes at −11.2 °C. Thus, the calcium chloride solution will freeze at a temperature 5.5 °C *colder* than a solution of sodium chloride.

12.17 Osmotic Pressure

Peel a fresh potato and slice it into rectangles the size of French fries. Put these immediately into a strong sugar solution. What happens?

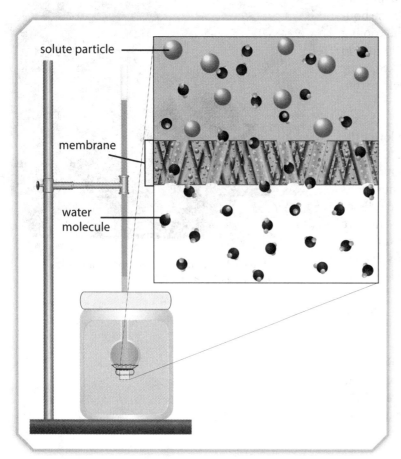

12-17 Osmosis is clearly demonstrated with this osmometer assembly.

The semipermeable membrane that surrounds each cell in the potato pieces allows the small water molecules to pass through it, but it prevents the larger sugar molecules from entering the cells. Only a few water molecules leave the sugar solution to enter the cells. Many more enter the solution from the water within the cells. As a result, the potato slices will shrink as they lose water. In fact, if you wait about thirty minutes, remove the potato slices, pat them dry, and put them on a balance, you can determine exactly how much water has been lost. The change in mass is due to osmosis. **Osmosis** is the process in which a solvent moves across a semipermeable membrane. In this case, the solvent is water and the semipermeable membrane is the cell membrane. Osmosis occurs as the water molecules attempt to reach an equilibrium on each side of the membrane.

The **osmotic pressure** of a solution is the amount of pressure required to prevent osmosis from occurring. The more solute particles there are in a solution, the higher the osmotic pressure will be. The osmotic pressure of a solution is a colligative property because it depends on the number of particles in solution.

Osmosis is important not only in chemistry but also in biology. For example, a cell could be thought of as a small amount of an aqueous solution enclosed by a semipermeable membrane. The solution surrounding the cell should be kept at the same osmotic pressure as the solution inside the cell. If this equilibrium is disturbed, water could leave the cell in sufficient quantities to cause dehydration or could enter the cell and cause the membrane to burst! The concentration of solutes inside a cell is approximately 0.3 M. If a cell is placed in a solution that is more concentrated—*hypertonic*—than the inside of the cell, water will flow out of the cell into the surrounding solution. If the cell is placed in a solution that is less concentrated—*hypotonic*—than the inside of the cell, water will flow into the cell, causing it to swell and possibly burst. When concentrations on both sides of the membrane are equal, it is said to be *isotonic*. When health care providers are administering intravenous fluids, the concentration of the solution must be adjusted to prevent harm to the patient.

12C Section Review

1. What determines how much a solute affects the colligative properties of a solution?
2. If one mole of each of these substances was added to an equal amount of water, which substance would decrease the vapor pressure the most?

$$KCl, NaCl, Na_3PO_4, PbSO_4$$

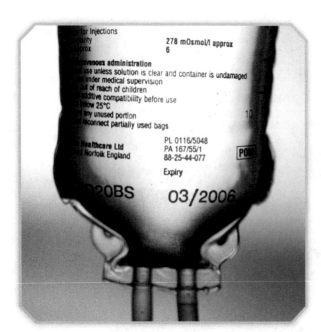

12-18 Intravenous fluids used to hydrate hospital patients are formulated to be isotonic to the concentrations in their cells.

3. Which measure of concentration is used in calculating boiling-point elevation and freezing-point depression?

4. What is the freezing point of an aqueous solution of 15.5 g glucose ($C_6H_{12}O_6$) dissolved in 150 g of water?

5. What is the boiling point of an aqueous solution of 15.5 g glucose ($C_6H_{12}O_6$) dissolved in 150 g of water?

6. What would happen to a lab culture of cells that was submerged in a solution whose concentration was 3.0 M?

12D Colloids

12.18 Introduction

Some mixtures cannot be classified as solutions or suspensions. Solutions are homogeneous mixtures that do not separate. *Suspensions*, on the other hand, contain larger particles that eventually settle out of a mixture. You learned about those types of mixtures in Section 2A. Some mixtures, however, do not fit neatly into one of these classes. What about fog? Is it a solution? Not really. The water molecules have not been completely separated from each other. Groups of molecules float through the air without becoming dissociated. Is fog a suspension? Again, the answer is no. Fog has a uniform consistency and does not readily separate like suspensions do.

12.19 What Are Colloids?

Colloids are mixtures that contain small particles dispersed in a medium. The particles are sometimes called the **dispersed phase**, and the medium is called the **dispersing medium**. Fog is a colloid. Fog consists of water droplets (dispersed phase) in air (dispersing medium). Since colloids are different from solutions, terms such as *solute*, *dissolved*, and *solvent* do not apply.

The dispersed phase particles usually exceed 1 nanometer (1 nm = 10^{-9} m), which is the approximate size limit on dissolved particles. The size of particles in a colloid cannot exceed 1000 nm,

> *12D Section Objectives*
> After finishing this section, you should be able to
> - describe colloids using the terms *dispersing medium* and *dispersed phase*.
> - distinguish between suspensions, solutions, and colloids on the basis of particle size.
> - recognize common colloid examples.
> - describe an apparatus to demonstrate the Tyndall effect in colloids.

> The intermediate size of colloidal particles allows them to pass through most filtering systems. Most membranes can catch suspended particles, but special membranes are needed to catch particles smaller than 1000 nm.

12-19 Fog is a colloid of tiny water particles in the air.

since particles any larger would quickly settle out from their medium. Intermolecular collisions buffet particles smaller than 1000 nm with sufficient force to counteract the constant tug of gravity.

We see and use colloids every day. They can be solids, liquids, or gases. Eight of the nine possible combinations of solids, liquids, and gases can form colloids. Mixtures of gases with gases are the exception; they are always solutions. Foams, aerosols, and gels are types of colloids.

12-7 Particle Size Comparison

Mixture	Particle size
solution	<1 nm
colloid	1–1000 nm
suspension	>1000 m

12-6 Types of Colloids

Particles	Medium	Common name	Examples
liquid	gas	aerosol	fog, clouds, mist
solid	gas	aerosol	smoke, dust in the air
gas	liquid	foam	shaving cream, whipped cream
liquid	liquid	emulsion	mayonnaise (oil dispersed in water)
solid	liquid	sol, gel	paint, pudding, blood, gelatin
gas	solid	solid foam	plastic foam, marshmallows, pumice
liquid	solid	solid emulsion	cheese, butter
solid	solid	solid sol	ruby glass (glass with dispersed metal)

John Tyndall (1820–93) was an Irish scientist and a persuasive and popular public speaker who convincingly supported evolution. He influenced many people to reject the authority of Scripture.

12.20 Properties of Colloids

In 1869, an Irish physicist named John Tyndall demonstrated that the particles in colloids were large enough to scatter light waves. This **Tyndall effect** is often used to distinguish colloids from solutions. A beam of light will pass through a solution without being seen because particles that are small enough to be dissolved are too small to disturb light waves. When a beam of light goes through a colloid, the beam's outline shows up distinctly. This effect can be seen when automobile headlights shine through air into a patch of fog. The outline of the beams does not show up in the air because air is a solution. However, the beams show up clearly in fog, since it is a colloid.

A British botanist named Robert Brown discovered major proof for the kinetic theory by observing

12-20 Can you tell which jar contains a colloid and which one contains a solution?

a colloid. In 1827, he watched pollen particles that were dispersed in water. He saw that the pollen particles moved slightly, as if they were being jostled by many small collisions. This movement, called **Brownian motion**, results from colloidal particles being buffeted by molecular collisions. This observation led scientists to conclude that matter is made of moving particles.

12.21 Using the Properties of Solutions to Solve Problems

Disease-causing bacteria and viruses, or *pathogens*, are the biggest problem in polluted water. Thus, any water treatment that would directly solve the problem of polluted water must focus on removing or killing these microorganisms. Water treatment units must be mobile to be useful in unstable or remote areas. Water purification must also be rapid to be really useful.

Practical water treatment involves three steps. First, pathogens are killed by chlorinating the water with chlorine dioxide or a similar compound. Then other compounds are added to clump impurities together, causing *coagulation*. This step, also called flocculation, makes impurities bigger so that a lower grade filter can be used, speeding up the process. Finally, the water is passed through a filter under pressure to quickly remove impurities. All of these things can be done in a smaller unit that will allow it to be mobile. Scientists and engineers are continuing to use technology to develop a solution to the problem of impure water.

12-21 In many parts of the world, treated water is unavailable.

12D Section Review

1. What determines whether a mixture is a solution, colloid, or suspension?
2. What are the terms used for colloids that are equivalent to the terms *solvent* and *solute* used for solutions?
3. What is the only combination of phases that cannot form a colloid? Why?
4. What is the Tyndall effect, and how is it used?
5. What is the relationship between Brownian motion and colloids?

DS 6. How is the process of water treatment discussed in the facet on page 207 (Ch. 8) different from the mobile water treatment described in Subsection 12.21?

Coming to Terms

Term	Page
solution	300
solvent	300
solute	300
miscible	300
immiscible	300
solvation	302
hydration	302
dissociation	302
ionization	302
enthalpy (heat) of solution (ΔH_{soln})	302
soluble	303
solubility	303
insoluble	303
unsaturated	304
saturated	304
dynamic equilibrium	304
supersaturation	305
Henry's law	307
effervescence	308
concentration	308
percent by mass	310
percent by volume	310
molarity (M)	311
molality (m)	312
colligative property	313
boiling-point elevation (ΔT_b)	313
molal boiling-point constant (K_b)	313
freezing-point depression (ΔT_f)	315
molal freezing-point constant (K_f)	316
osmosis	318
osmotic pressure	318
colloid	319
dispersed phase	319
dispersing medium	319
Tyndall effect	320
Brownian motion	321

Chapter Review

Chapter Summary

- Solutions are homogeneous mixtures of solids, liquids, or gases in a single phase. The major substance is the solvent, and it contains one or more solutes which are dissolved into it.

- Substances able to dissolve in a given solvent are said to be soluble. Liquids that are mutually soluble are miscible.

- Solvation is the dissolving process in liquids. First, solvent particles surround and separate particles of the solute, an endothermic process. Next, the relationships that form between solvent and solute particles release energy, an exothermic process. The net gain or loss of heat is measured as the enthalpy (heat) of solution.

- A solution is saturated when it can dissolve no more solute at its current temperature and pressure.

- An increase in the temperature of a liquid solvent generally increases the solubility of solid and liquid solutes but decreases the solubility of gas solutes.

- An increase in pressure has little effect on the solubility of solids and liquids but increases the solubility of gases.

- The two most useful measures of concentration are molarity (moles of solute per liter of solution) and molality (moles of solute per kilogram of solvent).

- Colligative properties are physical properties of a solution that are related to the number of particles in solution.

- Some colligative properties include lowered vapor pressure, elevated boiling point, depressed freezing point, and osmotic pressure.

- The main factor determining whether substances will form a suspension, colloid, or solution is particle size.

- Colloids have particles of intermediate size, but they do not settle out as suspensions do.

- The Tyndall effect is the scattering of light by tiny colloidal particles.

Review Questions

Concept Review

1. Identify the solvent and solute in the following solutions:
    a. salt water used for gargling
    b. carbonated water
    c. air
    d. carbon steel (1% Mn, 0.9% C, 98.1% Fe)

2. Predict whether each of the following pairs of liquids is miscible or immiscible:
   a. popcorn oil and vegetable oil
   b. gasoline and water
   c. iced tea and salt water
   d. popcorn oil and tea

3. Will the following substances dissolve in water or in oil?
   a. carbon tetrachloride (CCl_4)
   b. lithium chloride (LiCl)
   c. sodium bromide (NaBr)
   d. methanol (CH_3OH)

4. Using the kinetic theory, explain each of the following observations:
   a. Sugar dissolves more quickly in hot coffee than in cold coffee.
   b. Finely ground salt dissolves more quickly than large chunks of salt.
   c. Powders dissolve more quickly when stirred.
   d. Popcorn oil and vegetable oil mix more quickly than do vegetable oil and sugar.

5. Will a carbonated drink go "flat" faster if it is heated? Why?

6. Explain why
   a. red blood cells that are placed in pure water absorb water until they explode but do not when they are surrounded by blood plasma.
   b. red blood cells shrink when they are in a very concentrated salt solution.

7. How can you determine whether a substance is a solution or a colloid? Why does this method of determination work?

DS 8. Why do you think water contains so many contaminates?

DS 9. Why can't contaminates like salt and microorganisms simply be filtered out of contaminated water?

True or False

10. No gas dissolves in a solid.

11. Dissociation, ionization, and hydration are all types of solvation.

12. The easiest way to create a supersaturated solution is to take a saturated solution and carefully add more solute.

13. Temperature, surface area, and movement of solute particles all affect the number of molecular collisions upon which solvation depends.

14. The escape of a gas from a liquid is called effervescence.

15. Percent by mass is the measure of concentration most often used by chemists.

16. Colligative properties are based on whether the particles are polar or nonpolar.

Application

17. A 9.168 M aqueous solution of H_2SO_4 has a density of 1.4987 g/mL at 20 °C. How many grams of H_2SO_4 are in 50.0 g of this solution?

18. An aqueous solution of borax, or anhydrous sodium tetraborate ($Na_2B_4O_7$), is sometimes used to fireproof wood.
    a. What is the molarity of 2.50 L of solution that contains 1.85 mol of anhydrous sodium tetraborate?
    b. What is the molarity of 45.0 L of solution that contains 6.78 kg of anhydrous sodium tetraborate?
    c. How many moles of anhydrous sodium tetraborate are in 600.0 mL of a 1.57 M sodium tetraborate solution?
    d. A chemist needs 50.8 g of anhydrous sodium tetraborate for a reaction. How many milliliters of a 1.87 M solution contain this mass?

19. What is the molality of the following aqueous sucrose ($C_{12}H_{22}O_{11}$) solutions?
    a. a 0.100 M solution with a density of 1.0119 g/mL
    b. a sucrose solution with a boiling point of 100.100 °C
    c. a sucrose solution with a freezing point of –0.100 °C

20. Sodium monofluorophosphate (Na_2PO_3F) (144.0 g/mol) is the fluoride component in some modern toothpastes. What is the molality of a solution containing
    a. 1.85 kg of H_2O and 1.00 mol of sodium monofluorophosphate?
    b. 125.0 g of H_2O and 0.356 g of sodium monofluorophosphate?
    c. 500.0 g of solution and 12.0 g of sodium monofluorophosphate?

21. From each set, choose the solution that has the lowest vapor pressure.
    a. 1.8 m CH_3OH, 0.7 m CH_3OH, 2.9 m CH_3OH, 0.2 m CH_3OH
    b. 0.5 m Na_3PO_4, 0.5 m $MgCl_2$, 0.5 m $NaCl$, 0.5 m $Al_2(SO_4)_3$

22. An aqueous solution of an unknown compound boils at 101.00 °C (760 torr).
    a. What is the molality of this solution?
    b. What is the freezing point of this solution?
    c. What is the molar mass of this unknown compound if 7.80 g of this solute was added to 100.0 g of water in order to make this solution?

23. A chef reads that water will cook eggs faster if salt is added to the hot water.
    a. Suggest an explanation for this fact.
    b. What is the boiling point of 1.00 L of water in which 10.0 g of NaCl have been dissolved?
    c. How many grams of NaCl must be added to a liter of water to raise its boiling point to 105.00 °C?
    d. In light of the answers to Questions (b) and (c), do you think that adding salt to water is a convenient method for decreasing the cooking time of eggs?

24. Automobile antifreeze consists primarily of ethylene glycol ($C_2H_6O_2$). Small amounts of dye and anticorrosion substances are also added.
    a. The coldest temperature of the year in Augusta, Maine, is expected to be –26.0 °C. What must the molality of a solution of water (solvent) and ethylene glycol (solute) for the radiator fluid be to prevent it from freezing?
    b. How many grams of ethylene glycol must be added to 4.00 L of water to attain this molality?
    c. How many quarts of antifreeze is this? Assume that the density of antifreeze is 1.11 g/mL and ignore the fact that there are small amounts of other substances present.
    d. Why is it important to have a solution of antifreeze in the radiator during the summer?

CHEMICAL THERMODYNAMICS

CHAPTER 13

13A Thermochemistry	327
13B Reaction Tendency	337
Facet	
The Bible and the Second Law of Thermodynamics	347

DOMINION SCIENCE PROBLEM

Instant Ice

Have you ever had a sports injury? Sports injuries are becoming commonplace as more people participate in organized sports. Also, the growing popularity of wilderness excursions increases the frequency of hiking injuries in remote places. For many ligament and muscle strains, prompt chilling reduces swelling, promoting cell repair and relieving pain. Is there a portable, rapid way to provide cooling relief for people with these kinds of injuries?

13A Thermochemistry

13.1 Defining Thermochemistry

In Chapter 2, we introduced the laws of thermodynamics. Those laws describe energy changes in matter. In summary, they state that matter and energy can neither be created nor destroyed but that it can be converted from one form to another. During these conversions, matter naturally becomes more disordered as some energy becomes unusable. **Thermochemistry**, the branch of science that studies the transfer of energy during chemical reactions or phase changes, is an important application of these laws.

13.2 Measuring Heat and Temperature

In scientific investigations involving heat, scientists rely on measurements of two different properties: temperature and heat. Remember that *temperature*, measured in degrees Celsius or in kelvins, expresses the average kinetic energy in the particles of a sample. The temperature reading is directly related to the kinetic energy of the particles. *Heat*, on the other hand, measures the total amount of thermal energy transferred from one substance to another. Heat is typically expressed in *joules* (J). The joule is the standard SI unit of work and can be used to measure any kind of energy. Because a joule is a very small amount of heat, chemists usually express heat in *kilojoules* (kJ).

Because it is impossible to determine the absolute amount of heat in a substance, chemists measure the change in energy that occurs during a chemical reaction. This measurement is more useful to them anyway. Scientists have defined a property of matter called enthalpy to help them quantify the change in energy. **Enthalpy (H)** is the heat content of a system at a constant pressure. For the most part, chemical reactions in nature or the laboratory occur at a constant pressure.

Changes in enthalpy can be measured in a reaction vessel called a calorimeter. A **calorimeter** is an insulated container, similar to a thermos, in which a thermometer detects the temperature change that occurs during a chemical reaction. In one common kind of calorimeter, a chemical reaction occurs in a sealed

13A Section Objectives
After finishing this section, you should be able to
- define *thermochemistry*.
- distinguish between temperature and heat and give the standard units for each.
- relate the states of matter and the concepts of sensible heat and latent heat to a warming curve graph.
- predict the temperature changes that occur in a substance after the application of a given amount of thermal energy by using specific heat values.
- determine whether a given reaction is endothermic or exothermic by calculating the change in enthalpy from a table of standard enthalpies of formation.

Subduing the Earth
Scientists who responded to the difficulties of measuring heat in substances by defining enthalpy and devising ways of measuring the changes of energy in a reaction are subduing the earth as God commanded (Gen. 1:26–28).

enthalpy (EN THAL pee)

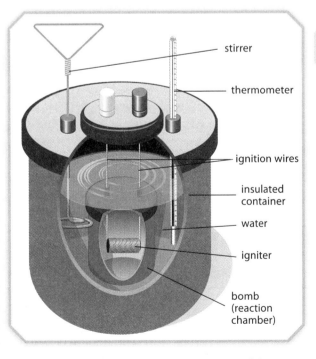

13-1 In this bomb calorimeter, the enthalpy of the reaction is captured by the water and calculated from the temperature change.

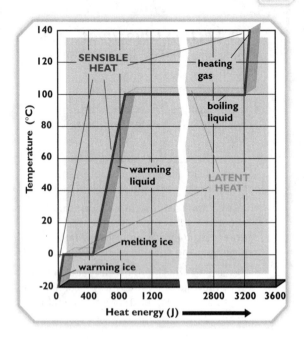

13-2 This warming curve for water shows all three states of matter and how the application of heat affects those changes.

reaction chamber submerged in a measured quantity of water. The change in the water temperature indirectly indicates how much energy the reaction released or absorbed.

13.3 Enthalpy of Phase Changes

The thermal energy, or heat, applied to a substance can be divided into two categories. **Sensible heat** is heat that produces a temperature change in the substance when applied. We can sense this type of heat by a change in temperature. The second category is called latent heat. **Latent heat** is heat that produces a phase change. No temperature change is observed. Examine Figure 13-2 to see where sensible and latent heat affect water's phase on the warming curve.

The melting of ice involves both sensible heat and latent heat. Ice, at a temperature that is below its melting point, can be heated until it reaches a temperature of 0 °C (sensible heat). Any thermal energy added at this point does not raise the temperature. Instead, it serves to supply the energy needed to overcome the intermolecular forces of attraction (latent heat). Not until the last piece of ice has melted will the temperature of the water rise above 0 °C.

A significant amount of heat must be added to a solid at its melting point to overcome intermolecular forces. That amount of heat differs for each substance. This latent heat is called the *heat of fusion*, or the **molar enthalpy of fusion** (ΔH_{fus}). It is defined as the

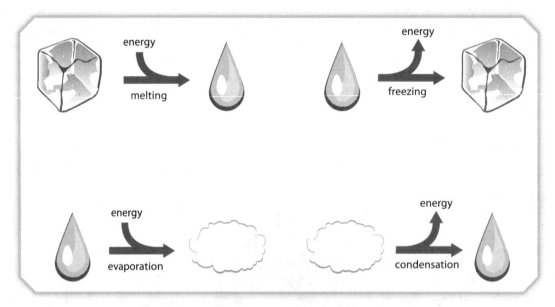

13-3 Note when thermal energy is absorbed and released during these changes.

13-1	Molar Enthalpies of Fusion and Vaporization			
Substance	Formula	ΔH_{fus} (kJ/mol)	ΔH_{vap} (kJ/mol)	
ammonia	NH_3	5.66	23.3	
ethanol	C_2H_5OH	4.94	38.6	
methanol	CH_3OH	3.33	35.2	
water	H_2O	6.01	40.7	

quantity of heat required to melt one mole of a solid to a liquid with no temperature change. It is usually expressed in units of kilojoules per mole. The same amount of energy is released when one mole of liquid freezes.

The amount of heat required to convert a mole of a liquid at its boiling point to its vapor at the same temperature is called the *heat of vaporization*, or the **molar enthalpy of vaporization** (ΔH_{vap}).

Enthalpies of vaporization differ widely because the strength of intermolecular attractions in different liquids varies. For instance, it takes 40.7 kJ to vaporize 1 mol of water at 100 °C and 1 atm. A mole of ammonia at its boiling point can be vaporized by adding 23.3 kJ of energy. Table 13-1 shows the molar enthalpy of vaporization for several other liquids.

13.4 Specific Heat

Have you ever noticed that an aluminum pan on a stove heats up faster than the water inside it? Why does this happen? You might explain this by observing that the pan receives the heat before the water, but there is a better explanation.

Not all substances heat at the same rate. The amount of thermal energy that raises 1 g of water by 1 °C will raise the temperature of 1 g of aluminum by 4.54 °C. Given the same amount of heat, the aluminum will have a greater temperature change. The **specific heat** (c_{sp}) of a substance is the amount of heat required to raise the temperature of 1 g of the substance by 1 °C. The usual unit for this quantity is joules per gram · kelvin (J/[g · K]). Degrees Celsius may be substituted for kelvins, since the degree sizes are equal. For example, 1 g of water can be raised 1 °C with the input of 4.18 joules of energy. Thus, the specific heat of water is 4.18 J/(g · °C). Specific heats of some common substances are listed in Table 13-2. Substances with high specific heats require large amounts of heat for a given temperature change and will change temperature only slowly. Substances with low specific heats require comparatively small amounts of heat for a given temperature change and will change temperature rapidly. The heat input is related to the mass of a substance (m), its temperature change, and its specific heat (c_{sp}) by the equation

$$Q = mc_{sp}\Delta T,$$

where Q is the thermal energy put into a substance and ΔT is the substance's change in temperature.

Rates of cooling are also related to specific heat. Substances with high specific heat values retain heat for longer periods than those with low specific heats. For example, the dry crust of an apple turnover may be slightly warm while the gooey filling is still hot enough to scorch your tongue. Like enthalpies of vaporization, specific heats serve as rough indicators of intermolecular attractions. Molecules that are strongly attracted to each other require extra thermal energy to increase their kinetic energies. Using the equation above for a cooling process would produce a negative change in temperature and a negative value for Q.

The warming curve of a substance reveals its relative specific heat. If the temperature rises quickly with the addition of heat, the specific heat must be low. In such a case, a small amount of heat results in a large change in temperature. A substance that has a gentle slope on its warming curve has a high specific heat because added heat does not change the temperature very much.

The reverse of the enthalpy of vaporization is called the enthalpy of condensation. This quantity is equal in magnitude but opposite in sign to the enthalpy of vaporization.

13-4 Aluminum and copper are both popular metals for the bottom of cooking pans because with their low specific heat they can heat up and cool down rapidly. Why do you think lead, with an even lower specific heat, is not used to make pans?

13-2	**Specific Heats**
Substance	Specific heat J/(g · °C)
mercury	0.14
lead	0.16
silver	0.23
copper	0.39
iron	0.45
chlorine	0.46
carbon	0.50
aluminum	0.92
oxygen	0.92
benzene	1.8
acetic acid	2.1
ethyl alcohol	2.4
water	4.18

Example Problem 13-1

Using the Specific Heat Equation to Calculate Temperature Change

If 942 J of thermal energy are added to 50.0 g of water at 25.1 °C, what will its final temperature be?

Solution

You must first determine the temperature change using the equation below.

$$Q = mc_{sp}\Delta T$$

$$\Delta T = \frac{Q}{mc_{sp}} = \frac{942 \text{ J}}{(50.0 \text{ g})(4.18 \text{ J/g}\cdot°C)}$$

$$\Delta T = 4.51°C$$

Then add this value to the starting temperature of 25.1 °C.

$$25.1°C + 4.5°C = 29.6°C$$

Example Problem 13-2

Using the Specific Heat Equation to Calculate Thermal Energy

A 28 g sample of silver is heated from 15 °C to 85 °C. How many joules were added to the sample?

Solution

First, determine the rise in temperature.

$$85°C - 15°C = 70°C$$

Next, find the specific heat for silver from Table 13-2. Use that value in the specific heat equation.

$$Q = mc_{sp}\Delta T$$

$$Q = (28 \text{ g})\left(0.23 \frac{J}{g\cdot°C}\right)(70.°C)$$

$$Q = 450 \text{ J}$$

A Complex Thermodynamic Problem

The last two sections have described measurements of energy changes in substances. These include molar enthalpy of fusion, molar enthalpy of vaporization, and specific heat. How can this information be used? Suppose you have an ice cube with a mass of 5.00 g at a temperature of 273 K. If that ice cube is heated until it is completely converted to steam at 373 K, how many joules of heat must be added? Remember the previous sections regarding latent and sensible heat. Latent heat is heat added to make a phase change with no change in temperature, and sensible heat is heat added with a noticeable temperature change. In this example, ice is being converted to water and then to steam. The first step is to gather all of the needed values from the various tables in this chapter.

1. Molar enthalpy of fusion of water (ice ⟶ water): 6.01 kJ/mol (from Table 13-1)

2. Molar enthalpy of vaporization (water ⟶ steam): 40.7 kJ/mol (from Table 13-1)
3. Specific heat of water: 4.18 J/(g · °C) (from Table 13-2)

The second step is to convert 5.00 grams of water into moles of water.

$$\frac{5.00 \text{ g H}_2\text{O}}{} \times \frac{1 \text{ mol H}_2\text{O}}{18.02 \text{ g H}_2\text{O}} = 0.277 \text{ mol H}_2\text{O}$$

The third step is to determine the latent heat energy (J) needed to melt 0.277 mol of ice.

$$0.277 \text{ mol} \times 6.01 \text{ kJ/mol} = 1.66 \text{ kJ}$$

Next, determine the sensible energy needed to raise the temperature from 273 K to 373 K.

The temperature change is 100 K. Since the Kelvin and the Celsius scales use equal values, the 100 K can be changed to 100 °C. Remember that specific heat is the amount of heat required to raise 1 g of a substance 1 °C.

$$\frac{5.00 \text{ g}}{} \times \frac{4.18 \text{ J}}{1.00 \text{ g} \cdot °\text{C}} \times \frac{100 \text{ °C}}{} = 2.09 \times 10^3 \text{ J}$$

$$= 2.09 \text{ kJ}$$

Then, the latent heat of vaporization as the water is converted to steam is

$$0.277 \text{ mol} \times (40.7 \text{ kJ/mol}) = 11.3 \text{ kJ}.$$

Finally, add all of the previously determined energy amounts to obtain the total energy needed to completely convert the 5.00 g of ice to steam.

$$1.66 \text{ kJ} + 2.09 \text{ kJ} + 11.3 \text{ kJ} = 15.1 \text{ kJ}$$

13.5 Enthalpy (Heat) of Reaction

The concept of enthalpy is especially valuable to scientists studying the energy changes of chemical reactions. They call the change in enthalpy that occurs during a reaction the **enthalpy (heat) of reaction** (ΔH). The enthalpy change of a reaction is the difference between the enthalpy of the products and the enthalpy of the reactants.

$$\Delta H = H_{products} - H_{reactants}$$

The formation of water from its two elements illustrates this process. When 1 mol of hydrogen gas is ignited in the presence of $\frac{1}{2}$ mol of oxygen gas, an explosive reaction forms 1 mol of water. Careful measurements reveal that 285.8 kJ of heat are released in the explosion.

$$H_2(g) + \tfrac{1}{2}O_2(g) \longrightarrow H_2O(l) + 285.8 \text{ kJ}$$

Because the coefficients in these balanced equations represent moles of substances and not molecules, it is permissible to use fractions in order to show the formation of a single mole of a product.

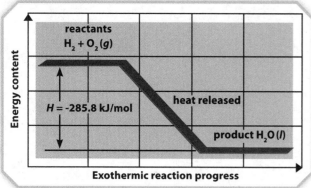

13-5 The products of exothermic reactions have less enthalpy than the reactants do.

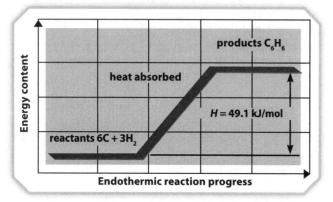

13-6 Endothermic reactions absorb heat, resulting in a greater amount of energy in the products.

Since there is a direct relationship between enthalpy and molar quantities, the standard balanced equation (which produces 2 mol of water) is as follows.

$$2H_2\,(g) + O_2\,(g) \longrightarrow 2H_2O\,(l) + 571.6 \text{ kJ}$$

A chemical equation, such as this one, that shows the reactants, products, and amount of energy that is released or absorbed as heat is called a **thermochemical equation**. These equations show the physical state of each substance because the physical states affect how ΔH for the reaction is calculated.

The products in an exothermic reaction have less enthalpy than the reactants because energy is released during the reaction. You can see from the enthalpy of reaction (ΔH) equation that exothermic reactions have negative values for ΔH. When the large enthalpy of the reactants is subtracted from the small enthalpy of the products, the difference is negative. Figure 13-5 illustrates the energy changes that are common for exothermic reactions. The products have less enthalpy than the reactants, and the difference is released as heat.

Products of endothermic reactions have more enthalpy than the reactants because energy has been absorbed. An endothermic reaction has a positive value for ΔH. Figure 13-6 illustrates the difference in energies between the reactants and the products of an endothermic reaction. The endothermic formation of 1 mol of benzene has the following thermochemical equation.

$$6C\,(s) + 3H_2\,(g) + 49.1 \text{ kJ} \longrightarrow C_6H_6\,(l)$$

If the ΔH of a reaction is known, the ΔH of the reverse reaction can be easily calculated. The ΔH of the reverse reaction has the same magnitude but the opposite sign of the ΔH of the forward reaction. For example, the reaction that formed a mole of liquid water had a ΔH of –285.8 kJ. A reaction breaking down the same amount of water has a ΔH of 285.8 kJ.

13.6 Enthalpy (Heat) of Formation

Chemists could measure the change in enthalpy for thousands of known reactions, but measuring so many reactions would be difficult, and the data would fill several volumes of books. Therefore, a quick technique for calculating changes in enthalpy was devised. A table that lists enthalpies of formation for common compounds is used (see page 333). A compound's **molar enthalpy of formation** (ΔH_f), also called the heat of formation, is the change in enthalpy that occurs when one mole of the compound is formed from its elements. The enthalpy of formation for a mole of liquid water is –285.8 kJ.

$$H_2\,(g) + \tfrac{1}{2}O_2\,(g) \longrightarrow H_2O\,(l); \Delta H = -285.8 \text{ kJ}$$

The amount of enthalpy in a substance varies with temperature and pressure, so these conditions must be specified. In thermodynamics the **standard state** is defined as 25 °C (298 K) and 1 atm of pressure. The **standard molar enthalpy of formation** ($\Delta H°_f$) is defined as the enthalpy change for the reaction that produces 1 mol of a compound in its standard state from its elements in their standard states. The degree symbol in $\Delta H°_f$ signifies that this ΔH refers to standard conditions. Table 13-3 lists several standard molar enthalpies of formation. Table 13-7 contains a more extensive listing.

13-3 Standard Molar Enthalpies of Formation

Compound	$H°_f$ (kJ/mol)
Fe_2O_3 (s)	−824.2
NH_4NO_3 (s)	−365.6
NO (g)	91.3
NO_2 (g)	33.2
$PbCl_2$ (s)	−359.4

13-7 The enthalpy of formation for Fe_2O_3 in a hand warmer is exothermic enough to warm your hands on a cold winter day.

As you will see later in Table 13-7, most values for $\Delta H°_f$ are negative. Thus, the reactions that produce most of the common compounds are exothermic. Compounds with large negative numbers release large amounts of energy when they form. Molecules such as Cl_2, H_2, and O_2 have no enthalpies of formation because these substances are the reference point to which the enthalpies of compounds are compared. This is true for all elements in their naturally occurring form.

EXAMPLE PROBLEM 13-3

Calculating $\Delta H°_f$

What is the $\Delta H°_f$ for $N_2 + 2O_2 \longrightarrow 2NO_2$?

Solution

Table 13-3 shows that the standard molar enthalpy of formation of nitrogen dioxide is 33.2 kJ/mol. The given reaction forms 2 mol of nitrogen dioxide.

$$\frac{2 \text{ mol NO}_2}{} \times \frac{33.2 \text{ kJ}}{1 \text{ mol NO}_2} = 66.4 \text{ kJ}$$

13.7 Enthalpy (Heat) of Combustion

Combustion reactions are a class of reactions that produce energy in the form of heat and light. They all involve a complete reaction with oxygen. The most significant combustion reactions occur with organic compounds containing carbon, hydrogen, and oxygen. Carbon dioxide and water are the typical products. The **standard molar enthalpy of combustion ($H°_c$)** is defined as the energy released as heat by the complete burning of one mole of a substance at standard conditions. For example, propane, the gas used for most outdoor grills, is useful because of the heat released during combustion. The thermochemical equation for the complete combustion of propane is as follows.

$$C_3H_8 (g) + 5O_2 (g) \longrightarrow 3CO_2 (g) + 4H_2O (l); \Delta H°_c = -2219.2 \text{ kJ/mol}$$

13-4 Standard Molar Enthalpies of Combustion

Substance	Formula	$\Delta H°_c$ (kJ/mol)
carbon monoxide (g)	CO	−283.0
graphite (s)	C	−393.5
hydrogen (g)	H_2	−285.8
methane (g)	CH_4	−890.8
propane (g)	C_3H_8	−2219.2
sucrose (s)	$C_{12}H_{22}O_{11}$	−5640.9

A list of values for the enthalpy of combustion for some other common compounds is found in Table 13-4. All of the values are negative because these reactions are all exothermic. As a rule, the larger the molecule, the more energy is released.

13.8 Calculating Enthalpies of Reaction

A chemical reaction involves two chemical processes. In the first step, each reactant breaks down to its elements in their standard states. In the second step, the elements in their standard states combine to form the products. Remember that the formation and breakdown of a particular substance are opposite processes and must therefore have opposite signs. If a reaction is endothermic in one direction, it will be exothermic in the other.

The following reaction between lead (II) oxide and carbon monoxide illustrates these two parts.

$$PbO\ (s) + CO\ (g) \longrightarrow Pb\ (s) + CO_2\ (g); \Delta H° = -65.7 \text{ kJ}$$

The first step in the reaction is the breakdown of lead (II) oxide and carbon monoxide. The $\Delta H°_f$ for the formation of lead (II) oxide at standard conditions is −217.3 kJ/mol. The breakdown of lead (II) oxide, which is the reverse reaction, requires 217.3 kJ.

$$PbO\ (s) \underset{-217.3 \text{ kJ}}{\overset{217.3 \text{ kJ}}{\rightleftarrows}} Pb\ (s) + \tfrac{1}{2}O_2\ (g)$$

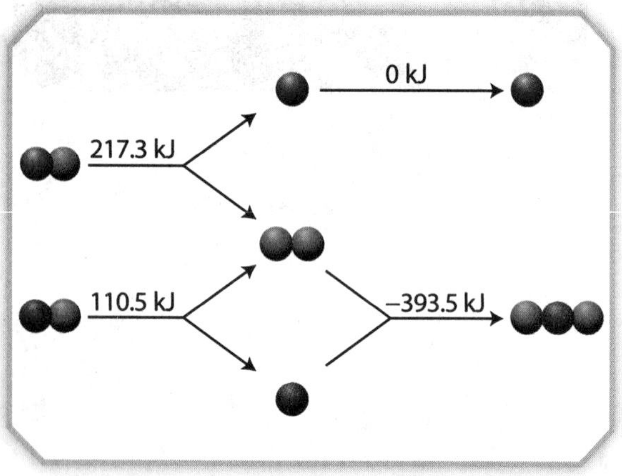

The $\Delta H°_f$ for carbon monoxide is −110.5 kJ/mol, so the breakdown of carbon monoxide to its elements requires 110.5 kJ/mol.

$$CO\ (g) \underset{-110.5 \text{ kJ}}{\overset{110.5 \text{ kJ}}{\rightleftarrows}} C\ (s) + \tfrac{1}{2}O_2\ (g)$$

The $\Delta H°$ for the first part of the reaction is obtained by adding together the two quantities of energy required to break the compounds apart. Their sum is 327.8 kJ.

$$PbO + CO \longrightarrow Pb + \tfrac{1}{2}O_2 + C + \tfrac{1}{2}O_2 \longrightarrow Pb + O_2 + C;$$
$$\Delta H = 327.8 \text{ kJ}$$

The second part of the reaction is the formation of products. The $\Delta H°$ for each of these changes can be easily found in the table listing values of $\Delta H°_f$. Since lead is an element in its standard state, its $\Delta H°_f$ is zero. The $\Delta H°_f$ for the formation of 1 mol of carbon dioxide from its elements at standard conditions is −393.5 kJ. The total $\Delta H°$ for the formation of products is −393.5 kJ.

13-8 This schematic diagram shows the energy changes for the reaction PbO (s) + CO (g) $\longrightarrow$ Pb (s) + CO_2 (g).

$$C\,(s) + O_2\,(g) \xrightleftharpoons[393.5\text{ kJ}]{-393.5\text{ kJ}} CO_2\,(g);\ \Delta H = -393.5\text{ kJ}$$

By adding the enthalpy for the breakdown of reactants to the enthalpy for the formation of products, we get the $\Delta H°$ for the reaction, which is -65.7 kJ. This result shows that the reaction is exothermic.

In this problem, calculating the $\Delta H°$ of the whole reaction involved combining the values for $\Delta H°$ in two partial reactions. **Hess's law** states that the enthalpy change of a reaction equals the sum of the enthalpy changes for each step of the process. The example below shows how one reaction could be thought of as the sum of two other reactions. The first reaction breaks the reactants apart, and the second puts them back together in a different arrangement.

$$\begin{aligned}
PbO + CO &\longrightarrow \cancel{Pb} + \cancel{O} + \cancel{C};\ \Delta H° = 327.8\text{ kJ} \\
+\ \cancel{Pb} + \cancel{O} + \cancel{C} &\longrightarrow Pb + CO_2;\ \Delta H° = -393.5\text{ kJ} \\
\hline
PbO + CO &\longrightarrow Pb + CO_2;\ \Delta H° = -65.7\text{ kJ}
\end{aligned}$$

The enthalpy change of the net reaction equals the sum of the enthalpy changes for the two steps. Notice how the common reactants and products cancel each other. The mathematical version of this statement is as follows.

$$\Delta H°_{reaction} = \Delta H°_{reactant\ decomposition} + \Delta H°_{product\ formation}$$

Because the $\Delta H°_f$ values in the table refer to the formation of compounds from elements, the values for reactant-breakdown enthalpies will be opposite in sign to those listed. Consequently, the equation can be changed to a form that uses the data in Table 13-7 without changing the signs of the numbers. The symbol Σ in this equation is the Greek letter *sigma*. It signifies a summation of numbers.

$$\Delta H°_{reaction} = \Sigma \Delta H°_{f(products)} - \Sigma \Delta H°_{f(reactants)}$$

There are two general principles for combining thermochemical equations.

1. When using Hess's law, you may use any valid balanced equation, but remember to change the sign on the ΔH if you reverse the equation.

2. Multiply the coefficients of a balanced equation to get the desired equation when the coefficients are added together, but remember to also multiply the ΔH by the same number.

EXAMPLE PROBLEM 13-4

Estimating $\Delta H°$

Estimate the $\Delta H°$ for the combustion of methane, using standard enthalpies of formation.

$$CH_4\,(g) + 2O_2\,(g) \longrightarrow CO_2\,(g) + 2H_2O\,(l)$$

Solution

The equation requires the use of the following enthalpies of formation from Table 13-7 on page 344.

$\Delta H°_f\,(CO_2) = -393.5$ kJ/mol
$\Delta H°_f\,(H_2O) = -285.8$ kJ/mol

(continued)

$\Delta H°_f$ (CH$_4$) = −79.6 kJ/mol
$\Delta H°_f$ (O$_2$) = 0 kJ/mol

Two moles of water and oxygen participate in the reaction. This is included when calculating the enthalpy change of the entire reaction.

$$\Delta H°_{reaction} = \Sigma \Delta H°_{f\,(products)} - \Sigma \Delta H°_{f\,(reactants)}$$

$$\Delta H°_{reaction} = [\Delta H°_f (CO_2) + \Delta H°_f (H_2O)] - [\Delta H°_f (CH_4) + \Delta H°_f (O_2)]$$

$$\Delta H°_{reaction} = [(1\text{ mol})(-393.5\text{ kJ/mol}) + (2\text{ mol})(-285.8\text{ kJ/mol})] - [(1\text{ mol})(-79.6\text{ kJ/mol}) + (2\text{ mol})(0\text{ kJ/mol})]$$

$$\Delta H°_{reaction} = -885.5\text{ kJ}$$

This reaction is highly exothermic.

> Notice that all of the mol units cancel each other.

13-9 A highly endothermic reaction in this cold compress reduces tissue swelling and eases pain.

13.9 Using Enthalpy to Solve Problems

Subsection 12.3 introduced the enthalpy (heat) of solution (ΔH_{soln}). This term was defined as the net change in energy that occurs during the dissolving process. As with all of the other enthalpy changes in this chapter, the end result may be endothermic (absorbing energy) or exothermic (releasing energy). A solvation process that is strongly endothermic can absorb enough thermal energy to produce a noticeable cooling effect. In fact, instant cold compresses use a common endothermic reaction. The dissociation of ammonium nitrate is a decomposition reaction that is strongly endothermic.

$$NH_4NO_3\,(s) \quad NH_4^+\,(aq) + NO_3^-\,(aq)$$

The enthalpy of solution is calculated the same way as the enthalpy of reaction. Simply subtract the heat of formation of the reactants from the heat of formation of the products.

$$\Delta H_{soln} = \Sigma \Delta H°_{f\,(products)} - \Sigma \Delta H°_{f\,(reactants)}$$

$$\Delta H_{soln} = [\Delta H°_f (NH_4^+) + \Delta H°_f (NO_3^-)] - [\Delta H°_f (NH_4NO_3)]$$

$$\Delta H_{soln} = [(-132.5\text{ kJ/mol}) + (-205.0\text{ kJ/mol})] - (-365.6\text{ kJ/mol})$$

$$\Delta H_{soln} = 28.1\text{ kJ/mol}$$

Instant ice packs are surprisingly simple in design. They typically contain a small packet of water which ruptures when squeezed, mixing with ammonium nitrate salt. If 75 mL of water completely dissolves 25.0 g of ammonium nitrate, the temperature may drop from standard-state temperature (25 °C) to −3 °C, producing ice to be applied to an injury.

By using their knowledge of enthalpy, scientists were able to develop a portable and easy-to-use way of treating injuries. God originally commanded that humans tame an unfallen creation. After Adam's sin, humans must also work on mitigating the effects of the Fall on creation to obey God's command to subdue the earth.

13A Section Review

1. What is the difference between heat and temperature? What units are used to measure each?
2. What would a scientist use a calorimeter for?
3. How does the specific heat of a substance relate to its ability to change temperature?
4. What is the difference between standard temperature and pressure (STP) and standard state?
5. Define the following symbols:
   a. H
   b. $\Delta H°_f$
   c. ΔH
   d. $\Delta H°_{vap}$
   e. $\Delta H°_{fus}$
   f. $\Delta H°_c$

⊙6. Calculate the standard change in enthalpy (ΔH) for each reaction and tell whether the reaction is endothermic or exothermic.
   a. $N_2 (g) + 3H_2 (g) \longrightarrow 2NH_3 (g)$
   b. $Ca(OH)_2 (s) \longrightarrow CaO (s) + H_2O (g)$

DS ⊙7. Some instant ice packs contain ammonium chloride (NH_4Cl), which dissolves to produce two ions in a reaction similar to that involving ammonium nitrate. Given that the ammonium ion has a $\Delta H°_f$ of −132.5 kJ/mol and that the $\Delta H°_f$ of the chloride ion is −167.2 kJ/mol, calculate the enthalpy of solution.

DS ⊙8. Compare the cooling efficiency of ice packs that use ammonium chloride to those that use ammonium nitrate, assuming each reaction involves the same volume of water and the same molar quantity of salt.

13B Reaction Tendency

13.10 Breaking Chemical Bonds

Chemical bonds store energy. As long as the bonds do not break or change, the energy that they contain cannot be released. When the bonds are broken, however, their energy takes a more observable form. Chemical bonds can be broken through *vibration*. In some cases, this vibration may be indirectly caused by the *rotation* or *translation* of the molecule. An outside source of energy, such as a laser, can cause the vibration. The atoms can be vibrated so strongly that the attraction holding them together breaks. If the nuclei are far enough from each other, the atoms will fly apart.

13.11 Chemical Bonds and Enthalpy

In Section 13A, you calculated the enthalpy of reactions by using enthalpies of formation. Another method of analyzing the enthalpy changes of chemical reactions is to focus on the bonds between pairs of atoms. The strength of the bond between any two atoms depends on the elements that are bonding and on the length of the bond. Because bonded atoms are more stable than unbonded atoms, energy is always required to break bonds. Bond breaking is thus an endothermic process. Conversely, the formation of bonds always releases

13B Section Objectives

After finishing this section, you should be able to

- recognize that exothermic reactions that increase entropy are the most favorable.
- define *entropy* and give some common examples.
- calculate the change in entropy of a chemical reaction.
- calculate the free-energy change in a chemical reaction and relate the results of the four possible combinations of enthalpy/entropy combinations.

13-10 Though it is not immediately apparent, there is more energy stored in the pile of gunpowder (left) than in the pile of sand (right).

energy and is thus exothermic. The stronger the bond is, the more energy is released.

These relationships can help explain why some reactions give off energy and others require energy. Strong, stable bonds require large amounts of energy to break them. Strong bonds also give off large amounts of energy when they form. Weak bonds can be broken with small amounts of energy. Predictably, they give off small amounts of energy when they form.

A reaction that breaks strong bonds and forms weak bonds requires energy. More energy is required to break the strong bonds than is released when the weak bonds form. As a result, reactions that produce compounds with weaker, less stable bonds are endothermic.

stronger bonds $\longrightarrow$ weaker bonds; $\Delta H > 0$ (endothermic)

A reaction that forms stable, low-energy compounds from compounds with high-energy, weak bonds releases energy. The amount of energy that must be used to break the weak bonds is small compared to all the energy that is given off when the strong bonds form.

weaker bonds $\longrightarrow$ stronger bonds; $\Delta H < 0$ (exothermic)

In most cases, exothermic processes are thermodynamically favorable and endothermic processes are not. But this rule does not always hold true. A negative ΔH does not always predict that a process is favorable.

13.12 Entropy and Reaction Tendency

We are surrounded by chemical processes that occur *spontaneously*, yet those processes absorb rather than give off energy. In other words, they are endothermic. One example is the release of compressed gas into the atmosphere. As gas under high pressure is released, it gets colder. You may have noticed this phenomenon when dispensing whipped cream from a can. In Chapter 12, we learned that the formation of many solutions causes the temperature of the solvent to decrease. Ice cubes absorb heat from their surroundings to melt into water. All of these are endothermic processes. Clearly, something in addition to enthalpy influences spontaneity.

Each of these examples shows us that chemical processes tend to proceed naturally in the direction that increases the disorder of a system. The *second law of thermodynamics* relates this tendency to a very large system when it states that natural processes decrease the order found in the universe. Gas molecules are less ordered after being released from the pressure of a can. Solute particles are more random after solvation and when dispersed through a solution. Water molecules in their

Endothermic reaction—energy required

$\Delta H_{reaction} = \Sigma \Delta H°_{f\,products} - \Sigma \Delta H°_{f\,reactants}$
= big number − small number
= positive number

Exothermic reaction—energy released

$\Delta H_{reaction} = \Sigma \Delta H°_{f\,products} - \Sigma \Delta H°_{f\,reactants}$
= small number − big number
= negative number

13-11 Endothermic reactions (top) form unstable, high-energy bonds. Exothermic reactions (bottom) form stable, low-energy bonds.

liquid state show less order than when they are locked in the crystal lattice structure of ice.

Entropy (S) is a measure of randomness or lack of orderliness in a system. Disordered substances have high entropies; very ordered substances have low entropies. Therefore, the greater the randomness, the greater the entropy. Gases, with their free-flying molecules, have high entropies. Crystalline solids have particles arranged in definite, repeating, organized systems; therefore, they possess low entropies. However, even orderly crystals eventually decay.

Because entropy is related to the kinetic energy of a substance, it is directly proportional to temperature. At absolute zero, theoretically all random movement of molecules or atoms ceases and the entropy equals zero. The entropy of substances increases with temperature. As the temperature becomes greater, the increased kinetic energy gives the particles greater motion. For example, when ice (a crystalline solid) melts, entropy has increased. Because the amount of entropy in a substance varies with temperature as well as pressure, these conditions must be specified. Table 13-5 lists the entropies per mole of several common substances at standard conditions. Although enthalpy values for pure elements are not listed in the extensive table, entropies are included because all substances have some entropy (see Table 13-7).

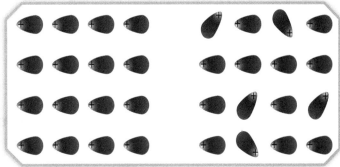

13-12 A perfect (theoretical) crystal of HCl has zero entropy ($S = 0$) at 0 K (left). When the temperature rises, vibrations within the lattice allow the dipolar molecules to begin to shift their orientations, thus increasing the entropy ($S > 0$) (right).

13-5 Standard Molar Entropies

Substance	Formula	$\Delta S°$ (J/mol · K)
hydrogen bromide (g)	HBr	198.7
hydrogen chloride (g)	HCl	186.9
ammonia (g)	NH_3	192.8
ammonium chloride (s)	NH_4Cl	94.6

13.13 Calculating Entropy Changes

From the data in Table 13-7, the entropy changes of many reactions can be calculated. An entropy change is the difference between the sum of the entropies of the products and the sum of the entropies of the reactants, similar to how the enthalpy of a reaction is calculated.

$\Delta S°$ = total entropy in products − total entropy in reactants

$$\Delta S° = \Sigma S°_{products} - \Sigma S°_{reactants}$$

A reaction with a positive $\Delta S°$ increases entropy or disorder. The products have more entropy than the reactants. Negative $\Delta S°$ signifies a decrease in entropy—an increase in order.

> **EXAMPLE PROBLEM 13-5**
> *Evaluating $\Delta S°$*
>
> Does the reaction $NH_3 (g) + HCl (g) \longrightarrow NH_4Cl (s)$ produce an increase or decrease in entropy?
>
> **Solution**
> $S° (NH_3) = 192.8$ J/(mol · K)
> $S° (HCl) = 186.9$ J/(mol · K)
> $S° (NH_4Cl) = 94.6$ J/(mol · K)
>
> $\Delta S° = \Sigma S°_{products} - \Sigma S°_{reactants}$
>
> $\Delta S° = \Delta S° (NH_4Cl) - [\Delta S° (NH_3) + \Delta S° (HCl)]$
>
> $\Delta S° = \left[1 \text{ mol} \times \dfrac{94.6 \text{ J}}{\text{mol} \cdot \text{K}} \right] - \left[\left(1 \text{ mol} \times \dfrac{192.8 \text{ J}}{\text{mol} \cdot \text{K}} \right) + \left(1 \text{ mol} \times \dfrac{186.9 \text{ J}}{\text{mol} \cdot \text{K}} \right) \right]$
>
> $= 94.6 \text{ J/K} - 379.7 \text{ J/K}$
>
> $= -285.1 \text{ J/K}$
>
> Since the value of $\Delta S°$ is negative, the amount of entropy has decreased. The product (a solid) is more ordered than the reactants (gases).

Notice that the mol units all cancel.

All processes, whether natural or unnatural, increase the entropy of the universe. A positive $\Delta S°$ means a reaction is more favorable. A reaction with a negative $\Delta S°$ goes against this tendency and attempts to decrease entropy. What can force a reaction to violate the trend toward disorder?

13.14 Free-Energy Change

Two tendencies drive reactions. The first is the tendency to decrease enthalpy (exothermic), and the second is the tendency to increase entropy. Reactions can be either exothermic or endothermic, and they can either increase or decrease entropy. Four possible enthalpy/entropy combinations exist.

1. Exothermic and increasing entropy ($-\Delta H$, $+\Delta S$)
2. Exothermic and decreasing entropy ($-\Delta H$, $-\Delta S$)
3. Endothermic and increasing entropy ($+\Delta H$, $+\Delta S$)
4. Endothermic and decreasing entropy ($+\Delta H$, $-\Delta S$)

Reactions are driven by the combined effects of change in enthalpy (ΔH) and change in entropy (ΔS). A single expression that relates ΔH and ΔS is needed to determine how they combine. J. Willard Gibbs, an American physicist and mathematician of the 1800s, formulated a single criterion of favorability. He combined the change in enthalpy and the change in entropy into a single term called **free energy**, which is sometimes called **Gibbs free energy (G)** in his honor. The **free-energy change (ΔG)** for a reaction can be calculated by the following equation,

$$\Delta G = \Delta H - T\Delta S,$$

where T is the temperature in kelvins.

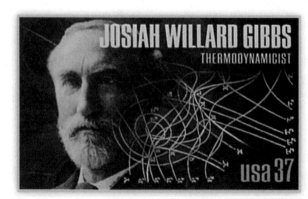

13-13 J. Willard Gibbs was a mathematical engineer, theoretical physicist, and chemist who established the theoretical groundwork for thermodynamics with the publication of two key papers in 1876 and 1878. He received the first engineering PhD in America from Yale in 1863. Other than the three years he studied in Europe, he spent his entire career at Yale.

ΔG is the change in free energy. It is the difference between the free energy of the products and the free energy of the reactants. A negative ΔG indicates a decline in free energy and signifies that a reaction may occur naturally and is spontaneous. A positive ΔG indicates a net increase in free energy and signifies that a reaction is not favorable, or nonspontaneous. The free-energy change can be negative under several conditions. A negative ΔH and a positive ΔS always contribute to a negative ΔG. When the two tendencies oppose each other, the temperature at which the reaction takes place often determines the sign of ΔG. Table 13–6 shows the conditions at which ΔG is negative and the reaction is spontaneous.

13-6 When ΔG Is Negative

Case	ΔH	ΔS	ΔG
1.	−	+	$-\Delta H - T(+\Delta S) = -\Delta G$ at all Ts
2.	−	−	$-\Delta H - T(-\Delta S) = -\Delta G$ at low Ts
3.	+	+	$+\Delta H - T(+\Delta S) = -\Delta G$ at high Ts
4.	+	−	$+\Delta H - T(-\Delta S) = +\Delta G$ at all Ts

13.15 Calculating Free-Energy Change

The ΔG of a reaction can be calculated in two different ways. First, we may calculate ΔG from the values for ΔH, ΔS, and T by substituting them into the free-energy equation. For example, for the reaction below, the free-energy change can be calculated from its enthalpy and entropy changes.

$$Mg(OH)_2 \,(s) \longrightarrow MgO\,(s) + H_2O\,(l) \text{ at } 298 \text{ K}$$

$$\Delta H = (-601.6 \text{ kJ} - 285.8 \text{ kJ}) - (-924.5 \text{ kJ}) = 37.1 \text{ kJ}$$

$$\Delta S = (27.0 \text{ J/K} + 69.95 \text{ J/K}) - (63.2 \text{ J/K}) = 33.75 \text{ J/K}$$

Note that the two values are not expressed in similar units. Converting the ΔS value from 33.75 J/K to 0.03375 kJ/K can remedy the problem.

$$\Delta G = \Delta H - T\Delta S$$
$$= 37.1 \text{ kJ} - 298 \text{ K} \;(0.03375 \text{ kJ/K})$$
$$= 27.0 \text{ kJ}$$

Calculating ΔG by the second method, where values from Table 13-7 are used, gives a result very close to the first value.

$$\Delta G = (-569.3 \text{ kJ} - 237.1 \text{ kJ}) - (-833.5 \text{ kJ}) = 27.1 \text{ kJ}$$

Since the ΔG is positive, the reaction is nonspontaneous at 298 K. At a higher temperature, the positive ΔS has a greater effect. This effect could be seen by calculating ΔG at 1000. K with appropriate values for $\Delta H_f °$ and $S°$.

Case 1: Decrease in Enthalpy, Increase in Entropy

Case 1 shows that exothermic reactions that increase entropy ($-\Delta H$, $+\Delta S$) are always favorable. They release energy and they increase disorder. ΔG for such reactions is always negative.

Example Problem 13-6

ΔG, Case 1

Calculate the Gibbs free-energy change for the following reaction:

$$2Ag\,(s) + S\,(s) \longrightarrow Ag_2S\,(s) \text{ at 298 K.}$$

Solution

$$\Delta H°_f = -32.6 \text{ kJ}$$

$$\Delta S = \left[1 \text{ mol} \times \frac{144.0 \text{ J}}{\text{mol} \cdot \text{K}}\right] - \left[\left(2 \text{ mol} \times \frac{42.6 \text{ J}}{\text{mol} \cdot \text{K}}\right) + \left(1 \text{ mol} \times \frac{32.1 \text{ J}}{\text{mol} \cdot \text{K}}\right)\right]$$

$$= 26.7 \text{ J/K}$$

$$\Delta G = \Delta H - T\Delta S = -32.6 \text{ kJ} - (298 \text{ K})(0.0267 \text{ kJ/K}) = -40.6 \text{ kJ}$$

Case 2: Decrease in Enthalpy, Decrease in Entropy

Case 2 shows that exothermic reactions that decrease entropy ($-\Delta H$, $-\Delta S$) may or may not be favorable. The two tendencies oppose each other. In cases like this, the temperature determines whether the reaction is favorable. The entropy change hinders the reaction less at low temperatures than it does at high temperatures, so ΔG is negative at low temperatures.

Example Problem 13-7

ΔG, Case 2

Calculate the Gibbs free-energy change for the following reaction:

$$2Fe\,(s) + \tfrac{3}{2}O_2\,(g) \longrightarrow Fe_2O_3\,(s) \text{ at 298 K.}$$

Solution

$$\Delta H°_f = -824.2 \text{ kJ}$$
$$\Delta S = [(1 \text{ mol})(87.4 \text{ J/mol} \cdot \text{K})]$$
$$\quad - [(2 \text{ mol})(27.3 \text{ J/mol} \cdot \text{K}) + (\tfrac{3}{2} \text{ mol})(205.2 \text{ J/mol} \cdot \text{K})]$$
$$= -275 \text{ J/K}$$

$$\Delta S = \left[1 \text{ mol} \times \frac{87.4 \text{ J}}{\text{mol} \cdot \text{K}}\right] - \left[\left(2 \text{ mol} \times \frac{27.3 \text{ J}}{\text{mol} \cdot \text{K}}\right) + \left(\frac{3}{2} \text{ mol} \times \frac{205.2 \text{ J}}{\text{mol} \cdot \text{K}}\right)\right]$$

$$= -275 \text{ J/K}$$

$$\Delta G = \Delta H - T\Delta S = -824.2 \text{ kJ} - (298 \text{ K})(-0.275 \text{ kJ/K}) = -742.3 \text{ kJ}$$

Case 3: Increase in Enthalpy, Increase in Entropy

Case 3 represents endothermic reactions that increase entropy ($+\Delta H$, $+\Delta S$). These reactions can be favorable if the entropy change is greater than the change in enthalpy. High temperatures magnify the effect of the entropy change and make ΔG negative. As a result, these reactions can proceed. At lower temperatures, however, they have positive ΔG values, making them unfavorable.

EXAMPLE PROBLEM 13-8

ΔG, Case 3

Calculate the Gibbs free-energy change for the following reaction:

$$2C\,(s) + H_2\,(g) \longrightarrow C_2H_2\,(g) \text{ at 298 K.}$$

Solution

$$\Delta H°_f = 227.4 \text{ kJ}$$

$$\Delta S = \left[1 \text{ mol} \times \frac{200.9 \text{ J}}{\text{mol} \cdot \text{K}}\right] - \left[\left(2 \text{ mol} \times \frac{5.7 \text{ J}}{\text{mol} \cdot \text{K}}\right) + \left(1 \text{ mol} \times \frac{130.7 \text{ J}}{\text{mol} \cdot \text{K}}\right)\right]$$

$$= 58.8 \text{ J/K}$$

$$\Delta G = \Delta H - T\Delta S = 227.4 \text{ kJ} - (298 \text{ K})(0.0588 \text{ kJ/K}) = 209.9 \text{ kJ}$$

Case 4: Increase in Enthalpy, Decrease in Entropy

Case 4 shows that endothermic reactions that decrease entropy ($+\Delta H$, $-\Delta S$) do not occur naturally. To proceed naturally, these reactions would have to store energy in chemical bonds and so increase order. Reactions that exhibit positive enthalpy changes can be forced if other types of energy are used. Otherwise, they are unfavorable at any temperature.

EXAMPLE PROBLEM 13-9

ΔG, Case 4

Calculate the Gibbs free-energy change for the following reaction:

$$2C\,(s) + 2H_2\,(g) \longrightarrow C_2H_4\,(g) \text{ at 298 K.}$$

Solution

$$\Delta H°_f = 52.4 \text{ kJ}$$

$$\Delta S = \left[1 \text{ mol} \times \frac{219.3 \text{ J}}{\text{mol} \cdot \text{K}}\right] - \left[\left(2 \text{ mol} \times \frac{5.7 \text{ J}}{\text{mol} \cdot \text{K}}\right) + \left(2 \text{ mol} \times \frac{130.7 \text{ J}}{\text{mol} \cdot \text{K}}\right)\right]$$

$$= -53.5 \text{ J/K}$$

$$\Delta G = \Delta H - T\Delta S = 52.4 \text{ kJ} - (298 \text{ K})(-0.0535 \text{ kJ/K}) = 68.3 \text{ kJ}$$

13-7 Standard Thermodynamic Property Values

Compound	$\Delta H°_f$ (kJ/mol)	$S°$ (J/mol·K)	ΔG (kJ/mol)	Compound	$\Delta H°_f$ (kJ/mol)	$S°$ (J/mol·K)	ΔG (kJ/mol)
Ag (s)	0	42.6	0	Cl_2 (g)	0	165.2	0
AgBr (s)	−10.4	107.1	−96.9	CO (g)	−110.5	197.7	−137.2
AgCl (s)	−127.0	96.3	−109.8	CO_2 (g)	−393.5	213.8	−394.4
AgI (s)	−61.8	115.5	−66.2	Co (s)	0	30.	0
Ag_2O (s)	−31.1	121.3	−11.2	Cr (s)	0	23.8	0
Ag_2S (s)	−32.6	144.0	−40.7	Cr_2O_3 (s)	−1139.7	81.2	−1058.1
Al (s)	0	28.3	0	Cu (s)	0	33.2	0
Al_2O_3 (s)	−1675.7	50.9	−1582.3	CuO (s)	−157.3	42.6	−129.7
Ba (s)	0	62.5	0	Cu_2O (s)	−168.6	93.1	−146
$BaCl_2$ (s)	−855.0	123.7	−806.7	CuS (s)	−79.5	120.9	−86.2
$BaCO_3$ (s)	−1213.0	112.1	−1134.4	$CuSO_4$ (s)	−771.4	109.2	−662.2
$BaSO_4$ (s)	−1473.2	132.2	−1362.2	F_2 (g)	0	158.8	0
Br_2 (l)	0	152.2	0	Fe (s)	0	27.3	0
C (s)	0	5.7	0	Fe_2O_3 (s)	−824.2	87.4	−742.2
Ca (s)	0	41.6	0	H_2 (g)	0	130.7	0
$CaCl_2$ (s)	−795.4	108.4	−748.8	HBr (g)	−36.3	198.7	−53.4
$CaCO_3$ (s)	−1207.6	917	−1129.1	HCl (g)	−92.31	186.9	−95.3
CaO (s)	−634.9	38.1	−603.3	HF (g)	−273.3	173.8	−275.4
$Ca(OH)_2$ (s)	−985.2	83.4	−897.5	HI (g)	26.50	206.6	1.7
$CaSO_4$ (s)	−1434.5	56.5	−477.4	HNO_3 (l)	−174.1	155.6	−80.7
CCl_4 (l)	−128.2	214.4	−65.3	H_2O (g)	−241.8	188.8	−228.6
CH_4 (g)	−79.6	186.3	−50.5	H_2O (l)	−285.8	69.95	−237.1
C_2H_2 (g)	227.4	200.9	209.9	H_2S (g)	−20.6	205.8	−33.4
C_2H_4 (g)	52.4	219.3	68.4	H_2SO_4 (l)	−814.	156.9	−690.
C_2H_6 (g)	−84.0	229.2	−32.0	Hg (l)	0	75.90	0
C_3H_8 (g)	−103.8	270.3	−23.4	HgO (s)	−90.79	70.25	−58.5
C_6H_6 (l)	49.1	173.4	124.5	HgS (s)	−58.2	82.4	−50.6
CH_3OH (l)	−239.2	126.8	−166.6	I_2 (s)	0	116.14	0
C_2H_5OH (l)	−277.6	160.7	−174.8	K (s)	0	64.7	0

(continued)

13-7 Standard Thermodynamic Property Values

Compound	ΔH°$_f$ (kJ/mol)	S° (J/mol·K)	ΔG (kJ/mol)	Compound	ΔH°$_f$ (kJ/mol)	S° (J/mol·K)	ΔG (kJ/mol)
KBr (s)	−393.8	95.9	−380.7	NO (g)	91.3	210.8	87.6
KCl (s)	−436.5	82.6	−408.5	NO$_2$ (g)	33.2	240.4	51.3
KClO$_3$ (s)	−397.7	143.1	−296.3	O$_2$ (g)	0	205.2	0
KF (s)	−567.3	66.6	−537.8	O$_3$ (g)	142.7	238.9	163.2
KOH (s)	−424.6	81.2	−379.4	Pb (s)	0	51.9	0
Mg (s)	0	32.7	0	PbBr$_2$ (s)	−278.7	161.5	−261.9
MgCl$_2$ (s)	−641.3	89.6	−591.8	PbCl$_2$ (s)	−359.4	136.0	−314.1
MgCO$_3$ (s)	−1095.8	65.7	−1012.1	PbO (s)	−217.3	66.5	−187.9
MgO (s)	−601.6	27.0	−569.3	PbO$_2$ (s)	−277.4	68.6	−217.3
Mg(OH)$_2$ (s)	−924.5	63.2	−833.5	Pb$_3$O$_4$ (s)	−718.4	211.3	−601.2
MgSO$_4$ (s)	−1284.9	91.6	−1170.6	PCl$_3$ (l)	−319.7	217.1	−272.3
Mn (s)	0	32.0	0	S (s)	0	32.1	0
MnO (s)	−385.2	59.7	−362.9	Si (s)	0	18.8	0
MnO$_2$ (s)	−520	53.1	−465.1	SiO$_2$ (s)	−910.7	41.5	−856.3
N$_2$ (g)	0	191.6	0	Sn (s)	0	51.2	0
Na (s)	0	51.3	0	SnCl$_4$ (l)	−511.3	258.6	−440.1
NaBr (s)	−361.1	86.8	−349.0	SnO (s)	−280.7	57.2	−251.9
NaCl (s)	−411.2	72.1	−384.1	SnO$_2$ (s)	−577.6	49.0	−515.8
NaF (s)	−576.6	51.1	−546.3	SO$_2$ (g)	−296.8	248.2	−300.1
NaI (s)	−287.8	98.5	−286.1	SO$_3$ (g)	−395.7	256.8	−371.1
NaOH (s)	−425.8	64.4	−379.7	Zn (s)	0	41.6	0
NH$_3$ (g)	−45.9	192.8	−16.4	ZnO (s)	−350.5	43.7	−320.5
NH$_4$Cl (s)	−314.4	94.6	−262.9	ZnS (s)	−206.0	57.7	−201.3
NH$_4$NO$_3$ (s)	−365.6	151.1	−183.9				

> **EXAMPLE PROBLEM 13-10**
>
> *Calculating ΔG at Different Temperatures*
>
> Determine whether the reaction between ammonia and hydrogen chloride is probable at 298 K and at 1000. K according to Table 13-7.
>
> $$NH_3 \,(g) + HCl \,(g) \longrightarrow NH_4Cl \,(s)$$
>
> **Solution**
>
> $$\Delta H = -314.4 \text{ kJ} - (-45.9 \text{ kJ} - 92.31 \text{ kJ}) = -176.2 \text{ kJ}$$
> $$\Delta S = 94.6 \text{ J/K} - (192.8 \text{ J/K} + 186.9 \text{ J/K}) = -285.1 \text{ J/K}$$
>
> To find the ΔG at 298 K, plug the appropriate values into the free-energy equation. Remember to convert the units of ΔS from J/K to kJ/K.
>
> $$\Delta G = \Delta H - T\Delta S$$
> $$\Delta G = -176.2 \text{ kJ} - (298 \text{ K})(-0.2851 \text{ kJ/K}) = -91.2 \text{ kJ}$$
>
> Since the ΔG is negative, the reaction is favorable at this temperature.
>
> Even though the values of $\Delta H°_f$ and $S°$ at 1000. K are not known for this reaction, the value of ΔG can be estimated by using the 298 K figures in the free-energy equation and solving for ΔG.
>
> $$\Delta G = \Delta H - T\Delta S$$
> $$\Delta G = -176.2 \text{ kJ} - (1000. \text{ K})(-0.2851 \text{ kJ/K}) = 108.9 \text{ kJ}$$
>
> The positive ΔG shows that the reaction is not favorable at 1000. K.

13B Section Review

1. Evaluate why the laws of thermodynamics are excellent scientific arguments against the theory of evolution.
2. Explain the difference between enthalpy and entropy.
3. Arrange the following examples in order of increasing entropy. (Begin with the most ordered and work to the least ordered.)
    a. 1 mole of $H_2O \,(g)$ at 125 °C
    b. 1 mole of $H_2O \,(g)$ at 140 °C
    c. 1 mole of $H_2O \,(s)$
    d. 1 mole of $H_2O \,(l)$
4. Define the following symbols:
    a. S
    b. $\Delta S°$
    c. T
    d. G
    e. ΔG
5. Calculate the standard entropy change ($\Delta S°$) for each reaction.
    a. $2KClO_3 \,(s) \longrightarrow 2KCl \,(s) + 3O_2 \,(g)$
    b. $SnO_2 \,(s) + 2H_2 \,(g) \longrightarrow Sn \,(s) + 2H_2O \,(l)$
6. Calculate the change in free energy (ΔG) for each reaction from Question 5 at 25 °C, and tell whether the reactions are spontaneous or nonspontaneous at this temperature.

THE BIBLE AND THE SECOND LAW OF THERMODYNAMICS
Facets of Chemistry

Two natural tendencies govern all thermodynamically favorable processes. First, matter tends toward stable, low-energy states. Second, natural processes decrease the order of the universe. That second principle is a statement of the *second law of thermodynamics*, also known as the law of increasing entropy.

This law applies to natural processes unaffected by outside manipulation by any intelligence. In other words, scientists can sometimes control conditions in a laboratory to temporarily overcome this limitation. The law of entropy also applies only to *closed systems*, that is, systems where no energy is allowed to enter or leave.

In some natural processes, entropy decreases, but only for a time. When energy enters from outside the system, it can counter the natural increase of disorder and allow an increase in available energy. Photosynthesis is a prime example of this phenomenon. The energy of the sun is captured in a cell's chloroplasts and used to build sugar molecules with high-energy bonds. Because smaller molecules and atoms are combined, entropy has decreased.

When a plant dies or is eaten, however, the energy is converted to other forms and sugar molecules are broken down to form carbon dioxide and water. Thus, entropy has increased overall.

Temporary and localized exceptions to the law of entropy do not threaten its reliability. Any of these contradictions is more than offset by any increase in the entropy of the surroundings. As a result, the entropy of the universe continues to increase.

You may wonder why the universe still has any energy if this law is true. The only possible answer is that God, the Creator, designed and brought forth a universe with an immense

supply of energy. The sun and stars contain vast stores of matter constantly being converted to energy, but even they will eventually expire. Isaiah 51:6 tells us that the heavens will eventually vanish and the earth will age like a garment. For one who does not acknowledge the Creator, this prophecy certainly sounds like bad news. But the verse goes on to promise that God's salvation will continue forever and His righteousness will never end.

For a Christian, the law of entropy is not a threat but a reminder that we live in a world that is broken. In addition to the moral brokenness evident in human actions, even the physical aspects of creation do not operate as God intended. When God finished His creative acts on the sixth day, He said that everything was very good (Gen. 1:31). While it is mere speculation to conjecture too much about how the laws of thermodynamics related to this newly created world, it is safe to say that the death and decay that surround us today did not exist before the Fall.

With Adam and Eve's decision to defy God by eating the forbidden fruit, everything changed. God's curse upon His perfect work of creation brought suffering, death, and decay. Certainly the second law of thermodynamics played some role in this radical transformation of nature.

Romans 8:20–21 teach that the decaying and groaning creation will one day be redeemed and re-created. The new earth will last forever. Will the laws of thermodynamics and entropy be changed on the new earth? Perhaps. Or perhaps the system will be open and God will continually supply sustaining energy.

Serving God as an HVAC Technician

Job Description

An HVAC technician performs a wide variety of skilled tasks necessary to install, repair, and maintain heating, ventilating, and air conditioning (HVAC) systems. He may also deal with water treatment and steam distribution systems in homes and industrial plants. These technicians use a variety of measuring and testing equipment, assemble and install metal ductwork, and plan and design new HVAC systems. HVAC technicians often analyze and solve complex technical problems.

The HVAC technician must demonstrate a professional demeanor and have excellent communication skills, be able to work without direct supervision, and maintain a positive, helpful attitude. Accurate record keeping is also a must.

The job's extensive lifting, climbing, and bending demand a worker who is physically fit. In addition, these technicians may be exposed to bad weather and potentially harmful chemicals and procedures.

Possible Workplaces

Many HVAC technicians serve customers as an employee of an HVAC company and are assigned spe-

cific jobs by supervisors or company owners.

All homes and businesses require some type of HVAC system, so there are practically unlimited workplaces to choose from.

Education

A well-qualified HVAC technician has at least a high-school education. In addition, certifications are required for certain procedures. Some HVAC positions require a four-year degree or equivalent experience in technical fields.

Dominion Opportunities

An HVAC technician may develop new methods or materials to improve working conditions and reduce exposure to dust, fumes, noise, high voltage, and toxic materials. Such inventions would be a great service to the Lord by showing care and love for his fellow man.

The technician could also assist in the design of more energy efficient and environmentally friendly heating, cooling, and ventilating systems. Understanding the principles of thermodynamics and the properties of materials would be essential. A cleaner environment would be safer for humans, animals, and plants.

Chapter Review

Chapter Summary

- Thermochemistry studies the transfer of energy during chemical reactions or phase changes.
- Temperature, measured in degrees Celsius or in kelvins, measures the average kinetic energy of the particles in a substance. Heat, the transfer of thermal energy between substances, is measured in joules.
- The heat content of a system at constant temperature is enthalpy (H). When scientists study the changes in a reaction, they measure the enthalpy of reaction (ΔH), which is defined as the total enthalpy of the products minus the total enthalpy of the reactants.
- The molar enthalpy of fusion (ΔH_{fus}) is the quantity of heat required to melt 1 mol of a solid at its melting point with no temperature change.
- The amount of heat required to convert 1 mol of a liquid at its boiling point to its vapor at the same temperature is called the molar enthalpy of vaporization (ΔH_{vap}).
- The specific heat of a substance is the amount of heat required to raise the temperature of 1 g of the substance 1 °C.
- The standard molar enthalpy of formation ($\Delta H°_f$), also called the heat of formation, is the change in enthalpy that occurs when 1 mol of the compound is formed from its elements at standard state, defined as 298 K and 1 atm of pressure.
- The molar enthalpy of combustion ($\Delta H°_c$) is defined as the energy released as heat by the complete burning of one mole of a substance.
- Hess's law states that the enthalpy change of a reaction equals the sum of the enthalpy changes for each step of the process. That law can be used to calculate the enthalpy of reaction.
- Entropy (S) is a measure of randomness or lack of orderliness in a system. The change in entropy ($\Delta S°$) is the total entropy in products minus the total entropy in reactants. ($\Delta S° = \Sigma S°_{products} - \Sigma S°_{reactants}$)
- The free-energy change (ΔG) for a reaction can be calculated by the equation $\Delta G = \Delta H - T\Delta S$, where T is the temperature in kelvins.

Coming to Terms

thermochemistry	327
enthalpy (H)	327
calorimeter	327
sensible heat	328
latent heat	328
molar enthalpy of fusion (ΔH_{fus})	328
molar enthalpy of vaporization (ΔH_{vap})	328
specific heat (c_{sp})	329
enthalpy of reaction (ΔH)	331
thermochemical equation	332
molar enthalpy of formation (ΔH_f)	332
standard state	332
standard molar enthalpy of formation ($\Delta H°_f$)	332
standard molar enthalpy of combustion ($\Delta H°_c$)	333
Hess's law	335
entropy (S)	339
free energy (G) (Gibbs free energy)	340
free-energy change (ΔG)	340

Review Questions

Concept Review

1. Why must a calorimeter be insulated?
2. What would happen to the temperature and physical state of a mole of liquid diethyl ether at its standard boiling point if its molar enthalpy of vaporization (26.52 kJ/mol) were added?
3. What is wrong with the statement, "As a substance freezes, it absorbs energy equal to its enthalpy of fusion"?

4. From the specific heat table (Table 13-2), what general conclusions can you draw about the relationship between specific heat values and the three states of matter?
5. Why is it necessary to designate the state of matter of each substance in a thermochemical equation?
6. Why are most of the standard molar enthalpies of formation ($\Delta H°_f$) negative?
7. What two tendencies influence all chemical reactions?
8. Which of the following situations are possible according to the laws of thermodynamics?
    a. insect larvae spontaneously forming in rotting meat
    b. the human body converting the energy in food to other forms of energy
    c. the energy and matter in the universe coming into being from nothing without any intervention from God
    d. the invention of an automobile engine that is 100 percent efficient

True or False

9. According to the laws of thermodynamics, matter can be destroyed but energy cannot.
10. A Celsius degree is the same size as a kelvin.
11. The heat measured in the enthalpy of vaporization of a substance is sensible heat.
12. Soup takes longer to cool than toast because water has a higher specific heat.
13. A compound with a negative molar enthalpy of formation value results from an endothermic reaction.
14. Hess's law allows you to combine the steps of a chemical reaction to compute change in entropy.
15. Exothermic reactions are more thermodynamically favorable than endothermic ones.

Application

16. If 2500. J of energy are added to 120. g of benzene at 30. °C, what will be its final temperature?
17. A 47.35 g mass of copper is heated from 20. °C to 85. °C. How many joules were added to accomplish this?
18. How much heat energy is lost by a 50.0 g piece of copper when it cools from 95.0 °C to 20.0 °C? (c_{sp} = 0.39 J/g · °C)
19. If 1000. J of heat energy are removed from 100. g of water at 35.0 °C, what is the final temperature of the water? (c_{sp} = 4.19 J/g · °C)
20. Suppose you have an ice cube with a mass of 8.00 g at a temperature of 295 K. If that ice cube is heated until it is completely converted to steam at 373 K, how many joules of heat must be added?

21. What is the enthalpy of formation (ΔH_f) for the following equation?

$$2C\ (s) + O_2\ (g) \longrightarrow 2CO\ (g)$$

22. For each of these reactions, find the enthalpy of reaction and tell whether the reaction is endothermic or exothermic.
    a. $Hg\ (l) + S\ (s) \longrightarrow HgS\ (s)$
    b. $HgS\ (s) \longrightarrow Hg\ (l) + S\ (s)$
    c. $2C\ (s) + 2H_2\ (g) \longrightarrow C_2H_4\ (g)$
    d. $C_2H_4\ (g) \longrightarrow 2C\ (s) + 2H_2\ (g)$

    For Questions 23–27, refer to these reactions:
    a. $4NH_3\ (g) + 5O_2\ (g) \longrightarrow 4NO\ (g) + 6H_2O\ (g)$
    b. $3NO_2\ (g) + H_2O\ (l) \longrightarrow 2HNO_3\ (l) + NO\ (g)$
    c. $2NH_4NO_3\ (s) \longrightarrow 2N_2\ (g) + O_2\ (g) + 4H_2O\ (g)$

23. Calculate the standard change in enthalpy ($\Delta H°$) for each reaction, and tell whether each reaction is endothermic or exothermic.

24. Calculate the standard entropy change ($\Delta S°$) for each reaction.

25. Calculate the change in free energy (ΔG) for each reaction at 25.0 °C, and tell whether each reaction is favorable (spontaneous) or unfavorable (nonspontaneous) at this temperature.

26. Calculate ΔG for each reaction at 1000. K, and tell whether each reaction is favorable (spontaneous) or unfavorable (nonspontaneous) at this temperature.

27. Calculate the temperature (if any) at which each reaction changes from being spontaneous to nonspontaneous (the temperature at which ΔG equals zero).

DS 28. How is developing an ice pack obeying two important commands from God?

Chemical Kinetics

CHAPTER 14

- 14A Reaction Rates   353
- 14B Reaction Mechanisms   361
- Facet
  - Spontaneous Combustion   356

Dominion Science Problem

Timely Medication

Have you ever taken pain reliever or cold medicine right before bed and then woke up in the middle of the night when the medication wore off? Sometimes medication needs to be much longer-lasting, especially for people who have problems with asthma, seizures, allergies, or severe pain. Ideally, medication should relieve symptoms for an extended period of time without building up a toxic amount of a drug in the body. How can this effect be accomplished?

14A Reaction Rates

14.1 Kinetics

What is the difference between an explosion and a minor reaction? Part of the difference is how much energy is released, as you learned when studying thermodynamics in Chapter 13. Another part of the difference is how fast the reaction takes place. The field of chemistry called kinetics answers questions about the speed at which reactions occur and the way in which they get started. While thermodynamics answers the basic question, "Can it react?" kinetics answers the more specific questions, "Will it react?" and "How fast will it react?" **Kinetics** is the study of the rates of reactions and the steps by which they occur.

The fact that a reaction is thermodynamically favorable does not mean that it will proceed automatically. Some reactions proceed at an extremely slow rate; others are rapid. Some need a push to get started; others proceed on their own. Consider three reactions. Each of them has a negative Gibb's free energy (ΔG), so they are all favorable (spontaneous) under the proper conditions of temperature and pressure.

1. The oxidation of a diamond: $\Delta G = -396.2$ kJ
2. The burning of methane: $\Delta G = -581.2$ kJ
3. The mixing and the neutralization of $Ba(OH)_2$ and H_2SO_4: $\Delta G = -131.0$ kJ

The first reaction will not proceed at a significant rate. The second reaction will not start unless it is given an energetic push (from a lighted match, for example). The third reaction proceeds as soon as the reactants are mixed together. Something besides thermodynamics must be used to explain why reactions have such a wide variety of rates. That "something" is kinetics.

14.2 Energy Diagrams

Thermodynamics relates only to the starting and ending points of a reaction and is therefore path-independent. For an analogy, consider a customer strolling through a multi-level mall, changing levels and going in and out of stores. Thermodynamics is concerned with only the levels at which he enters and exits the mall, that is, with his beginning and ending potential energy. On the other hand, kinetics is path-dependent. Kinetics addresses how a reaction proceeds from point A to point B. Are the energy changes during a reaction simple and direct, or must the reactants gain some energy before the reaction can proceed? What steps are involved? Kinetics seeks to determine what happens between the start and finish of a reaction. In the shopper illustration, kinetics would seek to determine which stores were entered and for how long, and when and where our shopper changed levels.

14A Section Objectives

After finishing this section, you should be able to

- compare and contrast thermodynamics and kinetics.
- explain how to read and use an energy diagram.
- explain how activation energy can prevent thermodynamically favorable reactions from occurring.
- explain the collision theory in your own words.
- interpret a reaction rate graph to determine the change in the concentration of reactants and products over time.
- describe how and why reactants, concentration, temperature, surface area, and the presence of catalysts can affect the rate of a reaction.
- recognize that enzymes belong to a large group of catalysts that control biochemical processes.

Loving Your Neighbor with Chemistry

The second greatest commandment is to love others. You can do this by general kindness to those around you, but you can also do this by becoming a scientist. By understanding the world God created, scientists are able to solve problems created by the Fall (as in the dominion science problem) or prevent problems before they happen (as in this chapter's facet).

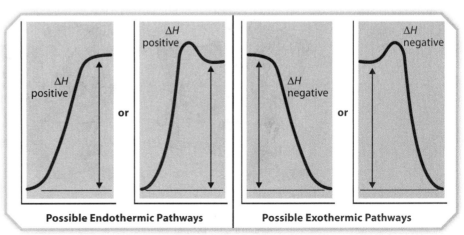

14-1 Possible reaction pathways

Energy changes during a reaction indicate that the reaction progresses from start to finish. These energy changes can be graphed. The horizontal scale represents the progress of the reaction being considered, and the vertical scale represents energy. The highest point on the energy diagram corresponds to the energy needed by the reactants to form products.

14.3 Collision Theory

A hydrogen gas molecule (H_2) can combine with an iodine gas molecule (I_2). A sufficiently forceful collision will disrupt the electron orbitals of these molecules. For a moment, four atoms form one highly energized group; a reaction then occurs to make hydrogen iodide (HI) molecules according to the equation

$$H_2 + I_2 \longrightarrow 2HI.$$

The **collision theory** of reactions outlines three conditions that must be met for any reaction to occur. The first is a fundamental principle of kinetics: (1) particles must collide before they can react.

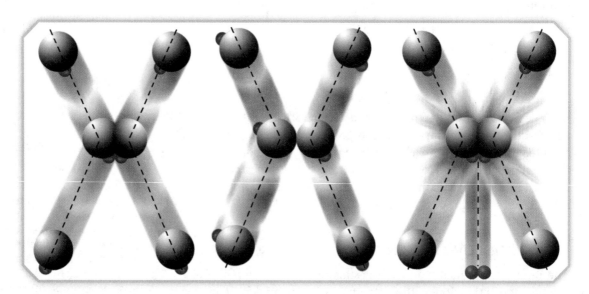

14-2 Collisions have three possible results. Only the third, with enough force and the proper orientation, produces a reaction.

14-3 Conditions for reactions: (1) proper orientation, not enough force, (2) enough force, improper orientation, (3) proper orientation and force

However, a collision alone does not guarantee a reaction. The collision must also be (2) properly oriented for the necessary rearrangement of atoms and electrons and (3) forceful enough—possessing enough energy to form products. Reactions follow only forceful, properly oriented collisions.

The collision theory explains why reactions occur at faster or slower rates, depending on reaction conditions. Any factor that increases the number of effective collisions increases the rate of a reaction. Therefore, the rate at which a chemical reaction proceeds is directly related to the frequency of effective collisions. Fast reactions occur when the three criteria are met easily. However, if even one criterion is difficult, the reaction occurs more slowly.

14.4 Activation Energy and the Activated Complex

Even though a specific reaction may be spontaneous, it will not proceed rapidly unless the initial energy barrier is overcome. The energy needed to jump-start a reaction, called the **activation energy (E_a)**, is the minimum amount of kinetic energy that must be possessed by the colliding molecules before they can react. As a rule, the lower the activation energy, the faster the reaction will occur. Each reaction has a specific activation energy.

Recall from Subsection 13.7 that a reaction can be pictured as a two-step process: (1) the breaking of bonds and (2) the formation of bonds. Although it is convenient to think of the reaction process in these terms, it is somewhat misleading. It is not necessary for all of the bonds to break before new bonds start to form. Those processes happen simultaneously.

Breaking bonds in molecules always requires energy. Consider the formation of liquid water from hydrogen and oxygen gases.

$$2H_2\ (g) + O_2\ (g) \longrightarrow 2H_2O\ (l)$$

If you mix hydrogen gas with oxygen gas in a tank, you won't produce any water, even though the standard enthalpy of formation ($\Delta H°_f$) is −285.8 kJ/mol and the free-energy change (ΔG) is −237.1 kJ/mol, both strong indicators of a favorable reaction. Why don't these diatomic molecules immediately react to form water?

Colliding reactants can form a theoretical, transitional structure between reactants and products, called an **activated complex**. The activated complex is an intermediate substance, but it does not appear as one of the substances shown in the steps of a chemical reaction. It appears briefly as old bonds are in the process of breaking and new bonds are in the process of forming. Its high energy content makes it extremely unstable, so it is very short-lived. The activated complex can break up to form the products of the reaction, or it can revert to the separate reactants. Because it can go in either direction, it is also called the *transition state*.

Although the net reaction in the formation of water is exothermic, the initial breaking of bonds within the hydrogen and oxygen reactants is endothermic. If the activated complex goes on to complete the reaction, a large amount of energy will be released. In exothermic reactions, this energy activates more reactants, and their reaction rate may accelerate. In the formation of water, the

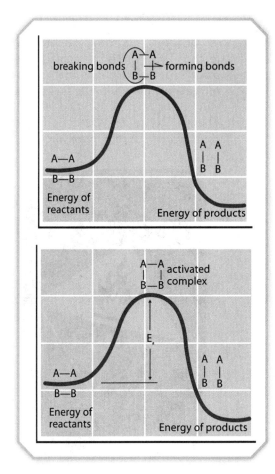

14-4 An activated complex is a transitional structure early in a chemical reaction.

> The activated complex is a stage of a reaction that exists for only a few femtoseconds. A femtosecond is a millionth of a billionth of a second (1 fs = 10^{-15} s). The activated complex lasts for only a short time because atoms move about 1000 m/s, and they have only a short distance to move when breaking and forming bonds.

Spontaneous Combustion
Facets of Chemistry

It was hot that June day in 1933 in North Arlington, New Jersey. Swimmers seeking relief from the oppressive heat swarmed a beach along the bank of the grimy Passaic River. Between them and the road was a storehouse full of pyroxylin—a highly combustible material used in lacquers, plastics, and artificial leather. Pyroxylin is the name for nitrocellulose compounds that can be formed into celluloid, the first pyroxylin plastic. Across the road was the Atlantic Pyroxylin Waste Company. Here, workers sorted pyroxylin scraps

collected from surrounding factories so that they could be used again.

At 9:12 p.m., lingering swimmers noticed fire coming from the roof of the factory. With a puff, both ends of the building blew out, showering the crowd with bricks. Burning celluloid fell on the people, setting many of them afire, and a vast flame swept over the surrounding area. The flames crossed the road and set fire to the small storehouse beside the beach. Another blast occurred, and flames engulfed the beach as well as nine nearby buildings. All told, ten people died and one hundred eighty were injured. The burning of the Atlantic Pyroxylin Waste Company is a classic example of spontaneous combustion.

Spontaneous combustion fires start without a flame or spark. They ignite because of the role that heat plays in speeding up reaction rates. The Atlantic Pyroxylin Waste Company stored pyroxylin scraps swept up from factory floors. These scraps often had machine oil on them. The heat necessary for the fire to start came from the reaction between this oil and atmospheric oxygen. The reaction does not normally generate great amounts of heat because it occurs slowly. Whatever heat is produced is released gradually. However, if the heat energy from this reaction cannot escape because the materials are too compact, it accumulates and the temperature increases. The increase in temperature causes the reaction to proceed faster and to produce more heat. This vicious cycle continues until the reaction proceeds fast enough to generate temperatures above the kindling point of the surrounding material.

Spontaneous combustion can occur in places other than factories. For example, it occurs in soft coal, especially if powdered coal covers a mass of lump coal. As the coal dust slowly reacts with oxygen, the heat cannot dissipate. The temperature increases until the coal ignites. Soft coal should be spread over a large area to allow the heat to escape.

Farmers must also be aware of the dangers of spontaneous combustion. If they pack freshly cut bales of hay into a barn before it is properly dried, oxidation will continue in the confined

space. When a certain temperature is reached, the hay will burst into flames. Hay should be stored in a cool, dry, well-ventilated barn. Compost piles containing leaves and other organic material can also spontaneously combust if not properly ventilated.

Oily, paint-saturated rags pose a threat to many homes and workshops. As paint dries, the oil it contains reacts with oxygen and forms an elastic solid. This reaction produces heat as the paint dries. A pile of oily rags provides the perfect conditions for a cycle of higher temperatures and faster reaction rates to be set up. Again the temperature increases until the rags catch fire. Paint rags should be hung outdoors where there is good air circulation, and then they should be stored in a metal can.

Preventing spontaneous combustion fires is a matter of controlling the rates of chemical reactions. Caution should be taken with combustible materials in powdered forms. The large amount of surface area allows reactions to proceed more quickly. More importantly, combustible material should be stored in a well-ventilated place where heat can be removed as fast as it is produced. This will prevent heat from accumulating in a place where it can increase the rate of reaction.

energy released when water molecules form is more than adequate to allow the reaction to continue without the further introduction of energy. Relating this concept to the energy diagrams discussed previously, you can see that the activation energy can also be defined in terms of the activated complex. It can be thought of as the minimum amount of energy needed to convert the reactants into the activated complex.

In conclusion, the activation energy plays a crucial role in determining how fast a reaction will proceed under given temperature and pressure conditions. A large activation energy can prevent an otherwise favorable reaction from proceeding quickly. Figure 14-5 shows energy diagrams for several reactions.

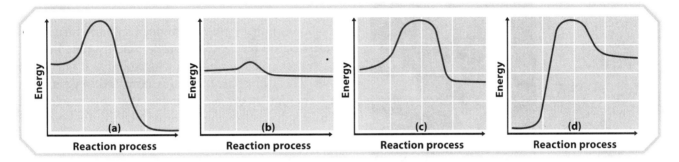

14-5 Which reaction is most likely to proceed rapidly? Reaction (d) requires a large initial input and a net gain in energy; it is doubtful that such a reaction will proceed rapidly on its own. The other reactions release energy but must overcome the activation energy barrier to do so. Reaction (b) has the smallest barrier to overcome, so it is the most likely to proceed rapidly as soon as reactants are mixed.

14.5 Rates of Reactions

During a reaction, the concentrations of reactants and products change constantly. At the beginning of a reaction, the reactants are in their highest concentrations. As the reaction progresses, product concentrations increase as the concentrations of reactants decrease by being consumed. Consider the reaction between iodine chloride and hydrogen.

$$2ICl\,(g) + H_2\,(g) \longrightarrow I_2\,(g) + 2HCl\,(g)$$

Figure 14-6 shows how the concentrations of the substances in this reaction vary with time. Notice that the reactants are being consumed as the products form. Also note that the changes are most rapid early in the reaction and then gradually slow.

Reaction rates tell how fast reactants change into products. They can describe how fast the reactants disappear or how fast the products appear. Although reaction rates may be measured in several different units, they usually tell how fast concentrations change with time. Units such as molarity per second or moles per liter per hour are common.

14.6 Factors That Affect Reaction Rates

Remember that reactions depend on forceful and properly oriented collisions. Anything that increases the frequency, strength, and efficiency of these collisions should increase the rate of a reaction. Five of the most important factors affecting reaction rates are explained here.

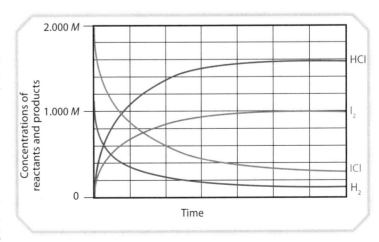

14-6 The concentration of reactants and products in the reaction of ICl with H_2

Nature of Reactants

The most obvious factor that controls the rate of a reaction is the chemical nature of the reactants. Logically enough, reactive substances react quickly. Phosphorus, in its white form, spontaneously bursts into flames when exposed to air. It is so reactive that it must be stored under water. At the other extreme, some of the noble gases rarely or never react. Another variable is the particular combination of elements. Hydrogen and chlorine, under certain conditions, react vigorously. Hydrogen and nitrogen will react only weakly under those same conditions.

Concentration

The typical test for the presence of oxygen gas relies on the fact that high concentrations of reactants increase the rate of a reaction. A glowing wooden splint is thrust into a gas-collection bottle. If the bottle contains pure or relatively pure oxygen gas, the glowing splint will burst into flames. The atmosphere contains only 21% oxygen. This small amount of oxygen supports the slow smoldering while the splint is glowing. An atmosphere of 100% oxygen increases the reaction rate; thus, the combustion proceeds more quickly.

The effect of concentration on the reaction rate varies from reaction to reaction. Doubling the concentration of a reactant in one reaction might not affect the rate at all. Doubling the concentration of a reactant in another reaction could double or even quadruple its rate. No rule can predict the effect of concentration changes on reaction rates. Thus, each reaction must be studied empirically.

14-7 A glowing splint tests for the presence of oxygen gas.

Temperature

A rule of thumb in kinetics is that reaction rates double for every 10 °C rise in temperature. Although there are many exceptions to this general rule, temperature does play a vital role in determining the rates of reactions. Higher temperatures increase reaction rates in two ways—by increasing both the number and the force of collisions between reactants. Higher temperatures increase the number of effective collisions. Inversely, lower temperatures decrease the number of effective collisions and thus decrease the rate of a reaction. However, the energy of the collisions must equal or exceed the activation energy for the activated complex for the reaction to proceed. As the temperature rises, more particles are likely to reach this level of energy.

For example, the decomposition of lactose (a sugar in dairy products) into lactic acid turns fresh milk into a foul-smelling fluid. The lactic acid that is produced gives old milk its distinctive sour taste. Can this reaction be stopped? Not totally, but keeping the milk cool in a refrigerator greatly slows the reaction.

Surface Area

In a mixture of gases or in a solute dissolved in a solvent, frequent collisions allow reactions, once started, to progress rapidly. If the reactants are in two different phases—a solid and a liquid, for example—the reaction rate is dependent on the area of contact

14-8 Magnesium does not noticeably react with water (above), but it reacts readily with steam to produce Mg(OH)$_2$ and H$_2$.

between the substances. Such reactions are called **heterogeneous reactions** because the reactants are in different phases.

Small particles have greater surface area per unit volume than large particles. The increased amount of exposed surface area affects the rate of a reaction by allowing more collisions to occur per unit of time. If a substance is broken up into pieces rather than being in one piece, more surface area is exposed and the reaction will occur faster due to an increase in collision rate. For example, since grain is a staple of our diet, we would not think of it as a hazardous, flammable substance. But when grain is in the form of a fine dust with a greatly increased surface area, it can be deadly. An open flame, a spark from static electricity or an electrical short, or even extreme heat can cause the airborne dust in a grain elevator to spontaneously ignite and explode with the force of a large bomb.

Presence of Catalysts

A **catalyst** is a substance that changes a reaction rate without being permanently changed or consumed by the reaction. It is present during the reaction, but it is neither a reactant nor a product. A catalyst provides an alternate route from reactants to products that is easier, that is, one that has a lower activation energy. However, a catalyst does not affect the concentration of products at the conclusion of a reaction, only the rate at which these products are formed. The catalyst for a given reaction accelerates both the forward and reverse reactions equally. Because it is balanced in its impact, a catalyst will not affect the enthalpy change (ΔH) of a reaction, either in its value or its sign. Neither will it enable a thermodynamically impossible reaction to occur.

How do catalysts work? One theory states that catalysts hold reactants in just the right positions for favorable collisions. For example, a mixture of hydrogen and oxygen gas does not react to form water at room temperature. But if a catalyst of powdered platinum is first introduced, the platinum causes an explosion as its surface becomes covered with adsorbed oxygen. The platinum atoms stretch and weaken the bonds of the O_2 molecules as they are held in place, lowering the activation energy required for the reaction. The oxygen atoms then react rapidly with the hydrogen molecules to form water.

There are several types of catalysts. A **homogeneous catalyst** is in the same phase as the reactants or in solution with a reactant. Homogeneous catalysts combine with one of the reactants to form an intermediate compound that will react more readily with the other reactant or reactants. A catalyst that is in a separate phase from the reactants is said to be a **heterogeneous catalyst**. Heterogeneous catalysts are materials capable of adsorbing molecules of gases or liquids onto their surfaces. Platinum, nickel, palladium, and other finely divided metals and metalloids are examples of heterogeneous catalysts.

14-9 The 46 cm (18 in.) thick concrete walls of this grain elevator were blown out by an explosion caused by the combustion of airborne grain dust.

The action of a catalyst is called *catalysis*.

Adsorption refers to the collection of one substance on the surface of another. It is different from absorption, which involves one substance being penetrated by another.

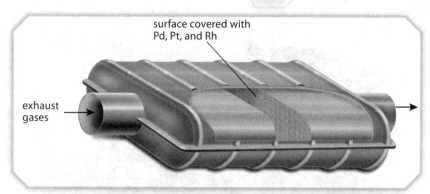

14-10 In the catalytic converter of a vehicle's exhaust system, harmful gases such as carbon monoxide, hydrocarbons, and nitric oxide are adsorbed on a surface of palladium, platinum, and rhodium particles and catalyzed to form harmless gases before emission.

One distinct class of catalysts comprises the naturally occurring biological substances known as **enzymes**. Enzymes are responsible for most of the essential biochemical reactions. More than a thousand enzymes have been identified, and each one is specific to a chemical reaction occurring within a living organism. The presence of an enzyme typically causes a reaction to occur millions of times faster than the same reaction when it is uncatalyzed. Without enzymes, these same processes would occur too slowly to be of any use to the organism. Enzymes also aid living things by providing a lower activation energy and thus a lower temperature at which chemical reactions can occur. The temperature that would be required for most reactions to proceed without the presence of an enzyme would cook the cells and result in tissue damage or the death of the organism.

Some catalysts, such as those that promote spoiling of food, are undesirable. An **inhibitor** is a substance used to reduce a catalyst's undesirable effects. Inhibitors work by bonding to the catalysts to slow the increase in reaction rate, or even to stop it completely. This effect, too, is the result of holding molecules in a certain position, but in this case to prevent collisions. Many important food preservatives act as inhibitors. Aspirin is another valuable inhibitor that works by blocking the action of two different enzymes that cause inflammation and pain.

> Enzymes are typically large proteins, ten to one hundred times larger than the largest molecules scientists have been able to synthesize in the lab. They are very specific to certain reactions, but only a small portion of the total molecule is the active site responsible for the catalysis. At this active site, the target compounds are held in the proper position to encourage collisions and chemical reactions. Secular scientists are baffled by the complexity and specificity of these important molecules. The whole process of enzyme catalysis is so complex that we are just beginning to understand it.

14A Section Review

1. How is the study of kinetics different from the study of thermodynamics?
2. Give two reasons that a collision between two reactive molecules might not result in a reaction.
3. Draw energy diagrams for both an endothermic and an exothermic reaction. Include the following labels in your diagram: reactants, products, activated complex, activation energy, and ΔH.
4. What are the five variables that can affect the rate of reaction?
5. How does a catalyst speed up a reaction rate?
6. Does a catalyst affect a reaction's activation energy? its enthalpy? Explain.
7. Write a sentence that properly relates these terms: *catalyst, biochemical, proteins, enzymes.*
8. A consumer product is available that, when applied to apple slices, prevents the enzymatic process that normally causes them to react with atmospheric oxygen and turn brown. Chemically speaking, what term describes this anti-browning agent?

14B Reaction Mechanisms

14.7 Mechanisms

Chemists can easily identify the reactants of a chemical process. They can also routinely analyze the products. However, the fragments of molecules that exist during reactions are much more difficult to pinpoint. For chemists, observing reactions is like studying a factory's processes from outside the building. They can see the raw materials going in and the finished products coming out, but they cannot see the individual steps involved in forming the products. The series of steps that make up a reaction is called a **reaction mechanism**.

Consider the reaction between hydrogen and iodine gases, one of the first to be studied in the field of kinetics.

$$H_2(g) + I_2(g) \longrightarrow 2HI(g)$$

But how does this reaction take place? Do the two molecules collide to form an activated complex that splits into two hydrogen iodide molecules? Do the atoms in both molecules split and then rearrange themselves? Does the iodine molecule split, surround the hydrogen molecules, and separate them? Or does something totally different occur? Theoretically, any one of these proposed mechanisms could produce hydrogen iodide from hydrogen and iodine. Often, a possible reaction mechanism is made of two or more possible steps, called **elementary steps**. The **rate-determining step** is the slowest elementary step in the reaction—the one that limits how fast the reaction will occur.

14B Section Objectives

After finishing this section, you should be able to

- recognize that a reaction mechanism describes the steps that make up a chemical reaction.
- determine which substances in a complex reaction are intermediates.
- state the purpose of rate laws and interpret a rate law as it relates to changes in reactant concentration.
- explain why some reactants in a chemical reaction do not appear in that reaction's rate law.
- state the connection between rate-determining steps and rate laws.
- formulate a rate law from a data table listing changes in reactant concentration.

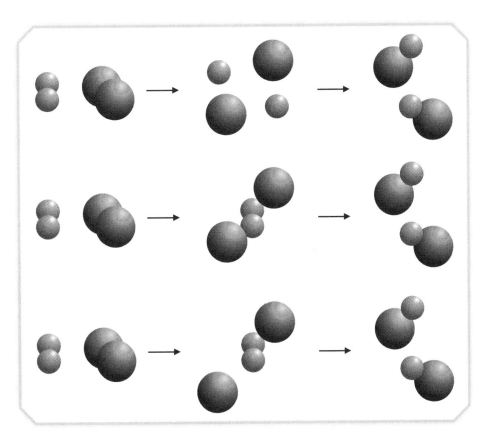

14-11 Three of the possible reaction mechanisms for the reaction between hydrogen gas and iodine gas

Compare the rate-determining step with the principle of an assembly line. Suppose that one job on an assembly line requires the worker to perform a task that is much more difficult than that of any other worker. The product can be produced only as fast as the worker with the difficult task completes his job.

Together, the individual elementary steps make up a **complex reaction**. During a complex reaction, there are substances that are formed in one step and consumed in the next step. These are the **intermediates**. Intermediates do not appear in the net chemical equation.

Here are some possible reaction mechanisms for the production of hydrogen iodide (HI). The phases have been omitted to make it easier to follow the steps. However, all of the reactants and products of this reaction are gases. Since they are all in the same phase, this is an example of a **homogeneous reaction**.

$$\text{Elementary Step 1:} \quad I_2 \rightleftharpoons 2I$$
$$\text{Elementary Step 2:} \quad 2I + H_2 \rightleftharpoons 2HI$$
$$\text{Net Chemical Equation:} \quad I_2 + H_2 \rightleftharpoons 2HI$$

Note how the iodine atoms cancel out. They are the intermediates.

A second possible reaction mechanism has three steps.

$$\text{Elementary Step 1:} \quad I_2 \rightleftharpoons 2I$$
$$\text{Elementary Step 2:} \quad I + H_2 \rightleftharpoons H_2I$$
$$\text{Elementary Step 3:} \quad H_2I + I \rightleftharpoons 2HI$$
$$\text{Net Chemical Equation:} \quad I_2 + H_2 \rightleftharpoons 2HI$$

Note that this reaction mechanism has an additional intermediate, H_2I.

Reaction mechanisms are temperature-dependent; therefore, temperature is the deciding factor in determining which of the possible mechanisms will actually occur.

14.8 Rate Laws and Reaction Orders

As stated before, the concentrations of reactants influence reaction rates. However, the effect of concentration varies from reaction to reaction. For example, the rate of this reaction,

$$H_2(g) + I_2(g) \longrightarrow 2HI(g),$$

is experimentally determined to be directly proportional to the concentration of hydrogen gas in the reaction vessel. The concentration of reactants in terms of moles per liter (molarity) is signified by square brackets. The expression $[H_2]$ represents the molar concentration of the hydrogen gas and means the same as $[H_2]^1$. It has been experimentally determined that doubling the concentration of hydrogen doubles the reaction rate. Tripling the concentration triples the reaction rate. The rate of this reaction is also experimentally determined to be directly proportional to the concentration of iodine gas, $[I_2]$. The **reaction order** for a particular reactant indicates how the rate of a reaction is affected by that specific reactant's concentration.

Because the reaction rate is directly proportional to the concentration of each of these gases, they have an assumed exponent of 1. The reaction is said to be first order for both of these reactants.

$$\text{rate} \propto [H_2][I_2]$$

This expression, called a rate law, can be converted to an equation by the insertion of a numerical constant k. A **rate law** is an equation that mathematically describes how fast a reaction occurs. Each reaction has its own experimentally determined **specific rate constant (k)**.

$$\text{rate} = k[H_2][I_2]$$

Table 14-1 gives several examples of rate laws. It is important to note that rate laws must be determined experimentally. If every chemical reaction occurred in a single step as shown in its balanced chemical equation, with no intermediates, it would be accurate to use the coefficients from the reactants as exponents for those substances. Single-step reactions are rare, however, so this is seldom an accurate method for writing rate laws.

14-1 Rate Laws for Several Reactions

Reaction	Rate =
$2H_2 + 2NO \longrightarrow N_2 + 2H_2O$	$k[H_2][NO]^2$
$2NO + Br_2 \longrightarrow 2NOBr$	$k[NO]^2[Br_2]$
$O_3 + NO \longrightarrow NO_2 + O_2$	$k[O_3][NO]$
$2NO + O_2 \longrightarrow 2NO_2$	$k[NO]^2[O_2]$
$C_4H_9Br + 2H_2O \longrightarrow C_4H_9OH + Br^- + H_3O^+$	$k[C_4H_9Br]$

In some cases, not all reactants appear in the rate law for the reaction. Changing the concentration of these reactants has no impact at all on the reaction rate. For that reactant, the reaction is zero order. The reactant would therefore not be included in the rate law. Can you find an example of such a reaction in Table 14-1?

Balanced chemical equations show only the reactants and the products they form, not the steps the reactants take to form those products. Reactions often have several possible mechanisms, and an equation alone cannot reveal which mechanism occurred. If there are several possible reaction mechanisms, the rate law can sometimes be used to eliminate one or more of the possibilities. A possible reaction can be eliminated if the experimentally determined rate law did not match the theoretical rate law derived from the proposed mechanism.

The reaction between nitrogen dioxide and carbon monoxide at low temperatures has the following rate law.

$$NO_2 (g) + CO (g) \longrightarrow NO (g) + CO_2 (g); \text{rate} = k[NO_2]^2$$

According to the rate law, changing the concentration of carbon monoxide (CO) has no effect on the rate of the reaction. Therefore, the concentration of carbon monoxide does not appear in the rate law. If this reaction is broken apart into the two most likely elementary steps, you can see why this is true.

$$\text{Elementary Step 1:} \quad NO_2 + NO_2 \rightleftharpoons NO_3 + NO$$
$$\text{Elementary Step 2:} \quad NO_3 + CO \rightleftharpoons NO_2 + CO_2$$
$$\text{Net Chemical Equation:} \quad NO_2 + CO \rightleftharpoons NO + CO_2$$

Scientists have experimentally determined that the first step, in which the nitrogen dioxide molecules collide, occurs at a slower rate than the second step. Consequently, the first step is the rate-determining step. Rate laws are always based on the rate-determining step. In reactions with a single step, the rate law would then be based on the original balanced equation. Because the carbon monoxide does not appear in the rate-determining step, the reaction rate does not depend on its concentration.

The exponent 2 on the $[NO_2]^2$ term of the rate law indicates that the concentration of nitrogen dioxide plays the crucial role in determining how fast the reaction occurs. If the concentration of nitrogen dioxide is doubled, we add a coefficient of 2 within the brackets, and the rate quadruples. This relationship occurs because the coefficient 2 in the brackets is raised to the second power, $[2NO_2]^2$. The reaction is said to be second order in NO_2.

Example Problem 14-1

Formulating a Rate Law

Nitrogen monoxide gas and chlorine gas combine as shown in the following equation.

$$2NO + Cl_2 \longrightarrow 2NOCl$$

If a laboratory scientist collects the following data on three trials of this reaction, state the reaction order for each reactant and determine an appropriate rate law.

Reaction Data for $2NO + Cl_2 \longrightarrow 2NOCl$			
Trial	Initial [NO] (M)	Initial [Cl_2] (M)	Initial rate [mol/(L · min)]
1	0.50	0.50	1.90×10^{-2}
2	1.00	0.50	7.60×10^{-2}
3	1.00	1.00	15.20×10^{-2}

Solution

Notice that when the concentration of NO was doubled from Trial 1 to Trial 2, the reaction rate went up fourfold. Therefore, the reaction is second order in NO.

When the concentration of Cl_2 was doubled from Trial 2 to Trial 3, the reaction rate also doubled. The reaction is first order in Cl_2. In light of this,

$$\text{rate} = k[NO]^2[Cl_2].$$

EXAMPLE PROBLEM 14-2

Interpreting a Rate Law

For the rate law

$$\text{rate} = k[A][B]^3,$$

what will happen to the rate of the reaction if
a. the initial concentration of A is doubled?
b. the initial concentration of A is cut in half?
c. the initial concentration of B is doubled?

Solution
a. The reaction rate will double.
b. The reaction rate will be cut in half.
c. The reaction rate will be eight times faster.

14.9 Using Kinetics to Solve Problems

Pharmacokinetics is the study of how the human body processes medication. Pharmacokinetics has many branches of study. The absorption, distribution, metabolism, and excretion of a drug can be studied to extend its beneficial effects. Absorption is the study of how the body takes a drug into the bloodstream from its point of entry. The distribution process is how the body spreads out the drug, which varies with the drug since different parts of the body are affected differently by a given drug. The body then metabolizes the drug, or breaks it down into different components. The liver is the organ that is chiefly responsible for metabolizing drugs. Finally, the drug is eliminated from the body, usually by means of the kidneys. Sometimes drugs simply build up in the body.

The metabolism process of a drug is crucial. The rate at which this process takes place depends on the quantity of the drug being metabolized and on the rate at which the liver can metabolize it. Blood flow and the number of enzymes available to process the drug also affect how fast it is metabolized. The metabolism process, a series of elementary steps, determines how fast the drug takes action to relieve symptoms. The intermediates in the metabolism process, called *metabolites*, may contribute to the side effects of a drug.

Pharmacokinetics can help people with chronic illnesses. For people who need long-term relief of certain symptoms, slowing down the body's absorption, distribution, metabolism, and excretion of medications can extend the duration of their relief from pain, seizures, asthma, and allergies. Time-release medications, which keep a more even concentration of medication in the body, can produce this effect. Sometimes derivations of medications can lessen the side effects and still provide relief from chronic illnesses.

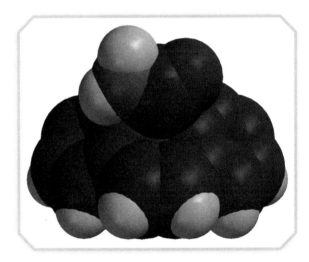

14-12 Carbamazepine is an anti-seizure medication that can be taken in a time-release form.

14B Section Review

1. Why do intermediates not appear in a net chemical equation?
2. There is usually one step in a multi-step reaction that is rate-determining. How does the rate-determining step affect the overall rate of the reaction?

3. What is the net chemical equation for the mechanism of this complex reaction?

   Elementary Step 1: $Cl + O_3 \longrightarrow O_2 + ClO$
   Elementary Step 2: $O_3 \longrightarrow O_2 + O$
   Elementary Step 3: $ClO + O \longrightarrow Cl + O_2$

4. What is the difference between a homogeneous catalyst and a homogeneous reaction?

5. In addition to the concentration of one or more of the reactants, what is found in all rate laws?

6. Why is it generally impossible to use only a balanced chemical equation to predict a rate law for a particular reaction?

7. What is the significance of the rate-determining step to the formulation of a rate law?

DS 8. The anti-seizure drug carbamazepine ($C_{15}H_{12}N_2O$) has been developed in the form of a time-release medication. What factors must pharmacologists consider when developing such a medication?

DS 9. Carbamazepine renders certain medications ineffective by binding with the enzymes that those medications need. What term describes carbamazepine's behavior in such cases?

Chapter Review

Chapter Summary

- Kinetics is the study of the rates of reactions and the steps by which they occur.
- Energy diagrams show the steps in a reaction, including the activation energy, rate-determining steps, the activated complex, and enthalpy changes.
- According to the collision theory, chemical reactions are caused by collisions that are properly oriented and forceful enough to form products.
- The activation energy (E_a) is the minimum amount of kinetic energy that must be possessed by the colliding molecules before they can react.
- The activated complex is a transitional structure between the reactants and the products. It is unstable and short-lived.
- Reaction rates tell how quickly reactants change into products. They are affected by the nature of the reactants, the concentration of the reactants, the temperature, the surface area, and the presence of catalysts.
- Enzymes are organic catalysts responsible for thousands of biochemical reactions.
- The series of steps that make up a reaction is called a reaction mechanism.
- Complex reactions have at least two elementary steps and form intermediate substances that do not appear in the net chemical equation.
- A rate law is an equation that mathematically describes how fast a reaction occurs. These equations take the form

$$\text{rate} = k[A]^m[B]^n,$$

where k is the specific rate constant, A and B are the molar concentration of reactants, and m and n are the reaction orders for A and B, respectively.

Coming to Terms

kinetics	353
collision theory	354
activation energy (E_a)	355
activated complex	355
reaction rate	357
heterogeneous reaction	359
catalyst	359
homogeneous catalyst	359
heterogeneous catalyst	359
enzyme	360
inhibitor	360
reaction mechanism	361
elementary step	361
rate-determining step	361
complex reaction	362
intermediate	362
homogeneous reaction	362
reaction order	362
rate law	363
specific rate constant (k)	363
pharmacokinetics	365

Review Questions

Concept Review

1. Contrast thermodynamics and kinetics by telling which questions each can answer.
2. The reactants for a reaction with a negative ΔG are mixed, but no reaction appears to occur. Suggest an explanation for the nonreaction.
3. According to the collision theory,
   a. why does an increased temperature increase the reaction rate?
   b. why does a greater concentration of reactants often increase the reaction rate?

c. why do powders react more quickly than crystals?
d. why will a reaction involving the collision of three molecules proceed more slowly than one involving the collision of two molecules (all other things being equal)?

4. Why do activated complexes and intermediates not appear in net chemical equations?

5. What property is usually measured in units such as mol/(L · min)?

6. Sugar needs temperatures much higher than 98.6 °F in order to burn. Yet sugar can be "burned" in your cells at this temperature. What is responsible for the ability of your body to burn sugar at this low temperature? Draw energy diagrams that illustrate the difference between the two situations.

7. In reactions that have several possible reaction mechanisms, what factor determines which path it will follow?

DS 8. Oxcarbazepine ($C_{15}H_{12}N_2O_2$) is a derivative of carbamazepine. This drug, which serves the same purpose as carbamazepine, has fewer and less serious side effects. How can this be, if it is so similar to carbamazepine?

True or False

9. Kinetics is path-independent.

10. The highest-energy step in an energy diagram corresponds to the rate-determining step.

11. The breaking of bonds within molecules is endothermic, even in exothermic reactions.

12. Refrigeration delays the souring of milk by slowing the chemical decomposition of lactose into lactic acid.

13. The surface area of a reactant influences reaction rates most in homogeneous reactions.

14. Inhibitors are substances that cause the formation of intermediates.

Application

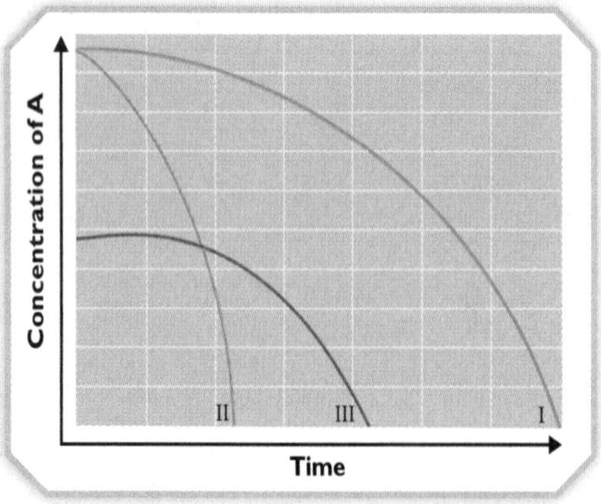

15. This diagram shows how fast an unnamed substance (A) reacts. Curves I, II, and III correspond to the reaction under different conditions.
    a. Under which set of conditions does the reaction proceed at the highest rate?
    b. Of reaction conditions I and III, which starts with the highest reactant concentration?
    c. If reaction conditions I and II occur without a catalyst, which one probably occurs at the higher temperature?
    d. If the temperatures for I and II are identical, which occurs in the presence of a catalyst?

16. The two curves in the energy diagram to the right represent the reaction 2KClO$_3$ (s) ⟶ 2KCl (s) + 3O$_2$ (g) occurring under two different sets of reaction conditions.

    a. Which set of conditions requires the lower activation energy?
    b. Which reaction pathway would more likely include a catalyst?
    c. Which reaction pathway requires higher temperatures?

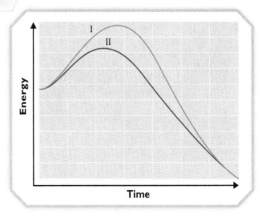

17. The reaction 2NO (g) + 2H$_2$ (g) ⟶ N$_2$ (g) + 2H$_2$O (l) occurs in two steps, which could be either

    A: (1) 2NO + H$_2$ ⟶ N$_2$ + H$_2$O$_2$
    (2) H$_2$O$_2$ + H$_2$ ⟶ 2H$_2$O

    or

    B: (1) 2NO + H$_2$ ⟶ N$_2$O + H$_2$O
    (2) N$_2$O + H$_2$ ⟶ N$_2$ + H$_2$O.

    a. What are processes A and B called?
    b. How can scientists decide whether A or B is correct?
    c. Suppose a scientist experimentally determines that the slower step for reaction mechanism A is step 1. He discovers that both gases influence the reaction rate and that the reaction goes four times faster when the concentration of nitrogen monoxide is doubled. What is the rate law for this reaction?

18. Propose a possible two-step mechanism for the decomposition of HgO according to this equation.

    2HgO (s) ⟶ 2Hg (l) + O$_2$ (g)

19. A scientist postulates that two possible rate laws for a reaction are as follows:

    A: rate = $k[NO_2][F_2]$

    and

    B: rate = $k[NO_2]^2[F_2]$.

    a. How could you tell which rate law is correct?
    b. What is k called?
    c. What is the reaction order of F$_2$ in rate law A?
    d. What is the reaction order of NO$_2$ in rate law B?
    e. In which of the rate laws would a change in the concentration of NO$_2$ have a greater influence?
    f. How would doubling the concentration of F$_2$ in rate law A affect the reaction rate?
    g. How would doubling the concentration of NO$_2$ in rate law B affect the reaction rate?
    h. From this information, can you tell whether the reaction is a complex reaction or a single-step reaction?

20. Acetone and bromine combine as shown in this equation.

$$C_3H_6O + Br_2 \longrightarrow C_3H_5OBr + HBr$$

If a laboratory scientist collects the following data on three trials of this reaction, state the reaction order for each reactant and determine an appropriate rate law.

\multicolumn{4}{c}{Reaction Data for $C_3H_6O + Br_2 \longrightarrow C_3H_5OBr + HBr$}			
Trial	Initial $[C_3H_6O]$ (M)	Initial $[Br_2]$ (M)	Initial rate [mol/(L·min)]
1	0.10	0.10	1.64×10^{-5}
2	0.10	0.20	1.64×10^{-5}
3	0.20	0.10	3.28×10^{-5}

CHEMICAL EQUILIBRIUM

CHAPTER 15

DOMINION SCIENCE PROBLEM

Hypoxia

It is frightening to watch someone struggling to breathe. Gasps, bluish skin, a heaving chest, and flared nostrils indicate that a person's body is deprived of something critical to life—oxygen. The lack of oxygen in the bloodstream is called *hypoxia*. If you have ever hiked at high elevations and experienced headaches, nausea, fatigue, or light-headedness, you probably were experiencing mild hypoxia. A person with severe hypoxia needs help *fast*. Untreated, it can cause a seizure, coma, and eventual death. What can you do to help someone with severe hypoxia?

15A Theories of Chemical Equilibrium 372

15B Applications of Equilibrium Chemistry 384

Facet

Fritz Haber (1868-1934): Gases for War and Peace 384

15A Theories of Chemical Equilibrium

15A Section Objectives

After finishing this section, you should be able to

- define *reversible reactions*.
- distinguish between a static and a dynamic equilibrium.
- write the equation for an equilibrium constant of a reaction when given the balanced chemical equation.
- calculate the equilibrium constant for a reaction when given concentrations of substances at equilibrium.
- calculate the concentration of a substance in an equilibrium mixture when given the equilibrium constant and the concentration of at least one of the other substances.
- list the four potential stresses that can affect reactions at equilibrium.
- predict the direction a reaction will shift when stressed.

When you think about chemical reactions, you probably assume that most reactions continue to completion. You assume that reactions completely change one or more reactants into one or more products. Reactions that conform to those assumptions are **irreversible reactions**. In reality, most natural processes are **reversible reactions**. Reversible reactions occur in both the forward and reverse directions. As products form, a portion of those products simultaneously reverts to reactants. In reversible reactions,

1. both reactants and products exist together as a mixture;
2. the ratio in which reactants and products are present in the mixture depends on the relative speeds of the forward and reverse reactions;
3. changing the rate of either the forward or the reverse reaction changes the ratio of products and reactants in the reaction mixture.

A chemical equilibrium results when forward and reverse reactions proceed simultaneously. This idea—like the kinetic theory and the atomic theory—is one of the big ideas of chemistry that explains many observations. While the study of thermodynamics determines whether the reaction can occur and kinetics determines how fast it will occur, the study of equilibrium determines how far the reaction will proceed.

15.1 Dynamic Equilibrium

When you think of the term *equilibrium*, you probably picture a balanced scale or a never-ending game of tug-of-war. Yet these examples are only one type of equilibrium—the static equilibrium of balanced physical forces. In chemistry, *equilibrium* refers to balanced changes, not balanced forces.

A **chemical equilibrium** exists when two opposing reactions occur simultaneously at the same rate, balancing each other. Unlike the examples of balanced forces, chemical equilibria are dynamic. The particles involved constantly move and react. Particle activity continues at a hectic pace even though no macroscopic changes seem to be happening.

Consider this reaction:

$$2C_2H_6 + 7O_2 \longrightarrow 6H_2O + 4CO_2.$$

As molecules of ethane (C_2H_6) and oxygen mix, they collide and produce some water and carbon dioxide. As molecules of water and carbon dioxide accumulate, the chances that they will collide with each other increase. Some of these collisions result in the re-formation of ethane and oxygen molecules. As more water and carbon dioxide molecules accumulate, more ethane and oxygen molecules re-form.

$$2C_2H_6 + 7O_2 \longrightarrow 6H_2O + 4CO_2$$

$$6H_2O + 4CO_2 \longrightarrow 2C_2H_6 + 7O_2$$

15-1 As long as this canoeist bails out water as fast as it enters the canoe, she maintains a dynamic equilibrium. What will happen if she bails at a slower rate?

Eventually the two reactions proceed at the same rate. To an observer it appears that the amounts of ethane, oxygen, water, and carbon dioxide remain constant; yet at the molecular level, the two opposing processes continue. The total system is designated by an equation with a special equilibrium sign.

$$2C_2H_6 + 7O_2 \rightleftharpoons 6H_2O + 4CO_2$$

For example, water molecules in a sealed jar continue to evaporate even after the air is completely saturated. The vapor molecules condense at the surface of the water at the same rate that others evaporate.

$$\text{liquid} \rightleftharpoons \text{vapor}$$

Dissolving processes also reach dynamic equilibria at their saturation points. Particles leave and rejoin an undissolved solid at the bottom of the solution at equal but opposing rates.

$$\text{solid particles} \rightleftharpoons \text{dissolved particles}$$

As you can see, chemical equilibria are vitally important. Many industrial processes increase their productivity by manipulating reversible reactions. But how can equilibrium reactions be controlled?

15.2 Equilibrium Concentrations

At equilibrium, the rate of the forward reaction equals the rate of the reverse reaction. However, at equilibrium the concentration of reactants does not necessarily equal the concentration of products. If a reaction is *product favored*, then the concentration of the products is higher than that of the reactants at equilibrium. In those reactions, the equilibrium is said to lie to the right. If a reaction is *reactant favored*, then the concentration of the reactants is higher than that of the products at equilibrium. In those reactions, the equilibrium is said to lie to the left.

The double-arrow equilibrium sign can be used to indicate the favorable direction in an equilibrium reaction. The side which is favored has a longer arrow pointing to it.

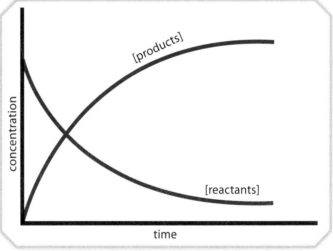

15-2 Graph of a product-favored reaction

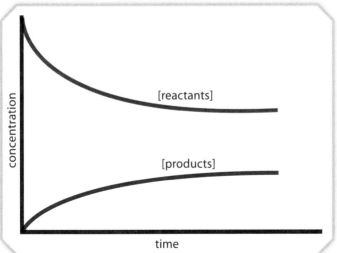

15-3 Graph of a reactant-favored reaction

For example, consider the reaction of hydrogen and iodine gas to form hydrogen iodide gas.

$$H_2\,(g) + I_2\,(g) \rightleftharpoons 2HI\,(g)$$

This reaction is reversible when it takes place in an enclosed flask. This chemical reaction has been well studied in experiments to determine the concentrations of reactants and products at equilibrium. For example, at 425 °C, the reaction begins with a 0.0175 M concentration of both H_2 and I_2 and with no HI. The concentrations of H_2 and I_2 both decrease by 0.0138 M, and the concentration of HI increases by 0.0276 M. At equilibrium, the concentrations of H_2 and I_2 are 0.0037 M and the concentration of HI is 0.0276 M. See this information summarized in the table below. From the graph in Figure 15-4 and from the table below, you can tell that there is more HI than H_2 and I_2 at equilibrium. Thus, this reaction is product favored, and it could be written as

$$H_2\,(g) + I_2\,(g) \rightleftharpoons 2HI\,(g).$$

> Because this reaction is reversible, it is also possible to reach equilibrium by starting with a concentration of HI and forming H_2 and I_2.

15-1 Equilibrium Concentrations of H_2, I_2, and HI at 425 °C

	$H_2\,(g)$	+	$I_2\,(g)$	$\rightleftharpoons$	$2HI\,(g)$
beginning concentration (M)	0.0175		0.0175		0
change in concentration (M)	−0.0138		−0.0138		+0.0276
equilibrium concentration (M)	0.0037		0.0037		0.0276

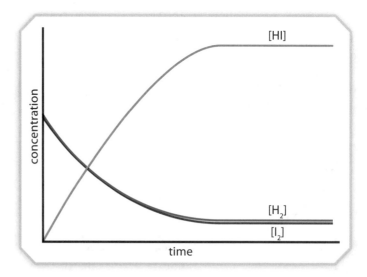

15-4 Graph of the reaction of H_2 and I_2 over time to form an equilibrium with HI

15.3 Equilibrium Constants

Chemists often describe an equilibrium using an equilibrium constant. Consider the reaction

$$A + B \rightleftharpoons C + D.$$

Suppose that experimentation shows that the rate law for the forward reaction (formation of products) is

$$R_f = k_f[A][B].$$

Also suppose that the rate law for the reverse reaction (formation of reactants) is

$$R_r = k_r[C][D].$$

Square brackets signify concentration in units of moles per liter, or molarity. By definition, chemical equilibrium is attained when the forward reaction rate matches the reverse reaction rate.

$$R_f = R_r$$

As a result, the two rate laws can be equated.

$$R_f = R_r$$

$$k_f[A][B] = k_r[C][D]$$

A few algebraic operations can rearrange this equation to a new form.

$$\frac{k_f}{k_r} = \frac{[C][D]}{[A][B]}$$

Both k_f and k_r are constants. Therefore, their quotient is also a constant. This quotient of rate constants is called an **equilibrium constant** and is denoted by K_{eq}.

$$\frac{k_f}{k_r} = K_{eq} = \frac{[C][D]}{[A][B]}$$

> Remember that square brackets in the rate law represent concentration in moles per liter, or molarity (*M*).

An equilibrium constant is a numerical expression of the ratio of the product concentration to reactant concentration. If the value of K_{eq} is greater than 1, the forward reaction predominates and the concentrations of products (in the numerator) are greater than the concentrations of the reactants (in the denominator). If the value of K_{eq} is less than 1, the reverse reaction predominates and the concentration of products is small compared to the concentration of reactants.

The equilibrium constant is determined from the stoichiometry of the reaction. The concentration of each substance is raised to a power that matches the coefficient of that substance in a balanced equation.

$$aA + bB \rightleftharpoons cC + dD$$

$$K_{eq} = \frac{[C]^c[D]^d}{[A]^a[B]^b}$$

It is possible to determine the equilibrium constant for a reversible reaction because of the **law of chemical equilibrium**. This law states that *at a fixed temperature, a chemical system may reach a point at which the ratio of the concentration of the products to the reactants is constant.*

15-5 Do you think the equation for this reaction is product or reactant favored?

Constants for Homogeneous Equilibria

In the equation $I_2 (g) + H_2 (g) \rightleftharpoons 2HI (g)$, the relationship for the equilibrium constant would be written

$$K_{eq} = \frac{[HI]^2}{[I_2][H_2]}.$$

This reaction is in **homogeneous equilibrium** because all of the reactants and products are in the same state. The notation (g) next to each reactant and product tells us that they are all gases.

> **EXAMPLE PROBLEM 15-1**
> *Formulating Equilibrium Constants*
>
> Write the equilibrium constant for the formation of ammonia from its elements. The reaction is $3H_2 (g) + N_2 (g) \rightleftharpoons 2NH_3 (g)$.
>
> **Solution**
>
> $$K_{eq} = \frac{[NH_3]^2}{[H_2]^3[N_2]}$$

Constants for Heterogeneous Equilibria

Not all reactions that reach equilibrium are limited to substances in the same state. Some involve a combination of gases and solids or liquids (solvents). These can reach a **heterogeneous equilibrium**, that is, one in which not all of the substances are in the same physical state.

Writing equilibrium constants for heterogeneous equilibria requires an additional bit of knowledge. The concentrations of solids and solvents (liquids) are both defined as 1; therefore, these substances do not affect equilibrium constants. For instance, the equilibrium between solid silver chloride and its dissolved ions in a saturated solution is not affected by the amount of undissolved solid present.

$$AgCl (s) \rightleftharpoons Ag^+ (aq) + Cl^- (aq)$$

> The concentrations of pure solids and liquids are not included in equilibrium constants. However, gases and aqueous substances are included.

The equilibrium constant for this reaction would normally include the AgCl in the denominator, but since the concentration of the solid [AgCl] is defined as 1, it drops out of the expression. This seemingly arbitrary rule makes sense because the amount of undissolved silver chloride that remains at the bottom of a saturated solution does not affect the equilibrium. The concentration of dissolved ions would remain the same even if an entire handful of silver chloride were added.

$$K_{eq} = \frac{[Ag^+][Cl^-]}{[AgCl]}$$

$$K_{eq} = [Ag^+][Cl^-]$$

The sublimation of solid iodine crystals into iodine gas in a closed system also reaches a heterogeneous equilibrium.

$$I_2 (s) \rightleftharpoons I_2 (g)$$

$$K_{eq} = [I_2]$$

Again, the equilibrium is not dependent on the amount of solid iodine that might be present, so it is not included in the expression.

When acetic acid molecules mix with water, some of them release hydrogen ions. The reaction adds hydrogen ions to some water molecules. Since only an insignificant amount of water is used up in the reaction, the concentration of water molecules is defined as 1. Although the equilibrium constant could include the concentration of water, it should be left out.

$$HC_2H_3O_2 (aq) + H_2O (l) \rightleftharpoons C_2H_3O_2^- (aq) + H_3O^+ (aq)$$

$$K_{eq} = \frac{[C_2H_3O_2^-][H_3O^+]}{[HC_2H_3O_2]}$$

Example Problem 15-2
Formulating Equilibrium Constants

Write the equilibrium constant for the precipitation of zinc hydroxide. The equation is

$$Zn^{2+} (aq) + 2OH^- (aq) \rightleftharpoons Zn(OH)_2 (s).$$

Solution

Since zinc hydroxide is a solid, its concentration is 1 in the equation for the equilibrium constant.

$$K_{eq} = \frac{1}{[Zn^{2+}][OH^-]^2}$$

15-6 After a period of time, a heterogeneous equilibrium occurs between the solid and the gaseous iodine.

Determining the Value of K_{eq}

The numerical value of an equilibrium constant must be determined experimentally. Suppose that 5 mol of sulfur dioxide and 4 mol of oxygen are placed in a container and allowed to react to form sulfur trioxide. At equilibrium with the sulfur trioxide, the various concentrations are measured. It is found that 1 mol of sulfur trioxide, 4 mol of sulfur dioxide, and 3.5 mol of oxygen are present.

$$2SO_2 (g) + O_2 (g) \rightleftharpoons 2SO_3 (g)$$

$$K_{eq} = \frac{[SO_3]^2}{[SO_2]^2[O_2]}$$

$$K_{eq} = \frac{[1]^2}{[4]^2[3.5]}$$

$$K_{eq} = 0.02$$

The equilibrium constant 0.02 describes the ratio between products and reactants at specific conditions.

15-7 Vinegar is a 4%–7% solution of acetic acid in water.

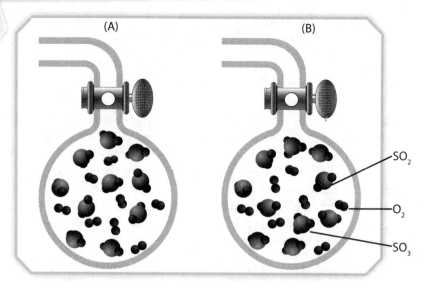

15-8 Flask A contains the reactant molecules, and Flask B holds molecules of the reaction at equilibrium. Notice that all of the reactants are still present with the product but in a different ratio.

Example Problem 15-3

Calculating the Value for K_{eq}

Calculate the numerical value of K_{eq} for the equilibrium involved in the formation of ammonia from nitrogen and hydrogen.

$$N_2(g) + 3H_2(g) \rightleftharpoons 2NH_3(g)$$

When the mixture of the gases reaches equilibrium, $[N_2] = 0.05\ M$, $[H_2] = 0.05\ M$, and $[NH_3] = 50\ M$.

Solution

Substitute the given concentration values into the equilibrium constant equation and solve for K_{eq}.

$$K_{eq} = \frac{[NH_3]^2}{[H_2]^3[N_2]}$$

$$K_{eq} = \frac{[50]^2}{[0.05]^3[0.05]}$$

$$K_{eq} = 4 \times 10^8$$

Significance of the Value of K

The value of the equilibrium constant can tell you much about the equilibrium reaction. If the value of K is much greater than 1, the reaction proceeds almost to completion as written, and the equilibrium lies far to the right. If the value of K is much less than 1, the reaction does not proceed very far at all before reaching equilibrium, and the equilibrium lies to the left. In other words, the larger the value of K, the more product favored the reaction is, and the lower the value of K, the more reactant favored the reaction is. If the value of K lies between these two extremes, then significant concentrations of both reactants and products exist at equilibrium.

15.4 Le Châtelier's Principle

We have discussed how the concentrations of reactants and products change before they reach dynamic equilibrium. But we've also mentioned that certain factors can control reactions to get a desired product. How can concentrations be changed? Is it possible to shift equilibrium from one side of an equation to the other?

In 1884, a French chemist named Henri (awn REE) Le Châtelier (luh SHAH tel YAY) questioned how equilibria behave when disturbed. His findings, called **Le Châtelier's principle** in his honor, state that *when a reversible process is disturbed, it will proceed in the direction that relieves the stress.* These stresses include concentration, pressure, temperature, and the action of a catalyst. Let's examine how these stresses influence equilibrium reactions.

15.5 The Effect of Concentration

Look back at Table 15-1 on page 374. What do you think would happen if the initial concentrations of H_2 and I_2 were doubled? More HI would be present at equilibrium. What would happen if HI were present initially? The reverse reaction would produce some H_2 and I_2. Changes in concentration definitely affect chemical equilibria. Consider the equilibrium equation

$$2SO_2\,(g) + O_2\,(g) \rightleftharpoons 2SO_3\,(g).$$

At one set of conditions, the K_{eq} has been measured to be 0.02.

$$K_{eq} = \frac{[SO_3]^2}{[SO_2]^2[O_2]} = 0.02$$

Suppose that more sulfur dioxide were added to the container. You learned in Chapter 14 that increasing the concentration of a substance increases the rate of the reaction involving it because there are more collisions per unit time. Thus, the system would be temporarily disturbed from equilibrium. Le Châtelier's principle predicts that the forward reaction would increase to reduce the extra sulfur dioxide—to reduce the stress of the additional sulfur dioxide. This reaction would increase the concentration of sulfur trioxide but decrease the concentration of oxygen. Despite all the

15-9 Henri Le Châtelier (1850–1936), a French chemist, was heavily influenced by his father, Louis Le Châtelier. Louis was instrumental in developing the French aluminum industry, improving iron and steel refining, and expanding rail transportation. Henri was trained as an engineer, but spent most of his life teaching. His research dealt mainly with metallurgy, salt solubility, and the principle that today bears his name.

The value of K_{eq} does not vary with changes in the concentration of reactants or products alone.

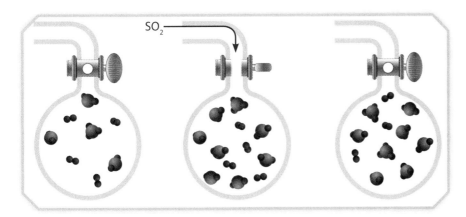

15-10 Additional SO_2 causes the equilibrium to shift forward.

changes, the original constant would remain unchanged once the rates again became equal.

If sulfur trioxide were added, the rate of the reverse reaction would increase; the effect would be the dissipation of the added substance. If some sulfur trioxide were removed, the reverse reaction involving it would decrease; in effect, the forward reaction would work to restore the lost amount because it would have a greater rate. Of course, the concentrations of sulfur dioxide and oxygen would decrease when the forward reaction predominated. This action would keep the value of the equilibrium constant the same, once equilibrium was reestablished.

15.6 The Effect of Pressure

Some gaseous reactions are affected by changes in pressure. Recall that 1 mole of any gaseous substance occupies 22.4 liters at STP. In the equilibrium

$$2SO_2\,(g) + O_2\,(g) \rightleftharpoons 2SO_3\,(g),$$

two moles of sulfur trioxide are produced for every three moles of gaseous reactants. Since there are fewer gas molecules in the product than in the reactants, the forward reaction decreases the pressure.

Suppose that the volume of a container was reduced, increasing the pressure inside it. This has the effect of increasing the *concentrations* of all the gases; thus, we can use the principle discussed previously to predict the effect. The reaction involving the larger number of molecules—the one whose concentrations are affected more—will increase more than the one with fewer, and the reaction will "shift" in the direction that has fewer molecules. Using Le Châtelier's principle, we can predict that the reaction that relieves the stress of greater pressure will be favored. The forward reaction would lower the pressure by producing 2 mol of sulfur trioxide from 2 mol of sulfur dioxide and 1 mol of oxygen. On the other hand, if the pressure on the reaction at equilibrium were decreased, the opposite would be true. The reverse reaction would have a greater rate and would act to "fill the void"—more sulfur dioxide and oxygen would be produced.

Some gaseous reactions are not affected by pressure changes. The reaction

$$H_2\,(g) + I_2\,(g) \rightleftharpoons 2HI\,(g)$$

has 2 mol of gas in the products for every 2 mol of gas in the reactants. The pressure of the entire system does not change with either the forward or the reverse reaction. Increased or reduced pressure on this equilibrium will not affect the direction of the reaction. Any reaction that has the same number of moles of *gas* as product and reactant will be unaffected by pressure changes that result from volume changes.

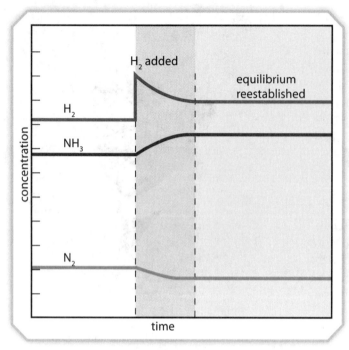

15-11 Notice how the concentrations change in the reaction $N_2 + 3H_2 \rightleftharpoons 2NH_3$ when hydrogen is added.

Equilibrium reactions involving solids and liquids are not significantly affected by a change in pressure because their particles are held more tightly.

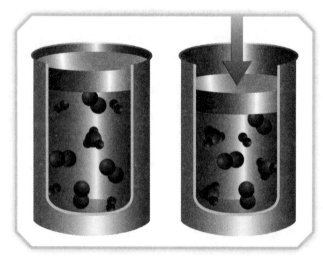

15-12 In the equation $N_2 + 3H_2 \rightleftharpoons 2NH_3$, additional pressure shifts the equilibrium forward.

Example Problem 15-4

Predicting the Effect of Pressure

Will the formation of ammonia be helped or hindered by high pressures?

$$O_2(g) + 2CO(g) \rightleftharpoons 2CO_2(g)$$

Solution

The balanced equation shows that 3 mol of gaseous reactants combine to form 2 mol of gaseous products. High pressures cause the reaction that produces the smallest volume to predominate. Thus, high pressures aid the formation of carbon dioxide gas.

> The value of K_{eq} does not vary with changes in pressure alone.

15.7 The Effect of Temperature

Heat may be thought of as a reactant or product in chemical reactions. The exothermic reaction between sulfur dioxide and oxygen produces sulfur trioxide and heat. If heat is applied to this reaction, the reaction will shift to the left, increasing the concentration of the reactants.

$$2SO_2(g) + O_2(g) \underset{\text{add heat}}{\overset{\longleftarrow}{\rightleftharpoons}} 2SO_3(g) + \text{heat}$$

> Temperature changes *do* affect the value of K_{eq}. Increases in temperature will lower the equilibrium constants of exothermic reactions and increase the equilibrium constants of endothermic reactions.

What would happen if you were to remove heat from the reaction faster than it is being produced? If you predicted that the reaction would shift to the right, you are correct.

Le Châtelier's principle predicts how temperature affects each of these equilibria. Since heat acts as a reactant in an endothermic reaction, it helps the forward reaction to occur. The equilibrium of hydrogen, iodine, and hydrogen iodide will shift to produce more hydrogen iodide if heat is added to the reaction.

$$H_2(g) + I_2(g) + \text{heat} \underset{\text{add heat}}{\overset{\longrightarrow}{\rightleftharpoons}} 2HI(g)$$

Example Problem 15-5

Predicting the Effect of Temperature

The formation of ammonia is exothermic. Will high temperatures help or hinder the forward reaction?

$$N_2(g) + 3H_2(g) \rightleftharpoons 2NH_3(g) + \text{heat}$$

Solution

Since heat is one of the products, additional heat would encourage the reverse reaction and would hinder the forward reaction. Again, when heat is added or removed, the value of the equilibrium constant will change. This relationship is why the temperature and pressure of the system must always be specified when the value of K_{eq} is considered.

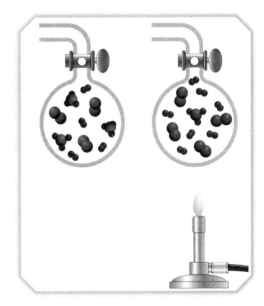

15-13 Additional heat shifts the equilibrium to the left in the equation $N_2 + 3H_2 \rightleftharpoons 2NH_3 + \text{heat}$.

15.8 The Effect of Catalysts

As discussed in Chapter 8, catalysts are substances that change the rate of a reaction without undergoing any permanent change themselves. A catalyst affects both the forward and the reverse reactions, causing the equilibrium to be reached sooner than if no catalyst were added. Although it causes equilibrium to be reached more quickly, the addition of a catalyst has no effect on equilibrium concentrations, or on the value of K_{eq}.

Table 15-2 summarizes the effects on equilibrium reactions and K_{eq} by changes in pressure, concentration, temperature, and addition of catalysts.

15-2 Effects of Stress on Equilibrium Concentrations and K_{eq}

Stress	Shift	Effect	Value of K_{eq}
Add catalyst	none	none	no change
Concentration increase			
of reactant	toward right	more products form	no change
of product	toward left	more reactants form	no change
Concentration decrease			
of reactant	toward left	more reactants form	no change
of product	toward right	more products form	no change
Pressure change (only for unequal numbers of moles of gaseous products and reactants)			
increase	toward side having smaller number of gaseous molecules	—	no change
decrease	toward side having larger number of gaseous molecules	—	no change
Temperature increase			
exothermic	toward left	more reactants form	decreases
endothermic	toward right	more products form	increases
Temperature decrease			
exothermic	toward right	more products form	increases
endothermic	toward left	more reactants form	decreases

15.9 Using Le Châtelier's Principle to Solve Problems

Oxygen is critical for the chemical reactions that make your body operate. Do you know how oxygen is delivered to your body's tissues?

Erythrocytes (red blood cells) in your bloodstream have an iron-containing protein called *hemoglobin* (Hb). Hemoglobin binds to oxygen received at the lungs. The heart pumps this oxygen-bound form of hemoglobin, called *oxyhemoglobin*, to all the parts of the body so that cells can receive the oxygen necessary to perform their functions. Circulation is an equilibrium process represented by the following reaction.

$$Hb\ (aq) + O_2\ (g) \rightleftharpoons HbO_2\ (aq)$$

So, a hypoxic person really needs more oxyhemoglobin to prevent serious medical problems. How can this be accomplished?

You can see from the equation above that an increased concentration of oxygen or hemoglobin will shift the equilibrium reaction to produce more oxyhemoglobin. When a person has hypoxia, pure oxygen gas can be administered, shifting the equilibrium to produce more oxyhemoglobin.

15-14 Without catalysts, refineries could not process crude oil into petroleum products.

15A Section Review

1. What generalizations about an equilibrium constant can be made if the value for K is large? If the value is small?
2. Is the following equilibrium reaction homogeneous or heterogeneous?

   $$CaF_2\ (s) \rightleftharpoons Ca^{2+}\ (aq) + 2F^-\ (aq) + heat$$

3. Write the equilibrium constant expression for the reaction

   $$CaF_2\ (s) \rightleftharpoons Ca^{2+}\ (aq) + 2F^-\ (aq) + heat.$$

4. When will a reversible reaction proceed to completion?
5. Predict the shift in equilibrium (to the right or to the left) under the following conditions for the following reaction.

   $$CaF_2\ (s) \rightleftharpoons Ca^{2+}\ (aq) + 2F^-\ (aq) + heat$$

   a. CaF_2 solid is added.
   b. Ca^{2+} aqueous ions are added.
   c. F^- aqueous ions are removed.
   d. Heat is added.
6. How can the chemical manufacturing industry utilize Le Châtelier's principle?

DS 7. Based on the reaction between oxygen and hemoglobin, why do you think people who live at high elevations don't experience hypoxia?

DS 8. Is the equilibrium in the production of oxyhemoglobin homogeneous or heterogeneous?

DS 9. Write the equilibrium constant for the production of hemoglobin.

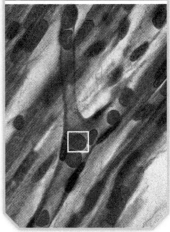

15-15 The hemoglobin molecule (top) found in erythrocytes (bottom) carries oxygen throughout the body.

15B Applications of Equilibrium Chemistry

15.10 Equilibria and Industry

> **15B Section Objectives**
>
> After finishing this section, you should be able to
> - recognize the relationship between the K_{sp} and the solubility of the salt.
> - convert between solubilities and K_{sp} values of solutes.
> - predict whether a precipitate will form when two given solutions are mixed, given the appropriate K_{sp}.

Esters

Chemical industries supply society with such diverse products as medicines, pesticides, fertilizers, paints, textiles, detergents, cosmetics, plastics, fuels, and building materials. Because the reactions used to make those products are usually reversible, the laws of equilibrium chemistry govern the processes. Methods of describing and controlling equilibria become indispensable tools in the fast, efficient, and economical production of consumer goods.

The characteristic flavors and odors of fruits are the results of naturally occurring compounds called *esters*. These compounds are parts of foods, cosmetics, and medicines. Because it is inefficient and expensive

FRITZ HABER (1868–1934): GASES FOR WAR AND PEACE

Facets of Chemistry

Though Fritz Haber was one of the twentieth century's most gifted and influential chemists and won the Nobel Prize in Chemistry in 1918, he died in despair. Who was this extraordinary scientist and what lessons can we learn from his life?

Fritz Haber was born December 9, 1868, in Breslau, Germany. His father was a prominent merchant and his mother died in childbirth. From an early age, Haber showed an interest and proficiency in chemistry and conducted many experiments during his school years.

From 1886 to 1891, he studied at three German universities under some of the preeminent chemists of his day. Before beginning his own career in academia, he worked briefly for three different companies, including his father's chemical business.

Haber accepted an assistantship in chemistry at Karlsruhe in 1894 and remained there until 1911. Here he was appointed as a professor of physical chemistry and electrochemistry in 1906. In addition, he became the director of an institute at Karlsruhe that studied those subjects.

During his time at Karlsruhe, Haber and fellow chemist Karl Bosch developed the Haber process. The Haber process is a catalytic procedure for reacting atmospheric nitrogen and hydrogen under high pressure and heat to produce synthetic ammonia. Ammonia is essential for

the production of nitric acid, which is used to make crop fertilizers. The experience that Bosch had in the chemical industry allowed the two of them to scale up the process, producing large amounts of ammonia from air at relatively low costs.

Prior to this, the world's supply of ammonia came almost entirely from Chile's Atacama Desert, where there are extensive beds of caliche, a mineral rich in sodium nitrate. Because it was being rapidly depleted to produce plastics, explosives, and fertilizer for a booming agriculture industry, many feared there would be worldwide famine and a population crash when when it ran out.

Haber's application of equilibrium chemistry to efficiently produce ammonia on a large scale was perhaps the most significant scientific breakthrough of his century. It greatly increased the efficiency of agriculture and made food more plentiful and affordable. Haber's work is an example of taking dominion over the earth for the benefit of one's neighbor.

But in a fallen world, inventions that can save life can also be used to take life. The same ammonia that saved the fertilizer industry was also valuable in making explosives. Haber's work helped the Germans produce war explosives that fueled both the first and second World Wars. Many historians believe that without the Haber process, Germany would have dropped out of World War I after just one year due to an inability to get Chilean caliche through the British naval blockades.

to extract esters from fruit, manufacturers of artificial flavorings form esters artificially by reacting alcohols and organic acids.

$$\text{alcohol} + \text{organic acid} \rightleftharpoons \text{ester} + \text{water}$$

The general reaction works well, but it is very easily reversed. Neither the forward nor the reverse reaction dominates. Thus efficient production of esters is difficult. Instead of going to completion, the reaction converts only a part of the reactants into the products. Rather than accept limited yields, manufacturers produce more ester per batch of reactants by forcing the forward reaction to prevail.

Manufacturers can shift the equilibrium by continually adding a reactant, removing a product, or both. If either the acid or the alcohol is inexpensive, huge quantities can be added to drive the equilibrium forward. Sometimes manufacturers can easily remove a product by boiling it off as it forms. Regardless of the technique that is used, ester production is increased.

15-16 All of these products are fruit flavored because of the presence of esters.

In 1911, the fiercely patriotic Haber became Director of the Kaiser Wilhelm Institute for Physical Chemistry and Electrochemistry at Berlin. Here, teams of scientists from universities, government, and industry conducted original research. Though much of that work improved the lives of others, Haber's studies also contributed to the German war effort.

In addition to giving us the Haber process, Fritz Haber is recognized as the father of chemical warfare for his efforts in developing and deploying chlorine and other gases. Mustard gas (bis[2-chloroethyl] sulfide) and phosgene ($COCl_2$) were among the most successful. Not only did he develop the gases, he personally supervised their first use at the western front in Ypres, Belgium, in 1915. Perhaps as many as 15,000 Allied troops were injured or killed in that single attack. Haber's enthusiastic support of chemical warfare caused his wife Clara, also a chemist, to take her own life.

Fritz Haber also supervised the production of other chemical warfare gases, including the notorious cyanide (HCN) gas formulation Zyklon-A. Although originally developed as a fumigant to kill insects in grain, it came to worldwide attention when modified to Zyklon-B—the poison used in

the Nazi gas chambers. In those gas chambers, millions of Jews, including members of Haber's own extended family, were murdered.

Although Haber pragmatically converted from Judaism to Christianity to be more accepted in German society, he was still a Jew in Nazi eyes. Not even his scientific contributions to the war effort were enough to earn him favor. He fled Germany in 1933, staying for a short time in England before deciding to take a professorship in Rehovot (present-day Israel). He died en route, in Basel, Switzerland, on January 29, 1934.

Haber's life and accomplishments are a testimony to the dangers science poses in a fallen world. Fallen humans' ability to exercise dominion over the world through science often leads to moral catastrophe. In our own day the issues of embryonic stem cell research, cloning, and genetic modification raise afresh the issue of what scientists *can* do versus what scientists *ought* to do. Without a scripturally tuned moral compass, scientists are left to make choices based on their impulses or the political sentiments of the moment.

Ammonia

Huge quantities of ammonia (NH_3) are used every year in fertilizers, cleaning compounds, and explosives. Until World War I, industrial countries obtained nitrogen atoms by importing mined saltpeter (KNO_3 and $NaNO_3$). All countries had access to nitrogen molecules (N_2) in the atmosphere, but no one knew how to liberate the atoms from the stable molecules. Rather than go to the expense of importing saltpeter, countries sought to form ammonia from atmospheric nitrogen.

Finding the appropriate reaction was no problem. The reaction

$$N_2\ (g) + 3H_2\ (g) \rightleftharpoons 2NH_3\ (g) + \text{heat}$$

seemed perfect. The relatively large equilibrium constant of this reaction (6.4×10^2 at 25 °C) indicated that the production of ammonia was the favored direction of the reaction. The problem was that nitrogen, hydrogen, and ammonia took a very long time to reach equilibrium. In 1909, a German chemist named Fritz Haber (1868–1934) found a way to manipulate the equilibrium to overcome this difficulty. He invented the Haber (HAH ber) process that supplied Germany's war machine with ammonia-based explosives. Today the Haber process provides ammonia for use in numerous industries.

The **Haber process** uses the following technique. Because 2 mol of gas are formed from 4 mol of gas, the products take up less space than the reactants. Le Châtelier's principle predicts that high pressures favor the production of ammonia. Because the forward reaction is exothermic, it proceeds best at low temperatures. Thus, engineers needed a process that used high pressure and low temperatures to make ammonia manufacturing easier.

In the Haber process, the reaction takes place under high pressures (100–200 atm), at high temperatures (350–500 °C), and with a catalyst. However, high temperatures significantly shift the equilibrium away from the formation of ammonia (at 500 °C, $K = 1.5 \times 10^{-5}$). To compensate, the Haber process manipulates other variables to increase the production of ammonia. A catalyst (mostly Fe_3O_4 with traces of K_2O and Al_2O_3) helps to speed up the reaction. To further increase the yield, ammonia is removed by liquefaction as it forms. This continual removal keeps the forward reaction going.

15-17 Plants like this use the Haber process to make ammonia. The annual worldwide production of ammonia exceeds 100 million tons.

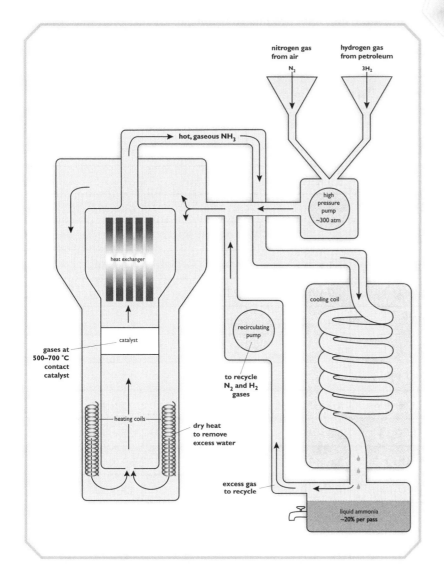

15-18 The Haber process

15.11 Ionic Equilibria

Saturated solutions are an equilibrium of undissolved solids and aqueous ions. Some salts, such as calcium sulfate, are slightly soluble. Others, such as lead (II) phosphate, dissolve so little that for all practical purposes they are insoluble.

$$CaSO_4\ (s) \rightleftharpoons Ca^{2+}\ (aq) + SO_4^{2-}\ (aq)$$

$$Pb_3(PO_4)_2\ (s) \rightleftharpoons 3Pb^{2+}\ (aq) + 2PO_4^{3-}\ (aq)$$

The forward reaction is dissolution, and the reverse reaction is precipitation. As with all equilibria, the rate of the forward reaction matches the rate of the reverse reaction at equilibrium. Since this reaction is an equilibrium, the relative amounts of dissolved ions and undissolved solids can be expressed with equilibrium constants. The equilibrium constant for the dissolution of $Pb_3(PO_4)_2$ is expressed as

$$K_{eq} = [Pb^{2+}]^3[PO_4^{3-}]^2.$$

As in other equilibrium expressions, solids are not included; thus, the $[Pb_3(PO_4)_2]\ (s)$ term is omitted in the expression of the equilibrium constant. The resulting solubility equilibrium constant is given a special name—the **solubility product constant** (K_{sp}). The equilibrium concentrations of the ions are raised to the appropriate powers as in other equilibrium constant expressions.

15-3 K_{sp} for Minimally Soluble Substances

Salt	Product	K_{sp} (at 25 °C)	Molar mass	Solubility (g/L)
silver chloride, AgCl	$[Ag^+][Cl^-]$	1.8×10^{-10}	143.35	1.9×10^{-3}
silver carbonate, Ag_2CO_3	$[Ag^+]^2[CO_3^{2-}]$	8.5×10^{-12}	275.81	3.6×10^{-2}
silver phosphate, Ag_3PO_4	$[Ag^+]^3[PO_4^{3-}]$	8.9×10^{-17}	418.67	6.4×10^{-2}
aluminum hydroxide, $Al(OH)_3$	$[Al^{3+}][OH^-]^3$	3.0×10^{-34}	78.00	4.1×10^{-7}
barium carbonate, $BaCO_3$	$[Ba^{2+}][CO_3^{2-}]$	5.1×10^{-9}	197.31	1.4×10^{-2}
barium sulfate, $BaSO_4$	$[Ba^{2+}][SO_4^{2-}]$	1.2×10^{-10}	233.37	2.6×10^{-3}
calcium carbonate, $CaCO_3$	$[Ca^{2+}][CO_3^{2-}]$	3.4×10^{-9}	100.09	6.6×10^{-3}
calcium sulfate, $CaSO_4$	$[Ca^{2+}][SO_4^{2-}]$	4.9×10^{-5}	136.15	2.1
iron (III) hydroxide, $Fe(OH)_3$	$[Fe^{3+}][OH^-]^3$	4.0×10^{-38}	106.87	2.1×10^{-8}
iron (II) hydroxide, $Fe(OH)_2$	$[Fe^{2+}][OH^-]^2$	4.9×10^{-17}	89.86	5.2×10^{-4}
magnesium carbonate, $MgCO_3$	$[Mg^{2+}][CO_3^{2-}]$	6.8×10^{-6}	84.32	1.8
magnesium hydroxide, $Mg(OH)_2$	$[Mg^{2+}][OH^-]^2$	5.6×10^{-12}	58.33	6.9×10^{-3}
lead (II) carbonate, $PbCO_3$	$[Pb^{2+}][CO_3^{2-}]$	7.4×10^{-14}	267.21	4.9×10^{-5}
lead (II) phosphate, $Pb_3(PO_4)_2$	$[Pb^{2+}]^3[PO_4^{3-}]^2$	3.0×10^{-44}	811.54	6.3×10^{-7}
lead (II) sulfate, $PbSO_4$	$[Pb^{2+}][SO_4^{2-}]$	2.5×10^{-8}	303.27	4.4×10^{-2}
zinc hydroxide, $Zn(OH)_2$	$[Zn^{2+}][OH^-]^2$	3×10^{-17}	99.43	4.2×10^{-4}

$K_{sp} < 1$ — insoluble
$K_{sp} > 1$ — soluble

15.12 Solubility and Equilibria

The K_{sp} constant quantifies salt solubilities. Large numbers of ions in solution will cause the solubility product to be relatively large (much greater than 1). Small numbers of ions in solution cause the K_{sp} to be small (less than 1). Thus, the larger the K_{sp} is, the more soluble the salt. Conversely, the smaller the K_{sp} is, the less soluble the salt. The K_{sp} of some slightly soluble salts are listed in Table 15-3.

K_{sp} values can be determined from measured solubilities. For example, barium sulfate has a solubility of 1.1×10^{-5} M. When barium sulfate dissolves, each formula unit releases one Ba^{2+} ion and one SO_4^{2-} ion.

$$1.1 \times 10^{-5} \text{ mol } BaSO_4\,(s) \rightleftharpoons 1.1 \times 10^{-5} \text{ mol } Ba^{2+}\,(aq) + 1.1 \times 10^{-5} \text{ mol } SO_4^{2-}\,(aq)$$

$$K_{sp} = [Ba^{2+}][SO_4^{2-}]$$

$$K_{sp} = (1.1 \times 10^{-5})(1.1 \times 10^{-5})$$

$$K_{sp} = 1.2 \times 10^{-10}$$

The K_{sp} of a substance can be used to determine the solubility in moles per liter. The K_{sp} of lead (II) phosphate is known to be 3.0×10^{-44} mol/L. The equilibrium for its dissociation can be written as follows:

$$Pb_3(PO_4)_2 \;(s) \rightleftharpoons 3Pb^{2+} \;(aq) + 2PO_4^{3-} \;(aq).$$

$$K_{sp} = [Pb^{2+}]^3[PO_4^{3-}]^2$$

$$3.0 \times 10^{-44} = [Pb^{2+}]^3[PO_4^{3-}]^2$$

Notice that when 1 mol of lead (II) phosphate dissolves, 3 mol of Pb^{2+} ions and 2 mol of PO_4^{3-} ions are released. The number of Pb^{2+} ions in a saturated solution must be three times the amount of $Pb_3(PO_4)_2$ units that were dissolved. Likewise, there must be twice as many PO_4^{3-} ions as the number of dissolved formula units. Suppose that S stands for the solubility of the salt, in this case $Pb_3(PO_4)_2$. The concentration of the Pb^{2+} ions can be expressed as $3S$, and the concentration of PO_4^{3-} ions can be expressed as $2S$. The concentrations of the ions can now be expressed in terms of the solubility of the salt.

$$K_{sp} = [Pb^{2+}]^3[PO_4^{3-}]^2$$

$$K_{sp} = (3S)^3(2S)^2$$

To find the solubility of the salt, substitute the value of the K_{sp} and solve for S.

$$3.0 \times 10^{-44} = (3S)^3(2S)^2$$

$$3.0 \times 10^{-44} = 27S^3 \times 4S^2$$

$$3.0 \times 10^{-44} = 108S^5$$

Divide by 108 and then take the fifth root of both sides to obtain $7.7 \times 10^{-10} = S$.

The solubility of lead (II) phosphate is 7.7×10^{-10} mol/L. This amount is so small that the salt is, for all practical purposes, insoluble.

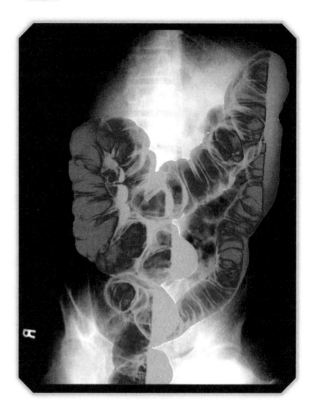

15-19 Barium ions from barium sulfate swallowed by this patient enhance the x-ray of the gastrointestinal tract.

EXAMPLE PROBLEM 15-6

Determining Solubility

Given that the K_{sp} of silver chloride is 1.8×10^{-10}, determine its solubility (*M*).

Solution

$$AgCl \;(s) \rightleftharpoons Ag^+ \;(aq) + Cl^- \;(aq)$$
$$K_{sp} = 1.8 \times 10^{-10}$$
$$K_{sp} = [Ag^+][Cl^-]$$

Since each silver chloride unit releases one Ag^+ ion and one Cl^- ion, the $[Ag^+]$ and the $[Cl^-]$ equal the solubility of the salt (S) in a saturated solution.

$$K_{sp} = [S][S]$$
$$1.8 \times 10^{-10} = S^2$$
$$1.3 \times 10^{-5} = S$$

Thus, 1.3×10^{-5} mol of silver chloride can be dissolved in 1 L of water.

The K_{sp} can be used to see whether a solution is unsaturated, saturated, or supersaturated. If the product of ion concentrations equals the K_{sp}, the solution is saturated. If the product of ion concentrations is less than the K_{sp}, the solution is unsaturated. A product greater than the K_{sp} indicates a supersaturated solution.

Another way to determine if a solution is unsaturated, saturated, or supersaturated is to compare the solubility of the substance in M to the concentration of the solution. Be sure, however, to compare it to the concentration in mol/L, not in g/L.

EXAMPLE PROBLEM 15-7
Determining Saturation State

Calculate the K_{sp} of a 2.5×10^{-6} M zinc hydroxide solution. Determine whether the solution is unsaturated, saturated, or supersaturated.

Solution
The equilibrium is

$$Zn(OH)_2\ (s) \rightleftharpoons Zn^{2+}\ (aq) + 2OH^-\ (aq).$$

When dissolved, 2.5×10^{-6} mol of zinc hydroxide produces 2.5×10^{-6} mol of Zn^{2+} ions and $2(2.5 \times 10^{-6})$ mol of OH^- ions. The product of these concentrations is as follows.

$$\begin{aligned} K_{sp} &= [Zn^{2+}][OH^-]^2 \\ &= [2.5 \times 10^{-6}][2(2.5 \times 10^{-6})]^2 \\ &= 6.3 \times 10^{-17} \end{aligned}$$

This product is greater than the K_{sp} of 3×10^{-17}, so the solution must be supersaturated.

15.13 Common-Ion Effect

Since salts dissociate when they dissolve, solutes that have a common ion can affect the equilibrium. For example, an unsaturated silver chloride solution has a concentration less than 1.33×10^{-5} M and a solubility product constant less than the K_{sp} of a saturated solution. What would happen if sodium chloride was added to the silver chloride solution, since they both have a common chloride ion?

Sodium chloride is very soluble. It dissolves and dissociates into sodium and chloride ions. Though the sodium ions have little or no effect on the original solution, the chloride ions do. The chloride ion is called the *common ion* because it is contained in both silver chloride and sodium chloride. The additional chloride ions from the sodium chloride cause the equilibrium to shift to the left in the reaction,

$$AgCl\ (s) \rightleftharpoons Ag^+\ (aq) + Cl^-\ (aq).$$

This shift will produce more solid silver chloride. The increased concentration of chloride ions raises the solubility product constant above its saturated value. The result of the **common-ion effect** is that the less soluble salt will precipitate out of the solution until the K_{sp} value is reached. Remember that the K_{sp}, like any other equilibrium constant, has a maximum value at a given temperature. That value cannot be exceeded without causing a shift to occur that will reestablish equilibrium conditions.

15-20 The addition of concentrated HCl to a saturated salt solution will cause NaCl to precipitate.

15.14 Precipitation Reactions

As we have seen, mixing two solutions can sometimes cause a precipitate to form. For instance, suppose that hydrochloric acid (HCl) and silver nitrate (AgNO$_3$) were mixed together. The resulting solution would contain H$^+$, Cl$^-$, Ag$^+$, and NO$_3^-$ ions. Hydrogen chloride and silver nitrate are quite soluble. However, an alternate combination of ions, Ag$^+$Cl$^-$, is not. A silver chloride precipitate will form if enough silver and chloride ions are in the solution. How much is enough? That depends on the K_{sp} of silver chloride. If the value of [Ag$^+$][Cl$^-$] in the solution exceeds the K_{sp} for silver chloride, a precipitate will form. How much precipitate will be deposited? Enough precipitate will form to lower the value of [Ag$^+$][Cl$^-$] to the K_{sp} for silver chloride. Thus, the K_{sp} of a salt can be used to predict whether precipitate will form and how much will form.

EXAMPLE PROBLEM 15-8

Predicting Precipitation

Will a precipitate form when equal volumes of a 5.0×10^{-5} M barium nitrate solution and a 1.0×10^{-3} M sodium carbonate solution are mixed?

Solution

Both barium nitrate and sodium carbonate are soluble salts. At this low concentration, they will both be completely dissolved. The first step is to calculate the theoretical concentration of each ion after both salts completely dissolve.

Ba(NO$_3$)$_2$ (s) $\longrightarrow$ Ba^{2+} (aq) + 2NO$_3^-$ (aq)
[Ba^{2+}] = 5.0×10^{-5} M
[NO$_3^-$] = 2(5.0×10^{-5}) M

Na$_2$CO$_3$ (s) $\longrightarrow$ 2Na$^+$ (aq) + CO$_3^{2-}$ (aq)
[Na$^+$] = 2(1.0×10^{-3}) M
[CO$_3^{2-}$] = 1.0×10^{-3} M

The next step is to check Table 15-3 to determine which combination of ions is most likely to precipitate. Both original combinations are very soluble, and so is sodium nitrate. Barium carbonate is the only salt formed by a combination of the ions in this solution that appears in Table 15-3 ($K_{sp} = 5.1 \times 10^{-9}$). The final step is to determine whether the value of [Ba^{2+}][CO$_3^{2-}$] exceeds the given K_{sp}. Divide their molarities in half since the combined solution has twice the volume.

[Ba^{2+}][CO$_3^{2-}$] = (2.5×10^{-5})(5.0×10^{-4}) = 1.25×10^{-8}

Since 1.25×10^{-8} is larger than the K_{sp} of 5.1×10^{-9}, a precipitate will form.

15B Section Review

1. An experimental solution has a K_{sp} value of 2.5×10^{-12}. That same solution has a known K_{sp} of 3.2×10^{-10}. Is the experimental solution saturated, unsaturated, or supersaturated? Explain your answer.

2. The solubility of copper (I) chloride is 1.08×10^{-2} g per 100 mL solution.
    a. Write the K_{sp} expression for CuCl.
    b. Find the K_{sp} value.

Coming to Terms

irreversible reaction	372
reversible reaction	372
chemical equilibrium	372
equilibrium constant (K_{eq})	375
law of chemical equilibrium	375
homogeneous equilibrium	376
heterogeneous equilibrium	376
Le Châtelier's principle	379
Haber process	386
solubility product constant (K_{sp})	388
common-ion effect	390

Chapter Review

Chapter Summary

- Many chemical reactions are reversible. The reactants and products can reach a dynamic equilibrium at which the ratio of the two remains constant.

- The ratios between products and reactants reflect the speed at which the forward and reverse reactions proceed.

- An equilibrium constant (K_{eq}) is a numerical expression of the ratio of products to reactants.

- An equilibrium constant expression is determined from the balanced equilibrium reaction. The equilibrium constant equals the product of the concentrations divided by the product of the reactants, all raised to the power of their stoichiometric coefficients. For the reaction

$$aA + bB \rightleftharpoons cC + dD,$$

$$K_{eq} = \frac{[C]^c[D]^d}{[A]^a[B]^b}.$$

- The law of chemical equilibrium states that at a fixed temperature, a chemical system may reach a point where the ratio between the concentration of the products and reactants has a constant value.

- In a homogeneous equilibrium, all of the reactants and products are in the same state. In a heterogeneous equilibrium, two or more states are represented.

- The numerical value of an equilibrium constant at a fixed temperature must be determined experimentally and is usually expressed without units.

- Le Châtelier's principle states that when a reversible process is disturbed, it will shift in the direction that relieves the stress. The stress may be a change in concentration, pressure, or temperature.

- Pressure changes act only on gaseous, homogeneous equilibrium reactions.

- A catalyst speeds up the forward and the reverse reactions, establishing equilibrium sooner. Catalysts do not affect equilibrium concentrations or the value of K_{eq}.

- Industries can use equilibrium chemistry to manufacture chemical substances more economically.

- The solubility product constant (K_{sp}) is a specific K_{eq} that expresses the solubility of a salt at a specific temperature. The maximum K_{sp} for a salt can be calculated from measured solubilities and vice versa.

- The K_{sp} can be compared with experimental solubility data to determine a solution's state of saturation.

- When two salts contain a common ion in the same solution, the common-ion effect can cause the less soluble salt (combination of ions) to precipitate out of solution.

Review Questions

Concept Review

1. What is the difference between a dynamic equilibrium and a static one?
2. When a bottle of soft drink is shaken violently, CO_2 gas escapes from solution and exerts increased pressure on the interior of the bottle. If the bottle were to be opened in this condition, the beverage would spurt out of the bottle. If the bottle remains capped and is allowed to sit for a short amount of time, the interior pressure subsides. Why does the pressure decrease with time?
3. A saturated solution of potassium chloride in water contains solid KCl and ions of K^+ and Cl^-. When this reaction is at equilibrium, is it homogeneous or heterogeneous?
4. Name three factors that can shift equilibria.
DS 5. Why do you think people with severe hypoxia receive oxygen and not hemoglobin?
DS 6. Sickle-cell anemia is a disease in which a person's hemoglobin experiences a mutation that often results in misshapen blood cells. This mutation affects the blood cell's ability to deliver oxyhemoglobin and reduces its ability to carry proper hemoglobin. What kind of serious effects could this have, based on the equilibrium reaction in Subsection 15.9?

True or False

7. Henri Le Châtelier developed a practical process for producing ammonia.
8. Some chemical reactions are irreversible.
9. According to the law of chemical equilibrium, in a reaction at equilibrium, the ratio between reactants and products is 1:1.
10. In an equilibrium reaction that is endothermic, the addition of heat will cause the equilibrium to shift toward the reactants.
11. K_{sp} values can be determined from measured solubilities.

Application

12. Write the equilibrium constant for each reaction.
    a. Laughing gas can decompose into nitrogen and oxygen.
       $2N_2O\ (g) \rightleftharpoons 2N_2\ (g) + O_2\ (g)$
    b. Carbon monoxide can be converted into methane in the process of converting coal into a gas.
       $CO\ (g) + 3H_2\ (g) \rightleftharpoons CH_4\ (g) + H_2O\ (g)$
    c. Methanol can be synthesized from carbon monoxide.
       $CO\ (g) + 2H_2\ (g) \rightleftharpoons CH_3OH\ (g)$
    d. Baking soda can extinguish fires because it can decompose to produce water and carbon dioxide, both of which smother combustion.
       $2NaHCO_3\ (s) \rightleftharpoons Na_2CO_3\ (s) + H_2O\ (g) + CO_2\ (g)$

13. Predict the effect of increasing the pressure on each of the four equilibria given in Question 12.

14. A scientist experimenting with the Haber process,
    $$N_2(g) + 3H_2(g) \rightleftharpoons 2NH_3(g) + \text{heat},$$
    seeks to determine the equilibrium constant at 450 °C.
    a. After permitting a reaction mixture of N_2 and H_2 to reach equilibrium, he finds that $[N_2] = 1.00$, $[H_2] = 0.0769$, and $[NH_3] = 0.000200$. Calculate K at this temperature.
    b. The scientist repeats the experiment under the same conditions with different amounts of gas in the reaction vessel. The scientist finds that $[H_2] = 0.0375$ and $[NH_3] = 0.000318$. Use the value of K already determined to calculate $[N_2]$.

15. Consider the reaction
    $$2NO(g) + O_2(g) \rightleftharpoons 2NO_2(g).$$
    a. What is the value of the equilibrium constant for this reaction if $[NO] = 0.890$, $[O_2] = 0.250$, and $[NO_2] = 0.0320$?
    b. If, under the same conditions, $[NO] = 1.00$ and $[O_2] = 0.500$, what will be the concentration of NO_2 in a reaction vessel?

16. Acetic acid ($HC_2H_3O_2$) is the compound that gives vinegar its distinctive smell and taste. When dissolved in water, it can ionize according to the reaction
    $$HC_2H_3O_2(aq) + H_2O(l) \rightleftharpoons C_2H_3O_2^-(aq) + H_3O^+(aq).$$
    a. Write the equilibrium constant for this reaction.
    b. A chemist working in a clinical research lab dissolves some acetic acid in water (at 25 °C) and finds that $[H_3O^+] = 1.01 \times 10^{-5}$, $[C_2H_3O_2^-] = 1.01 \times 10^{-5}$, and $[HC_2H_3O_2] = 5.67 \times 10^{-6}$. What is the value of K?
    c. If $[H_3O^+] = 3.6 \times 10^{-3}$ and $[C_2H_3O_2^-] = 3.6 \times 10^{-3}$, what is the concentration of un-ionized acetic acid?
    d. What equation gives the equilibrium constant for the reverse reaction?
    e. What is the value of the equilibrium constant for the reverse reaction?

17. Oxalic acid ($H_2C_2O_4$) experiences a two-step dissociation in an aqueous solution:
    $H_2C_2O_4 \rightleftharpoons H^+ + HC_2O_4^-$, where $K_{eq} = 5.90 \times 10^{-2}$ at 25 °C, and
    $HC_2O_4^- \rightleftharpoons H^+ + C_2O_4^{2-}$, where $K_{eq} = 6.40 \times 10^{-5}$ at 25 °C.
    If the equilibrium concentration of $[H^+] = 1.95 \times 10^{-3}$ M and $[HC_2O_4^-] = 1.95 \times 10^{-3}$, what are the equilibrium concentrations of $H_2C_2O_4$ and $C_2O_4^{2-}$? What is the equilibrium constant for the total reaction
    $$H_2C_2O_4 \rightleftharpoons 2H^+ + C_2O_4^{2-}?$$

18. Lithium carbonate (Li_2CO_3) is less soluble in hot water than in cold water.
    a. Do you think the dissolution of Li_2CO_3 is exothermic or endothermic? Explain.
    b. What will happen if a saturated solution of Li_2CO_3 at 25 °C is heated to 100 °C? Explain.

19. Write K_{sp} expressions for the dissolution of the following:
    a. AgBr
    b. MgF_2
    c. Ag_2S
    d. CaF_2
    e. $PbCl_2$

20. A scientist measures the solubility of $Mg(OH)_2$. The concentration of the Mg^{2+} ions in a saturated solution at 18 °C is 1.44×10^{-4} M, and the concentration of OH^- in the same solution is 2.88×10^{-4} M. What is the value of the K_{sp} at this temperature?

21. Using the solubilities and molar masses in Table 15-3, calculate the K_{sp} for the following:
    a. Ag_2CO_3
    b. $BaCO_3$
    c. $Fe(OH)_3$
    d. $PbSO_4$

    If the calculations are performed correctly, you should obtain a value close to the value for K_{sp} listed in the table.

22. A saturated solution of Li_2CO_3 has both Li^+ and CO_3^{2-} ions in it. What will happen if some solid LiCl is added to the solution? (LiCl is much more soluble than Li_2CO_3.)

23. Will a precipitate form if equal volumes of the following solutions are mixed? If so, what substance(s) will precipitate? (*Hint*: First determine whether each substance is soluble or only minimally soluble. If a salt is minimally soluble, refer to its K_{sp}.)
    a. 0.05 M NaCl and 0.05 M LiCl
    b. 6.3×10^{-4} M $MgCO_3$ and 1.0×10^{-4} M $Mg(OH)_2$
    c. 0.5 M NaOH and 2.0×10^{-5} M $Mg(OH)_2$

24. Barium sulfate ($BaSO_4$) is administered to people when x-rays of their digestive systems are taken, despite the fact that the Ba^{2+} ion is toxic to humans. Suggest the method by which $BaSO_4$ could be administered without the danger from the Ba^{2+} ions. Base your method on the common-ion effect.

25. The K_{sp} of AgCl at 50 °C is 1.32×10^{-9}. Predict whether AgCl will precipitate if $AgNO_3$ and NaCl solutions of the following concentrations are mixed in equal volumes:
    a. 1.00×10^{-5} M $AgNO_3$ and 1.00×10^{-5} M NaCl
    b. 1.00×10^{-5} M $AgNO_3$ and 1.32×10^{-4} M NaCl
    c. 1.00×10^{-5} M $AgNO_3$ and 4.00×10^{-4} M NaCl

26. Calculate the molar solubilities of each of the following:
    a. barium fluoride, BaF_2, $K_{sp} = 1.7 \times 10^{-6}$
    b. cadmium carbonate, $CdCO_3$, $K_{sp} = 2.5 \times 10^{-14}$
    c. lanthanum iodate, $La(IO_3)_3$, $K_{sp} = 6.0 \times 10^{-10}$
    d. silver arsenate, Ag_3AsO_4, $K_{sp} = 1.0 \times 10^{-22}$

Acids, Bases, and Salts

CHAPTER 16

16A Defining Acids and Bases	397
16B Acid-Base Equilibria	402
16C Neutralization	413
Facet	
Heartburn and Acid Reflux	414

Dominion Science Problem

Drunk Driving

Everyone knows that driving under the influence is a real problem on our highways. Deaths caused by driving under the influence are a significant percentage of all vehicular deaths. In 2006, the National Highway Traffic Safety Administration reported 42,642 deaths—41% because of alcohol or drugs. Data collected by the NHTSA shows that 77% of these deaths were caused by drivers who were legally intoxicated. Though the ideal solution is for people to avoid alcohol and drugs, how can we keep drunk drivers off the road?

16A Defining Acids and Bases

We live in a world full of chemicals. Some of these chemicals are compounds called acids and bases. How can we know if a chemical is an acid or a base? Since ancient times, people have identified acids and bases by their taste or other properties. As early as 1663, Robert Boyle classified many common substances as acids or bases based on their physical properties. It would be another two centuries, however, before scientists began to explain what causes these traits.

16.1 Observable Properties of Acids

Acids can be identified by their physical properties. Aqueous solutions of acids typically have a sour taste. For example, the citric acid in fruit juices has a tart, sour taste. Acetic acid puts the pucker in pickles. Aspirin tastes sour because of acetylsalicylic acid. Lactic acid's smell and flavor identify sour milk. In fact, sourness is so closely tied to acids that their name comes from the Latin *acidus*, meaning sour. Some acids dissolve in solution to conduct an electrical current because they ionize. For example,

> **16A Section Objectives**
> After finishing this section, you should be able to
> - describe acids and bases.
> - classify substances as acids or bases according to the Arrhenius, Brønsted-Lowry, and Lewis definitions.
> - compare and contrast the three acid-base models.
> - relate protonation and deprotonation to the formation of conjugate pairs.

16-1 The tart flavor of citrus fruits comes from citric acid.

> Don't ever attempt to taste acids, bases, or any other chemicals in the laboratory! You could severely damage your oral tissues.

hydrochloric acid (HCl) ionizes into hydrogen ions and chlorine ions. Any substance that ionizes to conduct electricity in a solution is called an **electrolyte**.

A simple test makes one of the chemical properties of acids observable. Litmus paper can be used to identify acids. Litmus paper contains wood cellulose like other paper, but during production it is mixed with a chemical extracted from certain lichens. Acids turn blue litmus paper red.

Acids can also be identified by their chemical properties. Acids react with active metals to produce hydrogen gas and a salt. If the metal is more active than hydrogen, it will replace the hydrogen ion of the acid, liberating hydrogen as its diatomic gas. The reaction between magnesium and hydrochloric acid illustrates this characteristic.

$$\text{Mg}\ (s) + 2\text{HCl}\ (aq) \longrightarrow \text{Mg}^{2+}\ (aq) + 2\text{Cl}^-\ (aq) + \text{H}_2\ (g)$$

16-2 The bubbles released from the surface of the magnesium ribbon are hydrogen gas formed by the reaction with HCl.

16.2 Observable Properties of Bases

Bases, like acids, can be identified by their physical properties. Most bases have a bitter taste, which you might have experienced by accidentally tasting soap. Thus, very few foods contain strong bases. Bases also feel slippery. The soapy feeling associated with bases results from a reaction between the base and the fatty acids and oils in your skin. Litmus paper can be used to identify stronger bases. These bases turn red litmus paper blue. However, some weak bases may not produce this color change. Some bases in solution also ionize to conduct an electrical current, making them electrolytes.

Bases can also be identified by their chemical properties. Strong bases and strong acids react to neutralize each other. A **neutralization reaction** between an acid and a base in an aqueous solution produces a salt and water. For example, see the reaction of hydrochloric acid with sodium hydroxide below to produce water and sodium chloride.

$$HCl\ (aq) + NaOH\ (aq) \longrightarrow NaCl\ (aq) + H_2O$$

Some common acids and bases are listed in Table 16-1.

16-3 These products contain bases.

16-1 Common Acids and Bases	
Product	Name
vinegar	acetic acid, $HC_2H_3O_2$
vegetables, vitamin C	ascorbic acid, $H_2C_6H_6O_6$
soft drinks	carbonic acid, H_2CO_3
citrus fruit	citric acid, $H_3C_6H_5O_7$
bee venom	formic acid, $HCHO_2$
sour milk, sore muscles	lactic acid, $H_2C_3H_4O_3$
rhubarb	oxalic acid, $H_2C_2O_4$
household ammonia	ammonium hydroxide, NH_4OH
battery acid	sulfuric acid, H_2SO_4
lime	calcium hydroxide, $Ca(OH)_2$
milk of magnesia	magnesium hydroxide, $Mg(OH)_2$
lye	sodium hydroxide, $NaOH$

16.3 The Arrhenius Model

Though we have described acids and bases, how do we determine whether a compound is an acid or a base? Since the physical properties of acids and bases can vary so much, scientists have focused on the chemical properties of acids and bases to define them.

The earliest of the modern acid-base definitions was proposed in the 1880s by the Swedish chemist Svante Arrhenius (uh REE nee us). Arrhenius classified acids and bases by the ions they produce in solution. **Arrhenius acids** ionize to release hydrogen ions (H^+) into aqueous solutions. Note that a hydrogen ion is just a proton, or a hydrogen nucleus. This hydrogen ion is not stable in solution, so it bonds to a water molecule to form the hydronium ion, H_3O^+. **Arrhenius bases** are substances that release hydroxide ions (OH^-)

16-4 Svante Arrhenius (1859–1927) was born near Uppsala, Sweden. In 1876, he began to study mathematics, chemistry, and physics at the University of Uppsala. His 1884 thesis, "Investigations on the Galvanic Conductivity of Electrolytes," was initially undervalued, but by 1903 the scientific community recognized his brilliance and he was awarded the Nobel Prize in Chemistry.

into aqueous solutions. According to the Arrhenius definitions, HCl, HCOOH, and H_2SO_4 are acids, and NaOH, $Mg(OH)_2$, and $Al(OH)_3$ are bases.

Hydrogen-containing compounds are not always acids. For example, methane (CH_4) has four hydrogen atoms, but none of these are ionizable hydrogens. Thus, methane is not considered an acid. Likewise, compounds with OH groups in their formulas are not always bases. Methanol (CH_3OH) does not normally release its OH group.

The Arrhenius model is useful but limited because it deals only with compounds in aqueous solutions. Normally this is not a great limitation, but modern chemistry is pioneering research in acid-base reactions that occur in nonaqueous solvents or even without solvents. The Arrhenius definition of bases does not recognize compounds such as ammonia as being bases because these compounds do not have an OH group. Nevertheless, ammonia exhibits properties long associated with bases. Although the Arrhenius model has some use, it is not as functional as other definitions. Such is the nature of science. As research continues, we must refine our models to explain new data.

16.4 The Brønsted-Lowry Model

In 1923, a Danish chemist named Johannes Brønsted and a British chemist named Thomas Lowry proposed new definitions for acids and bases. They separately formulated similar definitions but published their results within months of each other. A **Brønsted-Lowry acid** is a substance that donates a proton, and a **Brønsted-Lowry base** is a substance that accepts a proton. The process of losing a proton is called **deprotonation**, and the process of gaining a proton is called **protonation**.

The Brønsted-Lowry model encompasses all Arrhenius acids and bases, plus many others. Its definition of acids is essentially the same as the Arrhenius definition, but its definition of bases differs greatly. The Brønsted-Lowry definition greatly expands the number of substances called bases because it includes many substances without OH groups.

Brønsted-Lowry acids and bases exist in pairs called **conjugate pairs**. For example, when acetic acid mixes with water, it deprotonates in the presence of water to form an equilibrium with hydronium and acetate ions. The hydronium ion is capable of losing a proton, and the

16-5 Johannes N. Brønsted (1879–1947) was a Danish physical chemist who earned his PhD from the University of Copenhagen in 1908. His expertise was in the area of the catalysis of acids and bases. His interests ranged from playing piano and singing to painting and politics. He was elected to the Danish parliament in 1947 but died before he could assume office.

16-6 Thomas M. Lowry (1874–1936), son of a Methodist chaplain, was born in Bradford, England. He lectured in chemistry at several colleges before becoming the first professor of physical chemistry at Cambridge in 1920. Staunchly committed to his religious upbringing, he preached often at village chapels.

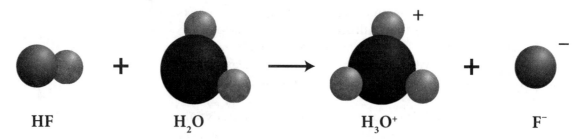

16-7 Can you identify the conjugate pair of acids and bases in this reaction?

acetate ion is capable of accepting a proton. The acetate ion is called the **conjugate base** of acetic acid. The hydronium ion is called the **conjugate acid** of water.

$$\underset{\text{acid}}{\underset{\text{proton donor}}{HC_2H_3O_2\ (aq)}} + \underset{\text{base}}{\underset{\text{proton acceptor}}{H_2O\ (l)}} \rightleftharpoons \underset{\text{conjugate acid}}{\underset{\text{proton donor}}{H_3O^+\ (aq)}} + \underset{\text{conjugate base}}{\underset{\text{proton acceptor}}{C_2H_3O_2^-\ (aq)}}$$

Notice that these conjugate pairs differ only by a single proton (H^+). Arrows connect the conjugate pairs on opposite sides of the equation. Examples of Brønsted-Lowry conjugate pairs are shown in Table 16-2.

16-2 Conjugate Pairs of Brønsted-Lowry Acids and Bases

Name of conjugate acid	Conjugate acid	Conjugate base	Name of conjugate base
hydrochloric acid	HCl	Cl$^-$	chloride ion
perchloric acid	HClO$_4$	ClO$_4^-$	perchlorate ion
sulfuric acid	H$_2$SO$_4$	HSO$_4^-$	hydrogen sulfate ion
hydrogen sulfate ion	HSO$_4^-$	SO$_4^{2-}$	sulfate ion
acetic acid	HC$_2$H$_3$O$_2$	C$_2$H$_3$O$_2^-$	acetate ion
ammonium ion	NH$_4^+$	NH$_3$	ammonia
water	H$_2$O	OH$^-$	hydroxide ion

(increasing acid strength →)

16.5 The Lewis Model

The Lewis theory of acids and bases is named after Gilbert Lewis, an American chemist. He published his ideas in 1923, the same year that Brønsted and Lowry reported their findings. His model is not used as commonly as the Brønsted-Lowry model. Lewis's model, based on bonding and molecular structure, is the broadest. It includes many more substances than either the Arrhenius or Brønsted-Lowry models. A **Lewis acid** is any substance that can accept a pair of electrons, and a **Lewis base** is a substance that can donate a pair of electrons. A Lewis acid has at least one empty orbital or bonding site, and a base has at least one lone (unbonded) pair of electrons. Notice that these definitions do not refer to specific atoms at all.

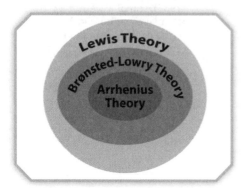

16-8 The Lewis definitions of acids and bases include more substances than do the other definitions.

The formation of ammonium (NH_4^+) ions in water shown below is a reaction between a Lewis acid and a Lewis base.

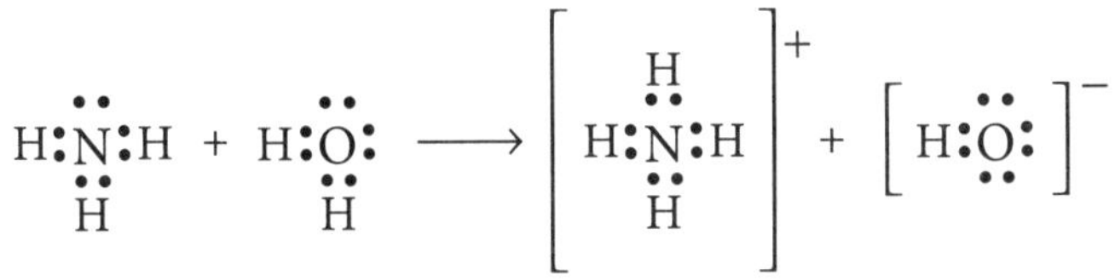

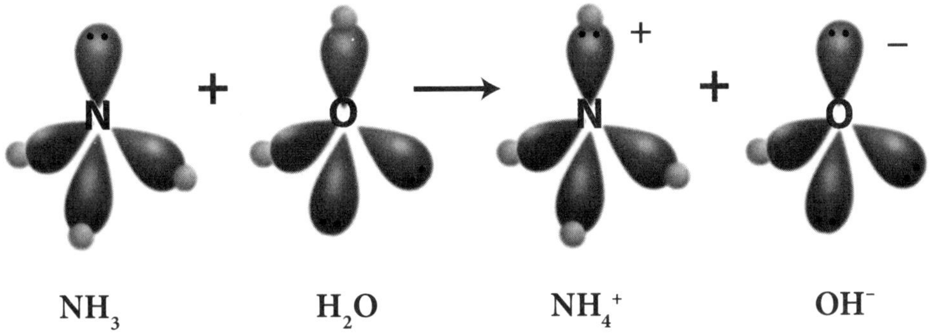

16-9 Ammonia is a Lewis base because it can donate a pair of electrons in the formation of a covalent bond.

The water molecule acts as a Lewis acid by accepting an electron pair to form a hydroxide ion. The electron pair comes from the lone pair on the nitrogen in the ammonia molecule, which acts as a Lewis base.

16-3 Acid-Base Model Summary

Type	Acid definition	Base definition
Arrhenius	releases H^+	releases OH^-
Brønsted-Lowry	donates protons (H^+)	accepts protons (H^+)
Lewis	accepts electron pairs	donates electron pairs

16A Section Review

1. Define the following:
   a. Arrhenius acid
   b. Arrhenius base
   c. Brønsted-Lowry acid
   d. Brønsted-Lowry base
   e. Lewis acid
   f. Lewis base

2. What is a neutralization reaction?

3. For each of the following reactions, label which substance is the Brønsted-Lowry acid and which is the Brønsted-Lowry base. Also, indicate the conjugate pairs.
   a. $H_2C_2O_4 + ClO^- \rightleftharpoons HC_2O_4^- + HClO$
   b. $HPO_4^{2-} + NH_4^+ \rightleftharpoons NH_3 + H_2PO_4^-$
   c. $SO_4^{2-} + H_2O \rightleftharpoons HSO_4^- + OH^-$

4. Identify each of the following as an Arrhenius acid or base, a Brønsted-Lowry acid or base, or a Lewis acid or base. Some substances may fit into more than one model.
   a. HCl
   b. Br^-
   c. CN^-
   d. H_2O
   e. $Ca(OH)_2$
   f. NH_4^+

16B Section Objectives

After finishing this section, you should be able to

- recognize the self-ionization of water as the basis for determining pH and pOH.
- calculate and convert between pH, pOH, $[H_3O^+]$, and $[OH^-]$ for solutions.
- use acid ionization constants to compare strengths of acids.
- choose the stronger of two acids on the basis of structural formulas.
- identify acids as monoprotic, polyprotic, diprotic, or triprotic on the basis of their chemical formulas.
- explain the chemical basis for the action of indicators.

16B Acid-Base Equilibria

Some acids are safe to eat and drink. But other acids are dangerous, even deadly. For example, perchloric acid, which explodes on contact with cloth or wood, would wreak havoc as soon as it touched your mouth! What's the difference? The difference is the degree to which they release or accept protons. Acid-base reactions are reversible processes. Equilibrium constants describe how readily acids deprotonate and bases protonate. These equilibrium constants give scientists a way to describe the acidity or alkalinity of a solution.

16.6 The Self-Ionization of Water

In addition to all its unique physical properties, water has an intriguing chemical property: it can react with itself. The reaction is an acid-base reaction in which one water molecule donates a proton to another water molecule. Electrical conductivity experiments which show that pure water is a weak electrolyte verify that the reaction involves acids and bases. As the two water molecules combine to form a hydroxide ion and a hydronium ion, one molecule acts as an acid and the other acts as a base. This reaction is called the **self-ionization of water**. It is also called the auto-ionization or autoprotolysis of water.

$$H_2O\ (l) + H_2O\ (l) \rightleftharpoons OH^-\ (aq) + H_3O^+\ (aq)$$

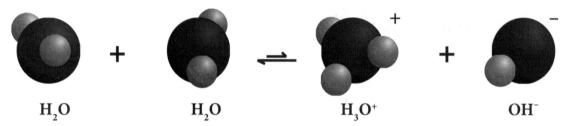

16-10 When a proton is transferred from one water molecule to another, self-ionization occurs.

The reaction is an equilibrium in which the reverse reaction predominates. Only a few molecules produce ions. This reaction can be described with an equilibrium constant.

$$K_{eq} = \frac{[H_3O^+][OH^-]}{[H_2O]^2} = [H_3O^+][OH^-]$$

Acids, Bases, and Worshiping God

As scientists study creation, they understand more and more about the complexity of God's creation. Anybody can taste when something is bitter or sour, but scientists who attempt to understand how this works realize the wonderfully complex world that explains our initial sense perception. God's world is marvelously engineered so that acids and bases react with each other the way they do. God reminds Job that the complexity of His creation should cause people to worship Him (Job 38–39).

Since water's concentration is nearly constant, it can be eliminated from the equation. The resulting constant is called the **ionization constant of water** (K_w). Experimental evidence has shown that the concentrations of hydronium and hydroxide ions are both 1.0×10^{-7} mol/L in pure water at 25 °C. Only 1 out of 555 million water molecules is ionized at any given instant. The product of these two concentrations gives K_w a value of 1.0×10^{-14}.

$$K_w = [H_3O^+][OH^-]$$
$$K_w = (1.0 \times 10^{-7}\ M)(1.0 \times 10^{-7}\ M)$$
$$K_w = 1.0 \times 10^{-14}$$

The value of K_w is a very important number. Whether a solution is acidic, basic, or neutral, the product of the concentration of hydronium ions and hydroxide ions always equals 1.0×10^{-14} at 25 °C.

Example Problem 16-1

Calculating Molarity from Hydronium Ion Concentration

The concentration of hydronium ions in a mild acid is found to be 5×10^{-7} mol/L. What is the concentration (molarity) of hydroxide ions?

Solution

Given that $K_w = [H_3O^+][OH^-]$, solve for $[OH^-]$.

$$[OH^-] = \frac{K_w}{[H_3O^+]}$$

Substitute the known values.

$$[OH^-] = \frac{1 \times 10^{-14}}{5 \times 10^{-7}} = 2 \times 10^{-8} \, M$$

16-11 For pool chemicals to work, the pH of the water must be between 7.0 and 7.6. The ideal pH is 7.2.

16.7 The pH Scale

You can see that the concentrations of hydronium and hydroxide ions in aqueous solutions are very small, frequently ranging between 1 and 1×10^{-14} mol/L. Scientists find it awkward to deal regularly with such small numbers. Søren P. L. Sørensen, a Danish chemist, proposed the pH scale in 1909 to provide a clear, concise, and convenient way to describe the concentrations of these ions without scientific notation. The **pH** of a solution is the negative logarithm of the molar hydronium ion concentration.

$$pH = -\log [H_3O^+]$$

For example, consider a sample of pure water. Its hydrogen ion concentration is 1.0×10^{-7} moles per liter. Let's calculate the pH of a sample of pure water.

$$pH = -\log [H_3O^+]$$
$$pH = -\log (1.0 \times 10^{-7} \, M)$$
$$pH = -(-7.0) = 7.0$$

Thus, the pH of pure water is 7.0. But what is the significance of this number?

The pH value also describes the acidity or alkalinity of a solution. All solutions with a pH less than 7 are called **acidic solutions** because the hydronium ion concentration is greater than the hydroxide ion

> The logarithm of a number is the power to which a base (in this case, 10) must be raised to produce that number. For example, because $10^2 = 100$, $\log_{10} 100 = 2$. In general, if $b^e = n$, then $\log_b n = e$.

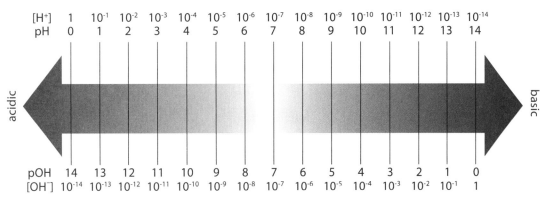

16-12 This diagram can help you understand the relationships among pH, pOH, and the concentrations of H^+ and OH^- ions.

concentration. All solutions with a pH greater than 7 are called **basic solutions**, or *alkaline solutions*, because the hydroxide ions outnumber the hydronium ions. When a solution's pH is 7, it is called a **neutral solution**. A pure sample of water is thus a neutral solution.

> For most scientific calculators, press the +/− key, press the log key, and then enter the $[H_3O^+]$, which is 2.0×10^{-5}. Different calculators have various methods for expressing exponents. Finally, press the enter key.

EXAMPLE PROBLEM 16-2
Calculating pH from Hydronium Ion Concentration

The hydronium ion concentration in a shampoo is 2.0×10^{-5} M. What is the pH of this shampoo?

Solution
$$pH = -\log [H_3O^+]$$
$$= -\log (2.0 \times 10^{-5} M)$$
$$= -(-4.7)$$
$$= 4.7$$

EXAMPLE PROBLEM 16-3
Calculating pH from Mass of Solute

What is the pH of an aqueous solution of 0.40 g of HI dissolved in 500. mL of water?

Solution
First, you must convert grams of HI to moles of HI.

$$\frac{0.40 \text{ g HI}}{} \times \frac{1 \text{ mol HI}}{127.9 \text{ g HI}} = 0.0031 \text{ mol HI}$$

Next, calculate the molarity of the HI solution in moles per liter.

$$\frac{0.0031 \text{ mol}}{500. \text{ mL}} \times \frac{1000 \text{ mL}}{1 \text{ L}} = 6.3 \times 10^{-3} M$$

$$pH = -\log [H_3O^+]$$
$$= -\log (6.3 \times 10^{-3} M)$$
$$= -(-2.2)$$
$$= 2.2$$

16.8 The pOH Scale

Just as pH describes the hydronium ion concentration, pOH describes the hydroxide ion concentration. The **pOH** of a solution is the negative logarithm of the hydroxide ion concentration.

$$pOH = -\log [OH^-]$$

> Assume that an acidic or basic solution is at 25 °C unless otherwise stated.

Because the hydronium and hydroxide ion concentrations are related, pH and pOH are also related. For a substance at 25 °C, the sum of the pH and the pOH of a solution is 14.

$$pH + pOH = 14.0$$

Thus, $[H_3O^+]$, $[OH^-]$, pH, and pOH are all related. For example, if the pH of a solution is known, the concentration of hydronium ions can be calculated. Because pH is a negative exponent, you can find $[H_3O^+]$ by raising 10 to a power given by −pH:

> Calculating the concentration from the pH value is the opposite of the process of calculating the pH from the concentration. The inverse of a logarithm is an antilogarithm. In this case, you are taking the antilogarithm of the negative pH to get the concentration.

$$[H_3O^+] = 10^{-pH}.$$

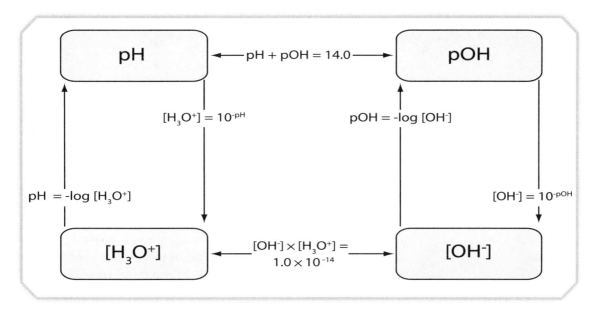

16-13 This chart shows how to convert between pH, pOH, and the different ion concentrations.

Consider an example of how these quantities are related. Household ammonia solutions used for cleaning are aqueous solutions of ammonia (NH_3). In water, ammonia produces ammonium hydroxide, which makes the solution basic (pH = 11.7). How can the pOH of this base be calculated? For the ammonia solution with a pH of 11.7, the hydronium ion concentration can be calculated using the antilog.

$$[H_3O^+] = 10^{-11.7}$$

$$[H_3O^+] = 2 \times 10^{-12} \ M$$

The pOH can be calculated directly from the pH.

$$pOH = 14.0 - pH$$

$$pOH = 14.0 - 11.7$$

$$pOH = 2.3$$

The hydroxide ion concentration can be calculated from the antilog of the pOH, just as the hydronium ion is calculated from the pH.

$$[OH^-] = 10^{-pOH}$$

$$[OH^-] = 10^{-2.3}$$

$$[OH^-] = 5 \times 10^{-3} \ M$$

Logarithms and Significant Digits

The number of significant digits in the ion concentration determines the number of decimal places in the pH or pOH. This is because the digits to the left of the decimal point in the pH or pOH value represent a power of ten—the exponent—and they are considered exact. For instance, if the pH equals 3.36 (2 decimal places), the hydronium ion concentration would be 4.4×10^{-4} (2 SDs).

EXAMPLE PROBLEM 16-4

Calculating pH and pOH from Concentration

A 0.10 *M* hydrochloric acid solution has a hydronium ion concentration of 1.0×10^{-1} mol/L. Calculate the pH, the hydroxide ion concentration, and the pOH of this solution. *(continued)*

Solution
The pH can be found first.
$$pH = -\log(1.0 \times 10^{-1})$$
$$pH = -(-1.00)$$
$$pH = 1.00$$
Next, find the pOH.
$$pOH = 14.00 - pH$$
$$pOH = 14.00 - 1.00$$
$$pOH = 13.00$$
The hydroxide ion concentration can now be calculated from the pOH.
$$[OH^-] = 10^{-pOH}$$
$$[OH^-] = 10^{-13.00}$$
$$[OH^-] = 1.0 \times 10^{-13}\,M$$

This example problem is pretty simple because the concentration was an integral power of ten. In these cases, you simply take the absolute value of the exponent as your pH or pOH value. The following example has values more typical of those you might encounter in the laboratory.

EXAMPLE PROBLEM 16-5
Calculating pH and pOH from Concentration

A sulfuric acid solution has a hydronium ion concentration of 3.2×10^{-2} mol/L. Calculate the pH, the $[OH^-]$, and the pOH.

Solution
The pH can be found first.
$$pH = -\log(3.2 \times 10^{-2})$$
$$pH = -(-1.50)$$
$$pH = 1.50$$
Next, find the pOH.
$$pOH = 14.00 - pH$$
$$pOH = 14.00 - 1.50$$
$$pOH = 12.50$$
The hydroxide ion concentration can now be calculated from the pOH.
$$[OH^-] = 10^{-pOH}$$
$$[OH^-] = 10^{-12.50}$$
$$[OH^-] = 3.2 \times 10^{-13}\,M$$

16.9 Acid-Base Strength

Though pH describes the concentration of a solution, it does not describe its strength. **Strong acids** give up protons easily and ionize completely. Surprisingly, only six common acids are classified as strong acids: $HClO_4$, HI, HBr, HCl, H_2SO_4, and HNO_3. **Weak acids** do not ionize completely because only a portion of their molecules lose protons. For example, only one in twenty-four molecules deprotonates in an acetic acid solution.

Similarly, the strength of a base depends on how easily it protonates. **Weak bases**, such as the Cl^- ion, are poor proton acceptors. **Strong bases**, such as the OH^- ion, accept protons readily. The hydroxides of active metals—$LiOH$, $NaOH$, KOH, $RbOH$, $CsOH$, $Mg(OH)_2$, $Ca(OH)_2$, $Sr(OH)_2$, and $Ba(OH)_2$—are the best sources of the hydroxide ion, so they are commonly called strong bases.

For solutions, *strong* is not the same as *concentrated*, and *weak* is not the same as *dilute*. The terms *concentrated* and *dilute* refer to the amount of acid in a given volume of aqueous solution. It is possible to have a dilute solution of a strong acid or a concentrated solution of a weak acid.

16-14 As electrolytes, both of these acids release ions and conduct electricity. Based on this photo, can you tell which one is stronger?

The conjugate pair of a strong acid or a strong base is always weak. For example, perchloric acid ($HClO_4$) is a strong acid that readily gives up a hydrogen ion. Once the hydrogen and the chloride ions are apart, they do not readily rejoin to form perchloric acid. Therefore, the perchlorate ion (ClO_4^-) is a weak conjugate base. Conversely, Cl^- is a weak base with a strong conjugate acid, HCl.

Acid and base strengths are described by the dissociation equilibrium constant. The **acid-ionization constant** (K_a) describes the extent of the forward equilibrium reaction in the formation of the hydronium ion. In this generalized formula, HA represents a typical acid where A is the anion.

$$HA + H_2O \rightleftharpoons H_3O^+ + A^-$$

$$K_a = \frac{[H_3O^+][A^-]}{[HA]}$$

Similarly, the **base-ionization constant** (K_b) describes the extent of the forward equilibrium reaction in the formation of the hydroxide ion. Here, B represents a base.

$$B + H_2O \rightleftharpoons OH^- + HB^+$$

$$K_b = \frac{[OH^-][HB^+]}{[B]}$$

Both of these constants are related to the constant for the self-ionization of water, K_w.

$$K_a \cdot K_b = K_w$$

Stronger acids and bases have higher K_a and K_b values because they ionize more readily. In fact, the forward reaction for strong acids and bases is so predominant that the reaction proceeds to completion because they completely dissociate. Table 16-4 lists conjugate pairs and their ionization constants, in order from strong to weak. Because the stronger acids have weaker conjugate bases, the bases

16-4 Relative Strengths of Some Acids and Bases

Acid		K_a	K_b		Base
perchloric acid	$HClO_4$	large	very small	ClO_4^-	perchlorate ion
hydriodic acid	HI	large	very small	I^-	iodide ion
hydrobromic acid	HBr	large	very small	Br^-	bromide ion
hydrochloric acid	HCl	large	very small	Cl^-	chloride ion
sulfuric acid	H_2SO_4	large	very small	HSO_4^-	hydrogen sulfate ion
nitric acid	HNO_3	large	very small	NO_3^-	nitrate ion
sulfurous acid	H_2SO_3	1.3×10^{-2}	8.3×10^{-13}	HSO_3^-	bisulfate ion
hydrogen sulfate ion	HSO_4^-	1.2×10^{-2}	8.3×10^{-13}	SO_4^{2-}	sulfate ion
phosphoric acid	H_3PO_4	6.9×10^{-3}	1.3×10^{-12}	$H_2PO_4^-$	dihydrogen phosphate ion
hydrofluoric acid	HF	6.8×10^{-4}	1.4×10^{-11}	F^-	fluoride ion
formic acid	$HCHO_2$	1.8×10^{-4}	5.6×10^{-11}	CHO_2^-	formate ion
acetic acid	$HC_2H_3O_2$	1.7×10^{-5}	5.6×10^{-10}	$C_2H_3O_2^-$	acetate ion
carbonic acid	H_2CO_3	4.3×10^{-7}	2.4×10^{-8}	HCO_3^-	hydrogen carbonate ion
hypochlorous acid	HClO	3.0×10^{-8}	2.9×10^{-7}	ClO^-	hypochlorite ion
boric acid	H_3BO_3	5.8×10^{-10}	1.4×10^{-5}	$H_2BO_3^-$	dihydrogen borate ion
ammonium ion	NH_4^+	5.7×10^{-10}	1.8×10^{-5}	NH_3	ammonia
hydrocyanic acid	HCN	4.9×10^{-10}	2.5×10^{-5}	CN^-	cyanide ion
hydrogen carbonate ion	HCO_3^-	4.8×10^{-11}	2.1×10^{-4}	CO_3^{2-}	carbonate ion
hydrogen peroxide	H_2O_2	2.6×10^{-12}	1.3×10^{-2}	HO_2^-	hydroperoxide ion
water	H_2O	1.0×10^{-14}	55.5	OH^-	**hydroxide ion**
ammonia	NH_3	very small	large	NH_2^-	**amide ion**
hydrogen	H_2	very small	large	H^-	**hydride ion**
methane	CH_4	very small	large	CH_3^-	**methide ion**

increasing acid strength →

← increasing base strength

*Acids and bases shown in **bold** are strong acids and bases.

16.10 Polarity, Charge, and Acid-Base Strength

Ionization constants describe acid-base strengths; they do not determine them. The location of shared electron clouds, the availability of a lone pair, and the charge on an acid or base determine how easily a proton (H^+ ion) can be released or absorbed. In other words, the strongest acids are more polar and have lower bond energies. The strongest bases are negatively charged and have available bonding sites. Strong bases are often soluble ionic compounds of hydroxides and alkali or alkaline-earth metals, like NaOH and $Mg(OH)_2$. These ionic compounds completely dissociate in water to form hydroxide ions, which are strong proton acceptors.

16-5 Causes of Strength

Compound	Electronegativity of central atom	Diagram
CH_4	2.4	
NH_3	3.0	
H_2O	3.5	
HF	4.0	

stronger →

Notice that the number of hydrogen atoms in a molecule has no bearing on whether it will be strong or weak. Some of the strongest acids have only one proton. Boric acid, with three hydrogen atoms, is one of the weakest.

To be easily ionizable, the nucleus of a hydrogen atom must be exposed. The negatively charged electron cloud of the covalent bond must be shifted away from the hydrogen toward a more electronegative atom in the center of the molecule. The series of compounds shown on Table 16-5 illustrates how a highly electronegative element can expose hydrogen nuclei. The hydrogen atoms in the methane molecule are virtually surrounded by negative charges. The highly electronegative fluorine atom pulls shared electrons to itself. Consequently, hydrofluoric acid is the strongest acid of this series.

The electrical charge of a molecule also determines the strengths of acids. The sulfuric acid molecule (H_2SO_4) has a neutral charge and is a strong acid. The hydrogen sulfate ion (HSO_4^-) has an ionizable proton, but this anion has little tendency to release another hydrogen ion because of electrostatic attraction. On the other hand, it will quite readily attract a proton because of its negative charge and available bonding site. Similarly, H_3PO_4, $H_2PO_4^-$, and HPO_4^{2-} have decreasing acidic strengths. Anions hold protons more strongly than neutral molecules.

16-15 Because boric acid is weak, it can be used in eye drops.

16-6 Acid/Base Strength Summary

Strong acids	Strong bases
exposed hydrogen nuclei	exposed electron pairs
minimal negative charges	possible strong negative charges

16.11 Amphoteric Substances

An **amphoteric** substance can act as both a Brønsted-Lowry acid and a Brønsted-Lowry base. Its behavior depends on reaction conditions. For example, the hydrogen carbonate ion accepts a proton (H^+ ion) when an acid is added to the solution. Under those circumstances, the hydrogen carbonate ion is a base.

$$HCO_3^- \,(aq) + H_3O^+ \,(aq) \rightleftharpoons H_2CO_3 \,(aq) + H_2O \,(l)$$

However, when a base is added to the solution, the hydrogen carbonate ion acts as an acid by donating a proton.

$$HCO_3^- \,(aq) + OH^- \,(aq) \rightleftharpoons CO_3^{2-} \,(aq) + H_2O \,(l)$$

Water is also an amphoteric substance, as demonstrated by its reactions with perchloric acid and ammonia.

$$H_2O \,(l) + HClO_4 \,(aq) \rightleftharpoons H_3O^+ \,(aq) + ClO_4^- \,(aq)$$

$$H_2O \,(l) + NH_3 \,(aq) \rightleftharpoons NH_4^+ \,(aq) + OH^- \,(aq)$$

In the first reaction, water acts as a Brønsted-Lowry base by receiving a proton from perchloric acid. In the second reaction, it acts as an acid by donating a proton to ammonia.

16.12 Polyprotic Acids

Monoprotic Brønsted-Lowry acids can donate only one proton. Examples of monoprotic acids include perchloric acid ($HClO_4$), hydrochloric acid (HCl), and nitric acid (HNO_3). **Polyprotic** acids can donate more than one proton per molecule. Each stage of ionization has a different equilibrium constant. A **diprotic** acid can donate two protons. For example, carbonic acid (H_2CO_3) is diprotic. The K_a for the first ionization is called K_{a1}, and that for the second ionization is called K_{a2}.

$$H_2CO_3 + H_2O \rightleftharpoons HCO_3^- + H_3O^+$$

$$K_{a1} = \frac{[HCO_3^-][H_3O^+]}{[H_2CO_3]}$$

$$= 4.3 \times 10^{-7}$$

$$H_2CO_3^- + H_2O \rightleftharpoons CO_3^{2-} + H_3O^+$$

$$K_{a2} = \frac{[CO_3^{2-}][H_3O^+]}{[H_2CO_3^-]}$$

$$= 4.8 \times 10^{-11}$$

For any polyprotic acid, all the possible stages of ionization occur in the same solution. For instance, a solution of phosphoric acid (H_3PO_4) would contain the ions $H_2PO_4^-$, HPO_4^{2-}, and PO_4^{3-} in addition to the hydronium ion H_3O^+.

All polyprotic acids have at least one amphoteric substance in their series of ionizations. In carbonic acid, the hydrogen carbonate ion is amphoteric. Furthermore, the K_{a2} is always smaller than the K_{a1}. The first ionization involves separation of a proton from an uncharged carbonic acid (H_2CO_3) molecule, and the second involves separation of a proton from the negatively charged hydrogen carbonate (HCO_3^-) ion. Because the strength of attraction between opposite charges depends on the magnitude of the charges, the second ionization is less favorable than the first. In general, $K_{a1} > K_{a2} > K_{a3}$. Acids with three ionizable protons are called **triprotic** acids. Can you name a triprotic acid? Several more examples of polyprotic acids are shown in Table 16-7.

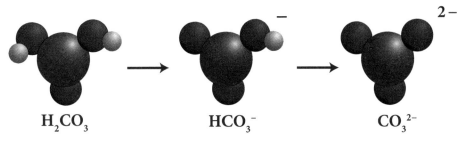

16-16 A proton is lost at each stage of ionization.

16-7 Ionization Constants of Polyprotic Acids

Polyprotic acid	Acid-base reaction	K_a
chromic acid	$H_2CrO_4 \rightleftharpoons H^+ + HCrO_4^-$	1.8×10^{-1}
	$HCrO_4^- \rightleftharpoons H^+ + CrO_4^{2-}$	3.2×10^{-7}
hydrosulfuric acid	$H_2S \rightleftharpoons H^+ + HS^-$	1.1×10^{-7}
	$HS^- \rightleftharpoons H^+ + S^{2-}$	1.0×10^{-14}
phosphoric acid	$H_3PO_4 \rightleftharpoons H^+ + H_2PO_4^-$	6.9×10^{-3}
	$H_2PO_4^- \rightleftharpoons H^+ + HPO_4^{2-}$	6.2×10^{-8}
	$HPO_4^{2-} \rightleftharpoons H^+ + PO_4^{3-}$	4.8×10^{-13}
sulfurous acid	$H_2SO_3 \rightleftharpoons H^+ + HSO_3^-$	1.3×10^{-2}
	$HSO_3^- \rightleftharpoons H^+ + SO_3^{2-}$	6.3×10^{-8}

16.13 Measuring pH

Indicators are substances whose colors are sensitive to pH. As a result, they change colors when the pH of a solution changes. At the beginning of this chapter you were introduced to litmus paper, a type of indicator. Indicators may be weak acids or bases whose conjugates are different colors. Typically, the anion has one color when it is combined with the hydrogen atom and a different color after ionization. Some indicators change colors at low pH values, some at high pH readings. Those that change at low pH values are the stronger acids because they ionize easily. The indicators that do not change color until the higher pH range are weaker acids. Some are polyprotic, so they may exhibit more than one color change.

Examples of indicators, their colors, and the pH range over which their colors change are given in Figure 16-17. Notice that each indicator has a range of pH values over which the color change occurs. This is known as the **transition interval**. Not all indicators have the same size transition interval.

Phenolphthalein, the first synthetic indicator, was introduced in 1877. Prior to that time, chemists used natural plant juices as acid-base indicators, some of which are shown in Figure 16-18.

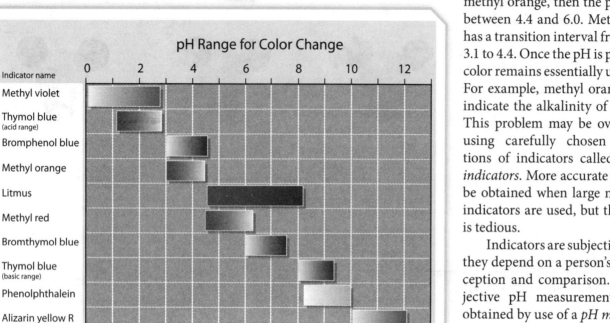

16-17 Color of some common pH indicators

Indicators give rough estimates of the pH value. For instance, if a solution is yellow when tested with both bromthymol blue and methyl orange, then the pH must be between 4.4 and 6.0. Methyl orange has a transition interval from a pH of 3.1 to 4.4. Once the pH is past 4.4, the color remains essentially unchanged. For example, methyl orange cannot indicate the alkalinity of a solution. This problem may be overcome by using carefully chosen combinations of indicators called *universal indicators*. More accurate values can be obtained when large numbers of indicators are used, but this process is tedious.

Indicators are subjective because they depend on a person's color perception and comparison. More objective pH measurements can be obtained by use of a *pH meter*. A pH meter indirectly determines the concentration of the hydronium ions by measuring the voltage they produce between two points and comparing the voltage to a standard. The instrument uses a *pH probe*, which is a sensor that responds only to hydronium ions. The meter also contains a reference probe and usually a temperature sensor or adjustment control, since the measurement of pH is temperature dependent. The voltage produced by the probe is directly proportional to pH. The slope of the pH probe's output is determined by temperature. The pH instrument converts the voltage and temperature information into a digital display of pH. Before

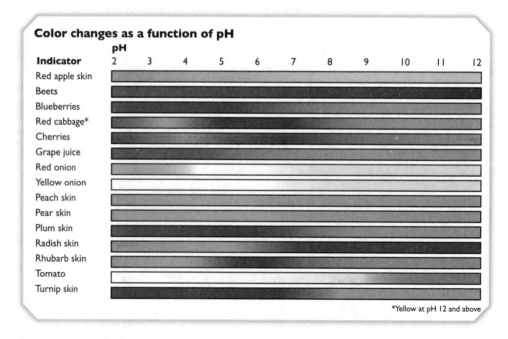

16-18 Natural indicators can be found in many fruits and vegetables.

measurements are taken, these meters are standardized against solutions of known pH to ensure accuracy.

16B Section Review

1. Find the pH of a solution whose [H_3O^+] equals 9.5×10^{-8}.
2. What is the [H_3O^+] of a solution with a pOH equal to 5.45?
3. Find the pOH of a solution whose pH is 3.52.
4. Using Table 16-4, indicate the stronger acid in each pair.
   a. HI, H_2SO_4
   b. HF, HCN
   c. H_2O_2, $HCHO_2$
   d. H_3BO_3, H_3PO_4
   e. HCl, $HClO_4$
5. A rainwater sample shows a yellow color when tested with both methyl red and bromthymol blue. What is the approximate pH of the sample? Is it acidic, basic, or neutral?
6. Calculate the pH of a solution with [OH^-] = 1.0×10^{-6}.

16-19 Exact pH measurements are possible with a pH meter.

16C Neutralization

The pain and discomfort of heartburn are usually associated with meals that were either too large or too spicy. However, the burning sensation comes not from oregano or green peppers but from excess hydrochloric acid. This acid, which is normally confined to the stomach, irritates the tissues of the esophagus if it rises past the upper part of the stomach. To relieve the pain, a person can neutralize the acid with an antacid (a base). Mixing these two active chemicals results in a harmless product, a salt.

16C Section Objectives

After finishing this section, you should be able to
- complete equations for neutralization reactions between acids and bases.
- calculate the concentration of a solution when given data from an appropriate acid-base titration.
- explain how a buffer system works and give examples of reactions in which they function.

16.14 Salts

As mentioned in Subsection 16.2, neutralization reactions are the reactions of a strong acid and a strong base to produce a salt and water. Antacids that contain magnesium hydroxide react to neutralize stomach acid.

$$Mg(OH)_2 \,(aq) + 2HCl \,(aq) \longrightarrow MgCl_2 \,(aq) + 2H_2O \,(l)$$

Sodium hydroxide is much too corrosive to be used as an antacid, but it too neutralizes acids.

$$2NaOH \,(aq) + H_2SO_4 \,(aq) \longrightarrow Na_2SO_4 \,(aq) + 2H_2O \,(l)$$

A **salt** is a substance formed when the anion of an acid and the cation of a base combine. In the antacid reaction, Cl^- (the acid's anion) combines with Mg^{2+} (the base's cation) to form $MgCl_2$ (the salt). Many different salts can be produced in neutralization reactions. Sodium chloride, zinc bromide, and potassium fluoride are all examples of salts. Most salts, like these three, are composed of a metal and a nonmetal, but this is not always true. Ammonium chloride (NH_4Cl), potassium sulfate (K_2SO_4), and aluminum phosphate ($AlPO_4$) contain polyatomic ions.

According to the Arrhenius definitions of acids and bases, the heart of every neutralization reaction is the formation of water from

16-20 Antacids are effective because they provide a base to neutralize the excess hydrochloric acid in your stomach.

hydronium and hydroxide ions. We can easily see this from the net ionic reaction of a neutralization reaction. The hydronium ions of acids and the hydroxide ions of bases combine to form two water molecules.

$$H_3O^+ (aq) + OH^- (aq) \rightleftharpoons 2H_2O (l)$$

Metal cations and nonmetal anions are present, but they are not involved in the neutralization reaction. Because the ions are found in the solution of the reactant and the product, they are called *spectator ions*. If the resulting solution evaporates, these ions will crystallize into a salt.

Acids, Bases, and Loving Your Neighbor

Scientists serve God by using His creation and the principles that govern it to serve others. Many lakes and ponds in the Northern Hemisphere are acidified due to excessive air pollution in the 1970s and 1980s. Although more stringent regulations limit further damage, the environment still suffers. Some companies have patented processes that adjust the pH of these bodies of water through the addition of bases such as lime and procedures that oxygenate the lakes by circulating the water.

Example Problem 16-6
Writing a Neutralization Reaction

Write a neutralization reaction to produce zinc chloride ($ZnCl_2$), a salt.

Solution

Salts are made from the anions of acids and the cations of bases. In this case, the most common acid that releases the chloride ion is hydrochloric acid. Zinc hydroxide is a likely source for the zinc ion.

$$2HCl\ (aq) + Zn(OH)_2\ (aq) \longrightarrow ZnCl_2\ (aq) + 2H_2O\ (l)$$

Heartburn and Acid Reflux

Facets of Chemistry

Imagine that you are asked to design a portable chemical reaction chamber whose walls are capable of withstanding liquids ten times more acidic than lemon juice. In fact, the chamber must manufacture its own acid and simultaneously protect itself against that same acid. Furthermore, it must be able to expand as much as 200% to contain added reactants, and it should have two valves that are able to open and close automatically. Would you believe you are already carrying this vessel everywhere you go?

Your stomach is a remarkable sac in which much of your chemical and mechanical digestion occurs. With nary a grumble, it churns away, day in, day out, processing everything that you swallow. Occasionally, a virus will make your stomach empty itself rapidly in the wrong direction, but otherwise you hardly give it a second thought.

Heartburn or acid indigestion is one common symptom that tells you your stomach is malfunctioning. The strong hydrochloric acid (HCl) in your stomach is unable to eat into your stomach wall because mucous glands keep the inner layer well coated. Above the valve where the

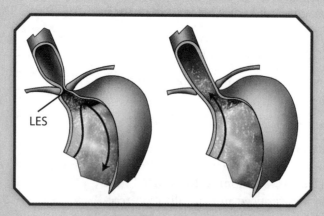

LES

esophagus empties into the stomach, however, there is less protection. If this ring-like valve, the *lower esophageal sphincter* (LES), does not close up tightly after food is dropped into the stomach, strong acidic fluids can move up into the esophagus. In some, but not all, cases, this triggers sharp, burning pain, which is sometimes mistaken for a heart attack.

Today, an increasing number of people are diagnosed with gastroesophageal reflux disease, commonly called GERD or *acid reflux*. In these people, the stomach contents regularly back up, or reflux, into the sensitive esophagus. In a minority of patients, the acid is strong enough or is present long enough to damage the wall of the esophagus. For reasons not fully understood, many people experience heartburn during these incidents, but others have no discomfort at all.

The same God Who designed the stomach to be a one-of-a-kind acid reaction chamber also made safeguards that prevent GERD for most people. Even in those cases where material is

EXAMPLE PROBLEM 16-7
Determining Salts

What salt is produced from the neutralization of barium hydroxide by acetic acid ($HC_2H_3O_2$)?

Solution

$$2HC_2H_3O_2\,(aq) + Ba(OH)_2\,(aq) \longrightarrow Ba(C_2H_3O_2)_2\,(aq) + 2H_2O\,(l)$$

Barium acetate forms from the barium and acetate ions.

16.15 Acid-Base Titrations

Acid-base **titrations** are controlled neutralization reactions in which scientists determine the unknown concentration of a solution by measuring its capacity to react with a solution of known concentration. This is all based on the concept of stoichiometry, which you studied in Section 9B.

Suppose that a chemist has 100.0 mL of a hydrochloric acid solution but that he does not know its exact concentration. Because he knows that NaOH neutralizes HCl, he can find the concentration by adding small volumes of a precisely measured 1.00 M sodium hydroxide solution until the pH of the solution rises to 7.0. At this point, there are as many moles of H^+ ions from the acid (HCl) as there are OH^- ions from the base (NaOH).

> The solution with a known concentration is sometimes called the *standard solution* or *titrant*.

---

refluxed into the esophagus, gravity usually causes it to flow back down into the stomach before it can cause problems. On top of this, the mouth continually produces saliva, which is subconsciously swallowed. This tends to wash any refluxed fluid back into the stomach. Also, bicarbonate ions (HCO_3^-) in the saliva react with the remaining HCl to neutralize it.

Some acid reflux sufferers have an LES that does not seal tightly. The loose seal can allow gastric fluid to leak into the esophagus and pool there for extended periods. In other cases, the valve opens up and stays open for minutes at a time. Because the valve is often open just after a meal, the strong stomach acid refluxes into the esophagus, where it begins to irritate the tender esophageal lining. Some other malfunctions associated with GERD include an esophagus that does not contract adequately to empty itself into the stomach and a stomach that is slow to empty itself into the small intestine. These disorders increase the time during which acid reflux can occur.

A number of complications can result from GERD. These include ulcers, a narrowing of the esophagus due to scar tissue, coughing, asthma, throat and larynx inflammation, lung infections, and even esophageal cancer.

How does a doctor determine whether a person has GERD that requires treatment or merely suffers from occasional acid reflux that is not threatening? The current diagnosis is usually made with a rather unusual pH meter. A tiny sensor that measures the pH level is attached to a long, thin catheter. This catheter is inserted into the nose and run down the throat, and the sensor dangles just above the stomach at the LES. The other end of the catheter goes from the nose, over the top of the ear, and down to a recording device at the waist. The time and duration of every acid reflux event is then accurately recorded. After the 24-hour test, the data is analyzed. The frequency, length, and severity of the acid reflux events help the doctor to make a diagnosis.

As you might expect, acid-base interactions play a role in some of the treatment options for GERD. Avoiding spicy or acidic foods such as citrus juices, tomatoes, and carbonated drinks can stave off potential problems. Chewing gum may stimulate increased saliva with its neutralizing bicarbonate ions. Antacids used for common heartburn may also help neutralize the hydrochloric acid before it damages the esophagus. Some other effective drugs relieve GERD by inhibiting or temporarily shutting down the acid-producing cells in the stomach wall. Another popular drug contains antacids combined with a foaming agent. When chewed and swallowed, these tablets react with the stomach contents to form a layer of foam that floats on top of the liquid. This chemical barrier neutralizes acids and at the same time serves as a physical barrier against reflux.

A graph called a **titration curve** shows how pH changes when an acid or base is added to a solution. Figure 16-22 shows how the pH of a solution of unknown concentration changes with the addition of 1.00 M sodium hydroxide.

Adding 50 mL of sodium hydroxide raises the pH of the acid to 7; so 50.0 mL of 1.00 M sodium hydroxide is chemically equivalent to 100 mL of the acid. With this information, the chemist knows that the 1.00 M sodium hydroxide is twice as concentrated as the hydrochloric acid. He can then calculate that the hydrochloric acid has a concentration of 0.50 M.

The **equivalence point** is the point in a titration at which an equivalent of titrant is added, that is, the point at which the number of H_3O^+ ions equals the number of OH^- ions. The equivalence point should not be confused with the **end point** of a titration. The end point is the point at which some change in a property of the solution is detected. For example, if phenolphthalein indicator is added to a solution, it remains clear until a pH reading of 8.5, at which a pink color appears. If a 0.1 M HCl solution is titrated with a 0.1 M NaOH solution, the equivalence point will occur at pH 7.0, but the end point will be detected at pH 8.5—the first noticeable pink tinge.

According to the rules of stoichiometry, it is possible to calculate the unknown molarity of a given amount of solution if we know the ratio between the reactants. The values of this ratio are obtained from the reaction's balanced equation. The following equation can be used to solve for unknown molarity concentrations in titration reaction calculations:

$$(M_k)(V_k) = (M_u)(V_u).$$

16-21 A buret releases precise amounts of solution to allow titration calculations.

In some lab titrations, the end point may be determined through a temperature change, precipitation of a salt, or a conductivity change.

The equivalence point is not always at a pH of 7. If a strong acid reacts with a strong base, the equivalence point is at a pH of 7. If a strong acid reacts with a weak base, the equivalence point is at a pH below 7. If a strong base reacts with a weak acid, the equivalence point will be at a pH greater than 7.

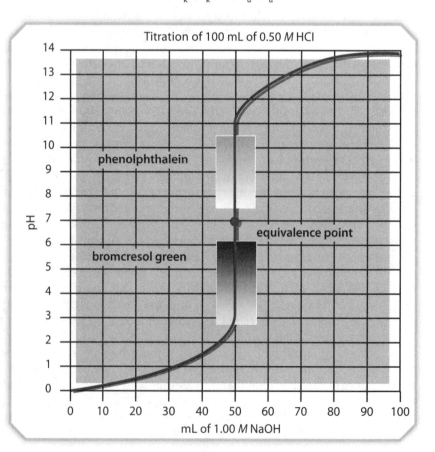

16-22 Chemical indicators denote a range of values. The end point is a specific value.

where k stands for "known" and u stand for "unknown." The V is the volume of each substance. This equation can be rearranged to solve for the unknown quantity.

EXAMPLE PROBLEM 16-8
Determining a Concentration

A volume of 108 mL of 5.00×10^{-3} M sodium hydroxide solution can neutralize 36.0 mL of a nitric acid solution. What is the concentration of the nitric acid?

Solution

$$M_u = \frac{(M_k)(V_k)}{(V_u)} = \frac{(5.00 \times 10^{-3} \text{ M NaOH})(108.0 \text{ mL NaOH})}{36.0 \text{ mL HNO}_3}$$

$M_u = 0.0150$ M solution of HNO_3

Since the concentration of H^+ and OH^- ions alone affects the neutralization reaction, the calculated or given molarity may need to be multiplied by an integer if the acid or base can accept or donate more than one hydrogen ion. This is called the number of *equivalents*. For instance, a solution of H_3PO_4 determined to be 0.306 M actually contributes 0.918 moles of H^+ ions (3×0.306) since there are three hydrogen atoms per molecule.

EXAMPLE PROBLEM 16-9
Determining a Concentration

Adding 38.0 mL of 1.50×10^{-2} M barium hydroxide solution can neutralize 20.0 mL of a hydrochloric acid solution according to the equation,

$$Ba(OH)_2 + 2HCl \longrightarrow BaCl_2 + 2H_2O.$$

What is the concentration of the hydrochloric acid?

Solution

First, note that each mole of barium hydroxide provides two moles of hydroxide ions in this neutralization, effectively doubling its concentration. Use the same neutralization formula from the previous example, multiplying barium hydroxide's concentration by 2 before substituting its value.

$$M_u = \frac{(M_k)(V_k)}{(V_u)}$$

$$M_u = \frac{(3.00 \times 10^{-2} \text{ M Ba(OH)}_2)(38.0 \text{ mL Ba(OH)}_2)}{20.0 \text{ mL HCl}}$$

$M_u = 0.057$ M solution of HCl

16.16 Buffers

In the course of a single day, you probably eat foods with very different pH values. It is obvious that the human stomach can handle a wide range of pH values. The bloodstream, however, cannot tolerate

> A buffer is a solution of weak conjugate acid-base pairs.

this wide pH range. Its pH must stay between 7.35 and 7.45. God has designed the bloodstream with an ingenious protective feature—a buffer system.

Buffers are solutions that resist pH changes. They "buffer" drastic pH changes when bases or acids are added. Buffer systems usually consist of a weak acid and its conjugate base or of a weak base and its conjugate acid. Water is not a buffer; a small amount of acid or base changes its pH dramatically. Adding 0.01 mol of hydrochloric acid or sodium hydroxide to a liter of pure water changes its pH by 5 units. The same substances added to a liter of blood change its pH only 0.1 units.

Buffers can be found in many chemical reactions in living things. Man has also harnessed the benefits of buffers to control chemical reactions for industry and to make safer medicines such as buffered aspirin. By properly selecting the acid-base pair, buffers can be used in almost any pH range. A given buffer acts to maintain the pH relatively constant within its working range. You can see several examples of buffers and their usable ranges in Table 16-8.

Buffers are most effective in regulating slight pH changes. If an enormous amount of acid or base is added to a buffered solution, the buffer will be depleted and the pH will change drastically. Keep in mind that the buffer does not keep the solution neutral (pH of 7.0), but rather maintains a constant pH for the given solution. For example, suppose that the pH of a given solution containing a buffer is 4.7. As a base is added to this solution, the buffer will maintain the pH at 4.7 until it has been depleted. Only then will the pH begin to rise with the addition of more base solution.

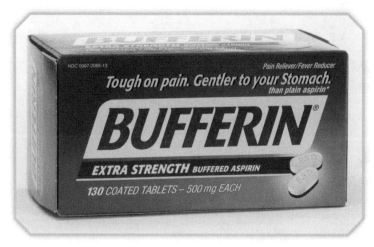

16-23 Some products are buffered to protect you from potentially harmful pH changes.

For example, blood is buffered by a mixture of carbonic acid (H_2CO_3), which is a weak acid, and hydrogen carbonate ions (HCO_3^-), which are its conjugate base. This combination of solutes keeps the pH relatively constant by reacting with hydronium or hydroxide ions. When an acid intrudes on this blood buffer system, hydrogen carbonate ions snare the hydronium ions.

$$H_3O^+ + HCO_3^- \longrightarrow H_2CO_3 + H_2O$$

When a base disturbs the equilibrium of the same buffer system, carbonic acid molecules spring into action and remove the hydroxide ions.

$$H_2CO_3 + OH^- \longrightarrow H_2O + HCO_3^-$$

Blood has a second buffer system that utilizes hydrogen phosphate ions (HPO_4^{2-}) and dihydrogen phosphate ions ($H_2PO_4^-$). If excess acid enters the bloodstream, it reacts with HPO_4^{2-} to form $H_2PO_4^-$. Bases react with $H_2PO_4^-$ to form HPO_4^{2-}. In this reaction, the H^+ ions released by the $H_2PO_4^-$ react with hydroxide from the base. Thus, the pH is stabilized in either direction.

16-8	Common Buffer Systems
Components	Usable pH range
formic acid + sodium formate	2.6–4.8
citric acid + sodium citrate	3.0–6.2
acetic acid + sodium acetate	3.4–5.9
sodium hydrogen carbonate + sodium carbonate	9.2–10.6
sodium hydrogen carbonate + sodium hydroxide	9.6–11.0

16.17 Using Acid Reactions to Solve Problems

The main ingredient in alcoholic beverages is ethanol, also called ethyl alcohol (CH_3CH_2OH). Ethanol is unique because it is absorbed unchanged into the bloodstream by the intestines, stomach, esophagus, and even the mouth because it does not readily dissociate in the presence of digestive acids. This is why you can tell a person has been drinking by how his or her breath smells. The alcohol in a person's breath is directly related to the amount in their blood, called the blood alcohol content (BAC). Is there an accurate way for police officers to measure ethanol in a person's breath right on the side of the road, deterring drunk driving?

Police officers can test a driver's BAC right on the side of the road using a device that analyzes a person's breath. The device has four main parts: a mouthpiece, two identical vials, and a photocell. The suspect breathes into a mouthpiece. The exhaled air is bubbled through one vial containing water, potassium dichromate (a green salt), sulfuric acid, and silver nitrate (a catalyst).

Evaporated ethanol in the breath sample reacts in this vial to produce a color change, as orange potassium chromate is used up to produce greenish chromium sulfate.

$$2K_2CrO_4\ (aq) + 3CH_3CH_2OH\ (aq) + 8H_2SO_4\ (aq) \xrightarrow{AgNO_3}$$
$$2Cr_2(SO_4)_3\ (aq) + 2K_2SO_4\ (aq) + 3CH_3COOH\ (aq) + 11H_2O$$

The intensity of this color change is directly related to a person's BAC. A light-sensitive device called a photocell compares this color change to another vial not exposed to the breath sample. In most states, if a driver's BAC exceeds 0.08 g of alcohol per 100 mL of blood, he or she is considered legally drunk and can be prosecuted.

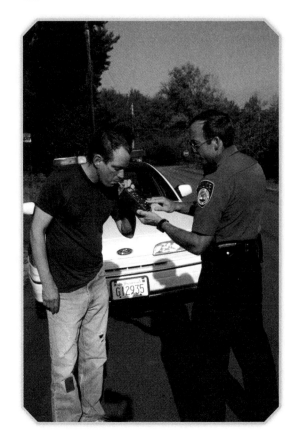

16-24 The level of ethyl alcohol in the blood is indirectly measured by its concentration in exhaled air.

16C Section Review

1. In titrations, what is the difference between the equivalence point and the end point?
2. Write a neutralization reaction to form the salt K_2SO_4.
3. What salt results from the neutralization of calcium hydroxide by hydrobromic acid?
4. In a titration reaction, 33.6 mL of 0.302 M RbOH neutralizes 20.0 mL of HF. Calculate the molarity of the acid solution.
5. The addition of 37 mL of 4.2×10^{-2} M HCl solution can neutralize 65 mL of NaOH solution. What is the concentration of NaOH?
6. A solution has a buffer that is effective in the 3.0–6.2 pH range. If the system is overloaded with acid, what will be the effect on the buffer system?
DS 7. Show how potassium dichromate is a salt produced by the neutralization reaction of chromic acid and potassium hydroxide.
DS 8. What other acids do you see in the reaction in Subsection 16.17?

Questionable Science?

Though such devices are a valuable tool to keep drunk drivers off the road, their accuracy can be significantly influenced by different factors. Other substances in the breath besides ethanol can produce a color change. Those substances cause the readings to be higher than a person's actual BAC. After studying these devices, researchers have found that they can produce readings at least 15% higher than a person's actual BAC. Thus, about one in four drivers can be unjustly accused of being legally drunk. This inaccuracy is especially significant if the BAC measurements are permitted as evidence in court cases.

Chapter Review

Chapter Summary

- Acids taste sour, turn blue litmus paper red, react with active metals to produce hydrogen and a salt, conduct an electrical current in solution, and neutralize bases.

- Bases taste bitter, turn red litmus paper blue, neutralize acids to form a salt and water, conduct an electrical current in solution, and feel slippery.

- Arrhenius defined acids as substances that ionize to release hydrogen ions into solutions, and bases as those that release hydroxide ions in solution.

- Brønsted and Lowry defined acids as substances that donate protons, and bases as those that accept protons. These substances exist as conjugate pairs. A conjugate acid is formed when the base protonates, and the conjugate base forms when the acid deprotonates.

- Lewis defined acids as substances that can accept a pair of electrons. Bases can donate a pair of electrons. His is the broadest definition of acids and bases.

- In self-ionization, two water molecules react to form a hydronium ion (H_3O^+) and a hydroxide ion (OH^-). The equilibrium constant of water (K_w) is the product of the two concentrations of these ions. It is 1.0×10^{-14} at 25 °C.

- The pH of a solution is the negative logarithm of the hydronium ion concentration. The pOH is the negative logarithm of the hydroxide ion concentration. A pH under 7 is acidic, a pH above 7 is basic, and a pH of 7 is neutral.

- The strength of an acid or base is not related to its concentration but instead to its ability to ionize in solution. Ionization constants quantify the degree to which these reactions proceed in the forward direction.

- Amphoteric substances are those which can act as either Brønsted-Lowry acids or bases, depending on the reaction.

- Monoprotic Brønsted-Lowry acids can donate only one proton, and polyprotic Brønsted-Lowry acids can donate more than one. Donations occur in steps of ionization, but all ions are found in the same solution.

- Indicators are substances (usually weak acids or bases) whose color is sensitive to pH changes.

- A strong acid and a strong base neutralize to form a salt and water.

- Titrations are controlled reactions in which the unknown concentration of a measured volume of a substance is determined by adding a known volume of a reactive substance of known concentration. A titration is expressed by the equation $(M_k)(V_k) = (M_u)(V_u)$.

Coming to Terms

Term	Page
electrolyte	397
neutralization reaction	398
Arrhenius acid	398
Arrhenius base	398
Brønsted-Lowry acid	399
Brønsted-Lowry base	399
deprotonation	399
protonation	399
conjugate pair	399
conjugate base	400
conjugate acid	400
Lewis acid	400
Lewis base	400
self-ionization of water	402
ionization constant of water (K_w)	402
pH	403
acidic solution	403
basic solution	404
neutral solution	404
pOH	404
strong acid	406
weak acid	406
weak base	406
strong base	406
acid-ionization constant (K_a)	407
base-ionization constant (K_b)	407
amphoteric	410
monoprotic	410
polyprotic	410
diprotic	410
triprotic	411
indicator	411
transition interval	411
salt	413
titration	415
titration curve	416
equivalence point	416
end point	416
buffer	418

- Buffers are solutions that resist pH changes despite small additions of hydronium or hydroxide ions. Many reactions in living things are possible because of buffer systems.

Review Questions

Concept Review

1. Fill in the blanks. (Use a separate piece of paper.)
   a. $Al(OH)_3$ is an Arrhenius _____ and therefore must be a Brønsted-Lowry _____.
   b. H_2SO_4 is an Arrhenius _____ and therefore must be a Brønsted-Lowry _____.
   c. NH_3 can be classified as a(n) _____ base and a(n) _____ base, but not as a(n) _____ base.
   d. The Cl^- ion can be classified as a(n) _____ base or a(n) _____ base, but not as a(n) _____ base.

2. Formic acid (HCO_2H) ionizes in water to form a formate ion and a hydronium ion.

   $$HCO_2H\ (aq) + H_2O\ (l) \rightleftharpoons H_3O^+\ (aq) + CO_2H^-\ (aq)$$

   a. What is the acid in the forward reaction?
   b. What is the conjugate base of this acid?
   c. What is the name of the process by which formic acid loses a proton?

3. Two reactions describe the stepwise ionization of H_2SO_4.

   $$H_2SO_4\ (l) + H_2O\ (l) \rightleftharpoons HSO_4^-\ (aq) + H_3O^+\ (aq)$$
   $$HSO_4^-\ (aq) + H_2O\ (l) \rightleftharpoons SO_4^{2-}\ (aq) + H_3O^+\ (aq)$$

   a. What is the acid in the forward direction of the first reaction?
   b. What is the conjugate base of this acid?
   c. What is the acid in the forward direction of the second reaction?
   d. What is the conjugate base of this acid?

4. Fill in the blanks. (Use a separate piece of paper.)
   a. Aqueous solutions of Arrhenius acids taste _____.
   b. Aqueous solutions of Arrhenius _____ are slippery.
   c. Arrhenius acids react with active metals to produce _____ gas.
   d. Aqueous solutions of Arrhenius bases taste _____.

5. What ion gives H_2SO_4 the ability to react strongly with many substances? What ion is responsible for the corrosiveness of strong bases like NaOH?

6. List four common properties of acids and four common properties of bases.

7. What particles, other than H_2O molecules, are always present in pure water?

8. Write the equation for the equilibrium constant that describes the reaction of water with itself. What is the numerical value of this constant?

9. Can water molecules act like acids, bases, both, or neither?

10. A young chemist measures $[H_3O^+]$ and $[OH^-]$ in an aqueous solution at 25 °C. He reports that $[H_3O^+] = 1 \times 10^{-8}$ and $[OH^-] = 1 \times 10^{-8}$. His lab instructor tells him to go back to the lab and make the measurements again. Why?

11. What advantage does the pH scale offer that scientific notation does not?

12. Choose the correct answer.
    a. (Strong / Weak) acids ionize incompletely.
    b. (Strong / Weak) acids have large K_as.
    c. Strong (acids / bases) accept protons easily.
    d. The conjugate base of a strong acid is a (strong / weak) base.

13. Referring to Table 16-4, tell which acid in each of the following pairs is the stronger one:
    a. $HClO_4$, H_3PO_4
    b. formic acid, acetic acid
    c. hydrocyanic acid, formic acid
    d. NH_4^+, HSO_4^-
    e. NH_4^+, H_2O

14. Choose the member of each pair that should be the stronger acid and then write a brief justification for your choice.
    a. $HBrO_4$, HBr b. PH_3, H_2S

15. Is the HCO_3^- ion an acid, a base, both, or neither? Explain.

16. Write the equation that shows how the ions in acids act to neutralize the ions in bases during a neutralization reaction.

DS 17. Though devices to analyze the breath of drivers for alcohol can help prevent or punish drunk driving, what is the ultimate solution to drunk driving?

True or False

18. All Lewis acids donate protons.
19. All Brønsted-Lowry acids accept protons.
20. All Arrhenius bases donate protons.
21. All Brønsted-Lowry acids are Lewis acids.
22. All Arrhenius acids are Lewis acids.

Application

⊙ 23. All the following substances can undergo deprotonation reactions. Write the equations for their ionization constants. If they go through a series of ionizations, include the equation for each.
    a. $HClO_4$
    b. H_2CO_3
    c. H_3BO_3

24. Calculate the [H₃O⁺] and the pH of each solution.
   a. 1×10^{-5} M HCl
   b. 5×10^{-3} M HCl
   c. 1×10^{-1} M H₂SO₄
   d. 1×10^{-5} M H₂SO₄

25. Calculate the [OH⁻] and the pOH of the four solutions in Question 24.

26. What is the [H₃O⁺] in each of the following solutions?
   a. orange juice, pH = 4.0
   b. black coffee, pH = 5.0
   c. pure water, pH = 7.0
   d. seawater, pH = 8.0
   e. phosphate detergent solution, pH = 9.5

27. Classify each of the solutions in the previous question as acidic, neutral, or basic.

28. A NaOH solution contains 1.000 M NaOH solution. If 25.00 mL of this solution neutralizes 27.00 mL of a HCl solution of unknown concentration, what is the molarity of the HCl?

29. How many milliliters of 2.00 M H₂SO₄ will be required to neutralize 45.0 mL of 3.00 M KOH?

30. A careless laboratory assistant attempted to determine the concentration of a perchloric acid (HClO₄) solution. He found that 50.00 mL of 1.000 M KOH neutralized 0.5000 L of the HClO₄ solution. His calculations were as follows:

$$M_u = \frac{(50.00 \text{ mL})(1.000 \text{ M})}{0.5000 \text{ L}} = 100.0 \text{ M HClO}_4.$$

Another laboratory assistant insists that the calculation is incorrect. Why? What is the correct answer?

31. A salad-dressing manufacturer wants to make tangy salad dressing by using vinegar with an acetic acid concentration of at least 1 M. The quality control department examines a sample of vinegar and determines that 300.0 mL of 0.100 M sodium hydroxide neutralizes the acetic acid in 25.00 mL of the vinegar. Does this vinegar meet the manufacturer's requirements?

32. How could you determine the point at which a chemical reaction raises the pH of a solution past 8.8?

33. What happens to excess H₃O⁺ ions when an acid is added to a solution buffered by a combination of acetic acid and sodium acetate?

34. What happens to excess OH⁻ ions when a base is added to a solution buffered by a combination of acetic acid and sodium acetate?

DS 35. The reaction in devices used to analyze breath for alcohol produces acetic acid as a byproduct from ethanol. Write the balanced equations and the equilibrium constant for acetic acid.

Oxidation-Reduction

CHAPTER 17

17A Redox Reactions	425
17B Electrochemical Reactions	432
Facets	
The Battle Against Corrosion	429
Oxidation, Reduction, and the Evolution of Life	433

Dominion Science Problem

Rusting

Cars, bicycles, ships, bridges, tools—what do all these things have in common? These everyday objects are made of metal, and they rust. Rusting eats away the metal in useful objects, or even forces apart their components, making them worthless junk. Despite man's advanced knowledge of chemistry, losses caused by rust in the United States alone run into the hundreds of billions of dollars annually. The amount of steel lost to rusting equals about 25% of the steel the United States produces annually. What can be done to prevent rusting?

17A Redox Reactions

Reactions involving transfers or shifts of electrons occur all around us. When you power on your cell phone, admire the chrome on your new car, or get a roll of camera film developed, you are directly benefiting from these kinds of reactions. Most chemical reactions involve electron shift and transfer, so the study of these reactions is an important part of chemistry. These reactions are called **oxidation-reduction reactions**, or **redox reactions**.

17.1 Oxidation: Loss of Electrons

Redox reactions occur in the formation of both ionic and covalent compounds. For example, when magnesium burns to form magnesium oxide, each magnesium atom transfers two electrons to an oxygen atom. Reactions between covalent compounds often end up shifting electrons toward or away from atoms, depending on their electronegativities. When sulfur dioxide forms, oxygen attracts the shared electrons closer to itself than sulfur does, thus gaining electrons. In both of these cases, one substance loses electrons and another substance gains electrons.

Oxidation takes place when an atom loses electrons. In the reaction between magnesium and oxygen previously mentioned, magnesium atoms are oxidized.

$$Mg \longrightarrow Mg^{2+} + 2e^-$$

Lost electrons (e^-) in the equation above are shown as reaction products. Electrons gained are shown as reactants. Because neutral magnesium loses electrons, it becomes a cation. When an atom is oxidized, its oxidation number always becomes more positive.

$$\overset{0}{2Mg} + \overset{}{O_2} \longrightarrow \overset{+2}{2MgO}$$

At this point it may seem logical to assume that all oxidation reactions involve oxygen. While oxygen usually causes atoms to oxidize because of its high electronegativity, other atoms can do the same thing. For example, when sodium reacts with chlorine, chlorine causes the sodium atoms to oxidize.

$$\overset{0}{2Na} + \overset{0}{Cl_2} \longrightarrow \overset{+1}{2Na^+} + \overset{-1}{2Cl^-} \longrightarrow \overset{+1\ -1}{2NaCl}$$

17.2 Reduction: Gain of Electrons

Electrons lost by atoms must go somewhere, so they are gained by other atoms. An atom that gains electrons undergoes a process called **reduction**. In the reaction between magnesium and oxygen, for example, each oxygen atom is reduced because it gains two electrons. Reduction makes oxidation numbers smaller (*reducing* the oxidation number) as more negative charges join atoms. The oxidation number of oxygen reduces from 0 to –2.

Is the reaction that forms magnesium oxide a reduction reaction? Is it an oxidation reaction? Actually, it is both. Oxygen atoms gain electrons and are reduced. But the magnesium atoms lose electrons

17A Section Objectives
After finishing this section, you should be able to
- identify redox reactions.
- name the oxidized and reduced substances in a redox reaction.
- name the oxidizing and reducing agents in a redox reaction.
- balance redox reactions.

17-1 Fireflies belong to a family of beetles that use flashes of light to attract mates. Light energy is released from special cells called photocytes when the molecule luciferin is oxidized.

Review of Oxidation Number Rules

Rule 1	free atoms = 0
Rule 2	ion charge = oxidation number
Rule 3	compound sum = 0
Rule 4A	Group 1 = +1
Rule 4B	Group 2 = +2
Rule 4C	H = +1 or –1
Rule 4D	O = –2 or –1
Rule 4E	Group 17 = –1
Rule 5	sum of ONs in polyatomic ion = charge

and are oxidized. Oxidation and reduction *always* occur simultaneously. When electrons leave one atom, they must join another.

For example, the development of photographic film is based on the reduction of silver from Ag^+ to Ag. Photographic film contains grains of a silver halide such as silver bromide (AgBr). When exposed to light, silver ions in the silver bromide grains are energized. These activated ions are more prone to be reduced during the developing process than those in areas not struck by light. The film is developed in a solution containing substances that cause the sensitized silver ions to gain electrons.

$$\overset{+1}{2Ag^+} + e^- \longrightarrow \overset{0}{Ag}$$

The film is then developed to produce a negative. The reduced silver appears as black areas on the negative, so light objects in the photographed scene appear dark. Conversely, dark objects in the scene appear light on the negative. To produce a positive, or print, that looks like the photographed scene, a light is projected through the negative to produce an image on paper treated with a silver halide. The paper is then developed with the same procedure used on the film. Though color photography is more difficult to explain, it is also based on reduction processes.

The only reactions that qualify as redox reactions are those in which the oxidation numbers of the atoms involved change. If oxidation numbers do not change, neither oxidation nor reduction has occurred.

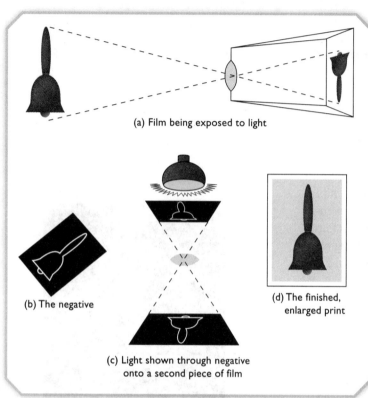

17-2 Redox reactions make photo developing possible.

(a) Film being exposed to light
(b) The negative
(c) Light shown through negative onto a second piece of film
(d) The finished, enlarged print

EXAMPLE PROBLEM 17-1

Identifying Redox Reactions

Is either of the following reactions a redox reaction?
a. $HCl + H_2O \rightleftharpoons H_3O^+ + Cl^-$
b. $2KClO_3 \longrightarrow 2KCl + 3O_2$

Solution

a. After assigning oxidation numbers to each atom, you can see that the oxidation numbers of all the atoms remain the same before and after the reaction. The reaction is not a redox reaction.

$$\overset{+1\ -1}{HCl} + \overset{+1\ -2}{H_2O} \rightleftharpoons \overset{+1\ -2}{H_3O} + \overset{-1}{Cl}$$

b. The oxidation numbers for both oxygen and chlorine change during the reaction. This reaction involves both oxidation and reduction.

$$\overset{+1\ +5\ -2}{2KClO_3} \longrightarrow \overset{+1\ -1}{2KCl} + \overset{0}{3O_2}$$

17.3 Oxidizing and Reducing Agents

A **reducing agent** is a substance that reduces another substance. Recall that developing photographs involves the use of one or more reducing agents. They cause silver ions to be reduced to silver atoms. Because reduction cannot occur without oxidation, reducing agents contain atoms that donate electrons. Because magnesium donates electrons, it is a reducing agent in the reaction between magnesium and oxygen.

$$2Mg + O_2 \longrightarrow 2MgO$$

Remember that oxidation and reduction always occur simultaneously. If magnesium acts as a reducing agent, it donates electrons and is oxidized. Reducing agents are always oxidized in redox reactions.

An **oxidizing agent** is a substance that oxidizes other substances. Oxidizing agents are commonly used as cleaners and disinfectants. Chlorine, for instance, is an oxidizing agent added to swimming pools to kill bacteria, algae, and fungi. Chlorine acts as an oxidizing agent by taking electrons from these microbes. In the process, chlorine atoms are reduced to chloride ions.

$$Cl_2 + 2e^- \longrightarrow 2Cl^-$$

Because chlorine gains electrons in this reaction, it is reduced. Oxidizing agents are always reduced in redox reactions.

17-3 Chlorine's strong oxidizing qualities make it useful for killing microorganisms in pool water.

EXAMPLE PROBLEM 17-2
Identifying Oxidizing and Reducing Agents

Tell which substance in the following reaction is a reducing agent and which is an oxidizing agent.

$$2Na + 2H_2O \longrightarrow 2NaOH + H_2$$

Solution

The oxidation number of sodium goes from 0 in sodium to +1 in sodium hydroxide. The oxidation number of hydrogen goes from +1 in water to 0 in hydrogen gas. Sodium was oxidized (lost electrons), and hydrogen was reduced (gained electrons). The hydrogen in water, therefore, is acting as an oxidizing agent, and sodium is acting as a reducing agent.

Household bleach is a 5% aqueous solution of sodium hypochlorite (NaOCl). Sodium hypochlorite is an effective oxidizing agent. It is reduced in redox reactions according to the following equation:

$$NaOCl + H_2O + 2e^- \longrightarrow NaCl + 2OH^-.$$

17-4 This mnemonic device can help you remember the relationship between reduction, oxidation, and the movement of electrons. **L**ost **E**lectrons = **O**xidation; **G**ained **E**lectrons = **R**eduction.

Bleach is normally used to eliminate stains and make dirty white clothes clean and white again. The molecules in stains possess very mobile electrons. These molecules are colored because they absorb visible light energy, energizing their electrons. When a dirty article of clothing is bleached, the sodium hypochlorite in the bleach grabs the less

17-5 The bleach this woman used on her hair and teeth is not sodium hypochlorite but hydrogen peroxide, another oxidizing agent.

tightly bonded electrons from the molecules in the stains. When these electrons are removed, the molecules no longer absorb visible light. Consequently, the color disappears, and the cloth is once again white. In effect, bleach does not remove stains; it merely decolorizes them.

17.4 Using Oxidation to Solve Problems

Rusting, a redox reaction, is a complex process. The formation of rust is similar to what happens chemically in a battery. Iron in steel loses two electrons. Water, a necessary component of the rusting process, gains those two electrons to form hydrogen gas and hydroxide ions in the presence of oxygen. Those hydroxide ions can combine with iron (II) ions to form iron hydroxide. The net reaction results in the production of iron hydroxide.

$$\text{Fe (s)} + 2\text{H}_2\text{O (l)} \longrightarrow \text{H}_2 \text{ (g)} + 2\text{Fe(OH)}_2 \text{ (s)}$$

This iron hydroxide then reacts with oxygen, again from the air, and produces the more familiar rust, or iron (III) oxide (Fe_2O_3).

$$4\text{Fe(OH)}_2 \text{ (s)} + \text{O}_2 \text{ (g)} \longrightarrow 2\text{Fe}_2\text{O}_3 \text{ (s)} + 4\text{H}_2\text{O (l)}$$

The formation of rust is much slower—about one hundred times slower—if no oxygen is present. Anything that pits or creates cracks, divots, and bends in the iron can aid the corrosion process. Another contributing factor is the presence of chloride ions, such as in salt. Knowing the conditions that encourage rusting can lead to its prevention. If iron can be coated to avoid contact with water and oxygen, rusting can be prevented.

Stainless steel is an iron-chromium alloy that is rust-resistant. It contains from 10.5%–28% chromium, which oxidizes in atmospheric oxygen to form a very thin layer of chromium (III) oxide (Cr_2O_3) on the surface of the alloy. This layer, which is thin enough to maintain the luster of the alloy, is impervious to water and oxygen. The layer is added in a process known as *passivation*. Stainless steel finds many uses in jewelry, cookware, surgical instruments, and building materials for skyscrapers and industrial machinery. However, even stainless steel can rust in extreme conditions, such as when it is gouged, pitted, stressed, or exposed to high concentrations of chloride ions.

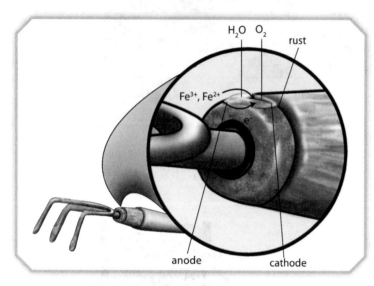

17-6 The process of rust formation occurs in the presence of iron, water, and oxygen and depicts the processes occurring in a voltaic cell.

17.5 Balancing Redox Reactions

Back in Chapter 8 you learned how to balance chemical reactions. Do you remember why chemical reactions need to be balanced? It is necessary for reactions to have the same number of each kind of atom on both sides of the chemical equation to maintain the law of the conservation of mass. Similarly, redox reactions must be balanced. Since electrons are being moved around, both the number of each kind of atom *as well as the overall charge on each side of the reaction* need to be balanced. The number of electrons on each side of the reaction must be conserved to maintain the first law of thermodynamics.

> Both atoms and charge need to be balanced in a redox equation.

Some redox reactions can be balanced like other reactions in the traditional way of identifying reactants and products, counting atoms of each element, and adjusting coefficients. For example, in Subsection 8.17, we considered this balanced equation:

$$2KClO_3\,(s) \rightleftharpoons 2KCl\,(s) + 3O_2\,(g).$$

This redox reaction can be balanced with little difficulty. There are two potassium atoms, two chlorine atoms, and six oxygen atoms on both sides of the equation. The overall charge is neutral on both sides of the equation.

In reactions where the same element appears in multiple reactants or in multiple products, however, a different method sometimes simplifies the process. Consider the following unbalanced reaction:

$$NH_3\,(g) + NO_2\,(g) \longrightarrow N_2\,(g) + H_2O\,(l).$$

17-7 Potassium chlorate ($KClO_3$) is found on matches and in fireworks and other explosives.

THE BATTLE AGAINST CORROSION

Facets of Chemistry

We've talked about how rust and corrosion happen and how stainless steel is rust-resistant. Other methods can be used to prevent metal from rusting.

As a general rule, when two metals are in contact with each other, the more active metal undergoes corrosion and the less active metal is protected. Iron may be protected from corrosion by electrically connecting it to a metal such as zinc or magnesium. The more active metal in such an arrangement more readily undergoes corrosion, saving the more desirable metal.

For example, the Alaskan oil pipeline is a large-diameter iron pipe that is protected by heavy zinc wires. Ships with exposed metal surfaces underwater are protected from the corrosive action of seawater by zinc anodes. Because these anodes are themselves consumed while protecting the less active metal, they are sometimes called sacrificial anodes.

Several materials are used to coat iron to protect it against corrosion. A "tin can" is actually an iron can coated with tin. Galvanized iron is iron coated with zinc. It is used for such items as trash cans and chain-link fences. Paints, lacquers, and varnishes are also used to protect the surfaces of iron and steel. Paints containing red lead or zinc chromate are especially effective for preventing corrosion.

Chemically altering the surface of iron can also form a protective coating. When red-hot iron is treated with steam, a thin coating of black iron oxide (Fe_3O_4) is formed. The black color of stovepipes is produced by this process. The iron oxide coating gives good protection to the metal even at high temperatures.

Is rust one of the effects of sin on the creation? Think of a junkyard of rusting cars or the streaks of rust on a tin roof. Rust seems to be a prime example of decay in our world. It may be that before the Fall rust did not occur. We do know that the Fall affected creation (Gen. 3:17–18; Rom. 8:22). Perhaps chemistry worked differently before the Fall. Perhaps the effect of rust was counteracted before the Fall by other chemical processes, or perhaps unfallen mankind would have developed uses of iron in ways that rust was

prevented. Today God has given us the ability to discover ways to counteract rust. More importantly, we are able to lay up treasures in heaven, where rust will not corrupt (Matt. 6:19–20).

With enough time and adjustment, you may be able to balance it by inspection, but another procedure might get you to the answer more quickly.

Step 1. Assign oxidation numbers to all atoms, and determine which atoms are oxidized and which ones are reduced. Since the oxidation numbers of hydrogen and oxygen remain unchanged in this particular reaction, ignore them for now and focus on the other atoms. As previously discussed, ion charges are written with the charge designation after the numeral (3–), and the oxidation numbers are written with the charge designation preceding the numeral (+5). This convention will help avoid confusion when dealing with reactions containing ions.

$$\overset{-3}{NH_3} + \overset{+4}{NO_2} \longrightarrow \overset{0}{N_2} + H_2O$$

Step 2. Determine which atoms are oxidized and which have been reduced. In this case, the nitrogen from the reactants is both oxidized and reduced. Nitrogen from ammonia (NH_3) is oxidized because its oxidation number increases from –3 to 0. The nitrogen in nitrogen dioxide is reduced from +4 to 0. Notice that hydrogen's and oxygen's oxidation numbers do not change.

$$\overset{\text{oxidation}}{\overset{-3}{NH_3} + \overset{+4}{NO_2} \longrightarrow \overset{0}{N_2} + H_2O}$$
$$\text{reduction}$$

Step 3. To help keep track of the oxidation and reduction halves of the reaction, draw a line connecting the atoms involved in reduction and those involved in oxidation. You should also write the net change in charge along each line.

$$\overset{+3}{NH_3 + NO_2 \longrightarrow N_2 + H_2O}$$
$$-4$$

Step 4. Balance the magnitude of the oxidation number changes by multiplying both pairs of reactants and products by a single coefficient for the reduction change, and another coefficient for the oxidation change. In some cases, you may need to multiply only one of the changes to balance the charge change. In this reaction, both changes must be multiplied to reach the least common multiple of 12. Notice that the N_2 on the product side has been multiplied by both 4 and 3. We add these multiples to get a coefficient of 7 on the right side of the equation.

$$\overset{+3 \times 4 = 12}{4NH_3 + 3NO_2 \longrightarrow (3+4)N_2 + H_2O}$$
$$-4 \times 3 = -12$$

Step 5. Now you can use the conventional inspection method to finish balancing the equation. Notice that you have fourteen nitrogen atoms on the right. There are half as many nitrogen atoms on the reactant side.

$$4NH_3 + 3NO_2 \longrightarrow 7N_2 + H_2O$$

Doubling the coefficients of each of the reactants would give you the right number of nitrogen atoms. But now you should recognize that the hydrogen and oxygen are still unbalanced.

$$8NH_3 + 6NO_2 \longrightarrow 7N_2 + H_2O$$

Since there are twenty-four hydrogen atoms and twelve oxygen atoms on the reactant side, the equation balances by simply adding 12 as a coefficient for the water product.

$$8NH_3 + 6NO_2 \longrightarrow 7N_2 + 12H_2O$$

Steps to Balancing Redox Reactions

- Step 1: Assign oxidation numbers.
- Step 2: Determine which atoms have been oxidized and which have been reduced.
- Step 3: Draw lines connecting atoms that have changed and write the magnitude of the charge change.
- Step 4: Multiply the reactant and product of each pair by a coefficient to balance the magnitude of the change.
- Step 5: Finish balancing the equation by the conventional method if necessary.

EXAMPLE PROBLEM 17-3
Balancing Redox Reactions

Balance the equation for the following redox reaction:

$$HCl + HNO_3 \longrightarrow HOCl + NO + H_2O.$$

Solution

Step 1. Assign oxidation numbers to all atoms whose charges have changed. Disregard hydrogen and oxygen in this reaction.

$$\overset{-1}{H}Cl + \overset{+5}{H}NO_3 \longrightarrow \overset{+1}{H}OCl + \overset{+2}{N}O + H_2O$$

Step 2. Determine which atoms are oxidized and which have been reduced.

(oxidation: Cl goes from −1 to +1; reduction: N goes from +5 to +2)

Step 3. Write the net change in charge along each line.

(+2 for oxidation, −3 for reduction)

Step 4. Balance the magnitude of the oxidation number changes by multiplying both pairs of reactants and products by a single coefficient for the reduction change and another coefficient for the oxidation change.

$$3HCl + 2HNO_3 \longrightarrow 3HOCl + 2NO + H_2O$$

(+2 × 3 = 6; −3 × 2 = −6)

Step 5. Now you can use the conventional inspection method to finish balancing the equation. In this case, the equation is balanced so no further coefficient changes are necessary.

$$3HCl + 2HNO_3 \quad 3HOCl + 2NO + H_2O$$

17A Section Review

1. In the redox equation $K + Br_2 \longrightarrow 2KBr$,
   a. what element is reduced?
   b. what element is oxidized?
   c. what is the reducing agent?
   d. what is the oxidizing agent?
   e. which element loses an electron or electrons?

f. which element gains an electron or electrons?
g. how many electrons are exchanged?

2. For each of these equations, determine whether it is a redox reaction. For each redox reaction, tell which element is oxidized and which is reduced.
   a. $2KBr + Cl_2 \longrightarrow 2KCl + Br_2$
   b. $2NaOH + CuCl_2 \longrightarrow 2NaCl + Cu(OH)_2$
   c. $Mg + I_2 \longrightarrow MgI_2$

DS 3. What substance is oxidized in the rusting process? How can you tell?

DS 4. What substance is reduced in the rusting process? How can you tell?

5. Balance the following equation using the oxidation-number method:
$$Cu + HNO_3 \longrightarrow Cu(NO_3)_2 + NO_2 + H_2O.$$

17B Electrochemical Reactions

The field of **electrochemistry** involves redox reactions and electricity. Redox reactions can be used to produce electricity, as in a battery, or electricity can trigger a redox reaction, as in the electroplating of metal objects.

17.6 Electrochemical Cells

Metals can conduct electricity because their bonds allow electrons to be mobile in a metal lattice. Some solutions can conduct electricity, but for a different reason. Water conducts electricity only when a sufficient number of ions are present. An **electrolyte** is any substance that, when dissolved in water, allows the resulting solution to conduct electricity. When an electrolyte is dissolved in water, anions (negative ions) and cations (positive ions) are formed. The ions move freely in the solution and therefore may carry a charge.

A substance's ability to act as an electrolyte depends on its ability to dissociate into ions. Solutions of strong electrolytes conduct electricity well because the electrolytes completely dissociate, producing many ions in the solution. Most salts, strong acids, and strong bases are strong electrolytes (NaCl, HCl, H_2SO_4, HNO_3, and NaOH). Substances that do not readily dissociate, such as weak acids, weak bases, and less soluble salts, are weak electrolytes. Solutions of weak electrolytes do not conduct electricity as well as solutions of strong electrolytes do. Covalent substances such as sugar, alcohol, and oxygen might dissolve in

17B Section Objectives

After finishing this section, you should be able to
- describe how an electrochemical cell works, correctly using the terms *anode*, *cathode*, *oxidation*, and *reduction*.
- describe how a voltaic cell works, correctly using the terms *anode*, *cathode*, *oxidation*, and *reduction*.
- give examples of voltaic cells.
- list several practical applications of electrochemistry.
- contrast rechargeable cells and fuel cells with conventional voltaic cells.

17-8 Solutions of a strong electrolyte (HCl), a weak electrolyte ($HC_2H_3O_2$), and a nonelectrolyte ($C_{12}H_{22}O_{11}$) have different electrical conductivities.

Oxidation, Reduction, and the Evolution of Life

Facets of Chemistry

"Mathematics and dynamics fail us when we contemplate the earth, fitted for life but lifeless, and try to imagine the commencement of life upon it. This certainly did not take place by any action of chemistry, or electricity, or crystalline grouping of molecules under the influence of force, or by any possible kind of fortuitous concourse of atmosphere. We must pause, face to face with the mystery and miracle of the creation of living things."

—Lord Kelvin*

Can you imagine the world on the seventh day of Creation, throbbing with life and brilliant in its perfection? Scripture reveals that everything necessary for life was available in abundance. At the end of the sixth day, God pronounced it all "very good."

However, evolutionists imagine the beginning of life much differently. Completely absent was the warmth of the Creator's loving touch.

In 1929, A. I. Oparin, from Russia, and John Haldane, from Britain, both biochemists, suggested that the earth's atmosphere was at one time without oxygen. They proposed that the earth's original atmosphere consisted primarily of methane, ammonia, hydrogen, and water vapor—completely devoid of oxygen. This supposed early atmosphere is described as reducing; it would have supplied electrons to substances and chemically reduced them to other forms. A reducing atmosphere was seen as necessary for the evolution of life because if sugars, amino acids, and the other necessary building blocks of cells were exposed to oxygen, they would quickly decompose to carbon dioxide and water.

Although Oparin and Haldane agreed that spontaneous generation was not possible in the current atmosphere, they speculated that in the reducing atmosphere the conditions favored spontaneous generation of life. It is interesting to note that the

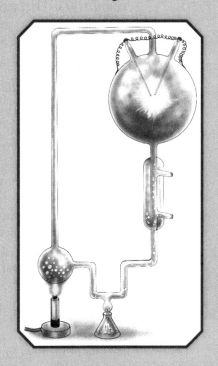

odds of a single cell's originating in a primitive environment, even given a 4.6 billion-year time frame, are one chance in $10^{40\,000}$.

American chemist Harold Urey (1893–1981) published an elaboration of the Oparin-Haldane theory in 1952. Working in Urey's laboratory as a graduate student, Stanley Miller (1930–2007) developed an experiment to test a hypothesis—could life have originated in the reduced atmosphere? His experiment simulated a reducing atmosphere high in methane, a gas thought to be abundant in the atmosphere of Jupiter and some planetary moons. Using electrodes to provide energy, Miller was able to produce some basic amino acids. Although the newspapers of that day claimed that he had "created life," those amino acids were no more living creatures than a pile of lumber, nails, and wire is a house.

Is there any evidence to support the idea of an early reducing atmosphere? Proponents of this theory point to geologic formations from the Precambrian era containing various sediments of reduced minerals. However, these same geologic formations that evolutionists use to support their claim for the reducing atmosphere contain large amounts of calcium carbonate ($CaCO_3$) and iron (III) oxide (Fe_2O_3), but these compounds could have formed only if large amounts of oxygen were present.

Two other problems face the idea of a reducing atmosphere. The first is that ultraviolet radiation from the sun in a reducing atmosphere without the protective ozone layer would have decomposed ammonia long before the envisioned life-forming reactions could have taken place. There couldn't be an ozone layer without oxygen! The second problem is explaining how oxygen was eventually introduced to the atmosphere. Although several explanations ranging from photosynthesis to volcanoes have been suggested, none are without serious problems.

Over the years, evolutionists have contrived many theories of earth history and the origin of life. However, none of the theories are more than speculation. Since theories have been known to change, it is possible that the reducing atmosphere may not be a permanent feature of evolutionary theory. Efforts will be made to patch it up for a time, but if it is finally deemed hopeless, it will be replaced by something else.

---

*Quoted in Barnes, Thomas G., "Physics: A Challenge to 'Geologic Time,'" in *The Battle For Creation: Acts/Facts/Impacts*, vol. 2, eds. Duane T. Gish and Henry M. Morris (San Diego: Creation Life Publishers, 1976), 232.

water, but they cannot conduct electricity. For this reason, such substances are called **nonelectrolytes**.

All electrochemical techniques rely on the electrical conductivity of metals and electrolyte solutions. To be useful, however, these substances must be assembled in just the right way. The fundamental apparatus used in electrochemistry is the **electrochemical cell**. An electrochemical cell consists of two electrical contacts, called **electrodes**, immersed in an electrolyte solution with a wire joining the electrodes. The electrodes are nothing more than metal rods or wires. They are commonly made of metals such as zinc, platinum, or copper.

There are two types of electrochemical cells—electrolytic and voltaic cells. **Electrolytic cells** use electrical energy to force a nonspontaneous chemical reaction to occur. **Voltaic cells** contain a chemical reaction that spontaneously occurs to produce electrical energy. All batteries consist of one or more voltaic cells.

In summary, all electrochemical cells contain a system of electrodes and electrolytes arranged so that electricity causes chemical changes or chemical reactions produce electricity.

17.7 Using Electrolytic Cells

Electrolysis is the process of forcing an otherwise nonspontaneous redox reaction to occur with the aid of an electrical current in an electrochemical cell. For example, normally stable water molecules can be pulled apart by an electrical current to produce hydrogen and oxygen gases. Current can be passed through an electrolytic cell when a source of electricity is connected to two electrodes immersed in an electrolyte solution.

What happens when electrons flow through the cell? First, electrons flow from the source of the electrical energy into the electrode called the **cathode**, giving it a negative charge. The electrons in the cathode become available for reduction reactions. A redox reaction cannot occur unless oxidation and reduction reactions occur simultaneously. The oxidation occurs at the **anode**, which is the positively charged electrode. Electrons then flow from the anode back to the source of electrical energy. Since any anions in the solution are negatively charged, they migrate to the positively charged anode. Similarly, positively charged cations migrate to the negatively charged cathode. That is why positively charged ions are called *cations* and negatively charged ions are called *anions*. If the electrical forces between the two electrodes are large enough, a redox reaction will occur.

Aluminum Manufacturing

Chemical engineers use electrolysis to purify active, valuable metals. Metals such as copper, tin, and iron, which are low on the activity series, can be freed from their natural compounds by chemical means. However, active metals such as sodium, lithium, and aluminum bond too strongly for those "mild" techniques to work, so engineers must separate the compounds in electrolytic cells. In nature, aluminum atoms are oxidized by oxygen atoms in an ore commonly known as bauxite. To get pure aluminum metal from the very stable aluminum oxide (Al_2O_3), the natural oxidation reaction must be reversed—electrons must be forced back into the aluminum ions.

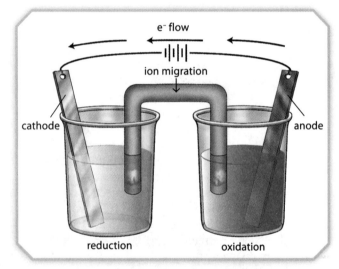

17-9 In an electrolytic cell, electrons flow in through the cathode and out through the anode.

In electrolytic cells, the cathode attracts cations and is negative. The anode attracts anions and is positive. Reduction takes place at the cathode, and oxidation takes place at the anode.

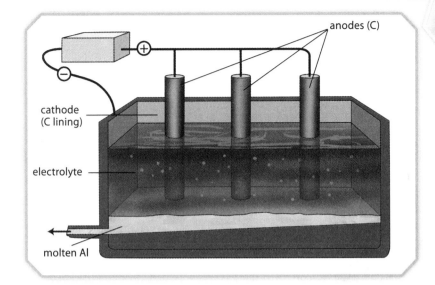

17-10 The Hall-Héroult process allows pure aluminum to be separated from the ore bauxite.

The electrolysis process used to produce aluminum in industry is called the Hall-Héroult process. This process is named after Charles Hall (an American) and Paul Héroult (a Frenchman), who each developed the process independently in 1886. Prior to this, aluminum was more precious than gold because, despite its abundance, there was no practical way to extract it. In the Hall-Héroult process, aluminum oxide is dissolved in molten cryolite (Na_3AlF_6) and electrolyzed with carbon electrodes.

$$3C + 4Al^{3+} + 6O^{2-} \longrightarrow 4Al + 3CO_2$$

The molten elemental aluminum falls to the bottom and is extracted. The oxygen reacts with the carbon of the anode and forms carbon dioxide.

Melting cryolite and bauxite requires great amounts of electricity. Although less than 6 V are needed, the current must be between 100 000 and 150 000 A. To lower operating costs, aluminum refineries are often located close to hydroelectric power stations, where energy is abundant and less expensive. It takes only one-twentieth as much energy to recycle aluminum as it does to extract it from its ore. That is why aluminum is recycled, despite the fact that it is so abundant in the earth's crust.

Electrolysis of Brine

Sodium hydroxide, chlorine gas, and hydrogen gas are produced by an electrolytic process from a concentrated sodium chloride salt solution called *brine*. These chemicals are valuable in industry. The most common type of cell used for this process uses a graphite anode and a steel cathode placed in the brine separated by a membrane. Although the membrane will allow cations to pass through it, it prevents substances produced at each electrode to mix. Oxidation at the anode produces chlorine gas from chloride ions and oxygen gas from water.

$$2Cl^- (aq) \longrightarrow Cl_2 (g) + 2e^-$$

$$2H_2O (l) \longrightarrow O_2 (g) + 4H^+ + 4e^-$$

17-11 The aluminum cap of the Washington Monument was placed at its completion in 1885, one year before Charles Hall perfected a method for extracting aluminum efficiently from bauxite.

Aluminum in History

When the Washington Monument was built, the pyramidal cap was made of aluminum, a costly, precious metal at the time. Napoleon is said to have had a set of aluminum dinner plates that he used only for his most honored guests.

Reduction at the cathode produces pure sodium from sodium ions and hydrogen gas from water molecules.

$$Na^+ (aq) + 2e^- \longrightarrow Na (s)$$

$$2H_2O (l) + 2e^- \longrightarrow H_2 (g) + 2OH^- (aq)$$

Electroplating

Sterling silver is at least 92.5% silver. Less expensive silverware is made of some common metal that is plated with a thin layer of silver. How is the thin silver layer deposited onto the inexpensive metal? One could pound, melt, or glue the silver onto the metal, but there is a much better way. Metallic ions in solution can be forced to cling to another metal through an electrochemical process that is called **electroplating**.

Electroplating is a type of electrolysis. The metal object to be plated is the cathode. The anode is made out of the metal to be plated onto the cathode. Silver plating is performed with a silver anode. The electrolyte solution contains silver ions. When current flows in the cell, electrons leave the anode, liberating more silver ions to maintain an adequate level of silver ions in the electrolyte solution. These silver ions migrate to the cathode and are reduced to metallic silver as electrons are attracted to the anode. The silver then plates onto the spoon.

$$Ag^+ + e^- \longrightarrow Ag$$

Zinc is sometimes electroplated onto steel to protect the steel from rust. The result is called galvanized steel. Zinc corrodes, but the product of this reaction protects the zinc from further oxidation. Therefore, the layer of zinc protects the iron in the steel from being oxidized. Even if a small crack forms in the zinc plating, the iron is still protected because zinc is more easily oxidized than iron. Any oxidation that occurs will be the oxidation of zinc. Zinc thus serves as a *sacrificial anode*.

Nonmetals can also be *electrodeposited* on metal surfaces. The automotive industry uses an electrodeposition process to paint car bodies. Positively charged paints are deposited onto the sheet metal. This technique more effectively covers the metal and gives better corrosion resistance than conventional spray painting.

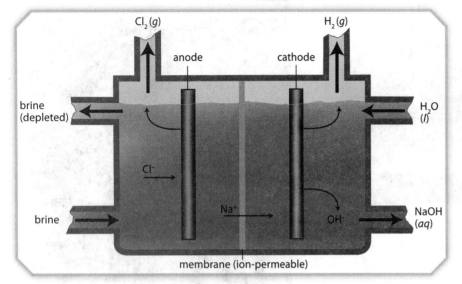

17-12 Many very useful elements and compounds can be separated from brine by electrolysis.

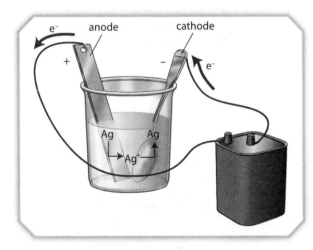

17-13 An electrolytic cell is used to electroplate metals.

In voltaic cells, the cathode attracts electrons and is positive. The anode repels electrons and is negative. Reduction takes place at the cathode, and oxidation takes place at the anode.

17.8 Voltaic Cells

Unlike electrolytic cells, voltaic cells use spontaneous redox reactions to produce electricity. They are called voltaic cells, in honor of Allessandro Volta (1745–1827), who invented the first battery in 1800. **Batteries** are collections of voltaic cells.

All voltaic cells have a negative electrode (anode) and a positive electrode (cathode). This designation appears to be opposite of the electrolytic cell convention in which the positive electrode is the anode and the negative electrode is the cathode. However, in both cases, oxidation occurs at the anode and reduction at the cathode. (The difference in sign convention will be explained later.) All voltaic cells contain an anode that loses electrons, a cathode that gains electrons, and an electrolyte between them. The electrolyte may be a liquid, such as the sulfuric acid solution in a car battery, or a moist paste, such as that found in most dry cell batteries.

In the voltaic cell diagrammed in Figure 17-15, the reaction occurring at the anode is the oxidation of zinc, and the reaction occurring at the cathode is the reduction of copper.

anode: $Zn \longrightarrow Zn^{2+} + 2e^-$

cathode: $Cu^{2+} + 2e^- \longrightarrow Cu$

The electrons from the oxidation *half-reaction* at the anode travel along the wire to the cathode.

At the cathode, copper (II) is reduced to metallic copper by the electrons from the anode. The steady flow of electrons from the anode to the cathode can be harnessed to produce electrical energy. As the reactions in each half-cell continue, the zinc solution builds up a positive charge from the accumulation of zinc ions. As the copper ions plate out as copper, that solution builds up a negative charge. The reactions will stop unless a mechanism is supplied to prevent the buildup of charges in the solutions. This is accomplished by the use of a salt bridge. A **salt bridge** is a tube of electrolytic gel that connects the two half-cells of a voltaic cell. The salt bridge allows the flow of ions but prevents two different solutions from mixing.

17-14 How many battery-operated devices can you find in this picture?

A salt bridge is illustrated in Figure 17-15. Current flows between the solutions as ions migrate. As zinc is oxidized, excess zinc ions accumulate in the solution around the anode. Chloride ions migrate from the salt bridge toward the concentration of positive charges and keep the solution close to neutral. As copper (II) is reduced to metallic copper at the cathode, positive charges are removed from the solution around the cathode. Potassium ions migrate from the salt bridge into the solution in order to keep the solution electrically neutral. If the salt bridge were removed, current would not flow.

The sign conventions of the anodes and cathodes are a common source of confusion. In one case, the anode has a positive (+) value, and in another, it has a negative (−) value. The confusion can be avoided if one remembers that the electrode that emits the electrons is always negative. Recall that oxidation always occurs at the anode. In the solution, the cations migrate to the cathode, where a reduction reaction occurs. The anions in the solution migrate to the anode—now positively charged—where oxidation takes place.

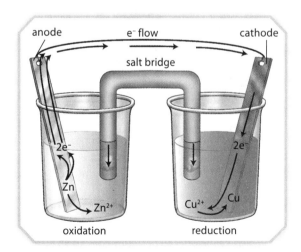

17-15 The salt bridge allows ions (K^+, Cl^-) to flow between the beakers, propagating the redox reaction that generates an electric current.

17-1 Electrode Conventions

Properties	Cathode	Anode
ions attracted	cations	anions
electron movement	into cell	out of cell
half-reaction	reduction	oxidation
sign of electrolytic cell	negative	positive
sign of voltaic cell	positive	negative

Zinc-Carbon Dry Cells

One common type of battery is the *zinc-carbon dry cell*. A dry cell consists of a zinc container filled with an electrolyte paste made of MnO_2, $ZnCl_2$, NH_4Cl, and a binder that keeps it all together. The zinc cylinder just inside the cardboard tube acts as the anode and loses electrons. Inserted into the electrolyte paste is a carbon graphite rod that acts as the cathode. The coated paper separator acts as the electrolyte layer and prevents a short circuit.

Although the reactions occurring at these electrodes are complicated, they can be summarized as follows:

$$\text{anode: } Zn \longrightarrow Zn^{2+} + 2e^-.$$

$$\text{cathode: } 2NH_4^+ + 2MnO_2 + 2e^- \longrightarrow Mn_2O_3 + H_2O + 2NH_3$$

In these half-reactions, zinc is oxidized from a charge of 0 to +2, and manganese is reduced from +4 to +3.

Alkaline Batteries

The most popular batteries in most consumer products today are *alkaline batteries*. They use an alkaline substance, potassium hydroxide. These batteries are a type of dry cell battery but have some chemical and structural differences. The reaction takes place in a paste of zinc metal and potassium hydroxide. They do not have a carbon rod cathode, nor do they have a solid metal liner serving as a cathode.

The reduction reaction occurring at the cathode is the same as the reaction that occurs in a zinc-carbon dry cell battery. The reaction at the anode oxidizes zinc from a charge of 0 to +2.

$$Zn + 2OH^- \longrightarrow Zn(OH)_2 + 2e^-$$

Some common types of alkaline batteries are nickel-cadmium (NiCad) and nickel-metal hydride (NiMH). Several other specialized alkaline batteries are listed in Table 17-2 on page 440.

Lead-Acid Storage Batteries

Automobiles first ignite their engines with a spark from a series of six lead storage cells linked together so that the voltages add to each other. These are called *lead-acid storage batteries*. The cathode of a lead storage cell is a series of lead-antimony alloy plates permeated with lead (IV) oxide (PbO_2). The anode is a series of lead-antimony alloy plates filled with spongy lead. The cathode and anode are immersed in sulfuric acid.

The oxidation half-reaction is

$$Pb + SO_4^{2-} \longrightarrow PbSO_4 + 2e^-.$$

17-16 Alkaline dry cell batteries contain an alkaline paste that serves as the anode.

The reduction half-reaction is

$$PbO_2 + 4H^+ + SO_4^{2-} + 2e^- \longrightarrow PbSO_4 + 2H_2O.$$

As strange as it seems, lead atoms are oxidized from 0 to +2 on one plate and reduced from +4 to +2 on another. As the battery discharges, the concentration of the sulfuric acid decreases as H_2SO_4 decomposes to form SO_4^{2-} ions and more water molecules.

Rechargeable Batteries

Reversing the polarity of the electrodes recharges batteries. This reverses the redox reaction and regenerates the cell. Some cells cannot or should not be recharged. Some manufacturers put a label on their alkaline batteries to warn the customers against recharging the batteries. They do not mean that alkaline batteries cannot be recharged; they mean that they should not be recharged because to do so would be dangerous. Even though the redox reaction in an alkaline cell is reversible, recharging produces some gases. Since some alkaline cells have no vents to release gases, they could explode if someone tried to recharge them.

Many types of batteries are designed to be recharged. The lead-acid batteries in automobiles and the nickel-cadmium batteries used in many electronic devices are good examples. Rechargeable cells combine aspects of voltaic and electrolytic cells. When they are discharged, they are voltaic cells, and when they are recharged, they reverse and act as electrolytic cells.

17-17 The white powder that sometimes collects on battery posts is evaporated sulfuric acid. What would be a good way to clean or neutralize this deposit?

Fuel Cells

Fuel cells are a special class of very efficient batteries. A fuel cell resembles a voltaic battery in that redox reactions occur, releasing electrons from one electrode to flow through a circuit to another electrode. However, there is one major difference. In the voltaic cell, the active ingredients are included within the cell and are depleted as the redox reactions occur. In a fuel cell, a gas or liquid fuel is supplied to one electrode and oxygen or air to the other from an external source. Fuel cells have a long life and have been used extensively in space vehicles since the 1960s. Figure 17-18 shows a schematic of an oxygen-hydrogen fuel cell. Current fuel-cell research is targeted at powering automobiles with hydrogen as a fuel.

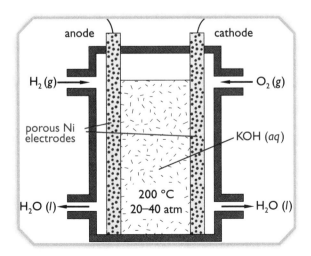

17-18 A fuel cell

Table 17-2 summarizes information on some of the most common types of batteries. All of these produce electricity through redox reactions.

17-2 Types of Batteries in Common Use

Type	Anode	Cathode	Electrolyte	Uses
zinc-carbon dry cell	Zn/Hg*	MnO_2**	NH_4^+ and Zn^{2+} chlorides	low-cost use
lead-acid	Pb	PbO_2	H_2SO_4	automotive and industrial uses
alkaline cells:				
nickel-cadmium	Cd	nickel oxides	NaOH or KOH	rechargeable batteries
nickel-metal hydride	rare earth or nickel alloys with many metals	$Ni(OH)_2$	KOH	cell phones, camcorders, laptops
silver-zinc	Zn/Hg*	Ag_2O	KOH	military applications
silver-cadmium	CdO	Ag	KOH	satellite uses
silver-iron	Fe	Ag	KOH and LiOH	emergency power uses

*Zn/Hg is zinc amalgamated with mercury.
**The cathode is a carbon rod, but it is inactive in the reaction. The reduction occurs in the paste surrounding the cathode and involves MnO_2.

17B Section Review

1. What determines the strength of an electrolyte?
2. What is the major difference between a voltaic cell and an electrolytic cell?
3. Will pure water conduct electricity? Why or why not?
4. What reaction occurs at the anode of a voltaic cell? at the anode of an electrolytic cell?
5. What is the main difference between a voltaic cell and a fuel cell?
6. What are some useful applications of electrochemical cells?
7. Describe how rechargeable batteries are both voltaic and electrolytic cells.
DS 8. What Scriptural command(s) do inventions like electroplating zinc onto steel or rechargeable batteries fulfill?
DS 9. Is everyone who develops technology that uses redox reactions pleasing God?

Chapter Review

Chapter Summary

- Oxidation-reduction, or redox, reactions are those chemical changes in which electrons are transferred from one atom or molecule to another.
- Oxidation involves the loss of one or more electrons, and reduction involves the gain or one or more electrons. These processes occur simultaneously.
- A reducing agent causes another substance to be reduced and is itself oxidized in the process. An oxidizing agent causes another substance to be oxidized and is itself reduced in the process.
- Equations of redox reactions must be balanced to maintain the law of the conservation of mass.
- An electrolyte is any substance that, when dissolved in water, allows the resulting solution to conduct electricity. Nonelectrolytes do not release ions and cannot conduct electricity in solutions.
- An electrochemical cell has two electrodes (electrical contacts) immersed in an electrolyte, with a wire joining the electrodes. The two main groups of electrochemical cells are electrolytic cells and voltaic cells.
- Electrolysis, a process occurring in electrolytic cells, requires electrical current to cause an otherwise nonspontaneous redox reaction to occur.
- In electrolytic cells, electrons flow from the source of electricity into the cathode, causing it to have a negative charge. Reduction occurs at the cathode. Oxidation occurs at the positively charged anode as electrons flow out of the anode and back to the source of electricity.
- In voltaic cells, spontaneous redox reactions produce electricity by causing electrons to flow. Batteries are collections of voltaic cells.
- In voltaic cells, oxidation occurs at the negatively charged anode. Reduction occurs at the positively charged cathode.

Coming to Terms

Term	Page
oxidation-reduction reaction (redox reaction)	425
oxidation	425
reduction	425
reducing agent	426
oxidizing agent	427
electrochemistry	432
electrolyte	432
nonelectrolyte	434
electrochemical cell	434
electrode	434
electrolytic cell	434
voltaic cell	434
electrolysis	434
cathode	434
anode	434
electroplating	436
battery	436
salt bridge	437
fuel cell	439

Review Questions

Concept Review

1. Does oxidation require that oxygen atoms be present?
2. In a voltaic cell, what is the electrical charge on the anode? What type of ion migrates toward the anode? What process (oxidation or reduction) occurs at the surface of the anode?
3. In a voltaic cell, what is the electrical charge on the cathode? What type of ion migrates toward the cathode? What process (oxidation or reduction) occurs at the surface of the cathode?
4. What is the difference between a voltaic cell and an electrolytic cell?

5. Describe how the Hall-Héroult process uses electrons to free aluminum atoms from bauxite.
6. Why is a special process necessary to purify active metals such as aluminum and sodium when other metals can be more easily purified from their ores?
7. Why is a salt bridge used in voltaic cells?
8. Why do some battery manufacturers recommend not to charge some batteries though they are based on the same reversible redox reactions as those in rechargeable batteries?
9. Explain why zinc plating will protect steel from corroding even if the plating is cracked.

True or False

10. Double replacement reactions are good examples of redox reactions.
11. In some redox reactions, the same element is both reduced and oxidized.
12. Oxidation always occurs before reduction.
13. The chlorine in swimming pools and bleach is a strong oxidizing agent.
14. Strong bases and acids can dissociate in water to form strong electrolytes.
15. In every type of electrochemical cell, oxidation always takes place at the anode and reduction always takes place at the cathode.
16. The use of a salt bridge improves the function of an electrolytic cell.
17. Sodium hydroxide, chlorine gas, and hydrogen gas can all be extracted from brine by electrolysis.

Application

18. Tell whether each reaction is a redox reaction.
    a. $2Fe\ (s) + 3Cl_2\ (g) \longrightarrow 2FeCl_3\ (s)$
    b. $CaO\ (s) + 2HCl\ (g) \longrightarrow CaCl_2\ (s) + H_2O\ (l)$
    c. $2C_2H_6\ (g) + 7O_2\ (g) \longrightarrow 4CO_2\ (g) + 6H_2O\ (l)$
    d. $Zn\ (s) + CuSO_4\ (aq) \longrightarrow Cu\ (s) + ZnSO_4\ (aq)$
    e. $H_2SO_4\ (aq) + KOH\ (aq) \longrightarrow K_2SO_4\ (s) + H_2O\ (l)$
    f. $Pb\ (s) + H_2SO_4\ (aq) \longrightarrow PbSO_4\ (s) + H_2\ (g)$
    g. $AgNO_3\ (aq) + HCl\ (aq) \longrightarrow AgCl\ (s) + HNO_3\ (aq)$

Questions 19–22 refer to the following redox reactions:
   a. $SO_4^{2-} + Zn + 4H^+ \longrightarrow Zn^{2+} + SO_2 + 2H_2O$
   b. $S_2O_3^{2-} + OCl^- + 2H^+ \longrightarrow Cl^- + S_4O_6^{2-} + H_2O$
   c. $8I^- + SO_4^{2-} + 10H^+ \longrightarrow 4I_2 + H_2S + 4H_2O$
   d. $3H_2S + 2CrO_4^{2-} + 10H^+ \longrightarrow 3S + 2Cr^{3+} + 4H_2O$
   e. $5SO_2 + 2MnO_4^- + 2H_2O \longrightarrow 5SO_4^{2-} + 2Mn^{2+} + 4H^+$
   f. $MnO_4^- + 5Fe^{2+} + 8H^+ \longrightarrow Mn^{2+} + 5Fe^{3+} + 4H_2O$
   g. $2SO_4^{2-} + C + 4H^+ \longrightarrow CO_2 + 2SO_2 + 2H_2O$

19. Identify the substance being oxidized in each reaction.
20. Identify the substance being reduced in each reaction.

21. Identify the oxidizing agent in each reaction.
22. Identify the reducing agent in each reaction.

DS 23. Assuming corrosion is similar to the chemical process of the battery, identify the anode and cathode in the corrosion process that occurs as a water drop rests on an iron surface.

DS ⊙ 24. How could stainless steel be further prevented from corroding, considering that the addition of chromium to iron results in significant corrosion resistance?

25. Balance each of these equations using the oxidation-number method.
    a. $SnCl_4 + Fe \longrightarrow SnCl_2 + FeCl_3$
    b. $SO_2 + Br_2 + H_2O \longrightarrow HBr + H_2SO_4$
    c. $CO + I_2O_5 \longrightarrow I_2 + CO_2$

⊙ 26. Would this pair of electrodes make a functional battery? Explain your answer.
$$Zn \longrightarrow Zn^{2+} + 2e^-$$
$$Cd \longrightarrow Cd^{2+} + 2e^-$$

⊙ 27. In the lead-acid storage battery of an automobile, you can determine the degree to which the power has been discharged by using a hydrometer to measure the density of the sulfuric acid solution. Explain why this is possible.

Serving God as a Groundskeeper

Job Description
A groundskeeper is responsible for the proper care and maintenance of residential or commercial lawns, flower beds, shrubs, and trees. He applies chemistry through the appropriate use of fertilizers, pesticides, and herbicides.

A groundskeeper must be dependable, willing to work flexible hours, and able to perform a full day of physical labor outdoors. The job requires skill at using and repairing hand tools and power equipment, good communication, and the understanding and consistent application of safety codes related to equipment and chemicals. An aptitude for design and layout and working with colors, shapes, and sizes of plants would be useful as well.

Possible Workplaces
There are myriad opportunities open for those interested in a career as a groundskeeper. Groundskeepers service athletic fields, cemeteries, golf courses, botanical gardens, parks,

schools, zoos, government buildings, apartment complexes, shopping malls, hotels, highways, farms, and, of course, residential properties.

Education
Many groundskeeping companies require a high-school education and prefer previous experience in the field. Some postings additionally ask for the candidate to possess a strong working knowledge of botany and of the structure, properties, proper uses, and effects of materials and chemicals on animals and other aspects of the environment.

Dominion Opportunities
A groundskeeper is someone who works hard to care for God's incredible creation and to beautify industrial, commercial, public, and private properties for the pleasure of mankind.

Genesis 2 tells us that after God created Adam and Eve, He placed them in the Garden of Eden as the groundskeepers—the very first profession. God must hold groundskeeping in high regard!

Organic Chemistry and Biochemistry

CHAPTER 18

18A Organic Compounds	445
18B Hydrocarbons	447
18C Substituted Hydrocarbons	455
18D Organic Reactions	463
18E Biochemistry	465
Facets	
Astrobiology and Polycyclic Aromatic Hydrocarbons	454
Trans Fats	460

Dominion Science Problem

Curing Cancer

Cancer. When you hear that word, you probably think of someone you know who is battling this disease. Over 1,500 Americans die of cancer every day. Cancer kills more Americans than any other cause except heart disease.

Scientists know that the cells that cause cancerous tumors divide rapidly. For many decades, biologists and biochemists have also understood the process of DNA replication that allows genetic material from cells to be copied so they can divide. This step is critical to cell division, allowing each daughter cell to contain the same genetic code as the parent cell. Could something inhibit the replication of DNA in cancerous cells, halting the growth of cancerous tumors? Could this idea be the key to the cure?

18A Organic Compounds

Carbon is a common element that is found in nature both as a pure element and in compounds. It can be found in all living matter, as well as in familiar fuels such as petroleum and natural gas. Truly, without carbon there would not be life.

By the early 1800s, chemists realized that a wealth of chemical compounds were produced by living organisms. They named these *organic* compounds since they were associated with living organisms. For many years, scientists struggled to produce even the simplest organic compounds in the laboratory, but none was successful.

In 1828, Friedrich Wöhler, a German chemist, unintentionally synthesized *urea* (NH_2CONH_2). Up to that time, urea was known only from the urine of humans and animals. Though Wöhler had not intended to create this compound, he recognized the significance of his accomplishment. It was the first synthetically produced organic compound.

This breakthrough initiated a series of similar experiments by other chemists, and a new field of study emerged. From these inauspicious beginnings, organic chemistry quickly rose to a place of prominence. Now, **organic compounds** are defined as covalently bonded carbon compounds, with the exception of carbonates, carbon oxides, and carbides. Read the labels of most packaged foods, cosmetics, or medicine, and you will see the impact organic chemistry has on your life.

By the twentieth century, scientists developed analytical techniques of such complexity that organic chemists could study the complex reactions taking place between organic compounds within living organisms. That field is called *biochemistry*.

18.1 The Unique Carbon Atom

Carbon is an element unparalleled in its ability to form a variety of compounds. Other elements can form several hundred thousand compounds, but carbon can form more than sixteen million compounds. What makes the carbon atom so versatile?

Carbon atoms form so many compounds because they have unique bonding abilities. There are three important properties of carbon that enable it to form large, stable molecules.

1. Carbon has four valence electrons, requiring four bonds to obtain an octet.
2. Carbon forms strong chemical bonds with other carbon atoms.
3. Carbon forms stable, essentially nonpolar bonds with hydrogen. The electronegativity difference between carbon (2.5) and hydrogen (2.1) is small, making it unreactive.

Carbon atoms can bond to a wide variety of atoms, including hydrogen, phosphorus, oxygen, nitrogen, sulfur, the halogens, and even metal atoms. Carbon atoms can bond together to form straight or branched chains of various lengths. Carbon atoms can even form rings. Carbon atoms can form double and triple bonds, further adding to the almost endless possibilities.

18A Section Objectives
After finishing this section, you should be able to
- give a brief history of organic chemistry.
- define *organic compound*.
- state three reasons why carbon forms so many compounds.
- write a compound's molecular and structural formula.
- distinguish between aliphatic and aromatic organic compounds.

18-1 Friedrich Wöhler, a German chemist, gave birth to a whole new field of chemistry when he synthesized urea.

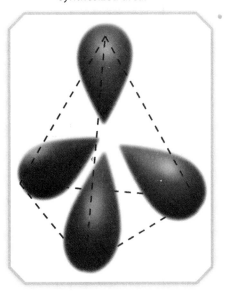

18-2 Carbon's valence electrons allow it to form covalent bonds with as many as four other atoms.

18.2 Structural Formulas

Organic chemists study more than the types of atoms in molecules. They also study how these atoms are arranged. If an analytical chemist announced that he had isolated a compound with the molecular formula C_2H_6O, his colleagues would not know which compound he was talking about. C_2H_6O could be ethanol, or it could be dimethyl ether. The arrangement of the atoms in the molecule makes the difference. For this reason, structural formulas are often used in organic chemistry. Many times, to make things simpler, hydrogen atoms are left out of these formulas. You can easily see that each carbon atom has four bonds, each oxygen atom has two bonds, and each hydrogen atom has one bond (Table 18-1).

18-3 Carbon can bond with itself in a great number of arrangements.

18.3 Classification

Organic chemistry nowadays almost drives me mad. To me it appears like a primeval forest full of the most remarkable things, a dreadful endless jungle into which one dare not enter for there seems to be no way out.

—Friedrich Wöhler, 1835

If Wöhler were alive today, he would be amazed to see how much larger the "jungle" has grown. Approximately 300 000 new organic compounds are synthesized for the first time each year. If a person

18-1 Structural Comparisons of Ethanol and Dimethyl Ether

Name	Molecular formula	Structural formula	Ball-and-stick model	Space-filling model
ethanol	C_2H_6O	H:C:C:O:H (with H's) or H-C-C-OH (with H's) or -C-C-OH		
dimethyl ether	C_2H_6O	H:C:O:C:H (with H's) or H-C-O-C-H (with H's) or -C-O-C-		

wishes to find his way through this ever-growing jungle, he must use some guidelines. A classification scheme that organizes compounds into easily identifiable groups serves as the map through the jungle.

Organic compounds can be divided into two large groups: aliphatic compounds and aromatic compounds. **Aliphatic compounds** are either straight-chain (open-chain) compounds or rings that can be formed by bending and connecting the ends of a straight chain. Cyclohexane is an example of a ring-shaped aliphatic compound. **Aromatic compounds** have ringed shapes, but they are unlike aliphatic rings because their electrons are not associated with a specific bond or atom.

18A Section Review

1. What do organic chemists do?
2. What three characteristics of carbon enable it to be found in millions of compounds?
3. Name several examples of atoms with which carbon can form bonds.
4. List and define the two large groups into which organic compounds are divided.

18B Hydrocarbons

The simplest aliphatic and aromatic compounds are hydrocarbons. As their name implies, **hydrocarbons** are organic compounds that contain only hydrogen and carbon. Some hydrocarbons contain carbon-carbon double bonds or even carbon-carbon triple bonds. We will also explore other elements or groups of elements that can substitute for the hydrogen atoms in hydrocarbons. Hydrocarbons are classified by the presence of these bonds and atoms (see Figure 18-4).

18.4 Alkanes

Our society relies on alkanes for fuel. Natural gas used to heat homes contains mainly methane (CH_4). Portable gas barbecue grills use bottles of pressurized propane, and automobiles burn a mixture comprised mostly of alkanes. All of these compounds contain only carbon and hydrogen. They have structural formulas that resemble open chains, and they contain only single bonds between their carbons. **Alkanes** are aliphatic hydrocarbons that contain only single bonds. Some are straight chains, while others have branches. The general formula for alkanes is C_nH_{2n+2}.

Methane (CH_4) is the simplest alkane. Its one carbon atom is bonded to four hydrogen atoms. Other alkanes are formed as additional carbons lengthen the chain. Ethane (C_2H_6) has two carbons, propane (C_3H_8) has three, and butane (C_4H_{10}) has four. Each carbon atom in an alkane is surrounded by

> **18B Section Objectives**
> After finishing this section, you should be able to
> - identify alkanes, alkenes, and alkynes from their formulas or structures.
> - draw structural formulas for compounds in common organic families.
> - use IUPAC rules to name simple alkanes, alkenes, and alkynes.
> - recognize the electron and structural differences between cyclic aliphatic compounds and aromatic substances.

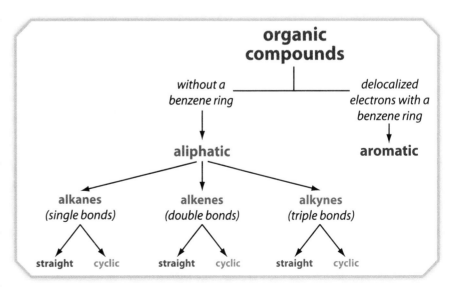

18-4 Classification scheme

four other atoms—the maximum number possible. For this reason, the molecules of alkanes are said to be **saturated**.

18-2	**Numerical Prefixes**
Number of C atoms	Prefix
1	meth-
2	eth-
3	prop-
4	but-
5	pent-
6	hex-
7	hept-
8	oct-
9	non-
10	dec-

18-5 While most school buses run on diesel, some operate on alternative fuels including compressed natural gas (CNG), bio-diesel, and gasoline engines modified to burn propane. This bus runs on propane.

The International Union of Pure and Applied Chemistry (IUPAC) is an organization that has devised and manages a system to accurately name organic compounds. This system relies on a series of prefixes to indicate the number of carbon atoms present. The prefix gives the number of carbons in the longest chain. The suffix identifies the type of compound. All alkanes have an *-ane* ending.

For example, a particular branched alkane has the following structure:

$$\begin{matrix} & -\overset{|}{\underset{|}{C}}- & \\ -\overset{|}{\underset{|}{C}}- & -\overset{|}{\underset{|}{C}}- & -\overset{|}{\underset{|}{C}}- \end{matrix}$$

There are three carbons in this alkane's longest chain, so it is a propane. Methyl is the name of the alkyl group CH_3, which is attached to the second carbon on the carbon chain, so this alkane is methylpropane. An **alkyl group** is similar to the structure of its straight formulas, and some properties of the first five alkanes are listed in Table 18-3.

Are alkanes polar or nonpolar? Bonds between two carbon atoms are not polar because both atoms have the same electronegativity. Bonds between carbon and hydrogen are arranged symmetrically, so their small polarities cancel. Alkanes dissolve well in nonpolar solvents such as carbon tetrachloride. However, in polar substances, such as water, liquid alkanes are immiscible and form an oily layer on top of water. Therefore, they are nonpolar.

Crude oil contains several types of alkanes. Chains from four carbons up to twenty carbons form naturally and are combined into one seemingly inseparable mixture. Chemical engineers take advantage of the properties of these alkanes to separate the mixture so that they can be used as lubricating oils, gasoline, and kerosene. How do chemical engineers separate the individual compounds? The boiling points of alkanes rise about 20–30 °C for each additional carbon added to the chain, so petroleum engineers can separate the alkanes in a process called *fractional distillation*.

18-3 Straight-chained Alkanes

Name	Structural formula	Melting point (°C)	Boiling point (°C)
methane	H–C–H with H above and below	−183	−164
ethane	H–C–C–H with H's	−172	−88.5
propane	H–C–C–C–H with H's	−187	−42
butane	H–C–C–C–C–H with H's	−138	−1
pentane	H–C–C–C–C–C–H with H's	−130	36

18-6 Methane gas forms in natural areas such as wetlands, but also in artificial environments such as landfills.

18.5 Structural Isomers

Some alkanes have branches off the main chain. For example, the molecular formulas of butane and isobutane are identical (C_4H_{10}), but they are obviously different. Compounds with the same molecular formulas but different structural formulas are called **structural isomers**. Any alkane with four or more carbon atoms can exhibit structural isomerism. Normal butane (*n*-butane, the straight-chained form) and its structural isomer, isobutane, have slightly different physical properties. Their melting and boiling points, density, and solubility reflect their different structures.

Isomer	Structural formula	Melting point	Boiling point	Density	Solubility
butane	–C–C–C–C–	−138	0	0.579 g/mL	1813 mL/100 mL ethanol
isobutane	–C–C–C– with –C– branch	−159	−12	0.549 g/mL	1320 mL/100 mL ethanol

18-7 Butane (top) and isobutane (bottom) are common structural isomers of C_4H_{10}.

> **EXAMPLE PROBLEM 18-1**
> *Drawing Structural Formulas*
> Draw structural formulas for all the structural isomers of pentane (C_5H_{12}).
> **Solution**
>
> dimethylpropane   2-methylbutane   n-pentane
>
> The 2 before methylbutane means that the alkyl group is attached to the second carbon from the nearer end of the carbon chain. There is no 2,2 before dimethylpropane because there is no other place that the alkyl groups could be attached and still yield propane.

18.6 Alkenes

Carbons can form double bonds quite easily. Hydrocarbons that contain double bonds are called **alkenes**. Alkenes use the same prefixes as alkanes but end with the suffix *-ene*. The simplest alkene is ethene. Because double bonds reduce the number of hydrogen atoms in this molecule, ethene and all alkenes are said to be **unsaturated**. Ethene contains only two carbons. When the carbon chain is longer than three carbons, the double bond could be in several locations. An alkene's name pinpoints the location of the double bond by numbering all of the carbons in the molecule and giving the number of the first carbon that is doubly bonded.

18-8 Ethene speeds up ripening. These tomatoes were picked and shipped green. Before sale, they were exposed to ethene for quick ripening.

18-4 *Alkenes*			
Name	Structural formula	Melting point (°C)	Boiling point (°C)
ethene	–C=C–	–169	–104
prop-1-ene	–C=C–C–	–185	–47
but-1-ene	–C=C–C–C–	–185	–6.3
but-2-ene	–C–C=C–C–	–139.3	—

The physical properties of alkenes are very much like those of alkanes. The first few are gases at room temperature. Pentene and larger compounds are liquids at room temperature because of greater intermolecular attractions. Alkenes are relatively nonpolar.

18.7 Alkynes: Chains with Triple Bonds

A triple bond between two carbon atoms identifies a member of the **alkyne** family. Alkynes use prefixes to indicate the number of carbons and end with the suffix *–yne*. The most common alkyne is also the simplest. When necessary, a number is used to tell where the triple bond occurs. Ethyne, commonly called acetylene, consists of two carbons joined by a triple bond. This compound is often used as a fuel for welding torches and as an ingredient for plastics.

$$H:C:::C:H$$
or
$$H-C\equiv C-H$$

Physically, alkynes are similar to other hydrocarbons. They are practically nonpolar, so they are insoluble in water and very soluble in nonpolar solvents. Their boiling points rise as the carbon chains get longer.

18-9 Acetylene (ethyne) gas, produced by the reaction of CaC_2 and H_2O, fuels this miner's helmet lamp.

18-5 Alkynes

Name	Structural formula	Melting point (°C)	Boiling point (°C)
ethyne	–C≡C–	–82	–75
propyne	–C≡C–C–	–101.5	–23
but-1-yne	–C≡C–C–C–	–122	9
but-2-yne	–C–C≡C–C–	–24	27
pent-1-yne	–C≡C–C–C–C–	–98	40
pent-2-yne	–C–C≡C–C–C–	–101	50
hex-1-yne	–C≡C–C–C–C–C–	–124	72
hex-2-yne	–C–C≡C–C–C–C–	–92	84
hex-3-yne	–C–C–C≡C–C–C–	–103	81
hept-1-yne	–C≡C–C–C–C–C–C–	–80	100
oct-1-yne	–C≡C–C–C–C–C–C–C–	–70	126
non-1-yne	–C≡C–C–C–C–C–C–C–C–	–65	151
dec-1-yne	–C≡C–C–C–C–C–C–C–C–C–	–36	182

18.8 Cyclic Aliphatic Compounds

Not all hydrocarbons are open chains of carbon atoms. Some hydrocarbons form a ring. Such compounds are called **cyclic aliphatic compounds**. Though many rings are possible, five- and six-carbon alkane rings are the most abundant. Simple alkenes and alkenes with more than one double bond multiply the number of possible structures. A few structures and their names are shown below.

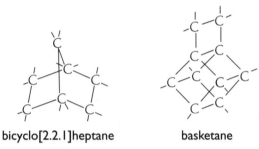

cyclohexane   cyclopentene   cyclohexa-1,3-diene

Some very unusual structures are possible when several rings combine.

bicyclo[2.2.1]heptane   basketane

The chemical activity of cyclic compounds is about the same as that of other members of their parent families. *Cycloalkanes* act like alkanes; *cycloalkenes* act like alkenes. Cyclic compounds have several unique uses, such as anesthetics, car maintenance, perfumes, and paint removers.

18.9 Aromatic Substances

Aromatic substances are an entirely different category of hydrocarbons. Aromatic substances all contain a form of the molecule benzene. These substances gained their name from the often pleasant-smelling odors that they produced.

Benzene was isolated in 1825 by Michael Faraday. An analysis of the elements in benzene and a determination of its molecular weight showed that the molecular formula was C_6H_6. This compound is the simplest aromatic compound known. Today it is one of industry's most important compounds, most often extracted from petroleum.

The structural formula of benzene puzzled scientists for thirty years after Faraday discovered the compound. Many clues were gathered, but they did not seem to fit together. The molecular formula C_6H_6 led chemists to believe that the molecule must have several double or triple bonds. Yet the chemical reactions of benzene did not support this idea. It behaved like an alkane, not an alkene or alkyne. When scientists determined the bond lengths between the carbon atoms, they found that the distances were not those of single or double bonds; they were in between. It was as if benzene used one and a half bonds. Furthermore, it became known that the carbons were arranged in a ring and that all the carbon atoms had identical bonds.

In 1865, August Kekule proposed a structure that could account for most of the observations. He described benzene in terms of two symmetrical ring structures.

18-10 Michael Faraday (1791–1867) was both a physicist and a chemist. Although probably best known for his work with electricity, he also invented an early form of the Bunsen burner, established the laws of electrolysis, and invented oxidation numbers.

The actual structure of a benzene molecule was thought to be a dynamic equilibrium of these two structures. Thus, the electrons in the double bonds were mobile and not "tied down" to specific locations. But the development of the valence bond theory modified this concept. The single bonds between the carbon atoms—sigma bonds—were due to head-on overlap of the carbon atom's hybridized orbitals. The double bond included the additional components of pi bonds—side-to-side overlap of additional orbitals between pairs of carbon atoms. These pi clouds overlap in the molecule to form doughnut-shaped areas of electron concentration above and below the plane of the ring. Therefore, the electrons are free to move throughout the entire "doughnut." For this reason, they are called **delocalized electrons**.

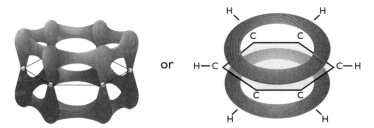

Because the pi electrons in the double bonds are not bound between any two carbon atoms, chemists often draw the structure of benzene like this:

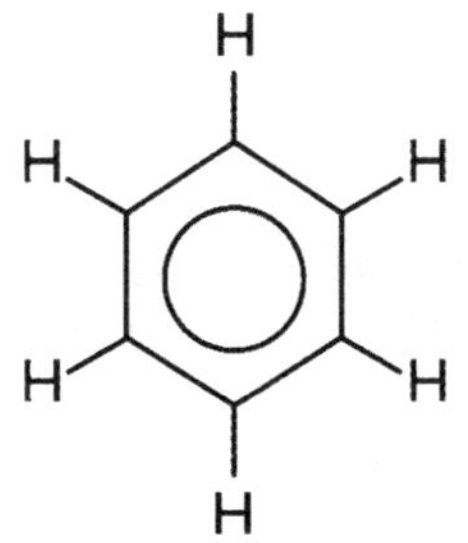

More simply, benzene can be depicted like this:

The modern definition of aromatic compounds has nothing to do with their smell. Rather, it deals with the nature of the bonds between atoms. All aromatic compounds have cyclic clouds of delocalized electrons. Benzene is the parent compound for a huge number of substances, many of which are used commercially. Figure 18-11 shows some of these commercial products.

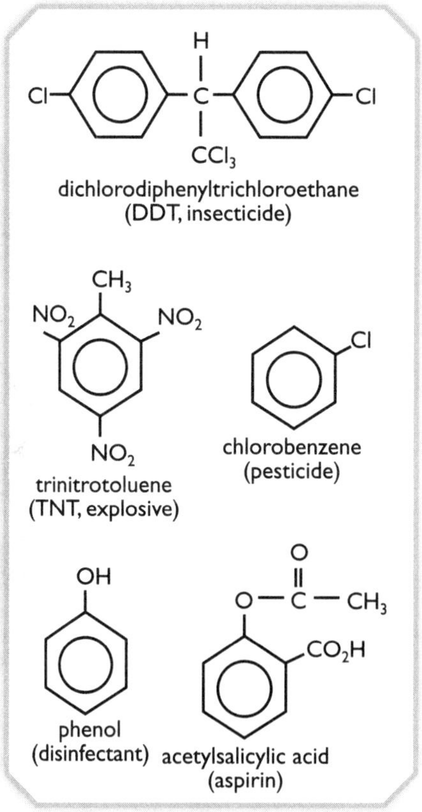

18-11 These benzene compounds are valuable in industry.

18B Section Review

1. What do hydrocarbons contain? Name some sources of hydrocarbons.

2. Explain the differences between saturated and unsaturated hydrocarbons.

3. What is a structural isomer?
4. Why are organic compounds mainly insoluble in water?
5. Compare and contrast alkanes, alkenes, and alkynes.
6. Define each molecular formula as an alkane, alkene, or alkyne.
   a. $C_{10}H_{22}$
   b. C_2H_4
   c. C_5H_8
   d. C_4H_6
   e. C_2H_2
7. What are carbon compounds called whose chains have been bonded into rings?
8. Describe the nature of the bonds between carbon atoms in aromatic compounds.

Astrobiology and Polycyclic Aromatic Hydrocarbons

Facets of Chemistry

Just as carbon atoms link readily to form chains, benzene molecules also combine to form plates of two or more of these aromatic rings. These larger molecules with joined rings are called *polycyclic aromatic hydrocarbons* (PAHs). The simplest one, naphthalene ($C_{10}H_8$), has only two rings, while others may contain ten or more. PAHs contain only carbon and hydrogen.

Recently, many secular scientists have suggested PAH molecules might help explain the origins of life. Frustrated by their inability to explain how life might have originated from non-living materials here on the earth, they have set their sights higher. The presence of PAHs in some meteorites and indirect evidence that PAHs and similar molecules might exist in the atmospheres of planets orbiting other stars has sparked their imaginations. This search for life in outer space is known as *astrobiology*.

If nitrogen could be substituted for some of the hydrogen atoms in PAHs, they would yield compounds essential for the production of hemoglobin. Hemoglobin is an important oxygen-carrying compound found in the blood of humans and many animals. These

modified PAH molecules are also found in chlorophyll, the primary pigment of photosynthesis. Some scientists speculate that the nitrogen for the replacement would have been available from the atmosphere on the earth, or in the form of frozen or gaseous methane on other planets. With this scenario, life might have originated on another planet, or a comet, and been delivered here. It could have also started here on the earth.

Other scientists have even claimed to find the origins of life in stacks of PAHs serving as the scaffolding for the production of the first simple RNA. Ribonucleic acid (RNA) is considered a necessary chemical of life since it and DNA carry the code to transfer traits.

This PAH-world hypothesis was proposed in 2005, but no laboratory work has shown that PAHs could ever form RNA. PAHs are actually harmful to human life, and many are carcinogenic. Those who reject the Scripture's account of the origin of life are forced to find natural explanations of God's supernatural working. Sadly, they are looking not to the God of heaven, but to the heavens themselves.

18C Substituted Hydrocarbons

Functional groups are the hot spots of chemical activity on an organic molecule. They greatly modify the behavior of hydrocarbons. These atoms or groups of atoms often substitute for hydrogen in a hydrocarbon, so the molecules they contain are called *substituted hydrocarbons*.

There are many kinds of functional groups. Some contain sulfur; some contain halogens. Most functional groups in organic chemistry contain arrangements of oxygen or nitrogen atoms. In each case the functional group greatly influences the molecule's properties, just as the number of carbons in a hydrocarbon or the presence of multiple bonds affects its properties.

> **18C Section Objectives**
> After finishing this section, you should be able to
> - explain how functional groups affect a compound's properties.
> - use IUPAC rules to name halides, alcohols, ethers, aldehydes, ketones, and carboxylic acids.
> - recognize the suffixes associated with each type of organic molecule.

18.10 Halides

When a halogen and an alkyl group combine, an alkyl halide forms. An **alkyl halide** is a combination of an alkyl group and a fluorine, chlorine, bromine, or iodine atom in place of one of the hydrogen atoms. The general formula for the whole family is R–X, where R represents an alkyl group and X represents a halide. When a halogen is attached to an **aryl group** (an aromatic group that lacks a hydrogen atom), the resulting molecule is called an **aryl halide**. Polytetrafluoroethylene (a non-stick coating for pans) and polyvinyl chloride (PVC plastic) are two alkyl halides you have used. Many other halides serve as intermediates in the synthesis of other compounds.

18-12 One alkyl halide is a plastic, non-stick coating on pans to aid cooking and cleaning.

18-6	Industrial Uses of Halides	
Name	Structural formula	Application
trichloromethane (chloroform)	H–C(Cl)(Cl)–Cl	early anesthetic
tetrachloroethylene	Cl₂C=CCl₂	nonpolar solvent (dry-cleaning agent)
triiodomethane (iodoform)	H–C(I)(I)–I	veterinary antiseptic

Use or Abuse?

To most people, the word *alcohol* evokes images of society's most-abused drug. Grain alcohol, or ethanol, has a reputation for destroying homes and lives and contributing to many highway fatalities. Yet the alcohol itself is not at fault. It is merely a combination of atoms that could just as easily be used to benefit man. Man is the guilty party when he drinks and abuses it. Even the process of fermentation, which yields alcohol, is a natural reaction allowed by God. Man, in his fallen state, though, is prone to take good or neutral things and misuse them. Noah proved this when he became drunk with wine shortly after God used him to preserve life during the Flood (Gen. 9:20–21).

18.11 Alcohols

Compounds that have a covalently bonded OH group attached to an alkyl group are classified as **alcohols**. The general formula for the whole family is R–OH, where R represents an alkyl group. The simplest alcohol is methanol. In this case the R group is the smallest one possible, CH_3 (methyl).

$$\begin{array}{c} H \\ | \\ H-C-OH \\ | \\ H \end{array}$$

Alcohol names consist of the standard prefixes that tell how long the carbon chain is plus an *-ol* ending in place of the *-e* ending.

18-13 Many common household products contain alcohols.

If the OH group is attached to a carbon other than the end carbon, its position is indicated by a numerical prefix. For instance, what is commonly called *rubbing alcohol* is a three-carbon chain with the OH group attached to the middle carbon. The IUPAC name is propan-2-ol.

The physical properties of alcohols are a result of the two following factors: a polar OH group and a nonpolar alkyl group. The combination of these two opposites determines the behavior of each specific molecule. If the hydrocarbon chain is relatively short, the OH group dominates the molecule. As a result, it behaves as a polar molecule. If the hydrocarbon chain is very long, the chain dominates and imparts nonpolar characteristics to the molecule. Small alcohols, under the influence of the OH groups, form hydrogen

18-7 Alcohols

Name	Structural formula	Application
methanol (wood alcohol)	—C—OH	solvent, fuel
ethanol (grain alcohol)	—C—C—OH	alcoholic beverages, engine fuel, solvent
propan-1-ol	—C—C—C—OH	solvent for making pharmaceuticals, resins
propan-2-ol (rubbing/ isopropyl alcohol)	—C—C(OH)—C—	sterilizing pads, solvent, gasoline additive
butan-1-ol	—C—C—C—C—OH	solvent, shellac, varnish

bonds, have high boiling points, and are soluble in water. Large alcohols, under the influence of their hydrocarbon chains, are insoluble in water and soluble in nonpolar solvents.

Alcohols that contain more than one OH group are called **polyhydroxy alcohols**. One common polyhydroxy alcohol is ethane-1,2-diol (also called ethylene glycol), which is used as an antifreeze in car radiators. Another common alcohol—propane-1,2,3-triol (glycerol)—is an ingredient for moisturizers in cosmetics.

ethylene glycol   glycerol

Industries use alcohols as solvents, paint thinners, antifreezes, and ingredients in after-shave lotions. In the United States, most gasoline is 10%–15% ethanol. Some vehicle engines are designed to run on E85 fuel, which is 85% ethanol and 15% gasoline. When ethanol (grain alcohol) is required, product manufacturers face the prospect of paying the same stiff taxes that are imposed on liquor. To avoid these extra costs, they denature the ethanol, or make it totally unfit to drink.

18.12 Ethers

Compounds that have the general formula R–O–R′ are called **ethers**. R′ (called R-prime) stands for a second alkyl group. It may be the same or different from the first group. Ethers are distinguished by the oxygen bridge between two carbon chains. The name of an ether includes the names of the alkyl groups on each side of the oxygen (smaller one first) and the word *ether* on the end.

diethyl ether   ethyl methyl ether

More complicated ethers have been synthesized, but diethyl ether is by far the most common. When people say "ether," they are usually referring to this compound. A doctor named Crawford Long made this ether famous when he painlessly removed a tumor from a patient's neck in 1842. The operation was painless because ether is an anesthetic; it puts a person "to sleep." Ether served the medical profession for many years until other anesthetics with fewer side effects were developed. Ethers are now used as solvents for perfumes, primers for engines, and reagents in the synthesis of organic materials.

18.13 Aldehydes and Ketones

When an oxygen atom forms a double bond with a carbon atom, it forms a **carbonyl group**. Carbonyl groups are associated with two major groups of organic compounds.

18-14 Many ethers are added to gasoline to increase engine performance.

Aldehydes

Aldehydes are organic compounds that contain an oxygen held with a double bond to the end carbon. Their general structure looks like this.

$$R-\overset{\overset{O}{\|}}{C}-H$$

All aldehydes contain a carbonyl group. According to the IUPAC rules, the name of an aldehyde is formed with an *-al* ending on the name of the corresponding alkane.

propanal   2-methylpentanal

The simplest aldehyde and one of the most important ones in industry is known by the common name formaldehyde. Its IUPAC name is methanal. This colorless gas with a piercing odor is often dissolved in water to make a 37% solution called formalin. Formalin has been used in the past to preserve frogs, fetal pigs, and other creatures dissected in biology labs. However, because of health concerns about its possible link to cancer and allergic reactions, the use of formalin has greatly declined. Not all aldehydes are potentially harmful. Some give unique flavors to foods.

benzaldehyde (almonds)   cinnamaldehyde (cinnamon)

Ketones

A carbonyl group in the middle of a carbon chain is a key feature of a compound called a **ketone**. The general formula of ketones is this.

$$R-\overset{\overset{O}{\|}}{C}-R'$$

The simplest ketone is commonly called acetone. Its IUPAC name is propanone. The name *ketone* is derived from the name of the alkane that has the same number of carbon atoms; the *-e* ending of the alkane is changed to *-one*. For chains over four carbons long, the location of the carbonyl group is shown by a numerical prefix.

acetone   butanone

18-15 The aroma of an iris is caused by irone, a ketone.

Acetone is an excellent solvent. While the molecule's methyl groups are nonpolar, the carbonyl group is polar. As a result, acetone dissolves most organic compounds but still mixes well with water. Acetone is widely used as a solvent for lacquers, paint removers, explosives, plastics, and disinfectants. It is also the active ingredient in some nail-polish removers.

Since both ketones and aldehydes have carbonyl groups, it is not surprising that they have similar physical and chemical properties. As a rule, though, aldehydes are more chemically reactive because their functional group is exposed on the end of the carbon chain.

18.14 Carboxylic Acids

Because of their abundance in nature, carboxylic acids were among the first organic compounds to be studied in detail. As a result, many of these acids acquired common names from their most familiar sources. For example, the Latin word for vinegar is *acetum,* so the acid in vinegar was called acetic acid. Butyric acid (from the Latin *butyrum*) gives rancid butter, aged cheese, and human perspiration their nasty odors. The Latin word for ant is *formica.* Consequently, the stinging acid of ants was called formic acid.

All **carboxylic acids** have this general formula.

$$R-C\overset{O}{\underset{OH}{\diagup}}$$

The COOH group is called the **carboxyl group**. Notice that a carboxyl group consists of both a carbonyl group and a hydroxyl group. This combination makes carboxyl groups quite polar. According to IUPAC nomenclature, the -*e* ending of the corresponding alkane is changed to -*oic,* and the word *acid* is added to form the name of a carboxylic acid.

CH_3COOH $CH_3(CH_2)_2COOH$

acetic acid (ethanoic acid)   butyric acid (*n*-butanoic acid)

Carboxylic acids with shorter alkyl groups are liquids at room temperature and have sharp or unpleasant odors. Acids with longer carbon chains are usually waxy solids. When the carbon chains are between twelve and twenty carbon atoms long, they are often called **fatty acids**. They can form hydrogen bonds between themselves and other molecules. These hydrogen bonds keep even the smallest members of the family liquid at room temperature.

Carboxylic acids are called acids because they contain an ionizable hydrogen. The ionizable hydrogen in these acids is the hydrogen in the carboxyl group. Since only a small fraction of carboxylic acid molecules ionize, these acids are weak. Only four percent of acetic acid molecules ionize. Although they are weak acids, carboxylic acids react quickly with strong bases to form salts. Many of these salts are commercially useful, such as sodium or potassium salts of fatty acids used to make soaps.

18-16 The citric acid found in this fruit contains three carboxyl groups.

Trans Fats

Facets of Chemistry

In December 2006, New York City became the first city in the United States to legislate a ban on artificial trans fats at all restaurants. They were required to eliminate this ingredient from all the foods they serve by July 2008. Why trans fats?

You may ask, isn't all fat bad for you? In a word, no. There are "good" fats and "bad" fats. Energy is stored in the body mostly in the form of fat. Dietary fat is necessary to supply essential fatty acids, substances required for growth but not produced by the human body. These substances also transport the fat-soluble vitamins A, D, E, and K and aid in their absorption in the small intestine.

There are three main types of fatty acids: saturated, monounsaturated, and polyunsaturated fatty acids. A saturated fatty acid has the maximum possible number of hydrogen atoms attached to every carbon atom. These fats are saturated with hydrogen atoms. Saturated fatty acids are long, straight molecules that line up beside each other easily. They pack together to form solid fats at room temperature.

$$H-\underset{H}{\overset{H}{C}}-\underset{H}{\overset{H}{C}}-\underset{H}{\overset{H}{C}}-\underset{H}{\overset{H}{C}}-\underset{H}{\overset{H}{C}}-\underset{H}{\overset{H}{C}}-\underset{H}{\overset{H}{C}}-\underset{H}{\overset{H}{C}}-COOH$$

Some fatty acids are missing one pair of hydrogen atoms somewhere in the molecule chain. This gap is a point of unsaturation (a carbon-carbon double bond). The fatty acid is referred to as "monounsaturated" because it has one gap. The gap at the point of unsaturation forms a "kink" in the molecule, so it won't line up and pack together with other molecules. This causes the substance to be a liquid at room temperature.

$$H-\underset{H}{\overset{H}{C}}-\underset{H}{\overset{H}{C}}-\underset{H}{\overset{H}{C}}-\underset{}{\overset{H}{C}}=\underset{}{\overset{H}{C}}-\underset{H}{\overset{H}{C}}-\underset{H}{\overset{H}{C}}-\underset{H}{\overset{H}{C}}-COOH$$

A polyunsaturated oil or fatty acid has many double or triple bonds in a molecule. This means that the molecule is missing more than one pair of hydrogen atoms.

$$H-\underset{H}{\overset{H}{C}}-\underset{}{\overset{H}{C}}=\underset{}{\overset{H}{C}}-\underset{H}{\overset{H}{C}}-\underset{}{\overset{H}{C}}=\underset{}{\overset{H}{C}}-\underset{H}{\overset{H}{C}}-\underset{H}{\overset{H}{C}}-\underset{H}{\overset{H}{C}}-COOH$$

Trans fats, or trans fatty acids, are created by adding hydrogen to vegetable oil through a process called *hydrogenation*. Trans fatty acids are

the building blocks of hydrogenated fats. Hydrogenating oils removes essential fatty acids, but it is used in the food industry to prolong the shelf life of processed foods.

Hydrogenation changes the molecular structure of a fat. This is one simple hydrogenation reaction:

$$-\underset{}{\overset{H}{C}}=\underset{}{\overset{H}{C}}- + H_2 \rightarrow -\underset{H}{\overset{H}{C}}-\underset{H}{\overset{H}{C}}-$$

The relatively high temperatures used in the hydrogenation process cause some of the carbon-carbon double bonds to change into the trans form. "Trans" and "cis" refer to the physical structure of the fat. "Cis" means that hydrogen atoms are on the same side of the unsaturated carbon atoms in a fatty acid; "trans" means that the hydrogen atoms are on the other side of the unsaturated carbon atoms.

$$-\underset{}{\overset{H}{C}}=\underset{}{\overset{H}{C}}- \quad -\underset{}{\overset{H}{C}}=\underset{H}{\overset{}{C}}-$$
(cis) (trans)

Trans fats are stiffer than cis fats because they are uniquely twisted, and this unusual structure allows them to pack together to form solid material needed for certain baking and cooking processes.

Small amounts of trans fat occur naturally in beef and dairy products. However, other trans fats can be produced in industry and are found in cookies, crackers, icing, potato chips, stick margarine, and microwave popcorn.

So are trans fats good or bad fats? Initially trans fats were thought to be a healthy alternative to animal fats (saturated fats) because they are chemically unsaturated and come mainly from plant oils. While unsaturated fats are beneficial when consumed in moderation, saturated and trans fats are not. Various studies have shown that a diet high in trans fatty acids may be linked to a greater risk of type 2 diabetes. A 1994 survey estimated that trans fats caused 30,000 deaths each year in the United States from heart disease.

Christians should be careful with their health and diet because they are stewards of the bodies God has given them. God made humans as physical beings, and He called that creation very good (Gen. 1:31). God calls His people's bodies the temple of God's Spirit (1 Cor. 6:19–20). Though the Fall has affected our bodies, God will one day give believers a glorified body.

18.15 Esters

If the hydrogen of a carboxyl group is replaced with an alkyl group, an **ester** forms. Esters have this general formula.

$$R-C(=O)-O-R'$$

Unlike their cousins the carboxylic acids, esters generally have appealing smells. These compounds are responsible for the flavors of many fruits and the fragrances of many flowers. In the naming of an ester, the R' group is indicated with its alkyl name, and the carboxylic acid part is given an *-oate* ending.

18-8 Esters

Name	Structural formula	Flavor or odor
2-methylpropyl methanoate		raspberry
pentyl ethanoate		banana
ethyl butanoate		pineapple
ethyl heptanoate		grape

18-17 A banana's distinct flavor comes from an ester that can be synthesized.

18.16 Amines and Amides

Nitrogen can bond into organic molecules in several different ways. **Amines** are a family of organic compounds with ammonia (NH_3) as their parent. Derivatives are formed when one or more of the hydrogen atoms are replaced with organic groups, such as alkyl groups. One, two, or even three hydrogens may be replaced. The names of the compounds that result commonly have the word *amine* as a suffix after the names of the alkyl groups.

methylamine

diethylamine

Other compounds that contain nitrogen are called **amides**. Amides are produced when an amine group takes the place of

18-18 This poison dart frog's toxin is an amine named batrachotoxinin A.

an -OH group in a carboxylic acid. All the members of this group have the following structure in common.

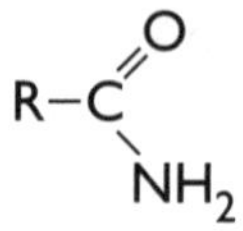

$$R-C\underset{NH_2}{\overset{O}{\lVert}}$$

This structure is especially important because it holds the amino acids in proteins together.

See Table 18-9 for a summary of the families of organic compounds we have discussed in this chapter.

18-9 Families of Organic Compounds

Hydrocarbons

alkane	alkene	alkyne	cyclic	aromatic
-C-C-C-	-C=C-C-	-C≡C-C-	cyclopropane ring	benzene ring
propane	propene	propyne	cyclopropane	benzene

Substituted hydrocarbons

Family	Structural formula	Suffix or group name	Typical compound	Name of compound
halide	R—C—X	chloro- bromo- fluoro- iodo-	F—C—C—C—	fluoropropane
alcohol	R—C—OH	-ol	—C—C—C—OH	1-propanol
ether	R—O—R′	ether	—C—O—C—	dimethyl ether
aldehyde	R—C=O	-al	—C—C—C(=O)	propanal
ketone	R—C(=O)—R′	-one	—C—C(=O)—C—	propanone
carboxylic acid	R—C(=O)—OH	-oic acid	—C—C—C(=O)—OH	propanoic acid
ester	R—C(=O)—O—R′	-oate	—C—C—C(=O)—O—C—	methyl propanoate
amine	R—N(R′)—R″	-amine	—C—N—C—C—	methylethylamine
amide	R—C(=O)—NH₂	-amide	—C—C—C(=O)—N—	propanamide

18C Section Review

1. What is the main difference between an alkyl halide and an aryl halide?
2. The physical properties of alcohol are a result of what two factors?
3. What influence does the length of the hydrocarbon chain have on an alcohol?
4. Which compounds are distinguished by an oxygen bridge between two carbon chains?
5. What is the main difference between an aldehyde and a ketone?
6. What is the main difference between an amine and an amide?
7. Identify the organic compounds that have the following IUPAC endings to their names:
   a. *-e* ending changed to *-one*
   b. ends in *-ol*
   c. ends in *-al*
   d. ends in *-oate*
   e. ends in *-oic acid*

18D Organic Reactions

The number of synthesized organic compounds is continually growing. These compounds can participate in many chemical reactions. Most biological processes rely on chemical reactions between organic molecules. These reactions are responsible for the movement of muscles, the digestion of food, the transmission of nerve impulses, and the sensing of light upon the retina. Industrial chemists manipulate molecules to make flavorings, plastics, fuels, synthetic fabrics, and a host of other products. Some of these reactions are quite complicated. This text will survey a few basic reactions in which organic compounds participate.

18D Section Objective
After finishing this section, you should be able to
- recognize and give examples of substitution, addition, and condensation reactions.

18.17 Oxidation-Reduction

When oxygen and carbon atoms bond, carbon atoms oxidize and oxygen atoms reduce. Since oxygen atoms have high electronegativities, they pull shared electrons away from carbon. This means that adding oxygen always makes the oxidation number of carbon more positive. Adding hydrogen atoms, on the other hand, reduces carbon atoms.

Combustion oxidizes all the carbon atoms in an organic molecule to form carbon dioxide. All hydrocarbons burn in oxygen to form carbon dioxide, water, and heat. The energy from these combustion reactions is used to produce motion, heat, and light.

The oxidation of methane, a simple alkane, is a combustion reaction.

$$CH_4 + 2O_2 \longrightarrow CO_2 + 2H_2O$$

Remember from Chapter 17 that oxidation is the loss of electrons, and reduction is the gain of electrons.

18-19 When a candle burns, alkanes found in paraffin wax are combusted in an oxidation reaction.

18-20 A halogen solution (bromine) is added to an alkane (left) and an alkene (right). The alkene, being unsaturated, reacts with the bromine in an addition reaction, and bromine's color disappears.

Fatty acid chains in food are oxidized by the human body similarly to how other hydrocarbons are burned. Providentially, the body controls the oxidation precisely so that only small amounts of energy are released at any instant. This regulation keeps the temperatures during oxidation tolerable and allows the body to capture and use most released energy.

18.18 Substitution

Substitution reactions replace one part of a molecule with another part. Typically, most of the reactions of chemically inactive compounds are substitution reactions. Alkanes are not very reactive, but when heated to high temperatures or exposed to energetic ultraviolet light, their hydrogen atoms can be replaced by other atoms. For example, methane and chlorine can react to form a variety of substitution products. In this example, a chlorine atom replaces a hydrogen atom when a mixture of the two gases is exposed to heat or certain types of light.

$$CH_4 + Cl_2 \xrightarrow{\text{heat or light}} CH_3Cl + HCl$$

Various groups can replace one or more of the hydrogens of aromatic compounds. Benzene can be nitrated, halogenated, or even alkylated.

18.19 Addition

Compared to carbon-carbon single bonds, double bonds and triple bonds are very reactive. Consequently, the double and triple bonds in a molecule react first and determine the molecule's behavior. An **addition reaction** is a reaction in which a multiple bond of a molecule is broken and two atoms or groups of atoms are added. This is a characteristic reaction of unsaturated molecules. Some of the most common addition reactions involve water, hydrogen gas, halogens, and hydrogen halides. In this example hydrogen converts an alkene (ethene) to an alkane (ethane).

$$C_2H_4 + H_2 \longrightarrow C_2H_6$$

18.20 Condensation

Reactions in which molecules combine with each other to lose a water molecule are called **condensation reactions**. Two identical alcohol molecules can be made to join together to form an ether under special reaction conditions.

$$R-OH + R-OH \xrightarrow{H_2SO_4} R-O-R + HOH$$

Esters form when carboxylic acids and alcohols go through a condensation reaction. This condensation reaction is called **esterification**. For example, an artificial banana flavoring can be made when ethanoic (acetic) acid and pentanol are mixed.

A condensation reaction is responsible for much of the clothing that people wear. **Polymers** are substances that consist of huge molecules that have repeating structural units. Polyester, one of the more common polymers, forms when ethylene glycol and terephthalic acid condense.

18D Section Review

Determine the most specific organic reaction that is taking place in the following examples:

1. Hydrogen atoms are replaced by chlorine atoms when an alkane is heated.
2. Unsaturated molecules react.
3. Carbon and oxygen atoms bond.
4. All of the carbon atoms in an organic molecule are converted to carbon dioxide.
5. A condensation reaction forms esters.
6. Molecules combine and a water molecule is lost.

18E Biochemistry

Biochemistry is the chemistry of life. The chemistry of genetics, metabolism, and other cell functions fall within this field of study. The marvelous complexities of life cannot be studied within the confines of a single subject area.

18.21 Biochemistry and Ultimate Questions

Because biochemistry deals with life, it leads to philosophical and spiritual issues. As technology develops and genetic engineering and cloning become a reality, scientists begin to encounter perplexing ethical predicaments. Should humans be cloned? Should human embryos be developed to harvest stem cells? Should human life be preserved at all costs? At the heart of these questions is the ultimate question, "What is the meaning of life? What gives it worth and value?" The Bible, not science, contains the answer to these questions, which is why Christians must engage in the field of biochemistry.

Biochemistry shouts out the infinite wisdom of the Creator. The complexity of this science forces those who study it to consider the existence of a purposeful, caring Creator. His handiwork is on display in the precise architecture of a protein molecule, the complexity of a metabolic pathway, and the efficiency of an enzyme. Every molecule bears witness to the power, ingenuity, and foresight of God. Man has been granted the privilege of making many discoveries in the field of biochemistry. These discoveries are exciting, but they are also humbling, for they reveal an omniscient, omnipotent Creator.

Let's embark on the study of biochemistry by studying some of the basic molecules of life—carbohydrates, lipids, proteins, and nucleic acids.

18.22 Carbohydrates

Carbohydrates are the most abundant biological compounds. Sugars and starches make up a large part of the human diet. Each year photosynthetic processes in plants convert water and carbon dioxide into one hundred billion tons of carbohydrates. The exoskeletons

---

18E Section Objectives

After finishing this section, you should be able to

- list the key functions of carbohydrates, lipids, proteins, and nucleic acids.
- identify carbohydrates, fats, oils, steroids, proteins, and nucleic acids when given their structural formulas.
- differentiate between saturated and unsaturated fats and their physical characteristics.
- describe the formation and structure of proteins using the terms *amino acids*, *peptide bonds*, and *polypeptide chains*.
- state the importance of the structure and shape of a protein molecule.
- recognize that an organism's traits are determined by genes, and that the genetic code is recorded as a sequence of nucleotides.
- list the three parts of a nucleotide and describe their arrangement.

Animals consume different carbohydrates than humans do. They store food in the form of glycogen, not starches. You can find glycogen in their liver and muscles. Glycogen is a branched polymer of glucose in which the branches occur more frequently than in starch.

of all the insects, crabs, and lobsters on this planet contribute even more carbohydrates to this already impressive amount.

Carbohydrate literally means "water of carbon." The name goes back to the days when these compounds were thought to be hydrates of carbon molecules, and the empirical formula of all carbohydrates was CH_2O. However, as larger and larger molecules were discovered, this formula did not hold true for all carbohydrate molecules. All carbohydrates have several of the −OH (hydroxy) groups common to alcohols. They also have the C=O (carbonyl) group of aldehydes and ketones. Concisely stated, **carbohydrates** are polyhydroxy aldehydes or ketones.

Carbohydrates have three primary functions: energy storage, an energy source for cellular functions, and structural elements in plants and animals. The carbohydrates cellulose and chitin form the structural support for plants and animals. Other carbohydrates, such as glucose, supply energy for cell activities. Carbohydrates can be classified into three groups based on the number of sugar units they contain.

Monosaccharides

Monosaccharides, or simple sugars, have one polyhydroxy aldehyde or ketone. All known monosaccharides—and there are many of them—are white, water-soluble, crystalline solids. Most have a sweet taste. They rarely occur in nature as free molecules but are usually

18-21 These mushrooms contain chitin. The tree on which they grow is made largely of cellulose. Both of these substances are carbohydrates.

18-22 Glucose, like many monosaccharides, exists in both open and ring forms in solution.

18-23 Two monosaccharides join in a condensation reaction to release a water molecule and form a disaccharide.

bonded to a protein, a fat, or another carbohydrate. Two significant exceptions are glucose and fructose.

Glucose is the most abundant sugar in nature. Ripe berries, grapes, and oranges contain 20 to 30 percent glucose. The human body maintains a reasonably constant level of 80 to 120 mg of glucose per 100 mL of blood. Hence, it is sometimes called blood sugar. Glucose is also the fundamental building block of the most common long-chain carbohydrates.

Although monosaccharides are often drawn as straight, chain-like molecules, they usually exist in rings that form when atoms near the end of the chain bond to atoms near the beginning of the chain. Glucose in an aqueous solution exists in an equilibrium between the ring form and the straight-chain form. The equilibrium lies strongly in favor of the ring form. Fructose also forms ring structures when it is in solution. Since the carbonyl group is not at the end of the carbon chain, it forms a five-member ring.

18-24 Which two disaccharides would you consume if you enjoyed this snack?

> Lactose and sucrose are too large to be absorbed by the human digestive system, so enzymes break them down into simpler monosaccharides. *Lactase* helps break down lactose into glucose and galactose. *Sucrase* converts sucrose to glucose and fructose.

Disaccharides

As their name implies, **disaccharides** contain two monosaccharide units. An oxygen bridge between the two monosaccharides holds the two units together. While many monosaccharides exist, even more disaccharides can be formed from combinations of monosaccharides.

Three disaccharides play an important part in the human diet—maltose, lactose, and sucrose. Maltose consists of two bonded glucose molecules. The bond forms between the first carbon of one molecule and the fourth carbon of the other. Maltose is found in germinating grain and is produced during the digestion of starches. Lactose is a disaccharide found only in milk. It consists of an isomer of glucose (galactose) joined to a glucose molecule. The sugar is not very sweet, but it is an important ingredient in milk. Milk sours when lactose breaks down into lactic acid. A glucose and a fructose molecule can bond in a condensation reaction, releasing a water molecule to form the disaccharide sucrose (table sugar). Sucrose occurs abundantly in fruits, sugar cane, sugar beets, and nectar.

Polysaccharides

Polysaccharides are molecules that contain many sugar units. This is evident from their large molecular masses (up to several million atomic mass units). Multitudes of sugar units are bonded into long chains. Polysaccharides may be built from several different monosaccharides, but this text will concentrate on the polymers of glucose.

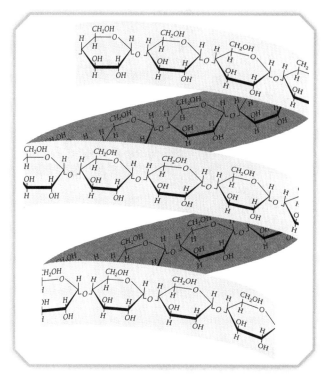

18-25 Straight-chained starch molecules coil into a helix.

Plants such as rice, potatoes, wheat, and oats store food in polysaccharide deposits called starch. **Starch** is a mixture of two glucose polymers, a straight chain and a branched chain. Together these polysaccharides supply nearly three-fourths of the world's food energy.

Humans do not eat some of the plants that animals can because these plants contain lots of cellulose. *Cellulose* is actually a glucose polysaccharide that is similar to starch and glycogen. In the polysaccharides that humans eat, the bonds between glucose

18-26 Bacteria in a cow's rumen digest the polysaccharide cellulose into absorbable sugars.

units—the oxygen bridges—point downward; however, the bonds in cellulose point upward. Thus, the difference between the starch in a tasty potato and the cellulose fibers in grass, cotton, or a splintery piece of wood is the orientation of the bonds. The digestive tracts of humans and animals contain different enzymes to process these saccharides.

18.23 Lipids

Although water is sometimes called the universal solvent, it does not mix with all the compounds in the human body. An entire class of compounds called **lipids** cannot dissolve in water. They are characterized by a high proportion of C–H bonds, making them nonpolar. Lipids play a key role in energy storage and the structure of cell membranes. Fatty acids, fats, oils, waxes, and steroids are examples of vital lipids.

Fats and Oils

Fats and oils are esters of glycerol and fatty acids. **Glycerol** is a three-carbon molecule that has three OH groups on separate carbons. Fatty acids are carboxylic acids with long hydrocarbon tails. Ester linkages join the COOH groups of acids to the OH groups in the glycerol to construct a fat or oil molecule and water. Fatty acids rarely occur alone in living things. Typically three of them bond with a glycerol molecule to form a triglyceride.

18-27 Three fatty acids and a glycerol molecule can combine to form a triglyceride.

The acids that bond to the glycerol usually contain a large hydrocarbon chain (12 to 18 carbons are common), and they usually contain an even number of carbons (the chains are made in two-carbon units). Some chains contain only single carbon-carbon bonds, and some contain a few double bonds. This difference leads to the distinction between fats and oils.

Fats exist as solids at room temperature. Because most of their fatty acid components have a large degree of saturation (single bonds), the chains tend to extend in straight lines from the glycerol backbone. They are considered saturated. As a result, they fit together well and have effective dispersion forces. Recall that intermolecular forces can have a great effect in determining the physical properties of matter. It is these intermolecular attractions that cause fats, such as lard, to be solids at room temperature.

Oils, on the other hand, are liquids at room temperature. Generally, their fatty acid components contain more double bonds than fats. Because of these double bonds, oils are considered to be unsaturated (recall that double bonds lessen the number of hydrogen atoms that are bonded). The double bonds in oils introduce bends in the chains. Because these irregularly shaped molecules do not fit together as well as the molecules in fats do, their dispersion forces are less effective. As a result, the oil molecules do not experience strong intermolecular forces—resulting in liquidity at room temperature.

18-28 The double bonds in unsaturated fats make them oils at room temperature. The saturated fats are solids.

Like carbohydrates and proteins, dietary fat is a vital nutrient to help promote a healthful lifestyle, especially as a source of energy for the body. It is the most concentrated source of energy in the diet, providing nine calories per gram compared to four calories per gram from either carbohydrates or protein. Fats also carry vitamins, cushion the body, and insulate nerves.

Steroids

Steroids are lipids, but their structures do not resemble triglyceride fats and oils in any way. They are the nonester lipids. The basic structure of a steroid is a combination of three six-membered rings and one five-membered ring. Functional groups attached to various points on the rings result in a wide variety of steroids. Cholesterol, vitamin D, cortisone, testosterone, and estrogen all use this common "chicken wire" frame.

The study of steroids ranks as one of the most active areas of chemical research. One steroid that has been studied much in recent years is *cholesterol*—a compound that, although vital to good health, can increase the risk of heart disease if excessive amounts are present in the blood. Contrary to popular belief, most of the cholesterol found in the body is manufactured by the body itself. The average total blood cholesterol in Americans is 203 milligrams per deciliter (mg/dL). The National Institutes for Health consider less than 200 mg/dL desirable and greater than 240 mg/dL high.

In addition to diet, blood cholesterol is influenced by a wide variety of factors, including heredity, smoking, age, race, gender, high blood pressure, obesity, and activity level. Researchers are also seeking to isolate a gene that may be used to identify individuals at risk for high cholesterol and apply early treatment.

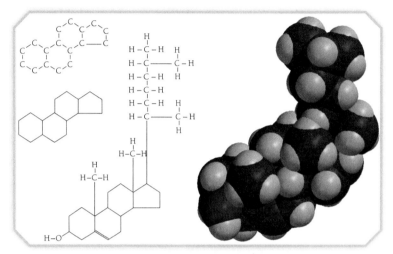

18-29 General structure of steroids (left); structural model (center) and space-filling model (right) of cholesterol

18.24 Proteins

Proteins are one of the most important substances in the bodies of living things. Proteins serve as the building blocks for muscles, hair, blood cells, skin, spider webs, silk, enzymes, insulin, and even snake venom. They are essential nutrients, and they have many life-sustaining functions. Protein chains exhibit an intricate three-dimensional architecture custom designed for their function.

Amino Acids

Amino acids are the building blocks of proteins. As their name suggests, they are molecules that contain an amine (NH_2) group and the carboxyl (COOH) group of carboxylic acids. Various side chains (R groups) attached to the carbon adjacent to the nitrogen atom result in twenty different common amino acids. It is the side chains that determine the properties of the protein.

Our bodies can synthesize 12 of the 20 amino acids. The other eight must come from proper diet. Those amino acids that cannot be manufactured by the body are called **essential amino acids**. Table 18-10 lists the essential amino acids.

> The Greek word *protos* means "first," and the French word *protéine* means "primary substance."

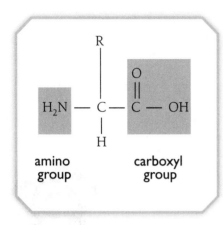

18-30 The general formula for amino acids

Name	R group	Adult requirements per day

isoleucine	—CHCH$_2$CH$_3$
	     |
	   CH$_3$
		1.4 g

leucine	—CH$_2$CH(CH$_3$)$_2$	2.2 g
lysine	—CH$_2$CH$_2$CH$_2$CH$_2$NH$_2$	1.6 g
methionine	—CH$_2$CH$_2$SCH$_3$	2.2 g
phenylalanine	—CH$_2$—C$_6$H$_5$	2.2 g

threonine	—CHOH
	    |
	   CH$_3$
		1.0 g

tryptophan	—CH$_2$-(indole)	0.5 g
valine	—CH(CH$_3$)$_2$	1.6 g

Table 18-10 Essential Amino Acids

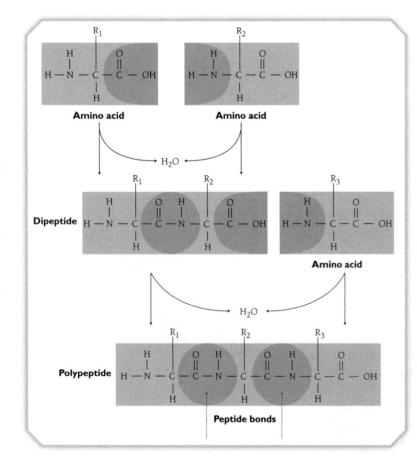

18-31 Polypeptides form when amino acids combine through peptide bonds.

Polypeptide Chains

Amino acids join together when the amine group of one amino acid reacts with the carboxyl group of another. The bond that links the two amino acids is called a **peptide bond**. Molecules that contain two amino acids are called **dipeptides**. Aspartame (aspartyl phenylalanine), a popular artificial sweetener, is a dipeptide. **Polypeptides** are polymers of many amino acids.

A **protein** is a polymer of one or more polypeptide chains. Some proteins contain only several hundred amino acids; others have many thousands of these building blocks. The basic order of the amino acids in a polypeptide chain is called its *primary structure*. A protein's molecular weight can range from 5000 to millions of atomic mass units. Even with a comparatively short chain, the number of possible different proteins is staggering. From the twenty common amino acids, a series of fifty amino acids could be arranged in 3×10^{64} different sequences, with each sequence forming a different protein. When one considers that most proteins are much longer, the complexity and variety of proteins in God's creation become apparent.

Like polysaccharides, polypeptides tend to form regular shapes. The exact form of the chain depends on the amino acids that are present and the order in which they appear. The order or primary structure contributes to the *secondary structure* of the protein. The shapes and electrical natures of amino acids cause many proteins to coil themselves into helixes. This tight spiral of amino acids is held in place by hydrogen bonds between certain atoms in the amino acid backbone of the protein. A hydrogen in one amino acid forms a hydrogen bond with an oxygen of the amino acid four positions away. The side chains of each amino acid extend like spokes from the surface of the helix. It is amazing that these intricate coils get their form from ordinary, but carefully arranged, hydrogen bonds. The structure truly is a masterpiece of art and engineering. Another preferred shape of polypeptide chains is a pleated sheet. The chains stretch out in an almost flat, but folded, pattern. Hydrogen bonds, this time between amino acids in different chains, hold the structure together. The same polypeptide chain may contain both helical and pleated-sheet sections. These forms influence a protein's *tertiary structure*.

When two or more polypeptide chains, each with its own unique shape, cluster together to form a protein, the *quaternary structure* is revealed. Each protein's distinctive shape (determined by the side chain) equips it to perform a unique function. For example, hemoglobin, the protein that carries oxygen in the bloodstream, consists of four interwoven chains. The shape of the protein provides perfectly formed crevices for oxygen molecules to nestle into.

18.25 Nucleic Acids

When cells reproduce, they pass genetic information to one another in long-chain molecules called **chromosomes**. Human body cells contain forty-six of these long molecules in their nuclei. Segments of chromosomes, called **genes**, carry the coded information that directs the production of specific polypeptide chains. Genes are constructed of many **nucleotides**, which are the building blocks of **nucleic acids**.

Each nucleotide consists of three units: a five-carbon sugar, a phosphate group, and a ring-shaped, nitrogen-containing base. The sugar and the phosphate groups alternate to form a long chain that supports the bases. In RNA (ribonucleic acid), the sugar is ribose, while in DNA (deoxyribonucleic acid), it is deoxyribose, which is a ribose with one less OH group.

There are five possible nitrogen-containing bases attached to a sugar: adenine, cytosine, guanine, thymine, or uracil (abbreviated A, C, G, T, and U). DNA strands contain only A, C, G, and T bases, and RNA strands have A, C, G, and U. Of these bases, C, T, and U are small, single-ringed structures, and A and G are larger, double-ringed structures.

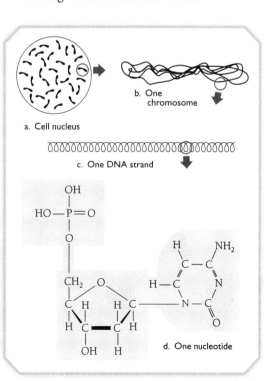

18-33 Notice the sugar, phosphate, and base in this nucleotide.

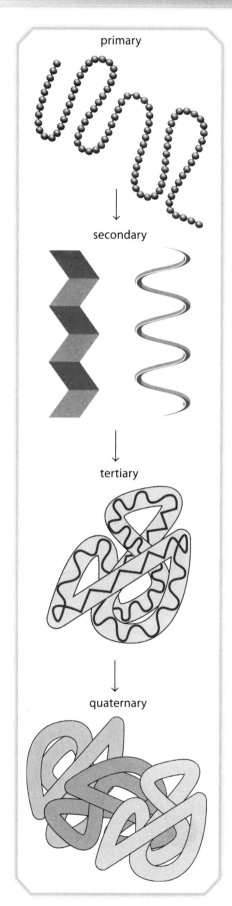

18-32 The structure of a protein is complex and can be described on several different levels.

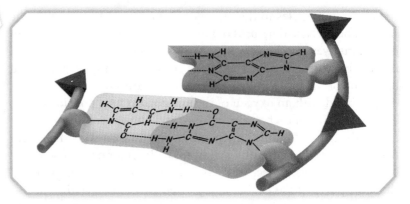

18-34 The dotted lines between cytosine and guanine in this base pair represent hydrogen bonds.

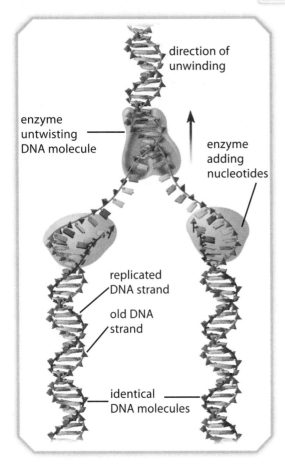

18-35 In DNA replication, an exact copy is made as the double helix strand separates and unravels.

DNA has been designed so that two strands coil around each other in a double helix. It looks somewhat like a ladder that has been twisted several times along its long axis. The backbones circle around the outside, and the bases mesh together inside the coil. Large, double-ringed adenine bases pair with smaller thymine bases, while large guanine bases fit neatly next to smaller cytosine bases. Hydrogen bonds hold the complementary bases snugly in position.

The way nucleic acids work is marvelous, almost miraculous. They are responsible for the faithful duplication of an organism's characteristics, and they perform this task with great precision and accuracy. In this regard, God's creation is astonishingly complex. Man has been privileged to unravel some of its workings, further verifying that he is "fearfully and wonderfully made" (Psalm 139:14). When cells divide, a DNA strand reproduces by first unraveling with the help of enzymes. Each of the two resulting strands serves as a template, or pattern, for a new complementary chain. Complementary bases on the newly forming strand match the now-exposed bases of the old strand. When the process of replication is completed, two identical double-stranded DNA molecules result. One strand goes to each half of the dividing cell.

DNA and RNA molecules instruct cells on how to make proteins. Because proteins are the building blocks of living things, as well as the controllers of all biological processes, these nucleic acids can be described as sets of "blueprints" that make each living thing unique.

18.26 Using Isomers to Solve Problems

In 1965, microbiologist Barnett Rosenberg was curious about how electrical current might affect the division of bacteria cells. To test this, he placed two platinum electrodes into a culture of *Escherichia coli*. He was startled to find that rather than dividing, the bacteria continued to grow, up to 300 times their typical length, without ever dividing. Through a series of follow-up experiments, he learned that the chemical responsible for inhibiting cell division was

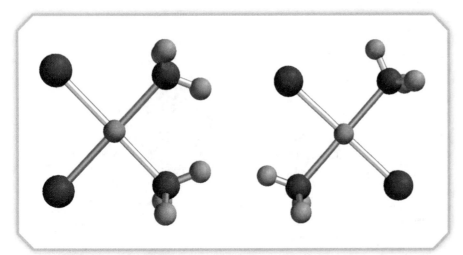

18-36 The isomer on the left, cisplatin, is an effective chemotherapy agent, while the other trans isomer is poisonous and cannot be used.

cis-diamminedichloroplatinum (II), cis-DDP or cisplatin for short. He had used platinum because it is unreactive with most chemicals, but in this case platinum reacted with chlorine ions and amine groups. The $PtCl_2(NH_3)_2$ that formed can take one of two isomer forms seen in Figure 18-36. Although the molecular formula is the same, trans-diamminedichloroplatinum (II) does not have the same effect.

Rosenberg conducted further tests to confirm that cisplatin did, indeed, inhibit the growth of certain malignant tumors. By 1978, the FDA had approved the compound for chemotherapy. It is now believed that cisplatin inhibits the division of cells by halting the replication of DNA. After hydrolysis, the chlorines of cisplatin are replaced by water molecules, allowing the molecule to attach to two separate DNA bases. This disturbance is enough to signal the DNA to try to repair itself by bringing in certain proteins. The bonding of these large proteins to the DNA-cisplatin complex causes the helical DNA molecule to kink, making replication impossible. Thus, a simple molecule is able to shut down the process of cell division and as a result prevent the growth of the tumor.

> Chemotherapy has helped save lives that would have otherwise been lost to cancer. However, chemotherapy has strong negative side effects, primarily the diminishing of red and white blood cells with the corresponding weakening of the immune system. In some cases the chemotherapy is life threatening. Why do you think chemotherapy attacks these cells in this way?

18E Section Review

1. Define *carbohydrate*.
2. List the three groups of carbohydrates.
3. What products result from the condensation reaction that forms sucrose?
4. List the functions of carbohydrates.
5. Explain the difference between saturated and unsaturated fats.
6. How does the difference in saturation affect physical properties?
7. What is the basic structure of a steroid?
8. List some factors besides diet that may influence a person's cholesterol level.
9. Define *protein*.
10. Which amino acids are essential? Why are they considered essential?
11. What is a polypeptide?
12. Compare the sugar unit in DNA with that in RNA.
⊙ 13. Research the etymology of the terms *cis* and *trans* to see how they relate to the two isomers pictured in Figure 18-36.
14. Why does chemotherapy often affect a patient's immunity by destroying blood cells?

Chapter Review

Chapter Summary

- Carbon is found in all living things, making it essential to life. All organic compounds contain carbon, and many of these can be synthesized in a laboratory.

- Carbon has four valence electrons and forms strong bonds with other carbon atoms and atoms of other elements. It can form into chains, rings, and other shapes with single, double, and triple bonds.

- Organic compounds are divided into two large groups: aliphatic compounds, which include those with straight chains or ring structures that can be formed by closing a straight chain; and aromatic compounds, which have ringed shapes with delocalized electrons.

- Hydrocarbons are a class of compounds containing only hydrogen and carbon. Some common hydrocarbon types are alkanes, which are open-chained and have only single bonds; alkenes, which include a double bond; and alkynes, which have at least one triple bond. Aromatic compounds are also hydrocarbons.

- Functional groups are atoms or groups of atoms that attach to organic compounds, giving rise to a great diversity of organic compounds with unique properties. Systematic names prescribed by the IUPAC are based on the nature and location of these functional groups.

- Some of the types of organic compounds that contain functional groups are halides, alcohols, ethers, aldehydes, ketones, carboxylic acids, esters, amines, and amides. Each of these is identified with one or more specific functional groups.

- Many reactions between organic compounds exhibit oxidation and reduction. Combustion is a common form of hydrocarbon oxidation.

- Both substitution and addition reactions add atoms to a molecule. In substitution reactions, part of the original molecule is replaced. In addition reactions, a multiple bond is broken and new atoms or groups of atoms are added.

- Condensation reactions join two molecules or two parts of the same molecule, releasing a water molecule.

- Carbohydrates are sources of energy and structure in living things and are polyhydroxy aldehydes or ketones. They may be simple sugars (monosaccharides), two simple sugars combined (disaccharides), or polymers of many sugars (polysaccharides). Starch and glycogen are important polysaccharides found in plants and animals, respectively.

- Lipids are a varied group of compounds that will not dissolve in water. Fatty acids are the simplest lipids. Several of these long-chain carboxylic acids can link to a glycerol backbone to form a triglyceride.

Coming to Terms

Term	Page
organic compound	445
aliphatic compound	447
aromatic compound	447
hydrocarbon	447
alkane	447
saturated	448
alkyl group	448
structural isomer	449
alkene	450
unsaturated	450
alkyne	450
cyclic aliphatic compound	451
delocalized electron	453
functional group	455
alkyl halide	455
aryl group	455
aryl halide	455
alcohol	456
polyhydroxy alcohol	457
ether	457
carbonyl group	457
aldehyde	458
ketone	458
carboxylic acid	459
carboxyl group	459
fatty acid	459
ester	461
amine	461
amide	461
substitution reaction	464
addition reaction	464
condensation reaction	464
esterification	464
polymer	464
carbohydrate	466
monosaccharide	466
disaccharide	467
polysaccharide	467
starch	467
lipid	468
glycerol	468
fat	468
oil	468
steroid	469
amino acid	469
essential amino acid	469
peptide bond	470
dipeptide	470
polypeptide	470
protein	470
chromosome	471
gene	471
nucleotide	471
nucleic acid	471

- Fats are lipids that exist as solids at room temperature. Oils remain in the liquid state because they are unsaturated and do not fit together as tightly as saturated fats.
- Steroids are nonester lipids that contain rings of carbon. Cholesterol is a well-known steroid.
- Proteins are large, organic polymers built from amino acids. They are essential for life due to their role in providing structure, acting as hormones, and serving as enzymes to control chemical reactions.
- Amino acids link through peptide bonds to form polypeptide chains. A protein is a polymer of one or more of these chains.
- The nucleic acids DNA and RNA are similar in structure and play related but unique roles in living things. DNA stores hereditary information, and RNA uses that code to assemble proteins.
- A chromosome is a long-chain molecule of nucleic acid. Sections of the chromosome that carry the code to direct the production of a certain polypeptide chain are called genes. Each gene is constructed of many units called nucleotides.

Review Questions

Concept Review

1. What is the modern definition of *organic compounds*?
2. Name a class of organic compounds that contains a substance that can be used as
   a. a fuel for welding.
   b. a fuel in automobiles.
   c. an antifreeze.
   d. a refrigerator coolant.
   e. an ingredient in soap.
   f. a paint remover.
3. Why do straight-chained alkanes have higher boiling points than alkanes with branched chains?
4. How can you tell when two compounds are structural isomers?
5. What is unique about the carbon-carbon bonds in benzene and other aromatic compounds?
6. Methane (CH_4) is a gas at room temperature, whereas methanol (CH_3OH) is a liquid. Aside from the difference in molecular masses, suggest an explanation for their different boiling points.
7. Methanol is soluble in water, but larger alcohols such as octanol are not. Why is this?
8. Alcohols and metallic hydroxides both have OH groups in their structural formulas. Why are metallic hydroxides such as NaOH caustic, but alcohols are not?
9. What is similar about the functions of HCl and CH_3COOH? How are they different?
10. Give one way in which NaCl and soap are similar and one way in which they are different.

11. List three reasons some scientists are looking to polycyclic aromatic hydrocarbons for secrets on the origin of life.
12. How could a Christian creationist respond to theories that polycyclic aromatic hydrocarbons were instrumental in the origin of life?
13. In most substitution reactions of organic compounds, what atom is usually replaced by another atom or group of atoms?
14. List three major functions of carbohydrates.
15. Describe what is meant by each term, and give one example of each.
    a. monosaccharide
    b. disaccharide
    c. polysaccharide
16. How are cellulose and glycogen similar? How are they different? How does the difference between cellulose and glycogen affect their nutritional value to man?
⊙ 17. Postulate a reason why termites can eat wood but people cannot.
18. Why are amino acids named amino acids? Draw the structural formula of an amino acid and identify the functional groups.
19. Polysaccharides and polypeptides are both polymers. What is the difference between them?
20. Why is the three-dimensional shape of a protein important?
21. Can a protein consist of more than one polypeptide? Explain.
22. What is the difference between a fat molecule and an oil molecule?
23. What distinguishes steroids from other lipids?
24. What is a nucleotide? What are nucleotides composed of? How do the nucleotides of DNA and RNA differ?
25. What is the general shape of a DNA molecule? What is the function of DNA?

True or False

26. Friedrich Wöhler synthesized the first organic compound, but not intentionally.
27. For simplicity, oxygen atoms are usually left off of structural formulas.
28. Benzene is a cyclic aliphatic compound.
29. Polymers form through condensation reactions.
30. Most of the cholesterol found in the body is manufactured by the body itself.
31. The eight essential amino acids are those that the human body is able to manufacture.

Application

⊙ 32. Modify the structural formula of the alkane 2-methylbutane to create the various kinds of compounds requested below. Remember that carbon atoms always have four bonds. Add or delete hydrogens when necessary.

$$-\overset{|}{\underset{|}{C}}-\overset{|}{\underset{|}{C}}-\overset{|}{\underset{\underset{|}{-C-}}{C}}-\overset{|}{\underset{|}{C}}-$$

a. an alkene
b. an alkyne
c. an alkyl halide containing one iodine atom
d. an alcohol
e. an aldehyde
f. a ketone
g. a carboxylic acid
h. an ester (Use an ethyl group to modify the carboxylic acid drawn in [g].)
i. an amine
j. an amide (Modify the carboxylic acid drawn in [g].)

⊙ 33. Classify each of the following compounds according to its general family:

a. $-\overset{|}{\underset{|}{C}}-\overset{|}{\underset{|}{C}}-Br$

b. $-\overset{|}{\underset{|}{C}}-\overset{|}{\underset{|}{C}}-\overset{O}{\overset{\|}{C}}-\overset{|}{\underset{|}{C}}-$

c. $-\overset{|}{\underset{|}{C}}-\overset{|}{\underset{\underset{|}{-C-}}{C}}-\overset{O}{\underset{OH}{\diagdown\diagup}}C$

d. $-\overset{|}{\underset{|}{C}}-\overset{|}{\underset{\underset{|}{-C-}}{C}}-\overset{|}{\underset{|}{C}}-\overset{|}{\underset{|}{C}}-\overset{O}{\underset{NH_2}{\diagdown\diagup}}C$

e. $-\overset{|}{\underset{|}{C}}-C\equiv C-\overset{|}{\underset{\underset{|}{-C-}}{C}}-\overset{|}{\underset{|}{C}}-$

f. $-\overset{|}{\underset{|}{C}}-\overset{|}{\underset{|}{C}}-\overset{|}{\underset{|}{C}}-NH_2$

g. $-\overset{|}{\underset{|}{C}}-\overset{|}{\underset{\underset{|}{-C-}}{\overset{\overset{|}{-C-}}{C}}}-\overset{|}{\underset{|}{C}}-\overset{|}{\underset{\overset{|}{-C-}}{C}}-\overset{|}{\underset{|}{C}}-$

h. $-\overset{|}{\underset{|}{C}}-\overset{O}{\underset{O-\overset{|}{\underset{|}{C}}-}{\diagdown\diagup}}C$

i. $-\overset{|}{\underset{|}{C}}-\overset{|}{\underset{\underset{|}{-C-}}{C}}-\overset{|}{\underset{\overset{|}{-C-}}{C}}-\overset{|}{\underset{|}{C}}-\overset{|}{\underset{|}{C}}-OH$

j. $-\overset{|}{\underset{|}{C}}-\overset{|}{C}=\overset{|}{C}-\overset{|}{\underset{|}{C}}-$

k. (benzene ring with Cl substituent)

l. $-\overset{|}{\underset{|}{C}}-\overset{|}{\underset{\overset{|}{-C-}}{C}}-O-\overset{|}{\underset{\overset{|}{-C-}}{C}}-\overset{|}{\underset{|}{C}}-$

⊙ 34. Draw the structural formula for each of the following compounds. Assume that the carbon chain is straight in each case.

a. hexane
b. hept-1-ene
c. oct-2-yne
d. pentan-1-ol
e. butan-2-ol
f. 1-chloropropane
g. ethyl butanoate
h. hexanoic acid
i. octan-3-one

Chapter Eighteen

35. Name each of the following compounds:

a. $-\underset{|}{\overset{|}{C}}-\underset{|}{\overset{|}{C}}-$

b. $-\underset{|}{\overset{|}{C}}=\underset{|}{\overset{|}{C}}-\underset{|}{\overset{|}{C}}-\underset{|}{\overset{|}{C}}-$

c. $-\underset{|}{\overset{|}{C}}-\underset{|}{\overset{|}{C}}-OH$

d. $-\underset{|}{\overset{|}{C}}-\underset{|}{\overset{|}{C}}-\underset{|}{\overset{|}{C}}-F$

e. $-\underset{|}{\overset{|}{C}}-\underset{|}{\overset{|}{C}}-\underset{\underset{OH}{|}}{\overset{|}{C}}-\underset{|}{\overset{|}{C}}-$

f. $-\underset{|}{\overset{|}{C}}-\overset{\overset{O}{\|}}{C}\underset{O-\underset{|}{\overset{|}{C}}-\underset{|}{\overset{|}{C}}-}{}$

g. $-\underset{|}{\overset{|}{C}}-\underset{|}{\overset{|}{C}}-\underset{|}{\overset{|}{C}}-\underset{|}{\overset{|}{C}}-\overset{\overset{O}{\|}}{C}-OH$

h. $\underset{|}{\overset{N}{\diagdown}}\,CH_3$

i. $-\underset{|}{\overset{|}{C}}-C\equiv C-\underset{|}{\overset{|}{C}}-$

j. $-\underset{|}{\overset{|}{C}}-\underset{|}{\overset{|}{C}}-\underset{|}{\overset{|}{C}}-O-\underset{|}{\overset{|}{C}}-\underset{|}{\overset{|}{C}}-\underset{|}{\overset{|}{C}}-$

k. $-\underset{|}{\overset{|}{C}}-\underset{|}{\overset{|}{C}}-\overset{\overset{O}{\|}}{C}-\underset{|}{\overset{|}{C}}-\underset{|}{\overset{|}{C}}-\underset{|}{\overset{|}{C}}-$

36. Draw an electron-dot structure of propyne. Would you expect this molecule to be polar or nonpolar? Why?

37. Match each of the following structural formulas to the general class of compound that it represents.

amino acid	monosaccharide	fat	polysaccharide
disaccharide	polypeptide	oil	steroid

a.
$$\begin{array}{c} CHO \\ -C-OH \\ HO-C- \\ -C-OH \\ H-C-OH \\ CH_2OH \end{array}$$

b. [steroid structure with HO- group and C_8H_{17} side chain]

c.
$$\begin{array}{c} H \\ H-C-COOH \\ NH_2 \end{array}$$

d. [repeating sugar polymer chain with CH₂OH, OH, H groups, marked "etc." on both ends]

e.

$\sim N-\underset{\underset{\underset{HN}{\overset{|}{C}}=C}{\underset{|}{C}H_2}}{\overset{H}{\underset{|}{C}}} - \underset{O}{\overset{\|}{C}} - N - \underset{\underset{O=C-NH_2}{\underset{|}{C}H_2}}{\overset{H}{\underset{|}{C}}} - \underset{O}{\overset{\|}{C}} - N - \underset{\underset{SH}{\underset{|}{C}H_2}}{\overset{H}{\underset{|}{C}}} - \underset{O}{\overset{\|}{C}} \sim$

f.

$$\begin{array}{l} H-\overset{H}{\underset{|}{C}}-O-\overset{O}{\overset{\|}{C}}-C_{17}H_{35} \\ H-\underset{|}{C}-O-\overset{O}{\overset{\|}{C}}-C_{15}H_{31} \\ H-\underset{|}{\underset{H}{C}}-O-\overset{O}{\overset{\|}{C}}-C_{15}H_{29} \end{array}$$

g. [structure of a disaccharide with two pyranose rings linked by O]

h.

$$\begin{array}{l} H-\overset{H}{\underset{|}{C}}-O-\overset{O}{\overset{\|}{C}}-(CH_2)_7-CH=CH-(CH_2)_7-CH_3 \\ H-\underset{|}{C}-O-\overset{O}{\overset{\|}{C}}-(CH_2)_7-CH=CH-(CH_2)_7-CH_3 \\ H-\underset{|}{\underset{H}{C}}-O-\overset{O}{\overset{\|}{C}}-(CH_2)_7-CH=CH-(CH_2)_7-CH_3 \end{array}$$

DS ⊙ 38. Although cisplatin is effective in halting cell division and the growth of tumors, it does have some side effects that make it a poor treatment choice for some people. Research to find out if there are any similar platinum-based compounds used in chemotherapy.

DS ⊙ 39. The compound cisplatin was discovered in the 1800s and approved for cancer treatment in the 1970s, yet it is still the leading cancer drug. What does this say about cancer research?

MODERN MATERIALS

CHAPTER 19

19A Ceramics	481
19B Polymers	484
19C Nanotechnology	492
Facet	
The Calling of Charles Goodyear	487

DOMINION SCIENCE PROBLEM

A Problem That Never Goes Away

Each American throws away more than 100 pounds of plastic packaging every year, much of it from food wrappers and containers. Most of this ends up in landfills where it will decompose for centuries, if it ever decomposes. In many communities, plastic recycling is unavailable or is not very accessible. Is there a way to reduce the environmental impact, perhaps by creating plastics that will break down in a reasonable amount of time?

19A Ceramics

Technology use has exploded in the last century. People carry around cell phones, MP3 players, and PDAs. The development of better computers, cars, airplanes, home appliances, and simply a better way of life has driven science to explore new frontiers. But do you know what these things are made of? This chapter explores modern materials that enable society to have these technologies.

19.1 Properties of Ceramics

Ceramics are nonmetallic and inorganic solid materials formed by shaping and firing a mixture until it hardens. Because ceramics can be created from a wide range of compounds, they can exhibit many types of chemical bonds. Ionic, covalent, and polar covalent bonds all exist in certain ceramics and contribute to their physical properties. When a ceramic contains a metal and a nonmetal (e.g., MgO), the bonds are primarily ionic. A ceramic that is comprised of a metalloid and a nonmetal is more covalent. An example is boron nitride (BN). Regardless of the bond type, most ceramics are crystalline, with regularly repeating unit cells. Their organized structure causes ceramics to respond to energy changes in predictable and useful ways.

> **19A Section Objectives**
> After finishing this section, you should be able to
> - distinguish between traditional and advanced ceramics.
> - recognize the physical properties shared by most modern ceramics.
> - describe the steps of the sol-gel process.
> - list some common examples of ways ceramics improve our lives.

19.2 The History of Traditional Ceramics

A discussion of ceramics would be incomplete without distinguishing between *traditional* and *advanced ceramics*. Traditional ceramics include pottery and artwork made from clay and mixed with other raw materials. The ratios and purities of these starting materials are significant but not meticulously measured. This material is shaped by a mold, by hand, or on a wheel. It is then sometimes glazed before being fired in a kiln. The firing process drives out the water molecules and forms strong bonds between the clay particles.

19-1 Traditional ceramics, such as this pottery, are made of clay and other raw materials that have been fired.

This use of clay in ceramic figures and useful objects is almost as old as civilization. By the time of ancient Egypt, ceramics had become so commonplace they were used as storage vessels, pots, and dishes. By 3000 BC the invention of glazing made these ceramic vessels even more serviceable.

19-2 During the Ming dynasty, the Chinese created beautiful porcelain objects from ceramics.

> The Greek word from which we get *ceramics* is *keramos*. It literally means "potter's earth."

White porcelain, a much finer ceramic, was developed around AD 600 during China's Tang dynasty. Porcelain is made from kaolin clay mixed with china stone. It is heated to a higher temperature than standard ceramics. Porcelain is less brittle and can therefore be fashioned into thinner objects. This material was perfected during the Ming dynasty, well known for its blue and white porcelain. This technology was carried by the Arabs to Spain and eventually throughout Europe.

19.3 Manufacturing Advanced Ceramics

Advanced ceramics emerged in the twentieth century as an outgrowth of refined laboratory techniques and experimentation with pure forms of elements and compounds.

Advanced ceramics fall into three general categories—oxides, non-oxides, and ceramic-based composites. Some common oxides are aluminum oxide (Al_2O_3) and zirconia (ZrO_2). Because they are already oxidized, these compounds are chemically inert. They have low thermal conductivity and are good electrical insulators. The non-oxide ceramics include a number of nonmetallic compounds such as carbides (e.g., SiC) and nitrides (e.g., BN). Non-oxide ceramics are known for their extreme hardness. Unlike most ceramics, they conduct heat and electricity. Ceramic-based composites are formulated to provide new or enhanced ceramic properties. Some common goals for composite ceramics include lighter weight, increased temperature stability, and shock resistance that overcomes the brittleness of traditional ceramics. **Ceramic-based composites** may be formed from oxide–oxide, non-oxide–non-oxide, or oxide–non-oxide combinations. Some may include other fillers such as ceramic fibers, polymers, or metal particles to impart specific properties.

While there are many variations on the manufacture of ceramics, most follow a similar process. Modern ceramics are usually manufactured using the **sol-gel process**.

Purified starting materials are finely but uniformly ground, based on the crystal size that is desired. Water is added along with binders to give stability. The raw material is then pressed or cast to the desired form and *sintered*, or fired to the appropriate temperatures. Although each type of ceramic requires specific sintering conditions, the temperature may be much lower than the high temperatures required for the production of traditional ceramics. Some ceramics also need to be fired in a vacuum or under high pressure conditions. All of these variables are controlled to affect the final crystalline structure of the ceramic because it is the crystalline structure that controls the material's properties. After cooling, the ceramic product must be finished to the right size by using lasers, grinding, machining, or drilling. The sintering process often causes shrinkage, so the size is imprecise without this final finishing stage.

19.4 Applications of Ceramics

Ceramics have been used for centuries, and they are part of an increasing number of advanced products. As scientists devise more precise methods to purify the starting materials and control their

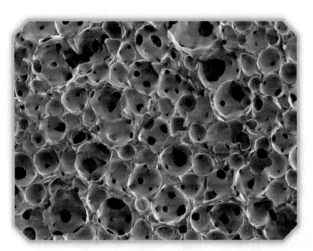

19-3 This artificial bone material, which is made from a ceramic-based composite, contains pores and tunnels through which living bone cells can grow and spread.

Organic polymers are added to some ceramic-based composites to increase flexibility.

manufacture, ceramics with new and more predictable properties are emerging. Ceramic products are used extensively in industry. You can also find ceramics in your car, home, electronic devices, and perhaps even your own body.

Many industries capitalize on the useful properties of certain ceramics. Due to their extreme hardness, some ceramics are used to make drill bits and other cutting tools. They wear much more slowly than typical metal tools. The hardness of ceramics, coupled with their low densities, makes them ideal for the nose cones of rockets and some aircraft. Nonconducting ceramics are great for electrical insulators and products such as light bulb sockets. The ability of most ceramics to withstand high temperatures makes them useful in metal refineries where they may line vats or pipes that carry molten metal. Diesel and jet engines may also include components made from or lined with ceramics.

For many years, ceramics have been used to make spark plugs for automobiles. Their nonconductivity of electricity and their heat resistance make them ideal for this role. Today, ceramics are finding even more applications in cars. Ceramic vibration sensors signal the car's computer system to make adjustments to the engine's firing. Ceramic products also serve as anti-theft sensors in some car doors and as rear sonar systems that warn you if you are about to back into something.

19-4 These silicon nitride ceramic ball bearings are used in race cars and motorcycles because they are lighter yet stronger and more durable than steel. In addition, they require less lubrication.

Most people are familiar with the types of ceramics that are used to make dishes, floor tiles, or wall tiles, but those are not the only places you might encounter ceramics at home. Your furnace may be lined with ceramic plates. Perhaps even your hair dryer has a ceramic heating element. Some high-quality paints today contain tiny ceramic microspheres as filler material. They do not add color, but they do enhance the paint's texture, shine, and resistance to microbes and marks. If you have an alarm system securing your home, the motion or sound detectors and the sensors at doors and windows may contain ceramics.

Many different electronic devices use advanced ceramics. A typical cell phone contains many ceramic capacitors. These miniaturized energy-storage devices are built up from many layers of ceramic material, but they are still as small as a grain of salt. The magnets found inside computers are often made of ceramic material rather than metal. The processor, which is the control center of the computer, also relies on ceramics. The compact media player in which you carry your music contains ceramics. The antenna on a GPS unit may be enhanced by ceramics. Even the beeping alarm in many digital watches is caused by a vibrating ceramic element.

Ceramics can also enhance your health. It has been more than two centuries since ceramics were first used to make artificial teeth. Today, dental ceramics are far more advanced than grandma's dentures. The old silver-tin-mercury amalgam fillings have been largely displaced by more aesthetically pleasing ceramic resin fillings. Crowns and veneers are also made from porcelain ceramics that more closely match the translucence and color of your own teeth. Ceramic dental braces are sometimes used in place of more conspicuous metal brackets.

19-5 Tiny ceramic capacitors, some as small as strawberry seeds, help provide power for cell phones.

19-6 Ceramics are used to make braces that are less conspicuous than the traditional metal ones. Ormco's Inspire ICE™ bracket, seen here, is a ceramic bracket manufactured from pure monocrystalline sapphire, which makes it virtually transparent.

Dentists are not the only medical professionals who recognize the usefulness of ceramics. Artificial bones and hip joints made from ceramics are proving just as strong as earlier metal alloy models. Stents used to keep coronary arteries open after the removal of blockages are often coated with ceramics. These coatings can even contain medicine that is released over time. Hollow ceramic microspheres, scarcely one-fourth the width of a human hair, can carry medicines to remote parts of the body where they dissolve and release their contents. Liver cancer patients may also profit from solid radioactive microspheres. In this targeted radiation therapy, the radioactive particles are injected into an artery that carries them directly to the liver, where their impact is more efficient with fewer side effects.

These advances in ceramics are fulfillments of God's mandate to subdue the earth (Gen. 1:26–28). Humans have taken God's creation and managed it in such a way that they have created many useful products. Creating products like those listed above is God-pleasing work for Christians. Jesus said the second greatest commandment was to love others as ourselves (Mark 12:31), and Christians have the opportunity to love other people by using chemistry to create helpful products for them.

19A Section Review

1. How do traditional ceramics differ from modern, advanced ceramics?
2. List four chemical or physical properties of most advanced ceramics.
3. Why are ceramics hard?
4. List four examples of technologies that use ceramics from which you might benefit daily.
5. Why should a Christian consider becoming a materials chemist?

19B Polymers

19.5 Structure and Properties

Polymers were introduced in Chapter 18. The term *polymer* was coined by Jons Jakob Berzelius in 1827 from the Greek *poly* meaning "many" and *meros* meaning "parts." He used the term to describe large molecules that were formed by the linkage of many lower-mass units called **monomers**. Some polymers are made of identical monomers. Starch is such a polymer because it contains only the monomer glucose. Other polymers are built from collections of two or more different monomers. We call these **copolymers**. Proteins are assembled from an assortment of different amino acids, so they would be considered copolymers. Starch and proteins are natural polymers found in much of our food and in our own bodies, respectively.

As a result of the development of organic chemistry, scientists have created a wide range of artificial or synthetic polymers. In fact, the opportunities are so vast that the polymer industry hires more chemists than any other single field. Through carefully controlled

---

19B Section Objectives

After finishing this section, you should be able to

- explain the relationships among the terms *polymer*, *monomer*, and *copolymer*.
- distinguish between addition and condensation polymers on the basis of their method of formation.
- recognize polyethylene, polystyrene, and rubber as examples of addition polymers.
- recognize nylon and polyester as examples of condensation polymers.
- contrast elastomers and plastics and give examples of each.
- explain how to practically distinguish a thermosetting polymer from a thermoplastic one.
- name two obstacles to plastic recycling.

reactions, chemists can polymerize monomers to create useful products. You are probably familiar with some of them, such as plastics and synthetic fibers like nylon and rayon. Many synthetic polymers owe their strength and usefulness to the strength of the carbon-carbon bonds that form their backbones.

19.6 Addition Polymers

Addition polymers form when monomers containing a double bond open up half of their double bond to create a new single bond to link them into chains. This change frees two electrons to contribute to single bonds on either end of the monomer.

Polyethylene

The formation of polyethylene, a common plastic, from the ethylene monomers is the clearest example of addition polymerization. The spaced dots represent the large number of units, often hundreds or even thousands of monomers, that form the polymer. The ends of the polymer are capped with hydrogen-carbon bonds or similar carbon bonds where carbon forms four bonds.

Polyethylene is one of the most used polymers due to the forms in which its molecules can be structured. More than 40 billion pounds of this polymer are produced annually in the United States. Though the molecule seems simple, its manufacturing process is delicate. The secrets of fashioning the different useful forms of polyethylene include controlling the temperature and the pressure and adding gases and catalysts.

High-density polyethylene (HDPE) is dense and rigid, allowing it to be used for whitewater kayaks, margarine containers, and bottles for milk, juice, and many household cleaning formulas. It is not used for carbonated beverages because it is porous to carbon dioxide and many other gases. The inner lining of cereal boxes is HDPE in a thin

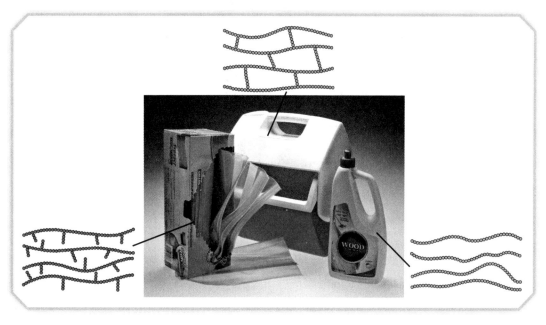

19-7 PEX polyethylene is very firm due to cross linking (center). HDPE, with its linear arrangement, is stiff and strong (right). LDPE is a branched polymer, making it less rigid (left).

film. This polyethylene is considered a **linear polymer** because the molecules are long and lack side chains. This characteristic allows them to pack tightly for more rigidity.

If polyethylene is formed by heating ethylene to 200 °C at a pressure of 200 atm or higher, molecules with side chains of ethylene will form. These branched-chain molecules cannot pack as tightly as HDPE, but instead form low-density polyethylene (LDPE). It is much less rigid, but it is useful for items such as dry cleaning and produce bags. LDPE is a **branched polymer** due to its side chains.

If hydrogen atoms are removed during the formation of polyethylene, side chains can form and link the strands together. Like HDPE, this polymer is tightly packed, but the connections make it even stronger. It is a **cross-linked polymer**. This strong plastic is used for pipes.

Polystyrene

Just as ethylene (ethene) monomers combine to form polyethylene, styrene monomers can link to form polystyrene, another versatile polymer. Like polyethylene, it can be created in different forms for various applications. Pure polystyrene, when melted, can be formed into transparent containers like clear cold drink cups. If instead the polystyrene is formed into beads which are then soaked in pentane, the mixture can be heated, releasing the absorbed pentane liquid and swelling the beads into tiny bubbles. This less dense form of polystyrene is commonly used in ice chests and foam cups for hot drinks.

19-8 These two cups are made from different forms of polystyrene.

Rubber

Natural rubber obtained from the rubber tree of the tropics is another polymer. The repeating monomer is 2-methyl-buta-1, 3-diene (isoprene). Natural rubber is therefore known as polyisopropene.

Although it had many uses among native peoples, it had little practical use in industry because, when heated, the rubber softens and becomes gooey. That changed in 1839 when Charles Goodyear perfected the process known as **vulcanization**. When natural rubber was heated to a high enough temperature with sulfur present, a pair of sulfur atoms would form cross links between adjacent polyisopropene molecules. This all-weather rubber was in great demand a few years later to make rubberized capes and tents for the Civil War.

Since then, political disturbances overseas have often limited the supply of available natural rubber. Organic chemists rose to the task by developing synthetic forms of rubber with similar properties. Styrene-butadiene rubber (SBR) and neoprene rubber are two commonly used substitutes. They are similar to polyisopropene in that they are polymers with cross links, but they are built from different monomers.

19-9 Natural rubber comes from processing the latex drained from the tropical rubber tree, *Hevea brasiliensis*.

THE CALLING OF CHARLES GOODYEAR

Facets of Chemistry

When you hear the name *Goodyear*, you probably think of automobile tires and blimps hovering over a football stadium. However, there are lessons to be learned in the story of the enigmatic man who birthed an industry through a combination of persistence, laboratory errors, and a sense of holy calling.

Charles Goodyear was born in New Haven, Connecticut, in 1800, the son of a hardware dealer. His father invented and marketed his inventions, so perhaps he was the source of Charles's talent for invention.

Goodyear showed strong religious conviction early in life and joined the Congregational Church at age sixteen. He seriously considered studying for the ministry but instead went to Philadelphia to learn the hardware business. His father's hardware firm began to prosper when he was admitted as a partner. However, his generosity led him to unwisely extend credit to the extent that both he and his father were financially ruined by 1834. Rather than declare bankruptcy, he went to the Philadelphia debtors' prison.

Earlier that same year, Charles visited the India Rubber Company in New York City. There he purchased a rubber life preserver and devised some modifications that he thought would make it a more valuable product. When he returned to offer them his invention, he learned that what was needed was a whole new process for manufacturing rubber. Though natural rubber was cheap and easy to form, it melted in the heat of summer and became brittle and hard in the winter cold. Rubber manufacturers were facing bankruptcy, left with warehouses full of useless materials.

Shortly after that, Goodyear felt the call to find a way to make rubber usable. He spent the next quarter of a century doing so, even at the risk of his own health and fortune. He spent much of that time in and out of debtors' prison, often selling off family possessions to buy more raw materials for his experiments.

In 1839, Goodyear accidentally learned how to make rubber usable. When he unintentionally exposed some rubber mixed with sulfur and lead (II) oxide to heat, the sample hardened and became durable. Goodyear had found the key to making rubber hard, yet elastic, at all temperatures. He called the process *vulcanization*, after Vulcan, the Roman god of fire and blacksmithing.

Many businesses used the vulcanization process without paying patent royalties to Goodyear. Though he sued often, Goodyear did not successfully win his patent claims until 1852. Even when he won his legal case, Goodyear still continued his spendthrift ways and continued in debt and debtors' prison.

Goodyear spent his final years promoting the use of rubber. He seemed genuinely convinced that rubber products would revolutionize society.

Charles Goodyear was motivated by an unswerving conviction that God had called him to serve his fellow man through scientific invention. In his own words, "He who directs the operations of the mind can turn it to the development of the properties of Nature in his own way, and at the time when they are specially needed. The creature imagines he is executing some plan of his own, while he is simply an instrument in the hands of his Maker for executing the divine purposes of beneficence to the race." Do you have that calling?

19.7 Condensation Polymers

Obviously, **condensation polymers** are those formed by condensation reactions. You may remember from Chapter 18 that condensation reactions occur when two molecules combine, releasing a smaller molecule, usually water, as the bond forms. Typically, condensation polymers are built from alternating monomers, making them examples of copolymers. Each monomer must have two functional groups so that it is able to link with other monomers at each end.

Nylon

Nylons are good examples of condensation polymers. They form from a diamine and a dicarboxylic acid. A *diamine* is an amine containing two amine groups. A *dicarboxylic acid* is characterized by two carboxyl groups. The numerical suffix in a nylon's name indicates the numbers of carbons donated by the monomers—the diamine first and the acid second. One of the most common forms is nylon 6,6, with the diamine (hexanediamine) and the dicarboxylic acid (adipic acid) each donating six carbons to the polymer chain.

$$n\text{H}-\underset{\text{hexanediamine}}{\text{N}(\text{H})-\text{CH}_2-\text{CH}_2-\text{CH}_2-\text{CH}_2-\text{CH}_2-\text{CH}_2-\text{N}(\text{H})-\text{H}} + n\underset{\text{adipic acid}}{\text{HO}-\overset{\text{O}}{\overset{\|}{\text{C}}}-\text{CH}_2-\text{CH}_2-\text{CH}_2-\text{CH}_2-\overset{\text{O}}{\overset{\|}{\text{C}}}-\text{OH}} \longrightarrow$$

$$\left(-\underset{\text{nylon 6,6}}{\text{N}(\text{H})-\text{CH}_2-\text{CH}_2-\text{CH}_2-\text{CH}_2-\text{CH}_2-\text{CH}_2-\text{N}(\text{H})-\overset{\text{O}}{\overset{\|}{\text{C}}}-\text{CH}_2-\text{CH}_2-\text{CH}_2-\text{CH}_2-\overset{\text{O}}{\overset{\|}{\text{C}}}}-\right)_n + n\text{H}_2\text{O water}$$

Nylon 6,6 was invented by Wallace Carothers at Dupont in 1935. Nylon was put to immediate use in WWII for parachutes, tires, tents, ropes, and other war supplies. In the years since then, nylon, in its various forms, has been fashioned into toothbrush bristles, hosiery, carpet fibers, and even machine parts.

Polyester

Another popular condensation polymer forms from monomers containing ester functional groups. Though the term *polyester* encompasses many different polymers, it is most often applied to polyethylene terephthalate (PET). This polymer contains monomers of terephthalic acid and ethylene glycol.

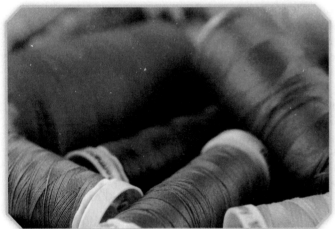

19-10 This strong thread is made of a polyester polymer.

$$n\text{OH}-\text{CH}_2-\text{CH}_2-\text{OH} + n\text{HO}-\overset{\text{O}}{\overset{\|}{\text{C}}}-\!\!\bigcirc\!\!-\overset{\text{O}}{\overset{\|}{\text{C}}}-\text{OH} \longrightarrow$$
$$\text{ethylene glycol} \qquad \text{terephthalic acid}$$

$$\left(-\text{O}-\text{CH}_2-\text{CH}_2-\text{O}-\overset{\text{O}}{\overset{\|}{\text{C}}}-\!\!\bigcirc\!\!-\overset{\text{O}}{\overset{\|}{\text{C}}}-\right)_n + n\text{H}_2\text{O}$$
$$\text{PET} \qquad\qquad \text{water}$$

Polyethylene terephthalate is widely used in many consumer products. In pure form or blended with cotton, this polyester is a fabric whose cross linking makes it wrinkle resistant. It is also used for many plastic films, soft drink bottles, pillow stuffing, and photographic film. Table 19-1 lists information about some common polymers.

19-1 **Common Polymers**			
Polymer	Starting materials	Formation	Uses
Dacron polyester	ethylene glycol and terephthalic acid	condensation	fabric
nylon	6-aminohexanoic acid	condensation	carpets, clothing, fishing line
polycarbonate	sodium bisphenol A	condensation	lenses, baby bottles, phosgene
polyethylene	ethylene	condensation	plastic bags, bottles, toys
polypropylene	propylene	addition	plastic milk containers
polystyrene	styrene	addition	molded objects, foam coffee cups
polyvinyl chloride (PVC)	vinyl chloride	addition	flooring, raincoats, pipes
polytetrafluoroethylene	tetrafluoroethylene	addition	cookware coating

19.8 Plastics and Elastomers

In addition to the reactions that form them, polymers can also be classified into one of two common groups based on the properties of the end product. In other words, the physical properties of the polymer itself allow it to be classified. **Plastics** are polymers that can be formed into different shapes, normally when heated or pressurized. It is nearly impossible to survey your surroundings in any civilized setting and not see plastics. Polymers that exhibit rubbery or elastic behavior after formation are appropriately named **elastomers**. Elastomers can be stretched but then resume their original shape. Rubber is a common example of an elastomer.

Because plastics display so many different properties, some materials scientists prefer to separate plastics into two subgroups. Those plastics that can be heated and then reshaped into other forms are

19-11 These rubber products are elastomers.

19-12 A thermoset plastic was used to make this durable countertop surface.

called **thermoplastic polymers** (or thermoplastics). Plastic milk jugs are made of a thermoplastic polymer because they can be shredded, melted, and then formed into pellets that can be melted again and formed to make other products. **Thermosetting polymers** (or thermosets), on the other hand, are created by irreversible chemical reactions that make them difficult or impossible to melt and reshape. Thermosetting polymers keep their original shape when heated. In fact, the energy provided during their formation usually causes them to cure and become stronger. The two-part epoxy glues and melamine plastic countertop surfaces are thermosets.

The terms from Subsection 19.6 describing the branching and cross linking of polymers relate directly to whether they are thermoplastic or thermosetting. All linear and most branched polymers are thermoplastic because the polymer chains are able to slide back and forth when heated. The cross-linked polymers, however, are thermosetting because they keep their original structure and shape.

19.9 Plastics and Recycling

You have probably recognized that many of the polymers described in this section are used to make containers for consumer products. These plastics are light, durable, and cheap. The drawback is that most are not biodegradable and so contribute to the trash problem. It is estimated that 80 million tons of plastic are used in the United States each year. Plastics are light, but they take up space, perhaps as much as 25% of what goes into landfills. With landfills filling up, recycling has become necessary for many types of polymers. The recycled plastic may then be used to make new containers, fiber filling for jackets, landscape timbers, or even carpeting.

Most plastic consumer products contain a number (1–7) within a recycling triangle. This resin identification coding system was introduced by the North American Society of the Plastics Industry in 1988. Table 19-2 provides some additional data on each of the polymers associated with these codes. These numbers are significant because when plastic is recycled, different types cannot be mixed within a batch. In some communities, only plastics with a 1 or a 2 can be recycled. Others may allow all plastics to be brought for recycling. In either case, these plastics must be sorted before processing.

19-13 Recycling numbers indicate the chemical composition of the material.

19-2 *Plastic Recycling Codes*			
Recycling number	Abbreviation	Polymer name	Density (in g/cm³)
1	PET(E)	polyethylene terephthalate	1.38–1.39
2	HDPE	high-density polyethylene	0.94–0.96
3	PVC or V	polyvinyl chloride	1.32–1.42
4	LDPE	low-density polyethylene	0.91–0.93
5	PP	polypropylene	0.90–0.92
6	PS or S	polystyrene	1.03–1.06
7	other	other plastics including polycarbonate, nylon, and acrylics	varies

Sorting is just the beginning of recycling plastics. The materials must then be separated from other contaminants like rock or glass. They are then washed and chopped into flakes. If mixed plastics are present, flotation can be used to separate the plastics due to their different densities. The flakes are then dried before being melted. While in the molten state, they can be filtered for one last attempt to remove contaminants. The molten plastic is formed into strands through a process called extrusion. After the plastic strands have cooled, they are chopped into pellets. These pellets are the starting material for new plastic products.

With proven technology like this, why do only 5% of plastic products get recycled? The answer to that question is based on economics and human nature. In some cases, the cost of processing post-consumer materials and recycling them costs more than making new plastics from raw materials. That is why in some communities only certain plastics are being recycled. Even in these cases, some of the plastics can be used as fuel for power plants. Since petroleum and natural gas are key ingredients for most plastics, the price of those commodities often dictates whether plastic recycling is financially viable. The other barrier to recycling is that many people do not find it convenient to participate. Recycling costs time and money. Those who participate must clean food containers and sort them from the other garbage. If curbside recycling is not available, they must also take the plastics to a recycling center or drop-off location. Sometimes there is extra cost involved. Nevertheless, Christians are charged with being good stewards of the planet, and plastic recycling may be one way you choose to conserve resources and help preserve the environment.

19-14 These "boards" are mostly recycled plastic. They will last for years without painting or any other treatment.

19B Section Review

1. What are the two main types of reactions through which polymers may be formed? Describe these two processes.
2. Explain the differences in structure seen in the three common polyethylene types: HDPE, LDPE, and PEX.
3. On the basis of molecular structure, explain why vulcanized rubber will not melt when heated though natural rubber will.
4. What type of polymer would a protein be, in addition to being a condensation polymer?
5. List three consumer products made from addition polymers and three products from condensation polymers.
6. Why are cross-linked polymers described as thermosetting materials?
7. How do the slightly different densities of certain plastics allow them to be sorted?
8. What are two significant barriers to plastic recycling?

19C Section Objectives

After finishing this section, you should be able to

- define *nanotechnology* and describe the size scale on which it is applied.
- describe the structure of fullerenes including buckyballs and carbon nanotubes.
- list four ways that fullerenes may be used to improve our lives.
- describe some potential risks of nanotechnology in relation to the environment and human health.

The History of Fullerenes

1985: Buckyballs were discovered.

1990: Researchers discovered an efficient way to manufacture buckyballs. C_{70} and C_{84} molecules were produced.

1991: Sumio Iijima discovered carbon nanotubes.

1992: Chemists found fullerenes in a meteorite and in a natural black rock called shungite. Fullerenes have since been found in a glassy rock known as fulgerite. Fulgerite forms when lightning strikes the ground.

1993: Researchers formed a fullerene lubricant.

1994: Scientists discovered that buckyballs were impervious (or impenetrable) to laser beams. This makes buckyballs of great significance to the military.

1996: The discoverers of buckyballs were awarded the Nobel Prize for Chemistry.

1999: Fullerenes were found within a meteorite that had hit the earth.

2000: Scientists speculated that fullerenes are abundant in the universe, particularly near red giant stars.

2003: Carbon nanotube transistor devices were developed.

19C Nanotechnology

19.10 Introduction

Imagine that you had the power to manipulate atoms or small molecules, one at a time, to construct useful objects. You could even design new devices that could control, carry, or manufacture other substances. That is what **nanotechnology** is all about. This cutting-edge science combines the disciplines of chemistry, engineering, and materials science to create useful new products to improve our lives. Much of the current work is focused on fabricating circuits and devices from individual atoms and molecules.

Exactly how small does something have to be to qualify as *nanoscale*? A nanometer is one billionth (10^{-9}) of a meter. For perspective, the diameter of a human hair averages between 80 000 and 100 000 nm. The width of a DNA helix is about 4 nm. Most manufactured molecules used in nanotechnology are smaller than 100 nm.

When materials can be manipulated and used on the atomic and molecular scale, some very interesting and valuable characteristics emerge, though the laws of physics still apply. Some of these altered properties concern strength, strain resistance, the ability to store other materials, and the conduction of thermal and electrical energy.

Nanotechnology is not limited to a particular element, but the most promising research and products involve carbon atoms. Because carbon has four valence electrons and tends to link to nearby carbon atoms, it is an ideal building material. Other materials under investigation include silicon and boron nitride.

19.11 Fullerenes and the Buckyball

Until 1985 there were only two known natural forms of pure carbon—graphite and diamonds. In that year, American chemist R. E. Smalley, British chemist H. F. Kroto, and their graduate students made a significant discovery while studying the nature of interstellar matter. While vaporizing carbon with a laser, they created sixty-carbon molecules in the shape of soccer balls. Geometrically, these carbon balls are truncated icosahedrons comprised of twenty hexagons and twelve pentagons. A carbon atom is found at each vertex, with a bond along each polygon edge. They called these molecules *buckminsterfullerenes,* or **buckyballs** (C_{60}). Buckminsterfullerenes are named in honor of R. Buckminster Fuller, who pioneered the use of light, strong, unusually shaped domes in architecture. Spaceship Earth at Epcot, Walt Disney World, is one of the most notable of the geodesic structures inspired by his designs.

When the buckyball was discovered, scientists already knew that long chains of carbon were present in space. These chains were identified using readings gathered from radio telescopes. The carbon chains exhibited characteristic readings, much like a fingerprint, on the radio telescope that could be compared to fingerprints of known molecules on the earth. These readings allowed the previously unknown carbon chains to be characterized. R. E. Smalley generated long-chain carbon molecules in order to measure their spectroscopic fingerprints. In the apparatus Smalley created, a laser was aimed at a rotating graphite disk in a helium-filled vacuum chamber. The laser delivered a short, high-energy burst that converted light energy to chemical energy. This rapid, intense heating of the graphite surface caused many carbon bonds in the graphite to break. As a result, carbon atoms were released from the graphite surface and collided in the helium vacuum. New bonding arrangements of carbon were produced, including the buckyball.

Buckyballs are part of a class of carbon molecules called fullerenes. **Fullerenes** are carbon molecules that are arranged in the form of a closed, hollow sphere, cylinder, or the like. Fullerenes have the ability to trap other atoms in the spaces between their carbon atoms. Structures have been created in many different shapes, such as fullerene "diamonds," monster fullerene balls, and honeycomb-shaped tubules. These are all various *allotropes* of carbon. Buckyballs are the only fullerenes to form a hollow sphere. They are electron-deficient alkenes, not aromatic hydrocarbons.

It has been estimated that buckyballs can contain between 30 and 980 carbon atoms in their stuctures. The C_{60} buckyball, however, is the most stable form and the largest possible symmetrical molecule. It was later found that C_{60} buckyballs can form outside of laboratory conditions, even in candle soot. Because of symmetry, the C_{60} buckyball is resistant to high-speed collisions. Although spinning at one hundred million times per second, C_{60} buckyballs can withstand a collision into a stainless steel plate at 15 000 mph.

Buckyballs are very useful because they form stable crystals that don't react. They are nonreactive because all the carbons are bonded to other carbons. When compressed to 70% of its original size, a pure mass of buckyballs becomes more than twice as hard as diamonds. When buckyballs are produced in large amounts, they produce a solid form called fullerite. Fullerite is a transparent yellow solid whose molecules are stacked together in a close-packed arrangement like a pile of cannon balls.

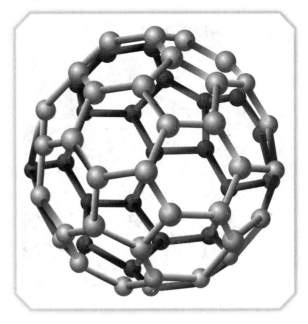

19-15 Buckyballs are composed of hexagons and pentagons of carbon atoms.

19.12 Carbon Nanotubes

In 1991, a Japanese scientist named Sumio Iijima noticed something bizarre in some soot he was viewing through an electron microscope—nanoscopic "threads" made of pure carbon allotropes. At one nanometer (nm) in diameter, carbon nanotubes are much smaller than living cells. Because these tubes are made purely of carbon atoms arranged into a hollow structure, they are also considered to be fullerenes. Originally called *buckytubes*, they are now usually called **carbon nanotubes**.

To understand a carbon nanotube, visualize a sheet of chicken wire and place a carbon atom at every angle in the chicken wire. Then roll up the sheet so that it closes upon itself at the edges seamlessly. You have just formed a long tube made solely of carbon atoms. Some nanotubes are open on both ends, while others are closed on one end like a test tube.

Another variation on carbon nanotubes is the formation of layers in the walls of the tube. Single-walled nanotubes (SWNT), described above, are the most common form, but multi-walled nanotubes (MWNT) also occur. These are concentric tubes of

Photo courtesy of NEC Corporation

19-16 Dr. Sumio Iijima helped to develop more powerful high-resolution electron microscopes and pioneered research in imaging individual atoms before he discovered the carbon nanotube in 1991 at NEC.

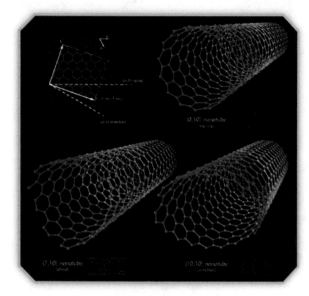

19-17 Carbon nanotubes are cylindrical fullerenes.

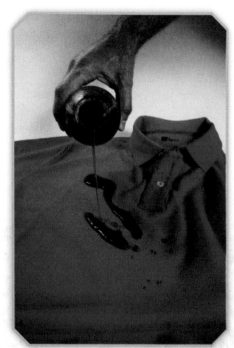

Photo courtesy of Nano-Tex

19-18 This cotton shirt has been transformed through nanotechnology to repel liquid spills.

different diameters. Although they are still formed of carbon atoms arranged in hexagons, MWNTs have some electrical conduction properties very different from SWNTs.

Scientists have learned to alter carbon nanotubes by causing deformations, such as junctions and bends. These irregularities introduce novel and useful electrical properties. Impurities such as catalysts may also be incorporated into the structure to impart other desirable properties.

Applications

Carbon nanotubes and other fullerenes can benefit man in many ways. Some of the products under development include highly efficient batteries; faster, smaller computer processors; brighter, more efficient computer and television displays; targeted medical treatments; biosensors that could detect disease, poisons, or radiation; and many more. Some futurists have even predicted that we could design nanorobots to perform surgery, rearrange molecules, or decontaminate the atmosphere or oceans. So much early research is underway in nanotechnology, especially with carbon nanotubes, that any list would only scratch the surface.

Because these nanotubes are usually about one thousand times as long as they are wide, they exhibit great flexibility. When their strong bonds are considered, they show great promise in the fabrication of lighter yet stronger materials. The interlocking structure of the carbon atoms provides great stiffness at a very low density. Carbon nanotubes can easily be added to polymers to make strong, durable composite materials. If used in vehicles, these could increase both safety, by increasing strength, and fuel efficiency, by decreasing weight.

19.13 Using Polymers and Nanotechnology to Solve Problems

Food packaging is a major component of the plastic entering our landfills. Effective packaging must not react with the oil or water in food, must keep the food fresh, must protect from contamination, and must appeal to the customer. Some packages must also be able to withstand the extreme temperatures of a freezer or a microwave. Many types of petroleum-based polymer plastics have been produced to meet these diverse needs. Regrettably, the needs of the food packaging industry are often at odds with the need for biodegradable materials.

Because traditional plastic polymers are not biodegradable, some materials scientists have engineered new *biopolymers* that will break down under natural processes in a relatively short period of time. Some completely decompose within forty-five days at a commercial composting facility. In the composting process, natural microorganisms such as bacteria and fungi break the bonds in a material, releasing smaller molecules to the environment for reuse. The three most common biopolymers are polyhydroxyalkanoates (PHA), polylactide (PLA), and thermoplastic starch (TPS).

PHA plastic is produced from polyhydroxyalkanoate granules formed inside bacteria. The tiny bacteria are grown in a culture, and the plastic is harvested from them. Starch from corn, wheat, sugar beets, or potatoes is the starting fuel for the PLA biopolymer. Microorganisms are used to change the starch to lactic acid monomers, which are chemically linked into long, polylactic-acid polymer chains. TPS polymers can be up to 100% starch, but smaller amounts of synthetic polymers can be blended in. This polymer is made directly from the plant starch without the need for a microorganism culture.

Although these three types of biopolymers are all biodegradable and reduce the use of fossil fuels, they lack some qualities of traditional packaging polymers. Some become brittle under certain conditions or melt when too warm, and others are too permeable to gases and water vapor. Materials scientists are now exploring the characteristics of biopolymer nanocomposite films. These biopolymers incorporate nanoparticles of natural clay particles as fillers. The tiny clay particles reinforce and strengthen the plastic without affecting its biodegradability.

19-19 The plastics most commonly used for food packaging are not biodegradable.

19.14 Risks

Nanotechnology is more than just speculation; it is already affecting our lives. More than three hundred consumer products contain ingredients created on the nanoscale. It is estimated that the industry will be worth $2 trillion by 2014. Many of these early products will be related to food and medicine. Consequently, there is a call for more research and regulation into any work that involves nanotechnology. Some of the most obvious areas of concern are human health and the environment.

Remember that nano-sized particles of elements or compounds often exhibit very different characteristics than they do on a macro scale. Some substances that are relatively inert become very reactive at the nano level. As a result, we have little research data to predict how our bodies or the environment might be affected. We do know that nanoparticles can enter the body by swallowing, inhalation, or even absorption through the skin. Once inside, they can easily move through tissues and even enter cells and certain cellular organelles. No one really knows the implications of how they could disrupt processes on the cellular level. Unprocessed nanotubes are very light and could become airborne, potentially reaching the lungs through inhalation. It is thought that the "free forms" of nanoparticles pose the greatest risks. Some researchers have concluded that carbon nanotubes are very toxic to laboratory animals—a serious health hazard to those working with them.

Carbon nanotube technology can be used to greatly benefit mankind in the fields previously mentioned but also must be used carefully and responsibly. Just because God has allowed man to create new materials does not mean we need to use them in every possible way. Because of the Fall, sometimes discoveries that at first appear to be helpful are discovered to be harmful. Remember that science cannot make value judgments. Christian chemists ought to be concerned that their inventions help and not harm other people. This need for discernment is an excellent reason for strong Christians to consider ministering to the Lord and to the people He created in the fascinating arena of materials chemistry.

19C Section Review

1. List several properties of matter that may be altered when the material is examined or used on the nanoscale.
2. List and describe two common forms of manufactured fullerenes.
3. List three potential benefits that could come from fullerenes, such as carbon nanotubes.
4. List some potential risks of using fullerenes.
DS 5. Although as a rule the biopolymers are more expensive to produce than traditional polymers, TSP is the least expensive of the three and is cost-competitive. Can you suggest an explanation why TPS would be lower in price?

Serving God as a Patent Attorney

Job Description
A patent attorney obtains and enforces intellectual property rights on behalf of scientists, other inventors, or organizations. Patents are granted by the federal government and allow inventors the right to prevent others from using their inventions for a specific time.

The term *intellectual property* conveys the idea that the subject matter is the result of the use of the mind or the intellect. The term implies that intellectual or academic theories and works (such as the development of a more efficient chemical test or apparatus) are the same as physical property (a car or house). This concept has caused quite a bit of controversy.

A patent attorney's responsibilities include keeping up to date with legal developments in the intellectual property field, researching and discussing inventions and processes with inventors or manufacturers, evaluating whether they are likely to be granted patents, studying and analyzing scientific or technical documents to assess in what ways an invention is new and innovative, writing detailed descriptions of inventions in precise legal and scientific terms, following the application through processing, and possibly tutoring and mentoring trainee patent agents.

Possible Workplaces
Patent attorneys may work in private practice, private firms of patent attorneys, or large manufacturing organizations that are involved with many areas of industry.

Education Needed
Prospective lawyers need not major in a specific field in their undergraduate years before going to law school. However, they should develop proficiency in writing and speaking, reading, researching, analyzing, and thinking logically. These skills are needed to succeed both in law school and in the profession. Regardless of a major, a variety of

humanities classes is recommended. Courses in English, foreign languages, public speaking, government, philosophy, history, economics, mathematics, and computer science, among others, are useful. Students interested in a particular aspect of law may find related courses helpful. For example, prospective patent lawyers need a strong background in engineering or science. As Christians, they need their concepts of science to be filtered by the lens of the Bible.

After earning a bachelor's degree, a student interested in serving the Lord as a patent attorney must prepare extensively to take the LSAT exam. Undergraduate grades and the LSAT exam score are the two strongest factors considered by law schools.

Dominion Opportunities
An experienced, dedicated Christian patent attorney with a biblical worldview serves inventors in the subject areas of chemistry and other disciplines with excellence and godly advice. He or she shows Christian love by explaining the complex patent procedure with kindness. A patent attorney must be honest and objective with facts discovered during research. Many scientists do not have a biblical worldview, which a godly patent attorney could influence. The attorney meets personally with his or her clients for extended periods of time to discuss details of their patent application. This frequent contact is a wonderful opportunity to build relationships, demonstrate the character and love of Christ, and perhaps have occasion to share the gospel!

Chapter Review

Chapter Summary

- Ceramics are solid materials that are nonmetallic and inorganic. Ceramics have several desirable characteristics including hardness and resistance to chemical reactions. Most are considered insulators because they do not conduct heat or electricity well.

- Traditional ceramics are made from clay mixed with other raw materials. Modern advanced ceramics may be classified as oxides, non-oxides, and ceramic-based composites.

- Most advanced ceramics are manufactured through the sol-gel process. The steps include precisely grinding purified materials, adding water and binders, pressing or casting, and sintering or firing the object. The final stage, finishing, may be necessary to properly size the object.

- Ceramics are valuable materials in automobiles, electrical and electronic devices, and even the human body.

- Polymers are large molecules formed by the linkage of many lower-mass units called monomers. If the polymer contains two or more different monomers, it is a copolymer. While there are some common, naturally occurring polymers such as proteins and starch, organic chemistry has yielded scores of useful synthetic polymers with many applications.

- Addition polymers form when monomers containing a double bond open up half of their double bond to create a new single bond that links them into chains.

- Linear polymers are long molecules that lack side chains. Branched polymers have long chains with shorter side chains. Cross-linked polymers have molecules that link together because their side chains bond to each other.

- Rubber is a natural addition polymer that was transformed into a more useful material through vulcanization. In this reaction, the addition of sulfur under high temperatures creates cross links that stabilize the rubber at high and low temperatures.

- Condensation polymers are those formed by condensation reactions. When two molecules bond, a smaller molecule, usually water, is released.

- Plastics are polymers that can be formed into different shapes, usually under heat or pressure. Those plastics that can be heated and reshaped are thermoplastic. The ones that will not melt and retain their shape are thermosetting. Elastomers are polymers that remain elastic or rubbery after formation.

- Plastics are common polymers that benefit our lives in many ways. Many of them are recyclable. A numerical coding system indicates the polymer type.

- Nanotechnology is focused on fabricating circuits and devices from individual atoms and molecules. At this scale, changes occur in some physical properties of the materials including

Coming to Terms

ceramics	481
ceramic-based composite	482
sol-gel process	482
monomer	484
copolymer	484
addition polymer	485
linear polymer	486
branched polymer	486
cross-linked polymer	486
vulcanization	486
condensation polymer	488
plastic	489
elastomer	489
thermoplastic polymer	490
thermosetting polymer	490
nanotechnology	492
buckyball	492
fullerene	493
carbon nanotube	493

strength, strain resistance, and conduction of thermal and electrical energy.
- Carbon, due to its electronic structure, is able to form some interesting and useful three-dimensional structures. These fullerenes include spherical buckyballs and cylindrical carbon nanotubes.
- Fullerenes are finding their way into products ranging from cosmetics to building materials. They may make their most significant impact on electronics and computers.
- Biodegradable biopolymers are being developed to reduce the problems of plastic trash. Most are formed from plant polymers.

Review Questions

Concept Review

1. Name two ways in which ceramics can be considered insulators.
2. What types of chemical bonds can be found in ceramic materials?
3. What are the three general categories of ceramics?
4. Which one of the ceramic types is typically able to conduct heat and electricity?
5. List five major steps of the sol-gel process in making advanced ceramics.
6. List at least three medical applications of ceramics.
7. Explain why proteins are considered copolymers but starch is not.
8. HDPE, LDPE, and PEX are different structure types of which common addition polymer?
9. How does the removal of hydrogen atoms during the formation of polyethylene affect its structure?
10. What is the significance of the number inside a triangle that is found on beverage or food containers?
11. What is the common name for a truncated icosahedron of carbon atoms?
12. What changes in carbon nanotubes are most likely to affect their electrical properties?
13. What feature of carbon nanotubes imparts their great flexibility and usefulness in composite materials?

DS 14. What are some of the weaknesses of the three most popular biopolymers?

True or False

15. Ceramics can be described as having a crystalline structure.
16. Porcelain is a special type of ceramic.
17. Sintering is a process in which ceramics are ground or machined to the desired size specifications.

18. Polystyrene can be clear and hard or white and soft.
19. Nylon is a very versatile addition polymer.
20. Because soft drink bottles can be heated, melted, and recycled, they are thermoplastic polymers.
21. Carbon is the only element used in nanotechnology.
22. Diamonds, coal, and fullerenes are simply different isotopes of carbon.

Application

23. Could you form an addition polymer from a monomer that had only single bonds? Explain your answer.
24. Name the element that is most significant in the structure of polymers and fullerenes and give several reasons for its usefulness.
25. Name two common polymers that this chapter mentions were used in wartime.
26. Would a polymer composed of isoprene monomers be an elastomer or a thermoplastic? Explain your answer.
27. If you ran a plastic recycling facility and were in charge of sorting mixed plastics after they had been cleaned and chipped, what process would you use to separate the PET, PS, and PVC plastic from the other types and why?
28. How many nanometers are in one centimeter?
29. Imagine that you need to produce plastic plates and cups that will break down quickly in a standard landfill situation without any special composting facility. You are trying to decide between PHA and PLA. Which would be the better choice?

Nuclear Chemistry

CHAPTER 20

20A Natural Radioactivity 501
20B Induced Reactions 512
Facet
Radioactive Age-Dating Methods 518

Dominion Science Problem

The Invisible Threat

Most of us feel safe in our own homes. With deadbolt locks, a security system, and smoke detectors, it doesn't seem like there is much to worry about. Believe it or not, many people have all of these safety precautions, yet they are still exposed to an invisible threat that could be greatly increasing their risk for lung cancer. In fact, Americans are seven times more likely to die from this in-home hazard than from a house fire. What is this dangerous substance and what can be done to remedy the situation?

20A Natural Radioactivity

20.1 Introduction

On August 6, 1945, the United States dropped the first nuclear weapon in the history of warfare on the Japanese city of Hiroshima. While no one doubts that nuclear power can be destructive, this same energy source can offer numerous peaceful uses. Radioactive tracers, for instance, are used in medicine to diagnose diseases. Properly constructed nuclear power plants offer a non-polluting, reliable source of electricity.

What causes radiation? How are nuclear reactions different from chemical reactions? Where does the awesome power come from? What makes a nuclear weapon different from a nuclear power plant? The answers to all these questions lie in the remarkable nature of the atomic nucleus.

In the chemical activities studied in the previous chapters, the atom's valence electrons play the crucial role. But this is not so in nuclear chemistry. **Nuclear chemistry** is concerned with **nuclear reactions**, those that take place within the nucleus of an atom. These nuclear reactions have nothing to do with the electrons in the atoms' energy levels. The nuclei have reactions all their own.

20.2 The Discovery of Radiation

Soon after the discovery of x-rays, a French physicist named Henri Becquerel (1852–1908) set out to determine whether some "glow-in-the-dark" crystals would emit x-rays. Becquerel gathered crystals of potassium uranyl sulfate that had been exposed to bright sunlight. These crystals have the strange ability to glow in the dark after they have been exposed to sunlight. To determine whether light from the crystals contained x-rays, he covered a photographic plate with black paper and exposed it to glowing crystals. He reasoned that only x-rays could pass through the paper to expose the film. As expected, the glowing crystals left marks on the plate. Becquerel correctly concluded that potassium uranyl sulfate emits some kind of penetrating ray. Yet Becquerel soon made an unexpected discovery that caused him to wonder just what kind of rays were coming out of the crystals.

Toward the end of his study, Becquerel determined that the crystals left marks on photographic plates even though they had not been exposed to the sun. Further investigations showed that other uranium compounds also darkened photographic plates. The crystals constantly gave off energetic rays despite being melted, dissolved, and recrystallized. No one knew where the energetic rays came from until two other French scientists made a startling discovery.

In 1898, two scientists in France, Marie Sklodowska Curie and her husband, Pierre, discovered another element, polonium, that seemed to defy the law of energy conservation. They coined the word **radioactivity** to describe the spontaneous emission of the penetrating rays from nuclei. The Curies correctly concluded that radioactivity is an atomic property that does not depend on how the atoms are chemically bonded. Becquerel's experimental results were then understood. The uranium atoms within the crystal, not the potassium uranyl sulfate crystals themselves, were responsible for the radiation.

20A Section Objectives

After finishing this section, you should be able to

- relate the events and scientists involved in the discovery of radioactivity.
- describe the charges, masses, penetrating power, and ionizing ability of alpha, beta, and gamma radiation.
- recognize the best applications for particular units of measuring radioactivity.
- complete nuclear equations for radioactive decays.
- calculate the amount of radioactive substance present after a given number of half-lives have passed.
- determine whether a given nuclide is stable or radioactive when given a graph of nuclide stability.

20-1 The French physicist Henri Becquerel (beh KREL) studied phosphorescence of crystals, and, with Pierre and Marie Curie, was awarded the Nobel Prize for Physics in 1903.

20-2 Pierre and Marie Curie studied Becquerel radiation. After Pierre's untimely death in an accident, Marie went on to earn a second Nobel Prize (1911), this time in chemistry for her discovery of radium and polonium. She is the only woman to have won two Nobel Prizes.

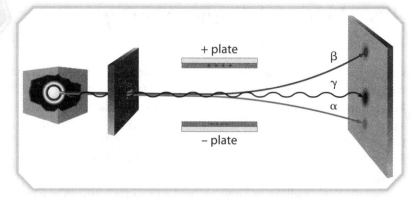

20-3 This experiment displays the charges and relative masses of radioactive products. Positively charged alpha particles are attracted to the negative plate. Beta particles, with their negative charge and low mass, are attracted to a greater degree toward the positive plate. Gamma rays, having no charge or mass, are unaffected.

While Becquerel had actually discovered radioactivity, the Curies correctly identified its source. Today, we use the term **radioisotope** to refer to any radioactive isotope of an element. Some elements are radioactive in their naturally occurring state, while others express this characteristic only in certain isotopic forms.

In time, scientists analyzed the rays from radioactive substances. They passed streams of radiation through powerful electrical and magnetic fields. By charting the deflections of the rays, they could determine the electrical charges of the different types of radiation. They found that radiation had three components: one with a positive charge, one with a negative charge, and one with no charge at all.

20.3 Alpha Particles

A stream of positively charged particles was deflected slightly toward the negative electrical plate. Because the particles swerved only slightly, scientists deduced that they must be more massive than the other types of radiation. They later found that these **alpha (α) particles** contained two protons and two neutrons. Interestingly enough, these were the same particles that Rutherford had used in his famous experiments with gold foil discussed in Chapter 4. Isotopic notation shows that alpha particles are actually $^{4}_{2}$He nuclei. The superscripted number shows that the mass number of the particles is 4, and the subscripted number gives the atomic number of the element. The alpha particle is also called the He^{2+} ion.

While alpha radiation will penetrate matter, it can be stopped easily. The alpha particles will interact quickly with the matter they strike to produce helium atoms. For instance, alpha particles cannot go through this page. If He^{2+} ions hit the page, they would immediately grab two electrons from the paper and turn into ordinary helium atoms. Although alpha particles cannot pass through the skin, they can be ingested or inhaled. Once inside the body, they can cause biological damage as they ionize molecules that are vital to biological processes. The alpha particles may cause burns, sickness, and even death. Some of the superficial burns that resulted from the nuclear weapons dropped on Hiroshima and Nagasaki during World War II were caused by the enormous number of alpha particles released by the bombs.

20.4 Beta Particles

Like alpha radiation, the second type of radiation is also composed of particles. **Beta (β) particles** carry a negative charge and have very little mass. Scientists noted that beta particles acted much like electrons when they swerved sharply toward a positively charged plate. The direction of the swerve identified the particle's charge, and the amount of the swerve indicated that the particle had negligible mass. Further studies gave a surprisingly simple identity to beta particles: they are electrons—electrons that have been formed in and emitted from the nucleus.

How can electrons be in the nucleus? Scientists theorize that neutrons produce beta radiation by breaking apart and then forming a proton and an electron. The proton stays in the nucleus, and the electron is emitted as a beta particle. Beta particles are represented as $_{-1}^{0}\beta$ in this notation. This convention shows both the negligible mass and the –1 charge of the particle.

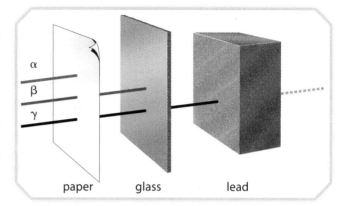

20-4 Alpha particles, beta particles, and gamma rays have different powers of penetration.

Beta radiation penetrates into substances deeper than alpha radiation. Beta particles can zip through sheets of paper, but wood, glass, or thin metal can stop them. Like alpha particles, they ionize atoms and molecules in the matter they hit, but not as readily as alpha particles do. Because they do not cause ionizations as readily, beta particles penetrate deeper and are more dangerous than alpha particles to the inside of the body.

20.5 Gamma Rays

Unlike alpha and beta radiation, gamma radiation is not composed of particles. It consists of electromagnetic waves similar to those of visible and ultraviolet light, but with more energy. **Gamma (γ) rays** are uncharged and are undeflected in an electrical field, as can be seen in Figure 20-3.

If no particles are formed by gamma emission, how are gamma rays produced? Gamma rays are emitted by nuclei in excited states. An excited nucleus is nothing more than a normal nucleus with

> The symbols α, β, and γ are the Greek letters alpha, beta, and gamma, respectively.

20-1 **Three Types of Radiation**						
Name	Symbol	Nuclear notation	Identity	Charge	Mass number	Penetration
alpha	α	$_{2}^{4}He$	helium nucleus	+2	4	low
beta	β	$_{-1}^{0}\beta$	electron	–1	0	medium
gamma	γ	$_{0}^{0}\gamma$	electromagnetic radiation	0	0	high

extra energy. Nuclei in excited states are sometimes produced as the result of alpha and beta emission. An excited nucleus returns to its ground state by releasing energy in the form of gamma rays. The symbol for gamma radiation is $^0_0\gamma$. The zeros indicate that there is no mass and no charge.

Of the three types of radiation, gamma rays are the most harmful. Partly because they have no electrical charge and partly because they are not particles, gamma rays have a lower ionizing ability than alpha or beta particles. As a result, gamma rays penetrate more deeply than either alpha or beta particles, doing more damage. They can, however, be stopped by several feet of concrete or a sheet of lead. Higher energy gammas can be stopped only by thicker layers of these materials.

20.6 Measuring Radiation

Measurement of nuclear radiation involves three types of units, depending on what is being measured—the radioactivity itself, the absorption of that radiation, or the biological effect caused by the radiation. Physical radiation units measure the activity of the radiation source, that is, the number of particles or rays produced per unit of time. This measurement is usually expressed using an SI unit called the **becquerel (Bq)**. A radiation source with an activity of one becquerel has one disintegration per second. Another unit, which is older but still used, is the curie. One **curie (Ci)** is the number of disintegrations per second in a one-gram sample of radium; one curie is 3.7×10^{10} Bq.

Other radiation units measure the amount of energy deposited by nuclear radiation in the absorbing material. The **gray (Gy)** is the SI unit of radiation absorption. A gray is equivalent to the transfer of one joule of energy to one kilogram of living tissue. The **rad** (*r*adiation *a*bsorbed *d*ose) is an older term, though it is still used, and is equal to 0.01 Gy. The **roentgen (R)** (RENT gin) was originally used to measure gamma and x-rays. It expressed the ionizing effect of electromagnetic radiation on a sample of air. A measure of 1 R is equal to 2.58×10^4 coulombs of charge per kilogram of air. Thus, 1 R is 0.0096 Gy—almost 1 rad.

The gray is a measurement of interest to physicists, but has little practical value to scientists concerned about the effect of radiation on living tissue. For that reason, another unit is necessary. Recall that the different types of radiation have different energies and penetration depths. Therefore, a 1 Gy dose of alpha radiation will have a different effect than a 1 Gy dose of gamma radiation. The portion of the body and the type of tissue that is irradiated also factor in to the radiation's effects. To overcome these differences, a unit was devised to measure the effects of different types of radiation on man. That unit, which takes into consideration both the dose and its impact on the tissue is called a **sievert (Sv)**. This unit is named for Rolf Sievert, a Swedish medical physicist who studied the biological effects of radiation. Table 20-2 shows the health effects of different doses of radiation measured in millisieverts.

20-5 Geiger counters detect radiation by the tendency of radiation to ionize gases, allowing them to conduct electricity. This change in electrical conductivity indicates the level of radioactivity.

20-2 Health Effects of Radiation

Dose	Health effects
0–250 mSv	No detectable clinical effect in humans.
250–1000 mSv	Slight short-term reduction in number of some types of red blood cells; severe sickness uncommon.
1000–2000 mSv	Fatigue, nausea; vomiting if >1500 mSv; longer-term decrease in some types of blood cells.
2000–3000 mSv	Nausea, vomiting first day of exposure; two-week latent period followed by malaise, appetite loss, sore throat, diarrhea. Recovery usually in 3 months unless complicated by infection.
3000–6000 mSv	Nausea, vomiting, and diarrhea in first few hours; after one week, fever, loss of appetite, and malaise; inflammation of the mouth and throat, diarrhea, emaciation follow. Some deaths in two to six weeks. Death likely for 50% if exposure was above 4500 mSv; others will recover in about six months.
6000+ mSv	Nausea, vomiting, and diarrhea in first few hours, followed by rapid emaciation and death possibly within two weeks. Death probable for 100%.

20.7 Nuclear Equations

Chemical equations describe the reactants and products in a chemical reaction. **Nuclear equations** describe what occurs when nuclei split, fuse, or release radiation. In some cases, the number of protons changes, so the atom becomes a different element. This change is called **transmutation**.

Nuclear equations are written not with chemical formulas but with an element's symbol preceded by a superscript and subscript. This identifies the particular nuclide. A **nuclide** is a unique atom of an element expressed with *isotopic notation* as $^A_Z X$, where X is the element's symbol, A is the mass number, and Z is the atomic number. Although we refer to atoms with identical atomic numbers but different mass numbers as isotopes of the same element, when discussing the nuclear characteristics of different elements we call them *nuclides*.

Nuclear equations identify the nuclides that react and the products of the reaction. They show whether gamma rays, alpha particles, beta particles, or other particles leave the nuclei.

Alpha Decay

For example, uranium-238 emits alpha particles. Because the alpha particle holds two protons, the atomic number decreases by two, and the nucleus becomes a thorium nucleus ($Z = 90$). Because the departed alpha particle had a mass number (A) of 4, the mass of the remaining nucleus is 234 instead of the original 238. A nuclear equation describes this process.

$$^{238}_{92}U \longrightarrow {}^{234}_{90}Th + {}^4_2He$$

20-6 In alpha decay, a helium nucleus, containing two protons and two neutrons, is released.

Remember that ^4_2He is an alpha particle. Two things can be noted about the nuclear equation for an alpha-emitting process. First, the sum of the atomic numbers of the reactants equals the sum of the atomic numbers of the products (92 = 90 + 2). Second, the sum of the mass numbers on one side of the equation equals the sum of the mass numbers on the other (238 = 234 + 4).

EXAMPLE PROBLEM 20-1
Alpha Decay

Write the nuclear equation that describes the alpha decay of uranium-234.

Solution

Alpha decay implies that an alpha particle is one of the products. Uranium-234 is the reactant.

$$^{234}_{92}U \longrightarrow ^A_ZX + ^4_2He,$$

where X is the unknown element, A is the mass number, and Z is the atomic number.

The sum of the mass numbers and the atomic numbers of the products must equal the mass number and the atomic number of the reactant. From the equation above, $A + 4 = 234$ and $Z + 2 = 92$. Therefore, the product nucleus must be a nuclide of thorium (atomic number = 90) with a mass of 230, or $^{230}_{90}Th$. Thus, the final equation is this:

$$^{234}_{92}U \longrightarrow ^{230}_{90}Th + ^4_2He.$$

Beta Decay

This type of nuclear decay is most likely to occur when the number of neutrons in an atom exceeds the number of protons. Nuclei are most stable when neutrons and protons are appropriately balanced.

A carbon nuclide with six protons and eight neutrons is prone to this kind of reaction.

$$^{14}_6C \longrightarrow ^{14}_7N + ^0_{-1}\beta$$

The mass numbers of carbon-14 and the product (nitrogen-14) are the same. The reacting nucleus does not lose nuclear particles when a neutron changes into a proton. The atomic number of the product nuclide is one greater than that of the reactant nuclide because of the extra proton.

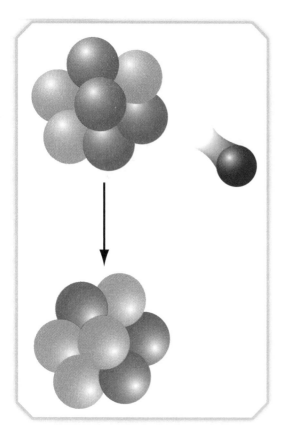

20-7 In the beta decay of C-14, a neutron is converted to a proton and a beta particle (electron), which is emitted.

EXAMPLE PROBLEM 20-2
Beta Decay

Write the nuclear equation for the beta decay of aluminum-28.

Solution

Since beta decay implies the emission of an electron ($^0_{-1}\beta$) from the nucleus, the equation can be set up as follows:

$$^{28}_{13}Al \longrightarrow ^A_ZX + ^0_{-1}\beta.$$

$A + 0 = 28$ and $Z + (-1) = 13$; therefore, $A = 28$ and $Z = 14$. The unknown element is silicon. The final equation is this:

$$^{28}_{13}Al \longrightarrow ^{28}_{14}Si + ^0_{-1}\beta.$$

Positron Emission and Electron Capture

In some cases, a nuclide may be unstable because there are more protons than neutrons. Nuclear chemists have offered two processes in explanation, each of which reduces the number of protons in the nucleus by one.

During **positron emission**, a proton is converted into a neutron and a positron. A **positron** ($^{0}_{+1}\beta$) has a positive charge but the same low mass as an electron. Positrons are emitted from the nucleus during positron emission. Potassium-38 undergoes positron emission in the following transmutation.

$$^{38}_{19}K \longrightarrow ^{38}_{18}Ar + ^{0}_{+1}\beta$$

Notice that the mass is unchanged because the neutron has approximately the same mass as the proton that was lost.

Electron capture occurs when the nucleus pulls in one of the closest electrons (usually one in the lowest level) and combines it with a proton to form a neutron. The process releases an x-ray photon. Silver-106 undergoes electron capture in the following transmutation to become palladium-106.

$$^{106}_{47}Ag + ^{0}_{-1}e \longrightarrow ^{106}_{46}Pd + \text{x-ray photon}$$

In electron capture, the mass again remains the same, but the atomic number is reduced by one.

Gamma Emission

Gamma emission is not accompanied by any changes in mass number or atomic number. The only change that occurs in gamma emission is a change in energy. For instance, an excited technetium-99 nucleus decays to form an unexcited technetium-99 nucleus and a gamma ray.

$$\text{excited } ^{99}_{43}Tc \longrightarrow ^{99}_{43}Tc + ^{0}_{0}\gamma$$

20-3 Radioactive Decay Processes

Process	Particle released	Mass number change	Atomic number change
alpha decay	$^{4}_{2}He$	−4	−2
beta decay	$^{0}_{-1}\beta$	no change	+1
positron emission	$^{0}_{+1}\beta$	no change	−1
electron capture	x-ray photon	no change	−1
gamma emission	$^{0}_{0}\gamma$	no change	no change

20.8 Radioactive Decay Series

Unstable nuclides attain stable states by emitting various types of radioactivity. Some release alpha particles and some release beta particles. Some radioactive nuclides, such as uranium-238, do not stop decaying after just one alpha emission. The thorium-234 that is produced by alpha emission is also an unstable nuclide, and it decays by beta emission.

$$^{234}_{90}\text{Th} \longrightarrow {}^{234}_{91}\text{Pa} + {}^{0}_{-1}\beta$$

The protactinium-234 produced by this reaction is still unstable, and it decays by beta emission.

$$^{234}_{91}\text{Pa} \longrightarrow {}^{234}_{92}\text{U} + {}^{0}_{-1}\beta$$

Uranium-234 is also unstable, and it too decays by a radioactive decay process. Do these decay reactions ever stop? Yes, but not until a stable nuclide is formed. When the product of a nuclear decay is stable, the radiation emissions cease. (They cease for that particular nucleus, not for the whole radioactive object.) Sequential alpha and beta emissions often form long series of nuclear reactions called **radioactive decay series**. When uranium-238 decays, nuclear decay reactions proceed until lead-206 is formed. Figure 20-8 shows the complete radioactive decay series of uranium-238. Depending on the pathway of the decay series, as many as fourteen decays will occur involving nine different elements.

All natural nuclides with more than eighty-three protons (all elements past bismuth on the periodic table) exhibit radioactivity. They all decay according to one of three natural radioactive decay series: the uranium series, the thorium series, or the actinium series. The uranium series begins with uranium-238 and ends with lead-206. The actinium series begins with uranium-235 and ends with lead-207. The thorium series begins with thorium-232 and ends with lead-208.

20-8 The radioactive decay series of U-238. Note the combination of alpa and beta decays.

20.9 Half-Life

How quickly do nuclear reactions occur? It might seem that individual decay steps occur instantaneously. But this is not necessarily so. While some radioactive nuclei release particles and waves readily, others release them only occasionally.

Scientists use the half-life concept to describe the decay rate of a particular isotope. The **half-life** ($t_{1/2}$) of a radioactive element is the amount of time that elapses while half of the nuclei in a sample of an element decay to form another specific nuclide. Each radioactive nuclide has a unique half-life. Short half-lives indicate that the nuclei decay quickly into other kinds of nuclei. Long half-lives show that radioactive decay proceeds more slowly.

The half-life of thorium-234 is 24.1 days. Of 40 g of Th-234, 20 g will decay into protactinium-234 within 24.1 days. How much thorium-234 will be left after two half-lives? It might seem logical to say none, but that is wrong. During each half-life, half of the *remaining* mass decays into another nuclide. There will be 20 g of thorium-234 remaining when the second half-life begins, and 10 g remain after 48.2 days.

20-4 Half-Lives	
Nucleus	**Half-life**
oxygen-13	0.0087 second
bromine-80	17.6 minutes
magnesium-28	21 hours
radon-222	3.8 days
thorium-234	24.1 days
hydrogen-3	12.26 years
carbon-14	5.73×10^3 years
uranium-238	4.51×10^9 years

After another 24.1 days, half of the 10 g will decay, leaving 5 g. Where is the mass going? It is not disappearing, but is being converted into another element. Which other element depends on the type of decay (α, β, or γ) that is occurring. There will always be approximately 40 g of material, but by the end most of it will be protactinium-234.

Example Problem 20-3
Half-Life Calculation

A 192 g sample of thorium-234 decays for 96.4 days. At the end of this time, how much thorium-234 is left?

Solution
Since the half-life is 24.1 days, you know that four half-lives have passed.

$$\frac{96.4 \text{ days}}{1} \times \frac{1 \text{ half-life}}{24.1 \text{ days}} = 4.00 \text{ half-lives}$$

During each half-life, the mass of the thorium-234 in the remaining sample is halved. Since there are four "halvings," one-half is multiplied by itself four times, or raised to the fourth power.

$$\left(\frac{1}{2}\right)\left(\frac{1}{2}\right)\left(\frac{1}{2}\right)\left(\frac{1}{2}\right) = \left(\frac{1}{2}\right)^4 = \frac{1}{16}$$

The amount of thorium-234 left after 96.4 days (four half-lives) is this:

$$\tfrac{1}{16}(192 \text{ g Th-234}) = 12.0 \text{ g Th-234}.$$

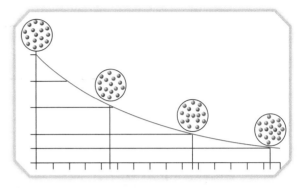

20-9 The mass of thorium-234 decreases exponentially with time.

20-10 Nuclear energy is efficient, but it creates radioactive waste that must be safeguarded.

Spent fuel rods from nuclear power plants contain plutonium-239, which has a half-life of 24 000 years. This raises the question of nuclear waste disposal. Special long-term nuclear waste facilities have to be constructed so that the waste is well contained. Part of the Christian's responsibility toward other humans and the planet is to make wise choices regarding long-term solutions for this material.

20.10 Nuclear Stability

As you learned back in Chapter 4, most of the atom's mass is packed into the incredibly dense nucleus. In fact, a sphere of nuclear material the size of a table-tennis ball would have a mass of 2.5 billion tons. One obvious question that has been ignored so far is "What keeps nuclei together?" Nuclei contain many protons that are packed together. The protons and neutrons found here are collectively called **nucleons**. Since protons are positively charged, they should electrostatically repel each other and immediately fly apart. Yet some nuclei are so stable that they exist for centuries without changing. Other nuclei are so unstable that their half-lives must be measured in units of seconds. Why are some nuclides stable (non-radioactive) and others unstable (radioactive)?

Scientists explain these facts by saying that **strong nuclear forces** hold the nucleons together. Nuclear forces are not well understood. Scientists postulate that these forces are effective only over small distances (less than 10^{-13} cm). Neutrons, being neutral in charge, do not repel each other, nor do they repel the positively charged protons. The strong nuclear forces allow the neutrons to balance the electrostatic repulsion between the protons and stabilize the nucleus.

Scientists say that stable nuclei have effective nuclear forces, while unstable nuclei have insufficient forces to keep them together.

To a certain extent, the stability can be predicted on the basis of the neutron to proton ratio within the nucleus. For smaller atoms ($Z < 20$), the most stable nuclei have a 1:1 ratio between neutrons and protons. As the number of nucleons increases, a surplus of neutrons with their strong nuclear forces is required to balance the repulsive forces between the protons. In some very large nuclei, the ratio may be as high as 1.5:1. The nuclide $^{206}_{82}Pb$ is one such example. With 124 neutrons and 82 protons, its ratio is 1.51:1.

A brief analysis of the stable nuclides reveals an interesting trend. Figure 20-11 represents the set of nuclides that researchers have observed to be stable. If the intersection point on the graph that corresponds to a particular nuclide is shaded in red, that form is stable and not radioactive. Yellow spots on the graph represent nuclides that are unstable. The curved band of points in this figure is called the **band of stability** because it identifies the nuclides that do not undergo radioactive decay. Notice the relationship between the band and the lines showing neutron:proton ratios. The smaller nuclei are stable when the ratio is nearly 1:1, but as the nuclei increase in size, they approach the 1.5:1 ratio. These ratios are indicated by the purple lines on Figure 20-11.

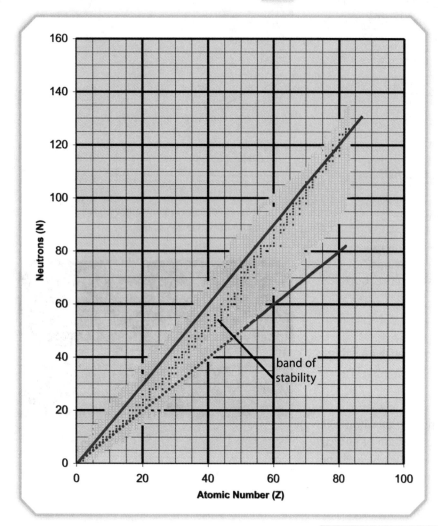

20-11 Stable nuclides

Example Problem 20-4

Nuclear Stability Prediction

Are $^{60}_{27}Co$ atoms radioactive or stable?

Solution

$^{60}_{27}Co$ atoms have twenty-seven protons and thirty-three neutrons. Referring to Figure 20-11, the point on the graph corresponding to the intersection of twenty-seven protons and thirty-three neutrons is not filled in. This represents an unstable, radioactive nuclide.

This same graph can also be used to predict how unstable nuclides will decay. Those with atomic numbers less than 83 that fall above the band of stability have more neutrons than their stable isotopes. It is reasonable, then, that they will transform one of these extra neutrons into a proton and will emit a beta particle. For instance, thallium-208 (eighty-one protons) decays by beta emission to form lead-208 (eighty-two protons), which has a stable nuclear structure. Nuclides with eighty-three or more protons frequently demonstrate alpha decay. In fact, all elements beyond bismuth on the periodic table are radioactive.

While unsaved scientists wonder what strong nuclear forces are, Christians have an authoritative source of information about this seeming mystery. Colossians 1:17 reveals that Jesus Christ "is before all things, and by him all things consist." God is ultimately responsible for holding all the protons in all the atoms of this world together.

Example Problem 20-5

Radioactive Decay Prediction

What type of radioactive decay (if any) are the following nuclides most likely to undergo?

a. $^{38}_{18}\text{Ar}$

b. $^{230}_{90}\text{Th}$

c. $^{35}_{13}\text{Al}$

Solution

a. This nuclide has eighteen protons and twenty neutrons. When plotted on Figure 20-11, the intersection of this example is on the band of stability; therefore, no radioactive decay occurs.
b. Alpha decay is most likely because the atomic number is greater than 82.
c. Beta decay is most likely because the nuclide has less than eighty-two protons, and when plotted on Figure 20-11, it falls above the band of stability.

The protons and neutrons in a nucleus seem to have energy levels similar to the energy levels in the electron shells. The **nuclear shell model** is a nuclear model in which nucleons exist in levels, or shells, that are analogous to the energy levels that exist for electrons. Just as "full" electron shells are more stable than those that are less full, there are "full" nuclear shells that appear to be more stable. The number of nucleons in a full shell is called the **magic number** for that shell. For protons, the magic numbers are 2, 8, 20, 28, 50, and 82. That explains why $_{50}\text{Sn}$ has ten different stable nuclides, while the elements on either side of it, $_{49}\text{In}$ and $_{51}\text{Sb}$, have only one and two, respectively. Neutrons have these same magic numbers, with the addition of 126. The nuclide $^{54}_{26}\text{Fe}$ is stable because it has a magic number of neutrons, even though its number of protons is not "magic." A few nuclides have stable numbers of both neutrons and protons and are considered "doubly magic." For example, $^{40}_{20}\text{Ca}$ is very stable.

20.11 Solving Problems with Radioactivity Detection

Radon-222, found in nearly every building, is a radioactive gas that forms as a decay product from uranium in the ground and water. This radioisotope of radon emits alpha particles as it decays:

$$^{222}_{86}\text{Rn} \longrightarrow \,^{4}_{2}\text{He} + \,^{218}_{84}\text{Po}.$$

Many studies have linked radon to an increased incidence of lung cancer, perhaps to as many as 21 000 cases annually in the United States. Radon gas is all around us. In outdoor air, its concentration is about 0.4 pCi/L (picocuries per liter). It can seep into basements or foundations of houses or even from some building materials. Because many homes have inadequate ventilation, its indoor concentration is usually much higher. The Environmental Protection Agency recommends that corrective measures be taken when the concentration exceeds 4 pCi/L and strongly suggests homeowners correct the problem if it falls between 2 and 4 pCi/L.

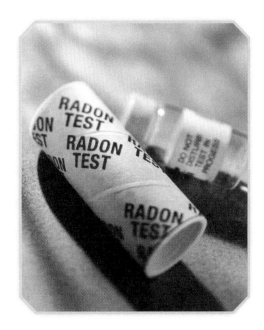

20-12 Homeowners can collect radon with this simple device and mail it to a laboratory for measurement.

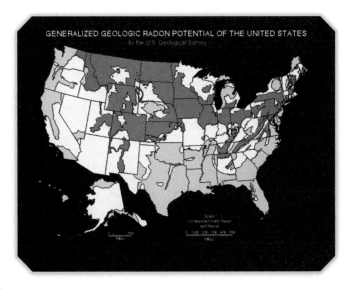

20-13 Radon levels vary from one region to another. Low levels are shown in blue, moderate levels in yellow, and high levels in magenta.

Several different types of radon detection devices are available to homeowners. Some are considered short-term and do their work in a few days, while long-term devices can give a more accurate picture of the radon exposure throughout a year's time. Most devices are purchased from a hardware or builders' store and placed in the home for a prescribed period. They are then sealed up and sent to a laboratory that is able to detect the number of radioactive particles that have been captured. Within a few weeks, the homeowner receives a report on the radon level and can take corrective action if necessary.

As many as one out of fifteen homes has unsafe levels of radon gas. This problem is simple to fix in most cases. Homeowners should improve ventilation, usually by ventilating the soil gases that accumulate beneath the house. The average cost for materials and the installation of this equipment is usually about $1200. In regions with high levels of radon, pipes and fans to ventilate this gas are often included in the original construction.

20A Section Review

1. List the types of natural radioactivity. Which are particles?
2. Fill in the missing nuclide in each of the following reactions:
    a. $^{150}_{64}\text{Gd} \longrightarrow \underline{\quad} + {}^{4}_{2}\text{He}$
    b. $^{245}_{96}\text{Cm} \longrightarrow \underline{\quad} + {}^{4}_{2}\text{He}$
    c. $\underline{\quad} \longrightarrow {}^{255}_{100}\text{Fm} + {}^{4}_{2}\text{He}$
    d. $^{141}_{56}\text{Ba} \longrightarrow \underline{\quad} + {}^{0}_{-1}\beta$
    e. $\underline{\quad} \longrightarrow {}^{32}_{16}\text{S} + {}^{0}_{-1}\beta$
    f. $^{239}_{94}\text{Pu} \longrightarrow \underline{\quad} + {}^{4}_{2}\text{He} + {}^{0}_{0}\gamma$
    g. $^{40}_{19}\text{K} \longrightarrow \underline{\quad} + {}^{0}_{-1}\beta + {}^{0}_{0}\gamma$
3. Barium-122 has a half-life of 2 minutes. A fresh sample weighing 80. g was obtained. If it takes 10 minutes to prepare the experiment, how much barium-122 remains when the experiment begins?
4. Iodine-131 is used in the treatment of a hyperactive thyroid gland. The half-life of iodine-131 is 8 days. If a hospital receives a 200. g shipment of iodine-131, how much will be left after 32 days?

20B Section Objectives

After finishing this section, you should be able to
- distinguish between nuclear fission and fusion.
- explain the relationship between mass of a fissionable substance and a chain reaction.
- compare the stabilities of nuclei when given a graph of nuclear binding energies.
- relate the mass defect of a nucleus to its binding energy.

20B Induced Reactions

20.12 Transmutation Through Nuclear Bombardment

Alchemists dreamed of transforming base metals into gold. Of course, they worked on that project before modern chemistry proved that transformations like this are impossible. Or are they? The transformation of lead into gold requires the removal of three protons and eight neutrons from a lead-208 nucleus. Ordinary chemical reactions cannot alter the nucleus, but particle accelerators can. Lead can be changed into gold, but only at a cost of many millions of dollars' worth of equipment and massive inputs of energy. In other words, it would cost more to make the gold than the gold itself would be worth.

Non-spontaneous nuclear reactions that are forced to occur are called **nuclear bombardment reactions**. In these reactions, a nucleus is struck by a high-energy particle or even another nucleus.

20.13 Fission

Extra neutrons induce some very heavy nuclei to split apart into smaller and more stable nuclei in a process called **nuclear fission**. For instance, uranium-235 can be bombarded with neutrons until the nucleus splits and releases various fragments, some liberated neutrons, and a tremendous amount of energy. There may be more than one way for the nucleus to split. Two of these ways are shown for uranium-235.

$$^{235}_{92}U + ^{1}_{0}n \longrightarrow ^{139}_{56}Ba + ^{94}_{36}Kr + 3\,^{1}_{0}n + \text{heat}$$

$$^{235}_{92}U + ^{1}_{0}n \longrightarrow ^{144}_{55}Cs + ^{90}_{37}Rb + 2\,^{1}_{0}n + \text{heat}$$

20-14 In the Trevaton accelerator at Fermi International Accelerator Laboratory in Illinois, nuclear bombardment occurs within a four-mile tunnel, thirty feet underground.

Even though the atomic and mass numbers all balance, empirical measurement of the products and reactants of fission reveals that some matter is missing. The energy released by nuclear fission is the result of the transformation of a small amount of nuclear mass into energy. According to Albert Einstein's theory of relativity, mass and energy are related by the equation $E = mc^2$, where E is energy, m is mass, and c is the speed of light. Since the speed of light is an astounding 3×10^8 meters per second, the energy released by the conversion of even a small amount of mass into energy is huge.

Chain Reactions

The fission of uranium-235 produces, among other things, three neutrons. What happens to these neutrons? They can either escape from the sample of uranium or they can collide with other uranium-235 atoms. If the three neutrons hit and are absorbed by three other uranium-235 nuclei, the resulting nuclei will immediately undergo fission, releasing a total of nine neutrons.

The neutrons produced by these reactions can, in turn, initiate more nuclear fission reactions, and so on. This ongoing fission process is called a **chain reaction**. It can only occur if the sample of uranium is large enough to intercept many of the released neutrons. If the mass of uranium is too small, that is, if it has a *subcritical mass*, many neutrons will escape without initiating more nuclear fission reactions, and a chain reaction does not occur. The smallest mass of a fissionable substance that can sustain a chain reaction is called the **critical mass** of the substance.

Chain reactions release large amounts of thermal energy because many nuclei are split apart. Each split converts a minute amount of matter into energy. If technicians control the reaction in nuclear reactors, the heat generated may be harnessed and transformed into useful forms of energy (such as electricity). Many industrialized countries have mastered the technology to harness this energy release

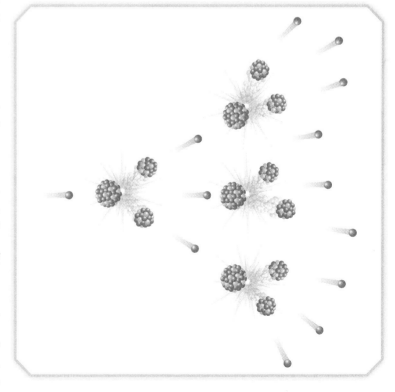

20-15 The fission of U-235 can trigger a chain reaction if a critical mass of uranium is present.

in nuclear power plants. Fission reactions at these plants produce about 20% of the electricity generated in the United States. Some other countries make even greater use of nuclear power for their energy needs. For example, France provides about 80% of its electricity through nuclear reactors.

If the mass of fissionable material is larger than the critical mass, many neutrons are produced. The chain reaction proceeds quickly, even explosively, if it is not controlled. The amount of fissionable material that can support an explosion is called the **supercritical mass**. An atomic bomb uses a supercritical mass of a fissionable substance to achieve extraordinarily large energy releases and great destruction. It is an uncontrolled chain reaction.

20.14 Fusion

Nuclear fusion is the transmutation process of directly combining several light (low atomic number) nuclei to form different, heavier elements. In other words, it is the opposite of fission. You might expect fusion to absorb energy, but instead it releases more energy. The characteristic that fission and fusion have in common is that both result in more stable materials. Thus, they are exothermic.

As with fission, there is a discrepancy of mass between the products of fusion and the reactants. The nuclei that form the larger nucleus have more mass than the product of the reaction. The nuclear

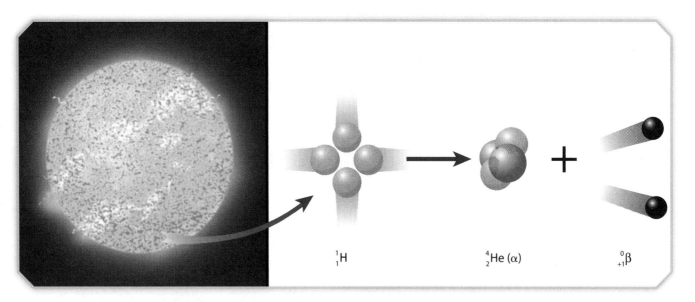

20-16 The fusion of hydrogen nuclei generates the energy released by the sun and other stars.

mass of helium is 4.0015 u. However, the combined masses of the two protons (2 × 1.0073 u) and the two neutrons (2 × 1.0087 u) equal 4.0320 u—a 0.0305 u difference. Just as in fission, this lost mass has been converted into an enormous amount of energy.

Nuclear fusion is important to all living things because it is believed to be the nuclear reaction occurring in our sun and other stars. In the sun, four hydrogen nuclei fuse to form a more stable helium nucleus (alpha particle), two positrons, and a great deal of energy.

Although the concept of fusing two atoms may appear simple, before nuclear fusion can occur, light nuclei must collide with enough kinetic energy to overcome the natural repulsion between nuclei.

Temperatures close to 100 000 000 K are required before nuclei have enough energy for fusion to proceed spontaneously. Because so much thermal energy is required, fusion reactions are sometimes called *thermonuclear reactions*. Needless to say, large amounts of energy must be expended to heat atoms to this temperature. In fact, in the first fusion reactions (hydrogen bombs), atomic bombs had to be used to supply the vast amount of kinetic energy required to initiate the fusion reaction.

Nuclear chemists and physicists hope to control nuclear fusion reactions and to convert their energy into electricity. The formation of helium by the fusion of two hydrogen isotopes (deuterium and tritium) offers particular promise.

$$^{2}_{1}H + {}^{3}_{1}H \longrightarrow {}^{4}_{2}He + {}^{1}_{0}n$$

However, using fusion to produce energy is a formidable problem for researchers because the 100 000 000 K temperatures must be contained. Currently there are no materials that can withstand such enormous temperatures. Researchers have devised two methods to confine the high temperatures. In inertial confinement, a small pellet of frozen hydrogen is compressed and heated by a laser so quickly that fusion occurs before the atoms can fly apart. Another approach is magnetic confinement, in which a strong magnetic field holds the ionized atoms together in a "magnetic bottle" while they are heated by microwaves, x-rays, or other energy sources. Most nuclear scientists agree that the magnetic bottle is the more viable method for commercial energy production. If research proceeds at its current pace, small-scale energy production by fusion could soon become a reality.

Why all the interest in fusion? If scientists are successful in harnessing fusion, it could provide an energy source that is almost inexhaustible. The fuel for fusion would be ordinary water. In every 6500 atoms of water there is one deuterium atom, giving a gallon of water the energy content of 300 gallons of gasoline. There are no hydrocarbon byproducts from fusion. The products of fusion (helium and a neutron) are not radioactive, and compared with the current fission reactors, there is minimal nuclear waste. All of these make fusion an attractive energy source.

20-17 The explosion of a hydrogen bomb demonstrates the great release of energy from nuclear fusion.

Cold Fusion?

Over the past few decades, there have been many reports of controlled nuclear fusion reactions at near-room temperatures and low pressures. These small-scale, tabletop experiments have been conducted by many different individuals using a variety of laboratory setups. If valid, cold fusion would offer enormous practical and financial advantages over the more traditional fission reactors. However, several reputable panels have evaluated the research and determined that the reported results are erratic or invalid. Attempts at replicating the most promising experiments have failed. For the time being, cold fusion remains on the fringe of reputable science.

20.15 Synthetic Elements

Before the atomic age began, scientists knew of no elements with atomic numbers higher than uranium's. The elements following uranium on the periodic table were synthesized with nuclear bombardment reactions. Scientists call elements with atomic numbers higher than ninety-two **transuranium elements**.

Ernest Rutherford, who discovered the nucleus, used nuclear bombardment to cause transmutation in 1919. He bombarded nitrogen with alpha particles and found that protons and an isotope of oxygen were produced.

$$^{14}_{7}N + {}^{4}_{2}He \longrightarrow {}^{17}_{8}O + {}^{1}_{1}H$$

His experiment was the first known induced transmutation. Transmutation can be performed on other light (low atomic number) elements, but (as with Rutherford's experiments) no new elements will be produced. If, however, heavier elements are bombarded, it is possible to produce an element with an atomic number greater than ninety-two.

Soon after discovering that beta decay increased the atomic number of an atom, scientists reasoned that this process could be the key to producing new elements. If uranium could be forced to emit an electron from one of its neutrons, the remaining part of the neutron would become a proton. The added proton would turn the uranium nucleus into element 93. To make a uranium atom emit a beta particle, scientists bombarded a sample of uranium with neutrons to produce uranium-239.

$$^{238}_{92}U + ^{1}_{0}n \longrightarrow ^{239}_{92}U$$

Neptunium was first synthesized in 1940 when uranium-239 atoms underwent beta decay.

$$^{239}_{92}U \longrightarrow ^{239}_{93}Np + ^{0}_{-1}\beta$$

A year later the same technique produced element 94 (plutonium). Larger elements resulted when alpha particles were used as projectiles. Einsteinium and fermium emerged from the fireball of an experimental hydrogen bomb in 1952. More sophisticated techniques and larger projectiles have since produced additional new elements. Scientists produced element 105 (dubnium) in 1970 when they bombarded californium-249 with nitrogen-15 nuclei, and element 112 (ununbium) was produced in a German laboratory in 1996 by bombarding lead atoms with zinc atoms. Presently, all elements up to atomic number 118 (ununoctium) have been produced, except element 117.

20.16 Mass Defect and Nuclear Binding Energy

All nuclei have slightly less mass than the protons and neutrons that comprise them. The idea that energy and mass are interconvertible explains this phenomenon. The difference between the mass of an atom and the total mass of all its components is called the **mass defect** of the nucleus.

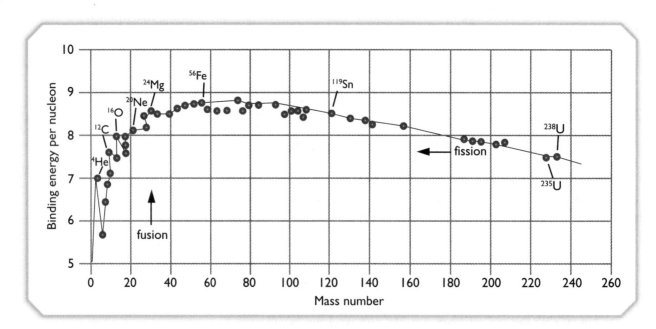

20-18 Binding energy per nucleon peaks near an atomic mass of 56 u.

While scientists do not fully understand the forces that hold a nucleus together, they do know that energy is required to separate all the protons and neutrons in the nucleus. This energy is the same as the energy that is released when a nucleus forms from its nucleons. This energy is called the **nuclear binding energy**. For the purpose of comparing the elements, researchers have determined the nuclear binding energies of various nuclei. They then divided this value by the total number of nucleons to obtain the **binding energy per nucleon**. Figure 20-18 shows the binding energy per nucleon plotted as a function of mass number. It shows that nuclei with mass numbers near 56 have the most binding energy per nuclear particle.

Nuclei with high binding energies per nucleon are more stable than nuclei with lower binding energies per nucleon. As expected, the observed mass defects for the various elements match the observed binding energies. Nuclei with mass numbers near 56 have the most mass defect per nuclear particle. The mass that has "disappeared" (transformed into energy) serves as the nuclear binding energy. These elements have more energy per nucleon holding their nuclei together, so they are more stable.

Example Problem 20-6

Nuclear Binding Energy Comparison

Which of the following nuclei is the most stable: hydrogen-2, arsenic-75, or uranium-235?

Solution

From Figure 20-18 you can see that of the three nuclei mentioned, arsenic-75 has the highest binding energy per nucleon. Therefore, it is the most stable (it is closest to 56 u).

Both fission and fusion will proceed in the direction that produces more stable nuclei as end products. Figure 20-18 helps us to understand this idea. The y-axis displays the binding energy per nucleon, and the x-axis shows the mass number of the elements. The graph peaks at a mass number of 56, demonstrating that the greater the binding energy released per nucleon is, the more stable the nucleus. Any element with the mass number of 56 will achieve the greatest stability possible. Thus, when bombarded by nuclear particles, elements higher than 56 fission into smaller elements to become more stable. Elements lower than 56 fuse when bombarded to become more stable. During both of these processes, vast amounts of energy are produced.

Example Problem 20-7

Fusion or Fission Prediction

Predict which of these are candidates for fusion and which are more likely to undergo fission: hydrogen-3, technetium-97, beryllium-7, and plutonium-239.

Solution

Because hydrogen-3 and beryllium-7 are found to the left of iron-56 on Figure 20-18, they are more likely to undergo fusion. The nuclides technetium-97 and plutonium-239 are strong candidates for fission because they are found to the right of iron-56.

20-19 The explosion of a reactor at the Chernobyl nuclear power plant reminded the world of the risks that must be considered.

20.17 Responsible Citizenship

The awesome power of the atom brings real risks. Nuclear weapons could spread large quantities of radioactive substances throughout the atmosphere. Nuclear fallout, as these airborne radioactive byproducts are called, would spread throughout the earth. Water and food supplies would be contaminated. Since some of the radioactive substances have long half-lives, health problems from the fallout (birth defects and cancer) would continue for years.

Similar difficulties could result if a massive accident occurred at a nuclear power plant. Such an accident did occur in 1986 at a nuclear reactor at Chernobyl in the Ukraine. Due to design and operator errors, a steam explosion caused a rupture and meltdown of one of the four reactors at the site. Although the groundwater was not significantly affected, an intense fire released a massive cloud of radioactive material into the atmosphere. This contamination was carried across the earth by the upper level winds.

A nuclear explosion at a nuclear power plant is highly unlikely because reliable designs and safeguards are used to keep the rate of

RADIOACTIVE AGE-DATING METHODS
Facets of Chemistry

News headlines from October 2007: "New 150 Million-year-old Crab Species Discovered," and "Fossilized Spider, 50 Million Years Old, Clear as Life"!

For more than a hundred years, evolutionists have insisted that the earth is millions or billions of years old and have used radioactive age-dating methods to support these claims. But these radioactive "clocks" do not always tell time accurately.

A "clock" is a device or system in which some component changes at a rate *assumed* to be constant. To be useful, the change must be observable so that one moment in time can be distinguished from another. The clock must also be "set," or calibrated, with another clock. A clock that is not set properly is worthless for telling time, even though it runs at a constant rate.

Radioactive dating techniques attempt to use nuclear reactions as clocks. An unstable radioactive substance, which is the "parent" of one chemical element, naturally decays to form a stable, non-radioactive isotope, or "daughter," of another element. The decay from parent to daughter occurs at a presumably constant rate. The quantities of each substance can be measured and compared.

Problems arise, however, because no one can calibrate the system. The amount of each material present at the beginning of the process cannot be known, and no one can definitely know if the rate of decay was constant. In the absence of these needed measurements, the investigator is forced to guess at the original quantities and assume the rate of decay. This guesswork is not based on observation and is therefore not science. Many of the "scientific" dating methods are actually unscientific.

There are three factors to keep in mind when reviewing age-dating methods:

- Age-dating methods are based on assumptions about the past that cannot be conclusively proven.
- Ninety percent of all age-dating methods give dates far *younger* than secular scientists require for their theories.
- Careful study of biblical chronology yields a date for the universe of only a few thousand years.

Three Commonly Used Age-Dating Methods

The uranium-lead method measures the radioactive decay of uranium-238

fission below the explosive level. However, technical considerations about nuclear plant safety and political concerns regarding the disposal of spent fuel rods have made nuclear energy very costly in the United States.

The storage of the spent radioactive control rods and byproducts of nuclear fission is an ongoing challenge. These **nuclear wastes** have very long half-lives and must be stored securely to prevent damage to the environment or human health. Most of the nuclear wastes of the nuclear power industry in the United States are currently in temporary storage while long-term solutions are being developed.

Considering the tremendous power and potential for evil in nuclear technology, some believe that we should avoid making use of atomic energy. As Christians, we must remember that God made this world originally good. This goodness extends to the physical properties of subatomic particles. God has also commanded us to manage this world—in all of its parts—in a good and wise manner.

There is, however, evil in this fallen world. Certain nuclear reactions produce radiation that may lead to cancer, burns, and death.

---

into lead-206. The greater the amount of lead compared to the amount of uranium, the older the sample is assumed to be. The obvious problem with this method is that much of the lead could already have been present in the original rock. Also, widely differing results are often obtained on different portions of the same sample.

The potassium-argon method is popular because it produces large numbers—older dates than those from the uranium-lead method. This method has the same major problem as the uranium-lead method; much of the argon could have already been present in the original rock.

Carbon-14 is found in essentially all fossilized organic material throughout the geologic column. The radiocarbon method, based on carbon-14, differs from those discussed above because a calibration curve has been set up according to measurements of the radioactivity in several samples of known age. Since it is calibrated against known dates, often using dendrochronology (tree rings), it has scientific legitimacy if it is used properly. In many cases, however, radiocarbon dating is stretched far beyond its reliable limits.

Radioisotopes and the Age of The Earth: The RATE Project

In 1997, a group of six young-earth creationist researchers met and formed a research initiative called the RATE (Radioisotopes and the Age of The Earth) Project. These scientists hypothesized that at some time in the past, accelerated rates of radioisotope decay may have occurred, leading to the production of large quantities of "daughter" products in a short period of time.

It has been suggested that these increased decay rates may have been part of the rock-forming process on the early earth or one of the results of God's judgment upon man following the Creation—the Curse or the Flood. The group's goal was to describe a physical process by which daughter isotopes were formed in a young-earth, biblical timeframe.

For five years beginning in 2000, the RATE group carefully investigated the current popular methods of radioisotope dating—methods that assume that the decay rate of a given isotope is constant, or that it has always been what it is today. A method can be used to estimate a reliable age only if nuclear decay rates have *always* been constant. The RATE researchers' goal was to explain why the radioisotope dating methods are giving such exaggerated ages.

One of the profound results of the RATE Project is that the researchers found strong evidence of accelerated radioactive decay in the past that supports an earth age of only *6000 years*! This evidence came from research into three different phenomena: the leaking of helium from zircon crystals, the presence of C-14 in coal and diamonds, and the presence of radiohalos and fission tracks in crystals, showing an accelerated rate of uranium decay.

The success of the RATE Project demonstrates the degree to which the worldview of the observer influences his research. One who holds an old-earth view would be much less open to exploring any research methods that might challenge that paradigm. Scientists must be willing to conduct accurate, honest research. But, they must also realize that a scientist cannot frame a question or seek to answer a question without appealing to his worldview. The RATE Project shows how careful scientific research combined with a biblical worldview may result in Christian scientists discovering things that other scientists miss.

Safeguards must be established and monitored to prevent this risk. More tragic is the fact that humans tend to make selfish use of the world God has made. Attempts to bypass safety control measures for financial gain show disregard for fellow human beings.

But the Christian is not to abandon the world. He must confront evil to the glory of God. This duty may require working to make nuclear energy more cost-effective or developing safe ways to reuse nuclear wastes. American technology is currently being used in Europe and Asia to safely reprocess spent nuclear fuel in order to make it reusable. It is estimated that the used fuel in storage in the United States, if reprocessed, could power every American home for 12 years. Federal governmental regulations currently prohibit nuclear fuel reprocessing.

Science alone cannot supply the answers to the problems associated with nuclear energy. Men need common sense, a code of ethics, and wisdom in order to apply science correctly. The best source of guidance is God's Word, the Bible. In it, Christians can find the truth and insight necessary to make wise decisions.

20B Section Review

1. Explain the difference between nuclear fission and fusion.
2. Explain the relationship between the terms *critical mass*, *supercritical mass*, and *chain reaction*.
3. Of the following pairs of nuclides, which will be more stable? (Use Figure 20-18.)
    a. oxygen-16, tin-119
    b. magnesium-24, carbon-12
    c. uranium-238, neon-20
    d. tin-119, helium-4
4. Would each of the following nuclides be more likely to undergo fusion or fission? (Use Figure 20-18.)
    a. carbon-12
    b. uranium-235
    c. neon-20
5. What are transuranium elements?

Chapter Review

Chapter Summary

- Nuclear chemistry studies the nuclear reactions that occur in an atom's nucleus.

- Henri Becquerel discovered radioactivity, the spontaneous emission of energy, by exposing photographic plates to crystals in the absence of light. Pierre and Marie Curie realized that the emissions were an atomic property and coined the term *radioactivity*. Isotopes that exhibit this property are called radioisotopes or nuclides.

- Alpha particles are composed of two protons and two neutrons, giving them a +2 charge. They ionize readily on contact with other atoms and have little penetrating power.

- Beta particles are electrons that form in and are released from the nucleus. They have a charge of −1 and very little mass. They have lower ionizing potential than alpha particles but greater penetrating ability.

- Gamma rays are the most energetic of the three types of radiation and are uncharged electromagnetic waves. They have the greatest power of penetration.

- The SI unit for measuring radiation is the becquerel (Bq). The gray (Gy) is an SI unit that quantifies the radiation absorption. Sieverts (Sv) are used to measure the biological effects on living tissue.

- Nuclear equations represent the nuclides that react and the particles or waves that are emitted. In transmutations, a change in the number of protons forms a new element.

- Alpha decay, beta decay, positron emission, electron capture, and gamma emission each cause predictable changes to the charge and mass of an atom. Table 20-3 summarizes these changes.

- Many nuclides undergo a series of nuclear reactions culminating in a stable form. These changes can be represented as a radioactive decay series.

- Half-lives measure the amount of time it takes for half of a sample of a nuclide to decay.

- Strong nuclear forces hold the nucleons (protons and neutrons) together within the nucleus. These forces, which operate over short distances, overcome the repulsive forces between the protons. Nuclear stability is related to the ratio between protons and neutrons.

- The nuclear shell model relates nuclear stability to magic numbers of protons and neutrons.

- Non-spontaneous nuclear reactions can be triggered through nuclear bombardment with particles or even other nuclei. This process can be used to create synthetic transuranium elements.

Coming to Terms

Term	Page
nuclear chemistry	501
nuclear reaction	501
radioactivity	501
radioisotope	502
alpha particle (α)	502
beta particle (β)	503
gamma ray (γ)	503
becquerel (Bq)	504
curie (Ci)	504
gray (Gy)	504
rad	504
roentgen (R)	504
sievert (Sv)	504
nuclear equation	505
transmutation	505
nuclide	505
positron emission	507
positron	507
electron capture	507
radioactive decay series	508
half-life ($t_{1/2}$)	508
nucleon	509
strong nuclear force	509
band of stability	510
nuclear shell model	511
magic number	511
nuclear bombardment reaction	513
nuclear fission	513
chain reaction	513
critical mass	513
supercritical mass	514
nuclear fusion	514
transuranium elements	515
mass defect	516
nuclear binding energy	517
binding energy per nucleon	517
nuclear waste	519

- In nuclear fission, atoms are split to produce more stable nuclides. In nuclear fusion, nuclei are fused to form larger, more stable nuclei. Both of these release a great amount of energy. A chain reaction of controlled fission is harnessed in nuclear power plants.
- The difference between the mass of an atom and the total mass of all its components is called the mass defect or nuclear binding energy.

Review Questions

Concept Review

1. Which types of radiation consist of particles and which type consists of waves?
2. What are the charges of the three types of radiation?
3. Rank the three types of radiation according to decreasing ionizing power and then according to decreasing penetrating power.
4. Explain why the different types of radiation have different ionizing capabilities.
5. Explain the relationship between ionizing power and depth of penetration.
6. Gamma radiation, visible light, and radio waves are all forms of electromagnetic radiation. Why are gamma rays much more dangerous than visible light and radio waves?
7. What are the advantages of fusion over fission?
8. Why do the decay processes in a radioactive decay series not continue indefinitely?
9. Scientific theories are based on observation. What observation led to the concept of strong nuclear forces?
10. Give two ways in which nuclear fission and nuclear fusion are similar and one way in which they are dissimilar.
11. Why do nuclear fission and nuclear fusion both release energy?
12. Why do some nuclei undergo fission but not fusion?

True or False

13. Henri Becquerel discovered radioactivity, but did not name it.
14. Alpha particles have more mass but less penetrating power than beta particles.
15. Fusion processes convert energy into matter.
16. The greater the binding energy is, the less stable the nucleus is.
17. Light nuclei tend to undergo fission in order to become more stable.
18. The mass needed to sustain an explosion is known as the critical mass.
19. The high temperatures of fusion can be confined within magnetic fields.

Application

20. Write the nuclear equation for each of the following:
    a. alpha decay of $^{217}_{86}Rn$
    b. alpha decay of $^{238}_{94}Pu$
    c. beta decay of $^{214}_{83}Bi$
    d. beta decay of $^{234}_{91}Pa$
    e. gamma decay of $^{152}_{63}Eu$
    f. gamma decay of $^{165}_{66}Dy$

21. Complete these nuclear reactions and tell what type of decay has occurred.
    a. $^{242}_{95}Am \longrightarrow {}^{242}_{96}Cm + \underline{}$
    b. $^{220}_{87}Fr \longrightarrow {}^{216}_{85}At + \underline{}$
    c. $\underline{} \longrightarrow {}^{126}_{54}Xe + {}^{0}_{-1}\beta$
    d. $\underline{} \longrightarrow {}^{216}_{85}At + {}^{4}_{2}He$

22. Referring to Table 20-4, which nucleus decays more quickly, bromine-80 or magnesium-28?

23. The half-life of hydrogen-3 is 12.26 years. If 35.8 g of hydrogen-3 are allowed to decay for 36.78 years, what mass of hydrogen-3 will be left?

24. Refer to Figure 20-11, and tell whether the following nuclei are stable or unstable. If they are unstable, predict whether the nucleus will undergo alpha or beta emission.
    a. $^{154}_{66}Dy$
    b. $^{61}_{26}Fe$
    c. $^{232}_{94}Pu$
    d. $^{200}_{80}Hg$
    e. $^{37}_{17}Cl$

25. Determine which nucleus is the more stable by referring to Figure 20-18.
    a. lithium-6, magnesium-24
    b. uranium-235, iron-56
    c. helium-4, uranium-235

Contents

APPENDIXES

A	Physical Constants	525
B	Unit Conversions	525
C	Commonly Used Abbreviations and Symbols	527
D	Element Data	529
E	Electron Configurations of the Elements	531
F	Names and Charges of Common Ions	534

GLOSSARY ... 535

INDEX ... 551

PHOTOGRAPH CREDITS ... 559

Appendix A—Physical Constants

Quantity	Symbol	Traditional units	SI units
atomic mass unit	u	1.66054×10^{-24} g	1.66054×10^{-27} kg
Avogadro's number	N_A	6.022×10^{23} particles/mol	6.022×10^{23} particles/mol
electron rest mass	m_e	9.11×10^{-28} g or 0.00055 u	9.11×10^{-31} kg
fundamental electrical charge	e	1.602×10^{-19} C	1.602×10^{-19} C
molar volume (STP)	V_m	22.4 L/mol	2.24×10^{-2} m^3/mol
neutron rest mass	m_n	1.67495×10^{-24} g or 1.00867 u	1.675×10^{-27} kg
normal boiling point of water	T_b	212 °F	373.15 K = 100 °C
normal freezing point of water	T_f	32 °F	271.15 K = 0.00 °C
proton rest mass	m_p	1.6726×10^{-24} g or 1.0073 u	1.673×10^{-27} kg
speed of light	c	2.9979×10^{10} cm/s	2.9979×10^8 m/s
universal gas constant	R	$\dfrac{0.0821 \text{ L} \cdot \text{atm}}{\text{mol} \cdot \text{k}}$ or $\dfrac{62.36 \text{ L} \cdot \text{torr}}{\text{mol} \cdot \text{k}}$	$\dfrac{8.3145 \text{ J}}{\text{mol} \cdot \text{k}}$

Appendix B—Unit Conversions

The conversion factor required to convert a dimension in the far left column to the dimension in the top row is found in the table where the two dimensions cross. For example, to convert centimeters to inches, you would multiply by 0.3937.

Length	cm	m	km	in.	ft	mi
1 cm =	1	10^{-2}	10^{-5}	0.3937	3.281×10^{-2}	6.214×10^{-6}
1 m =	100	1	10^{-3}	39.37	3.281	6.214×10^{-4}
1 km =	10^5	1000	1	3.937×10^4	3281	0.6214
1 in. =	2.540	2.540×10^{-2}	2.540×10^{-5}	1	8.333×10^{-2}	1.578×10^{-5}
1 ft =	30.48	0.3048	3.048×10^{-4}	12	1	1.894×10^{-4}
1 mi =	1.609×10^5	1609	1.609	6.336×10^4	5280	1

1 angstrom (Å) = 1.000×10^{-10} m   1 ly = 9.461×10^{15} m

Time	s	min	h	d	y
1 s =	1	1.667×10^{-2}	2.778×10^{-4}	1.157×10^{-5}	3.169×10^{-8}
1 min =	60	1	1.667×10^{-2}	6.944×10^{-4}	1.901×10^{-6}
1 h =	3600	60	1	4.167×10^{-2}	1.141×10^{-4}
1 d =	8.640×10^4	1440	24	1	2.738×10^{-3}
1 y =	3.156×10^7	5.260×10^5	8.766×10^3	365.2	1

(continued)

Volume	cm³	m³	L	in.³	ft³
1 cm³ =	1	1.000×10^{-6}	1.000×10^{-3}	6.102×10^{-2}	3.531×10^{-5}
1 m³ =	10^6	1	1000	6.102×10^4	35.31
1 L =	1000	1.000×10^{-3}	1	61.02	3.531×10^{-2}
1 in.³ =	16.39	1.639×10^{-5}	1.639×10^{-2}	1	5.787×10^{-4}
1 ft³ =	2.832×10^4	2.832×10^{-2}	28.32	1728	1

1 fl. gal = 4 fl. qt = 8 pt = 128 fl. oz = 231 in.³

Pressure	atm	mm Hg	torr	Pa	lb/in.²
1 atm =	1	760	760	1.013×10^5	14.70
1 mm Hg =	3.342×10^{-2}	1	25.4	3386	0.4912
1 torr =	1.316×10^{-3}	760	1	133.3	1.934×10^{-2}
1 Pa =	9.869×10^{-6}	101 325	7.501×10^{-3}	1	1.450×10^{-4}
1 lb/in.² =	6.805×10^{-2}	14.7	51.71	6.895×10^3	1

Mass	g	kg	u	oz	lb	t
1 g =	1	0.001	6.022×10^{23}	3.527×10^{-2}	2.205×10^{-3}	1.102×10^{-6}
1 kg =	1000	1	6.022×10^{26}	35.27	2.205	1.102×10^{-3}
1 u =	1.661×10^{-24}	1.661×10^{-27}	1	5.855×10^{-26}	3.660×10^{-27}	1.829×10^{-30}
1 oz =	28.35	2.835×10^{-2}	1.708×10^{25}	1	6.250×10^{-2}	3.125×10^{-5}
1 lb =	453.6	0.4536	2.732×10^{26}	16	1	5.0×10^{-4}
1 t =	9.072×10^5	907.2	5.465×10^{29}	3.2×10^4	2000	1

Density	g/cm³	kg/m³	lb/in.³	lb/ft³
1 g/cm³ =	1	1000	3.613×10^{-2}	62.43
1 kg/m³ =	0.001	1	3.613×10^{-5}	6.243×10^{-2}
1 lb/in.³ =	27.68	2.768×10^4	1	1728
1 lb/ft³ =	1.602×10^{-2}	16.02	5.787×10^{-4}	1

Energy	BTU	J	kcal	cal
1 BTU =	1	1055	0.252	252.0
1 J =	9.478×10^{-4}	1	2.390×10^{-4}	0.2390
1 kcal =	3.9868	4.184×10^3	1	1000
1 cal =	3.9868×10^{-3}	4.184	0.001	1

Appendix C—Commonly Used Abbreviations and Symbols

Unit Abbreviations

Unit	Abbreviation	Dimension	Unit	Abbreviation	Dimension
atmosphere	atm	pressure	kelvin	K	temperature
atomic mass unit (amu)	u	mass	kilogram	kg	mass
			meter	m	length
becquerel	Bq	nuclear activity	minute	min	time
coulomb	C	charge	molality	m	concentration
curie	Ci	nuclear activity	molarity	M	concentration
degree	°	temperature or angle	mole	mol	quantity (particles)
degree (Celsius)	°C	temperature	pascal	Pa	pressure
degree (Fahrenheit)	°F	temperature	percent	%	ratio (per hundred)
foot	ft	length (US)	revolutions per minute	rpm	rate of rotation
gram	g	mass	roentgen	R	nuclear activity
gray	Gy	nuclear activity	second	s	time
hour	h	time	sievert	Sv	nuclear activity
inch	in.	length (US)	yard	yd	length (US)
joule	J	energy or work	year	y	time

Other Abbreviations

Term	Abbreviation	Term	Abbreviation
acidity/alkalinity	pH	electronegativity number	EN
alpha particle	α	electronegativity numbers, difference in	ΔEN
aqueous	(aq)	fundamental electrical charge	e
beta particle	β		
Bose-Einstein condensate	BEC	gamma ray	γ
deoxyribonucleic acid	DNA	gas	(g)
electromagnetic	EM	International Union of Pure and Applied Chemistry	IUPAC
electron	e⁻	kinetic energy	KE

(continued)

Term	Abbreviation	Term	Abbreviation
liquid	(*l*)	significant digit	SD
National Institute of Standards and Technology	NIST	solid	(s)
		standard temperature and pressure	STP
neutron	N		
potential energy	PE	*Système International d'Unités*	SI
proton	p	ultraviolet	UV
ribonucleic acid	RNA	valence shell electron pair repulsion theory	VSEPR

Formula Quantities and Symbols

Quantity	Symbol	Quantity	Symbol
acid ionization constant	K_a	heat capacity	C
activation energy	E_a	ionization constant of water	K_w
area	A	Kelvin temperature	T
atomic number	Z	kinetic energy	KE
Avogadro's number	N_A	mass	m
base ionization constant	K_b	mass number	A
boiling-point elevation	ΔT_b	molal boiling-point constant	K_b
Celsius temperature	t_C	molal freezing-point constant	K_f
constant (general)	k	molar enthalpy of formation	ΔH_f
critical pressure	P_c	molar enthalpy of fusion	ΔH_{fus}
critical temperature	T_c	molar enthalpy of vaporization	ΔH_{vap}
density	D	neutron rest mass	m_n
distance	d	proton rest mass	m_p
electronic charge	e	specific heat	c_{sp}
electron rest mass	e_m	speed of light in vacuum	c
enthalpy of reaction	ΔH	speed (velocity)	v
enthalpy of solution	ΔH_{soln}	standard molar enthalpy of combustion	$\Delta H°_c$
entropy	S	standard molar enthalpy of formation	$\Delta H°_f$
Fahrenheit temperature	t_F	time interval	Δt
free-energy change	ΔG	universal gas constant	R
freezing-point depression	ΔT_f	volume	V
Gibbs free energy	G	wavelength	λ

Appendix D—Element Data

Name	Atomic number	Symbol	Atomic mass	Name	Atomic number	Symbol	Atomic mass
actinium	89	Ac	[227]	francium	87	Fr	[223]
aluminum	13	Al	26.98	gadolinium	64	Gd	157.3
americium	95	Am	[243]	gallium	31	Ga	69.72
antimony	51	Sb	121.8	germanium	32	Ge	72.64
argon	18	Ar	39.95	gold	79	Au	197.0
arsenic	33	As	74.92	hafnium	72	Hf	178.5
astatine	85	At	[210]	hassium	108	Hs	[277]
barium	56	Ba	137.3	helium	2	He	4.003
berkelium	97	Bk	[247]	holmium	67	Ho	164.9
beryllium	4	Be	9.012	hydrogen	1	H	1.008
bismuth	83	Bi	209.0	indium	49	In	114.8
bohrium	107	Bh	[264]	iodine	53	I	126.9
boron	5	B	10.81	iridium	77	Ir	192.2
bromine	35	Br	79.90	iron	26	Fe	55.85
cadmium	48	Cd	112.4	krypton	36	Kr	83.80
calcium	20	Ca	40.08	lanthanum	57	La	138.9
californium	98	Cf	[251]	lawrencium	103	Lr	[262]
carbon	6	C	12.01	lead	82	Pb	207.2
cerium	58	Ce	140.1	lithium	3	Li	6.94
cesium	55	Cs	132.9	lutetium	71	Lu	175.0
chlorine	17	Cl	35.45	magnesium	12	Mg	24.31
chromium	24	Cr	52.00	manganese	25	Mn	54.94
cobalt	27	Co	58.93	meitnerium	109	Mt	[268]
copper	29	Cu	63.55	mendelevium	101	Md	[258]
curium	96	Cm	[247]	mercury	80	Hg	200.6
darmstadtium	110	Ds	[281]	molybdenum	42	Mo	95.94
dubnium	105	Db	[262]	neodymium	60	Nd	144.2
dysprosium	66	Dy	162.5	neon	10	Ne	20.18
einsteinium	99	Es	[252]	neptunium	93	Np	[237]
erbium	68	Er	167.3	nickel	28	Ni	58.69
europium	63	Eu	152.0	niobium	41	Nb	92.91
fermium	100	Fm	[257]	nitrogen	7	N	14.01
fluorine	9	F	19.00	nobelium	102	No	[259]

(continued)

Name	Atomic number	Symbol	Atomic mass	Name	Atomic number	Symbol	Atomic mass
osmium	76	Os	190.2	strontium	38	Sr	87.62
oxygen	8	O	16.00	sulfur	16	S	32.07
palladium	46	Pd	106.4	tantalum	73	Ta	181.0
phosphorus	15	P	30.97	technetium	43	Tc	[98]
platinum	78	Pt	195.1	tellurium	52	Te	127.6
plutonium	94	Pu	[244]	terbium	65	Tb	158.9
polonium	84	Po	[209]	thallium	81	Tl	204.4
potassium	19	K	39.10	thorium	90	Th	232.0
praseodymium	59	Pr	140.9	thulium	69	Tm	168.9
promethium	61	Pm	[145]	tin	50	Sn	118.7
protactinium	91	Pa	231.0	titanium	22	Ti	47.87
radium	88	Ra	[226]	tungsten	74	W	183.8
radon	86	Rn	[222]	ununbium	112	Uub	[285]
rhenium	75	Re	186.2	ununhexium	116	Uuh	[291]
rhodium	45	Rh	102.9	ununoctium	118	Uuo	[294]
roentgenium	111	Rg	[272]	ununpentium	115	Uup	[288]
rubidium	37	Rb	85.47	ununquadium	114	Uuq	[289]
ruthenium	44	Ru	101.1	ununtrium	113	Uut	[284]
rutherfordium	104	Rf	[261]	uranium	92	U	238.0
samarium	62	Sm	150.4	vanadium	23	V	50.94
scandium	21	Sc	44.96	xenon	54	Xe	131.3
seaborgium	106	Sg	[266]	ytterbium	70	Yb	173.0
selenium	34	Se	78.96	yttrium	39	Y	88.91
silicon	14	Si	28.09	zinc	30	Zn	65.41
silver	47	Ag	107.9	zirconium	40	Zr	91.22
sodium	11	Na	22.99				

Appendix E—Electron Configurations of the Elements

Symbol	Atomic number	1s	2s	2p	3s	3p	3d	4s	4p	4d	4f	5s
H	1	1										
He	2	2										
Li	3	2	1									
Be	4	2	2									
B	5	2	2	1								
C	6	2	2	2								
N	7	2	2	3								
O	8	2	2	4								
F	9	2	2	5								
Ne	10	2	2	6								
Na	11		Neon core		1							
Mg	12				2							
Al	13				2	1						
Si	14				2	2						
P	15				2	3						
S	16				2	4						
Cl	17				2	5						
Ar	18	2	2	6	2	6						
K	19							1				
Ca	20							2				
Sc	21						1	2				
Ti	22						2	2				
V	23		Argon core				3	2				
Cr	24						5	1				
Mn	25						5	2				
Fe	26						6	2				
Co	27						7	2				
Ni	28						8	2				
Cu	29						10	1				
Zn	30						10	2				
Ga	31						10	2	1			
Ge	32						10	2	2			
As	33						10	2	3			
Se	34						10	2	4			
Br	35						10	2	5			

(continued)

Symbol	Atomic number	1s	2s	2p	3s	3p	3d	4s	4p	4d	4f	5s
Kr	36	2	2	6	2	6	10	2	6			
Rb	37											1
Sr	38											2
Y	39					Krypton core				1		2
Zr	40									2		2
Nb	41									4		1
Mo	42									5		1
Tc	43									6		1
Ru	44									7		1
Rh	45									8		1
Pd	46									10		
Ag	47									10		1
Cd	48									10		2

Symbol	Atomic number		4d	4f	5s	5p	5d	5f	6s	6p	6d	7s
In	49	Kr core	10		2	1						
Sn	50		10		2	2						
Sb	51		10		2	3						
Te	52		10		2	4						
I	53		10		2	5						
Xe	54		10		2	6						
Cs	55		10		2	6			1			
Ba	56		10		2	6			2			
La	57		10		2	6	1		2			
Ce	58		10	2	2	6			2			
Pr	59		10	3	2	6			2			
Nd	60		10	4	2	6			2			
Pm	61		10	5	2	6			2			
Sm	62		10	6	2	6			2			
Eu	63		10	7	2	6			2			
Gd	64		10	7	2	6	1		2			
Tb	65		10	9	2	6			2			
Dy	66		10	10	2	6			2			
Ho	67		10	11	2	6			2			
Er	68		10	12	2	6			2			
Tm	69		10	13	2	6			2			
Yb	70		10	14	2	6			2			
Lu	71		10	14	2	6	1		2			
Hf	72		10	14	2	6	2		2			
Ta	73		10	14	2	6	3		2			
W	74		10	14	2	6	4		2			

(continued)

			4d	4f	5s	5p	5d	5f	6s	6p	6d	7s
Re	75	Kr core	10	14	2	6	5		2			
Os	76		10	14	2	6	6		2			
Ir	77		10	14	2	6	7		2			
Pt	78		10	14	2	6	9		1			
Au	79		10	14	2	6	10		1			
Hg	80		10	14	2	6	10		2			
Tl	81		10	14	2	6	10		2	1		
Pb	82		10	14	2	6	10		2	2		
Bi	83		10	14	2	6	10		2	3		
Po	84		10	14	2	6	10		2	4		
At	85		10	14	2	6	10		2	5		
Rn	86		10	14	2	6	10		2	6		
Fr	87		10	14	2	6	10		2	6		1
Ra	88		10	14	2	6	10		2	6		2
Ac	89		10	14	2	6	10		2	6	1	2
Th	90		10	14	2	6	10		2	6	2	2
Pa	91		10	14	2	6	10	2	2	6	1	2
U	92		10	14	2	6	10	3	2	6	1	2
Np	93		10	14	2	6	10	4	2	6	1	2
Pu	94		10	14	2	6	10	6	2	6		2
Am	95		10	14	2	6	10	7	2	6		2
Cm	96		10	14	2	6	10	7	2	6	1	2
Bk	97		10	14	2	6	10	9	2	6		2
Cf	98		10	14	2	6	10	10	2	6		2
Es	99		10	14	2	6	10	11	2	6		2
Fm	100		10	14	2	6	10	12	2	6		2
Md	101		10	14	2	6	10	13	2	6		2
No	102		10	14	2	6	10	14	2	6		2
Lr	103		10	14	2	6	10	14	2	6	1	2

Appendix F—Names and Charges of Common Ions

1−	2−	3−
acetate, $C_2H_3O_2^-$	carbonate, CO_3^{2-}	arsenate, AsO_4^{3-}
amide, NH_2^-	chromate, CrO_4^{2-}	arsenite, AsO_3^{3-}
azide, N_3^-	dichromate, $Cr_2O_7^{2-}$	borate, BO_3^{3-}
bromate, BrO_3^-	hydrogen phosphate, HPO_4^{2-}	phosphate, PO_4^{3-}
bromite, BrO_2^-	oxalate, $C_2O_4^{2-}$	phosphite, PO_3^{3-}
chlorate, ClO_3^-	peroxide, O_2^{2-}	
chlorite, ClO_2^-	silicate, SiO_3^{2-}	

		1+
cyanate, OCN^-	sulfate, SO_4^{2-}	
cyanide, CN^-	sulfite, SO_3^{2-}	ammonium, NH_4^+
dihydrogen phosphate, $H_2PO_4^-$	tetraborate, $B_4O_7^{2-}$	hydronium, H_3O^+
hydrogen carbonate, HCO_3^- (bicarbonate)	thiosulfate, $S_2O_3^{2-}$	

		2+
hydrogen sulfate, HSO_4^- (bisulfate)		mercury (I), Hg_2^{2+}
hydrogen sulfide, HS^-		
hydrogen sulfite, HSO_3^-		
hydroxide, OH^-		
hypobromite, BrO^-		
hypochlorite, ClO^-		
hypoiodite, IO^-		
iodate, IO_3^-		
iodite, IO_2^-		
nitrate, NO_3^-		
nitrite, NO_2^-		
perbromate, BrO_4^-		
perchlorate, ClO_4^-		
periodate, IO_4^-		
permanganate, MnO_4^-		
thiocyanate, SCN^-		

Glossary

A

absolute zero (p. 37) 0 K; the theoretical temperature at which all molecular motion ceases; the coldest temperature possible.

accuracy (p. 55) A numerical evaluation of how close a measurement is to an accepted reference or theoretical value; the measurement of error.

acidic solution (p. 403) A solution with more H_3O^+ ions than OH^- ions, resulting in a pH less than 7.

acid-ionization constant (K_a) (p. 407) Related to the ionization constant of water, the value that describes the extent of the forward equilibrium reaction in the formation of the hydronium ion.

acoustic energy (p. 32) *See* **sound energy**.

actinide series (p. 112) A portion of the seventh series of the periodic table that includes the inner transition metals from actinium to lawrencium.

activated complex (p. 355) A theoretical, unstable, energetic group of reactants that forms as a transitional structure during a chemical reaction.

activation energy (E_a) (p. 355) The minimum amount of kinetic energy that must be possessed by colliding molecules before they can react.

activity series (p. 205) A table of metals or nonmetals arranged in order of descending chemical activity.

actual yield (p. 235) The measured amount of product at the end of a reaction; normally expressed in grams of the product.

addition polymers (p. 485) Form when monomers containing carbon-carbon double (or triple) bonds open their double bonds and create new single bonds available for the addition of new monomers; also called *chain-growth polymers*.

addition reaction (p. 464) An organic reaction in which one reactant joins another reactant at the site of a double or triple bond.

adhesion (p. 287) The attraction between the particles of a liquid and the particles of other materials.

alchemy (p. 10) The ancient study of transmutations between base metals and gold, sickness and health, age and youth, or even earthly and supernatural existence.

alcohol (p. 456) An organic compound of the general form

$R-\overset{|}{\underset{|}{C}}-OH$, having a covalently bonded –OH functional group attached to a nonaromatic group.

aldehyde (p. 458) Any organic compound of the general form

$R-\overset{|}{C}=O$, having an aldehyde group in its structure.

aliphatic compound (p. 447) An open-chain compound and any cyclic compound whose bonds resemble those of an open-chain compound.

alkali metal (p. 120) A Group 1 (1A) metal; has one valence electron.

alkaline-earth metal (p. 120) A Group 2 (2A) metal; has two valence electrons.

alkane (p. 447) An open-chain, aliphatic hydrocarbon that contains only single bonds.

alkene (p. 450) An open-chain, aliphatic hydrocarbon that contains at least one carbon-carbon double bond.

alkyl group (–R group) (p. 448) A group of bonded atoms that can be thought of as an alkane with one hydrogen atom missing.

alkyl halide (p. 455) An organic compound that contains an alkyl group and a halogen as a functional group.

alkyne (p. 450) An open-chain, aliphatic hydrocarbon containing at least one triple bond.

allotrope (p. 284) One of two or more forms of a polymorphic element that exists in the same physical state. For example, O_2 and O_3 are allotropic forms of oxygen.

alloy (p. 154) A mixture of the atoms of a metal with another element where the mixture has metallic properties.

alpha particle (α) (p. 502) The nucleus of the helium atom (two protons, two neutrons); represented by He^{2+} or $He\text{-}4^{2+}$.

amalgam (p. 301) An alloy of mercury with some other metal.

amide (p. 461) An organic compound of the general form

$R-\overset{O}{\overset{\|}{C}}-NH_2$, in which an amine is substituted for an –OH group in a carboxylic acid.

amine (p. 461) An organic compound which can be thought of as an ammonia molecule whose hydrogen atoms have been replaced by other atoms or groups of atoms.

amino acid (p. 469) The building block of a protein; a carboxylic acid containing an amino group.

amorphous (p. 127) Without shape, form, type, or structure; solid in which the particles occur in random positions with no orderly pattern, for example, asphalt and paraffin.

amphoteric substance (p. 410) A substance that can act as either a Brønsted-Lowry acid or a Brønsted-Lowry base.

analog instrument (p. 57) A nondigital metric instrument that measures by comparing the object or phenomenon being measured directly to a scale (e.g., a ruler or dial) or displays the reading using a pointer against a scale (e.g., a triple-beam balance).

analytical chemistry (p. 262) The techniques by which chemists devise equipment and methods to discover what is in a sample of material and to quantify its constituents.

anhydrous (p. 194) The form of a hydrate that has lost the water from its crystalline structure.

anion (p. 98) A negatively charged ion.

anode (p. 434) The electrode at which oxidation occurs during an electrochemical reaction; the electrode that attracts anions because of its positive charge.

antibonding orbital (p. 165) Orbitals that form when atomic orbitals combine in an unfavorable manner.

apothecary (p. 9) An early pharmacist who prepared and sold a wide variety of chemicals and herbs.

applied science (p. 12) Exploring natural products and processes for specific applications to benefit humans.

aromatic compound (p. 447) Benzene and any compound that has a structure resembling benzene's characteristic ring structure; aromatic compounds are all cyclic compounds with clouds of delocalized electrons.

Arrhenius acid (p. 398) A substance that releases hydrogen ions (H^+) into aqueous solutions.

Arrhenius base (p. 398) A substance that releases hydroxide ions (OH^-) into aqueous solutions.

aryl group (–Ar group) (p. 455) A group of bonded atoms that can be thought of as an aromatic compound with one hydrogen atom missing.

aryl halide (p. 455) An organic compound that contains an aryl group and a halogen as a functional group.

atom (p. 28) A neutral particle with a centrally located nucleus consisting of protons and neutrons with electrons around it; the smallest representative unit in an element.

atomic (p. 39) Pertaining to whole atoms.

atomic number (Z) (p. 78) The number of protons in the nucleus of an atom.

atomic radius (p. 113) The distance from the center of an atom's nucleus to its outermost electron; measured using x-ray diffraction.

atomism (p. 10) Democritus's naturalistic theory that all matter is made up of particles.

Aufbau principle (p. 89) The principle stating that the electron configuration of an atom may be obtained by building on the electron configuration of an atom of lower atomic number; the electrons fill the sublevels in the order given by the diagonal rule.

auto-ionization (p. 402) *See* **self-ionization**.

autoprotolysis (p. 402) *See* **self-ionization**.

Avogadro's law (p. 259) A law of gas behavior stating that the volume of a gas, maintained at a constant temperature and pressure, is directly proportional to the number of moles of the gas.

Avogadro's number (N_A) (p. 216) The number of carbon atoms in exactly 12 g of ^{12}C; 6.022×10^{23}.

Avogadro's principle (p. 259) The principle stating that equal volumes of gases at the same temperature and pressure contain equal numbers of molecules.

azimuthal quantum number (l) (p. 83) The number that defines the symmetrical shapes arranged around the nucleus within which electrons may be found; sometimes called the *sublevel quantum number*.

B

balanced chemical equation (p. 197) A chemical equation in which coefficients are arranged to show the conservation of mass in a reaction.

band of stability (p. 510) The group of stable nuclei represented on a graph of atomic numbers versus number of neutrons.

barometer (p. 246) An apparatus that measures atmospheric pressure by allowing it to support a column of liquid, for example, mercury.

base-ionization constant (K_b) (p. 407) Related to the ionization constant of water, the value that describes the extent of the forward equilibrium reaction in the formation of the hydroxide ion.

base unit (p. 50) Seven units in the SI, each of which measures a basic dimension, from which all other metric dimensional units are derived, for example, meter, kilogram, second, kelvin, mole, ampere, and candela.

basic solution (p. 404) A solution with fewer H_3O^+ ions than OH^- ions, resulting in a pH greater than 7; also called an *alkaline solution*.

battery (p. 436) One or more voltaic (galvanic) cells arranged to produce electricity.

becquerel (Bq) (p. 504) The SI unit of measure describing the radioactivity of a substance; 1 Bq = 1 disintegration per second.

bent (p. 170) A structural arrangement in which two particles are bonded to a central particle nonlinearly.

beta particle (β) (p. 503) An electron that has been formed in and emitted from a nucleus.

bias (p. 5) A person's preference, whether intentional or unintentional, for certain conclusions over others.

big bang theory (p. 121) The most popular evolutionary theory for the origin of the universe; suggests that all matter in the universe was condensed to a tiny point and expanded rapidly in a giant explosion that started our universe some 13.7 billion years ago.

binary acid (p. 194) An acid molecule consisting of hydrogen and one other nonmetal in which hydrogen is the first element named, for example, HCl.

binary covalent compound (p. 188) A compound consisting of only two different types of usually nonmetal atoms.

binding energy per nucleon (p. 516) The nuclear binding energy of a nuclei divided by the total number of nucleons.

biomass (p. 177) Organic products of living and newly deceased plants and animals.

biopolymers (p. 494) Biodegradable polymers that may be used as plastics and packaging. Four main types are produced from starch, sugar, cellulose, or synthetic materials.

boiling (p. 289) A physical change from the liquid state to the gaseous state caused by the application of heat that occurs when the vapor pressure of a liquid equals the prevailing atmospheric pressure.

boiling point (p. 290) The temperature at which the vapor pressure of a liquid equals the applied atmospheric pressure.

boiling-point elevation (ΔT_b) (p. 313) A raising of the boiling point of a substance caused by the presence of solute particles.

bond axis (p. 161) A region of high electron density that forms on a line and connects two nuclei.

bonding orbital (p. 165) Orbitals that form when electrons reinforce each other.

bonding pair (p. 143) Two shared electrons that make up a pair.

Bose-Einstein condensate (BEC) (p. 40) The state of matter that exists at temperatures close to absolute zero.

Boyle's law (p. 249) A law of gas behavior stating that the volume of a fixed quantity of a confined gas is inversely proportional to its pressure when its temperature is held constant ($PV = k$).

branched polymer (p. 486) Low-density polyethylene with side chains formed when ethylene is heated to 200 °C above a pressure of 200 atm.

bridge notation (p. 53) A special notation for multiplying and dividing several measurements together at the same time; usually used when converting units.

Brønsted-Lowry acid (p. 399) A substance that donates protons.

Brønsted-Lowry base (p. 399) A substance that accepts protons.

Brownian motion (p. 321) The random, chaotic movements of microscopic particles in a colloidal dispersion.

buckminsterfullerene (p. 492) See **buckyball**.

buckyball (C_{60}) (p. 492) The most common and stable of the fullerene molecules, consisting of 60 carbon atoms that are arranged in the shape of a soccer ball.

buffer (p. 418) A solution that can receive moderate amounts of either acid or base without a significant change in pH.

C

calibrate (p. 49) To accurately subdivide into measurement units.

Calorie (Cal) (p. 38) Equivalent to the kilocalorie; used in reference to the energy content of foods.

calorie (cal) (p. 38) The amount of energy required to raise the temperature of 1 g of water 1 °C.

calorimeter (p. 327) An insulated container in which a thermometer detects the temperature change that occurs during a chemical reaction in order to calculate enthalpy.

capillary action (p. 288) The movement of a liquid up a narrow tube caused by the attraction of the molecules in the walls of the tube for the molecules of the liquid.

carat (p. 154) A unit of weight equal to 200 mg.

carbohydrate (p. 466) A polyhydroxy aldehyde or ketone; a compound that can be hydrolyzed to form a polyhydroxy aldehyde or ketone.

carbon nanotubes (p. 493) Nanoscopic fullerene tubes made of pure carbon atoms that have highly beneficial uses in technology; originally called *buckytubes*.

carbonyl group (p. 457) A carbon atom with a doubly bonded oxygen atom attached to it.

carboxyl group (p. 459) A carbon atom with a doubly bonded oxygen atom and a hydroxide group attached to it.

carboxylic acid (p. 459) An organic compound of the general form $R-\underset{\underset{O}{\|}}{C}-OH$, having a carboxylic acid functional group in its structure.

catalyst (p. 199) A substance that changes a reaction rate without being permanently changed by the reaction.

cathode (p. 434) The electrode at which reduction occurs during an electrochemical reaction; the electrode that attracts cations because of its negative charge.

cathode rays (p. 75) The stream of electrons emitted from the cathode in a cathode-ray tube.

cation (p. 98) A positively charged ion.

Celsius scale (p. 36) A temperature scale, proposed by Anders Celsius, that divides the range from the freezing point of water (0 °C) to the boiling point of water (100 °C) into 100 increments and labels absolute zero as −273 °C.

ceramic-based composites (p. 482) One of the three general categories of advanced ceramics formulated to provide new or enhanced ceramic properties—high strength and high temperature tolerance; may be formed from oxide and nonoxide ceramic combinations.

ceramics (p. 481) Used extensively in industry; nonmetallic and inorganic solid materials formed by shaping and firing a mixture until hard; crystalline in structure with regularly repeating unit cells.

chain reaction (p. 513) A self-sustaining fission process in which neutrons produced from fission reactions cause more fission reactions.

Charles's law (p. 251) A law of gas behavior stating that when the pressure on a sample of dry gas is held constant, the Kelvin temperature and the volume are directly proportional ($V/T = k$).

chemical bond (p. 140) A force of attraction that holds atoms together in compounds; an attraction produced by the transferring or sharing of electrons.

chemical change (p. 25) A change in the identity of a material that results in a different material with a different composition and properties. See **chemical reaction**.

chemical energy (p. 32) The potential energy stored in the chemical bonds between atoms that is released or absorbed during chemical reactions.

chemical equation (p. 196) An expression that represents a chemical reaction by using chemical formulas, chemical symbols, and coefficients.

chemical equilibrium (p. 372) The state of balance attained in a reversible reaction in which the forward and reverse reactions proceed at the same rate.

chemical formula (p. 29) Chemical symbols arranged to represent molecules or formula units that make up compounds. Formulas show both the identity and number of atoms from each element.

chemical property (p. 25) A property of matter that describes how one substance reacts in the presence of other substances.

chemical reaction (p. 196) A change in which a substance loses its physical characteristics and becomes one or more new substances. See **chemical change**.

chemical subscript (p. 29) A number written at the lower right of a chemical symbol, indicating the number of atoms or groups of atoms (within parentheses) in a chemical formula. For example, O_2 indicates that two oxygen atoms make up an oxygen molecule.

chemical symbol (p. 29) A one-, two-, or three-letter representation for an element.

chemistry (p. 8) The study of the composition and properties of matter and the energy transformations accompanying changes in the fundamental structure of matter.

chemoreception (p. 176) The physical and chemical response of a sensory organ to the presence of a chemical.

chitosan (p. 176) A long, complex, positively charged sugar molecule that can be extracted from the shells of shrimp and other crustaceans; accelerates erythrocyte clotting.

Christian worldview (p. 6) A perspective on viewing the world based on the teachings of the Christian Scriptures.

chromosome (p. 471) A strand of DNA combined with proteins; it is usually formed within the nucleus of the cell.

coagulation (p. 321) See **flocculation**.

coefficient (p. 30) A number that appears in front of a chemical formula and indicates how many atoms, ions, molecules, or formula units are present. For example, $2O_2$ has two oxygen molecules, and $4K_2SO_4$ has four potassium sulfate molecules.

cohesion (p. 287) The attraction between the particles of a liquid.

colligative property (p. 313) A physical property of a solution that depends only on the number of particles present, without regard to type.

collision theory (p. 354) The theory stating that molecules and atoms must undergo forceful, properly oriented collisions before they can react.

colloid (p. 319) A mixture of small particles (between 1 and 1000 nanometers in size) that remain dispersed in a medium, for example, fog, foam, aerosols, and gels.

combined gas law (p. 254) A law of gas behavior that combines Boyle's law, Charles's law, and Gay-Lussac's law, where the temperature, pressure, and volume are constant; $P_1V_1/T_1 = P_2V_2/T_2$.

combustion (p. 25) A chemical reaction that yields heat and occasionally light; also called *burning*.

commodity chemicals (p. 215) Relatively inexpensive yet important chemicals used in industry that are sold in bulk, for example, ethanol.

common ion effect (p. 390) An equilibrium phenomenon in which two or more substances dissolve and release a common ion, thereby decreasing the ionization of the weaker electrolyte.

complex reaction (p. 362) The combination of the individual elementary steps of a reaction.

compound (p. 29) A substance that consists of atoms of different elements chemically bonded together.

compressibility (p. 245) The physical ability of a substance (usually a gas) to decrease its volume to fit its container.

concentration (p. 308) A measure of the number of solute particles in a certain volume or mass of the solvent, or of the solution itself.

condensation (p. 42) A physical change from the gaseous state to the liquid state.

condensation polymers (p. 488) Also called *step-growth polymers*. Formed by condensation reactions where two molecules combine, releasing a smaller molecule, usually water, as the bond forms.

condensation reaction (p. 464) An organic reaction in which two compounds combine with each other by losing water; the opposite of hydrolysis.

conductivity (p. 25) A physical property of matter indicating the ability to transfer heat or electrons through a substance.

conjugate acid (p. 400) The structure formed when a base is protonated.

conjugate base (p. 400) The structure formed when an acid has donated a proton.

conjugate pair (p. 399) Two particles that differ from each other by only a hydrogen ion.

continuous spectrum (p. 78) A complete visual spectrum with no frequency gaps, emitted only by an ideal luminous object.

controlled experiment (p. 16) An experiment where only one condition is varied at a time to isolate and measure its effect on the outcome.

conversion factor (p. 53) A ratio that is constructed from the relationship between two units and is equal to 1, for example, 1 kg/1000 g.

copolymers (p. 484) Polymers formed from the bonding of two or more different monomers.

corrosion (p. 429) The chemical destruction of a metal by its immediate surroundings.

covalent bond (p. 141) Two atoms held together in an ion or molecule by their attraction for the same pair of shared electrons.

Creation Mandate (p. 3) God's first commandment to mankind: to exercise good and wise dominion over the earth by managing and using His world and its resources (Gen. 1:26, 28).

criss-cross method (p. 184) A technique for writing the formula for binary ionic compounds that uses the absolute value of the oxidation number of one ion as the subscript for the other element.

critical mass (p. 513) The smallest mass of a fissionable substance that can sustain a chain reaction.

critical pressure (P_c) (p. 294) The pressure required to liquefy a gas at its critical temperature.

critical temperature (T_c) (p. 292) The highest temperature at which a gas can be liquefied. Each gas has its own characteristic critical temperature.

cross-linked polymer (p. 486) A polymer that is covalently bonded at various positions to nearby linear polymers resulting in a stronger molecule. Also called *network polymers* or *thermosets*.

cryogenic (p. 292) Relating to the science or technology of very low temperatures, i.e., below −150 °C.

crystal (p. 280) A solid in which the particles occur in a regular repeating pattern, for example, salt, sugar, and monoclinic sulfur.

crystal lattice (pp. 151, 282) An extensive three-dimensional structure of points or objects that represents the regular alternating pattern of atoms or ions.

crystalline solid (p. 280) *See* **crystal**.

curie (Ci) (p. 504) A measure of radioactivity based on the nuclear decay of radium; 1 Ci = 1 disintegration/sec of 1 gram of radium.

cyclic aliphatic compound (p. 451) An aliphatic, organic compound whose carbon chains are bonded in ring shapes, for example, cycloalkanes and cycloalkenes.

D

Dalton's law of partial pressures (p. 255) A law of gas behavior stating that the total pressure of a mixture of gases equals the sum of the partial pressures of the constituent gases; $P_{total} = P_1 + P_2 + P_3 + \ldots P_n$.

data (p. 58) Information collected through observation; any recorded observation that is a detailed description or measurement.

de Broglie's hypothesis (p. 81) The concept that if waves can behave like particles, then particles can behave like waves.

decomposition reaction (p. 203) A chemical reaction of the general form AB ⟶ A + B in which a reactant breaks down into two or more simpler products.

deductive reasoning (p. 13) Arguments from general, accepted statements called premises to more specific conclusions.

delocalized electrons (pp. 151, 453) Electrons that can move between several different bonds. In benzene, delocalized electrons can move throughout circular spaces above and below the plane of bonded nuclei.

density (p. 14) A measure of the concentration of matter; it is expressed as a ratio of the object's mass to its volume.

deposition (pp. 42, 282) A physical change directly from a gaseous state to a solid state, for example, snowflakes and frost.

deprotonation (p. 399) The process of losing a proton; Example: $H_2SO_4 \longrightarrow H^+ + HSO_4^-$.

derived unit (p. 50) Any mathematical combination of the base units. The SI contains only twenty-two derived units, for example, square meter, cubic meter, hertz, and joule.

descriptive chemistry (p. 118) The study of elements and the compounds they form that stresses identification of properties rather than theoretical calculations.

diagonal rule (p. 88) A mnemonic device that gives the energy levels and sublevels in the order in which they are filled.

diamine (p. 488) An amine containing two amine groups.

diatomic element (p. 28) An element whose atoms bond into two-atom units. Some examples are N_2, O_2, H_2, Cl_2, F_2, and Br_2.

dicarboxylic acid (p. 488) A condensation polymer characterized by two carboxyl groups.

diffusibility (p. 288) The physical ability of a liquid to mix spontaneously with another similar liquid by particle motion as determined by intermolecular attractions.

diffusion (p. 34) Spontaneous, uniform mixing caused by particle motion.

digital instrument (p. 57) A measuring instrument that has a display consisting only of numerical digits, for example, a digital thermometer.

dilute (p. 308) A term referring to the relatively low amount of solute in a solvent.

dimension (p. 49) Any measurable property of the physical universe. Examples include length, mass, time, volume, and temperature.

dipeptide (p. 470) Two amino acids joined by a peptide bond.

dipole-dipole force (p. 276) An intermolecular force; the attraction of the positive end of one polar molecule to the negative end of another polar molecule.

dipole moment (μ) (p. 174) The vector sum of the distance between charges and the strength of the charges; a measure of all bond polarities that exist within a molecule.

diprotic acid (p. 410) An acid that can donate two protons.

disaccharide (p. 477) A carbohydrate composed of two monosaccharide units, for example, sucrose.

dispersed phase (p. 319) The particles in a colloidal mixture.

dispersing medium (p. 319) The medium in a colloidal mixture.

dispersion (London) force (p. 277) An electrostatic attraction that arises between atoms or molecules because of the presence of instantaneous and induced temporary dipoles; the only intermolecular force acting on nonpolar molecules.

dissociation (p. 302) The process in which a solvent disrupts the attractive forces in a solute that is an ionic compound and pulls the solute apart into ions.

distillation (p. 291) A technique used to separate a mixture by evaporating its components at their boiling points and then condensing and collecting the vapors.

dominion, have (p. 3) To reign, to rule over.

dominion science (p. 3) Scientific activity that seeks to obey the Creation Mandate for the glory of God and for the benefit of other humans.

double replacement reaction (p. 206) A chemical reaction of the general form AY + BZ $\longrightarrow$ AZ + BY, in which the cation of one compound combines with the anion of another compound, and vice versa.

ductility (p. 25) A physical property of matter describing its ability to be drawn into a wire.

dynamic equilibrium (p. 289) A chemical equilibrium in which there is a continuation of two or more opposing events occurring at the same rate but resulting in no net change. For example, the balancing processes of evaporation and condensation.

E

effervescence (p. 308) The escape of a gas from a liquid-gas solution.

effusion (p. 244) A process in which gas particles pass through a tiny opening into an evacuated chamber or space.

elastomers (p. 489) Polymers that exhibit rubbery or elastic behavior after formation.

electrical energy (p. 32) The ability to do work through the action of the electromagnetic force on and by electrical charges.

electrochemical cell (p. 434) An apparatus consisting of two electrical contacts (electrodes) that are immersed in an electrolyte solution and joined by a wire.

electrochemistry (p. 432) The field of chemistry that involves redox reactions and electricity; the interactions between chemical and electrical phenomena.

electrode (p. 434) A conductor that allows an electrical current to enter or leave an electrolytic cell (or other apparatus).

electrolysis (p. 434) A process that uses electricity to force an otherwise nonspontaneous chemical reaction to occur in an electrochemical cell; the separation of a compound into simpler substances by an electrical current.

electrolyte (p. 397) A substance that releases ions and conducts electricity when it dissolves in water.

electrolytic cell (p. 434) An electrochemical cell used to split compounds by electrolysis.

electromagnetic (EM) energy (p. 32) The combined action of electrical and magnetic energies in the form of wavelike, radiant energy, for example, visible light. (Electricity and magnetism are different forms of EM energy.) Also called *radiant energy*.

electron (e^-) (p. 75) A particle with a −1 charge and a mass of 0.00055 u found orbiting the nucleus in an atom.

electron affinity (p. 115) The amount of energy required or released when an electron is added to a neutral atom to form a negative ion.

electron capture (p. 507) A process that occurs when a nucleus pulls in one of its closest electrons and combines it with a proton to form a neutron, and an x-ray photon is released.

electron configuration (p. 89) A representation of how electrons are positioned in an atom: a number indicates the principal energy level, a letter indicates the sublevel, and a superscript denotes the number of electrons contained within the sublevel.

electron-dot notation (p. 97) A chemical symbol with surrounding dots representing the valence electrons of an atom.

electronegativity (p. 116) A measure of the tendency of bonded atoms to attract electrons; designated by a small decimal called the electronegativity number (EN).

electron-sea theory (p. 151) A theory that offers an explanation of how metals bond; the valence electrons of atoms are said to be freely shared among all atoms. Also called the *free-electron theory*.

electron-spin quantum number (m_s) (p. 84) The number that describes the two possible "spin" states of a pair of electrons in an orbital.

electroplating (p. 436) An electrochemical process where metallic ions are forced to cling to another metal by means of an electrical current.

electrostatic force (p. 144) The field force exerted by electrical charges. It may be repulsive or attractive depending on the kinds of charges interacting.

element (p. 27) A substance whose atoms all have the same atomic number and that cannot be broken down by ordinary chemical means into anything that is both stable and simpler.

elementary steps (p. 361) The two or more possible steps that make up a reaction mechanism.

empirical (p. 16) Describes an explanation determined or data gathered by experimentation, for example, empirical data.

empirical formula (p. 219) A formula that tells the types of atoms that are present in a compound and the simplest whole number ratio between the atoms. For example, the empirical formula of C_2H_4 is CH_2.

endothermic (p. 36) A term describing a process that absorbs heat energy.

end point (p. 416) In a titration, the point at which some change in a property of the solution is detected.

energy (p. 31) The ability to do work.

enthalpy (*H*) (p. 327) The energy (heat) content of a system at a constant pressure.

enthalpy (heat) of solution (ΔH_{soln}) (p. 302) The net energy change during the dissolving process; expressed in kJ/mol.

enthalpy of condensation (p. 329) A quantity equal in magnitude but opposite in sign to the enthalpy of vaporization.

enthalpy of reaction (ΔH) (p. 331) The change in enthalpy that occurs during a reaction; calculated by taking the difference between the enthalpy of the products and the enthalpy of the reactants.

entropy (*S*) (p. 339) The measure of randomness or lack of orderliness in a system; a measure of the increasing unavailability of energy to perform useful work.

enzyme (p. 360) A naturally occurring protein molecule that acts as a catalyst.

equilibrium (p. 372) In chemistry, referring to balanced changes.

equilibrium constant (K_{eq}) (p. 375) A numerical expression of the ratio of the products to reactants at equilibrium; each concentration in the expression is raised to the power that matches the substance's coefficient in the balanced chemical reaction.

equivalence point (p. 416) The point in a titration at which an equivalent amount of titrant has been added; the number of H_3O^+ ions equals the number of OH^- ions.

essential amino acid (p. 469) An amino acid required for life that cannot be manufactured by the body and must be obtained through diet.

ester (p. 461) An organic compound of the general formula
$$R-\overset{\overset{\displaystyle O}{\|}}{C}-O-R'$$, having a carbonyl-oxygen-carbon system.

esterification (p. 464) The formation of an ester through the condensation reaction of a carboxylic acid with an alcohol.

ether (p. 457) An organic compound of the general form $R-O-R'$, in which an oxygen atom links alkyl groups. Ethers differ from esters in that the carbons next to the oxygen link are not bonded to another oxygen atom.

evaporation (p. 289) A physical change from the liquid state to the gaseous state that occurs at the surface of a liquid when the temperature is between the substance's freezing and boiling point.

excess reactant (p. 233) A reactant remaining in its original unchanged state at the end of a reaction where a limiting reactant affected the yield of product.

exothermic (p. 36) A term describing a process that releases heat energy.

expansibility (p. 245) The physical ability of a gas to expand without limit in an environment of lower pressure.

experiment (p. 15) A repeatable procedure that involves observing a natural process, sometimes under controlled conditions, for the purpose of analysis. *See also* **controlled experiment**, **natural experiment**.

F

family (p. 111) A vertical column of elements in the periodic table with similar physical and chemical properties.

fat (p. 468) A molecule formed from glycerol and three carboxylic acids with mostly saturated carbon chains; a fat differs from an oil in that it is solid at room temperature, while an oil is liquid.

fatty acid (p. 459) A carboxylic acid with a long, aliphatic chain that can be obtained by the hydrolysis of animal fat or vegetable oils.

first ionization energy (p. 115) The minimum amount of energy required to remove the first electron from the outermost shell of a single neutral atom in its gaseous state.

first law of thermodynamics (p. 33) The physical law stating that energy can be neither created nor destroyed but can be converted from one form into another. *See also* **law of conservation of mass-energy**.

flocculation (p. 207) A process that causes solid particles (the dispersed phase) to coalesce into larger masses to be separated from a liquid.

fluid (p. 245) A substance that has the ability to flow and take the shape of its container; a liquid or a gas.

formula unit (p. 150) The simplest ratio of elements in an ionic compound that describes its composition.

fraction (p. 291) One of the pure samples separated out from liquid mixtures by the process of distillation.

fractional distillation (p. 291) The separation of a mixture of liquids with different boiling points by the process of distillation.

free-electron theory (p. 151) *See* **electron-sea theory**.

free energy (G) (p. 340) A term that includes the enthalpy and entropy of a substance. Also called *Gibbs free energy*.

free-energy change (ΔG) (p. 340) The driving force of a reaction and the indicator of spontaneity; $\Delta G = \Delta H - T\Delta S$. If ΔG is negative for a process, the process is energetically favorable.

freezing (p. 42) A physical change from the liquid state to the solid state.

freezing-point depression (ΔT_f) (p. 315) A lowering of the freezing point of a solvent caused by the presence of solute particles.

fuel cell (p. 439) A class of highly efficient batteries that operates with a continuous supply of reactants for fuel from an external source, and the products are continually removed.

fullerene (p. 493) A class of carbon molecules in which the carbon atoms are arranged in the form of a hollow sphere or cylinder.

functional group (p. 455) An atom or group of atoms that is common to the members of a family of compounds and imparts characteristic chemical properties to that family; often substituted for hydrogen in a hydrocarbon.

G

gamma ray (γ) (p. 503) Uncharged electromagnetic waves of very high frequency and short wavelength; the most harmful type of radiation.

gas (p. 39) A state of matter in which the particles have enough energy to overcome the attractive forces. A gas has no definite size or shape.

gastroesophageal reflux disease (GERD) (pp. 414–15) Sometimes called *acid reflux*, this is a medical condition in which hydrochloric acid from the stomach refluxes up into the sensitive esophagus, sometimes causing temporary discomfort or more extensive tissue damage.

Gay-Lussac's law (p. 252) A law of gas behavior stating that the pressure of a confined gas is directly proportional to temperature in kelvins, provided its volume is held constant ($P/T = k$).

gene (p. 471) A segment of DNA capable of producing a specific polypeptide that is responsible for a particular characteristic, for example, hair color.

Gibbs free energy (p. 340) *See* **free energy**.

glycerol (p. 468) A three-carbon molecule with three hydroxyl groups of the general form

$$\begin{array}{c} \text{OH} \quad \text{OH} \quad \text{OH} \\ | \quad\quad | \quad\quad | \\ -\text{C}-\text{C}-\text{C}- \\ | \quad\quad | \quad\quad | \end{array}$$

Graham's law of effusion (p. 245) A law stating that the rate of effusion for a gas is inversely proportional to the square root of its molar mass.

gray (Gy) (p. 504) The SI unit of biological radiation effect.

Greek prefix system (p. 188) A nomenclature that indicates how many atoms of each element are in a binary covalent compound or how many water molecules are in a hydrate, for example, dinitrogen pentoxide and sodium carbonate monohydrate.

ground state (p. 80) An atom's lowest energy state, containing as many as seven levels.

group (p. 111) *See* **family**.

H

Haber process (p. 386) The industrial preparation of ammonia from nitrogen and hydrogen gas that uses high temperatures, high pressures, and catalysts.

half-life ($t_{1/2}$) (p. 508) The amount of time required for one-half of the nuclei in a radioactive sample to decay into another kind of nucleus.

halogen (p. 131) A Group 17 (7A) element; has seven valence electrons, making it chemically reactive.

heat (p. 36) A measurement of the total amount of thermal energy transferred from one substance to another; expressed in joules (J) or kilojoules (kJ).

heat transfer (p. 38) The transfer of thermal energy between two objects or substances.

Heisenberg uncertainty principle (p. 81) The principle that it is impossible to know both the energy or momentum (velocity) and the exact position of an electron at the same time.

Henry's law (p. 307) The law stating that the solubility of gases is directly proportional to the partial pressures of the gases above the solutions.

Hess's law (p. 335) The law stating that the enthalpy change of a reaction equals the sum of the enthalpy changes for each step of the reaction. Also known as the *law of constant heat summation*.

heterogeneous alloy (p. 301) An alloy whose particle sizes exceed the size limits for solutions.

heterogeneous catalyst (p. 359) A catalyst that is in a separate phase from the reactants.

heterogeneous equilibrium (p. 376) A chemical system at equilibrium in which two or more states are represented.

heterogeneous mixture (p. 27) A mixture composed of two or more separate phases that have their own properties, for example, a suspension.

heterogeneous reaction (p. 359) A reaction where the reactants are in different phases.

homogeneous catalyst (p. 359) A catalyst that is in the same phase as a reactant or in solution with the reactant.

homogeneous equilibrium (p. 376) A chemical system in which all of the reactants and products are at equilibrium and are in the same state.

homogeneous mixture (p. 27) A mixture that shows only a single phase with uniform properties throughout, for example, a solution.

homogeneous reaction (p. 362) A reaction in which the reactants and products are all in the same phase.

Hund's rule (p. 90) The rule stating that as electrons fill a sublevel, all orbitals receive one electron before any receives two.

hybridization (p. 168) The process of forming new kinds of orbitals with equal energies from a combination of orbitals of different energies.

hydrate (p. 193) A compound that holds a characteristic number of water molecules within its crystalline structure.

hydration (p. 302) The process by which water molecules surround and interact with solute particles; solvation in water.

hydrocarbon (p. 447) An organic, nonpolar compound containing only hydrogen and carbon atoms.

hydrogen bond (p. 276) A special (intermolecular) dipole-dipole force involving hydrogen and a highly electronegative element (fluorine, oxygen, or nitrogen).

hypothesis (p. 16) A temporary, testable explanation of a phenomenon that stimulates and guides further scientific investigation.

I

ideal gas (p. 264) A hypothetical gas whose behavior is exactly predicted by the kinetic-molecular theory (the ideal gas law).

ideal gas law (p. 266) A law of gas behavior that relates pressure, volume, temperature, and amount for an ideal gas ($PV = nRT$).

image of God (p. 2) Bearing a likeness to God (Gen. 1:26–27).

immiscible (p. 300) A term describing two liquids that are not soluble in each other.

immunotherapy (p. 57) Treatment or prevention of disease by producing immunity.

indicator (p. 411) A substance that changes color when the pH of a solution changes.

inductive reasoning (p. 13) Arguments that begin with known facts and proceed to general conclusions.

inhibitor (p. 360) A substance used to slow a reaction or reduce a catalyst's undesirable effects.

inner transition metal (p. 124) A member of the lanthanide or actinide series.

insoluble (p. 303) A term describing a solute that is unable to dissolve in a certain solvent.

instrument (p. 49) Any artificial device made for the purpose of refining, extending, or substituting for the human senses when measuring.

intermediate (p. 362) A substance formed in one step and consumed in the next step during a complex reaction; does not appear in the net chemical equation.

intermolecular forces (p. 276) An electrostatic attraction between molecules; it is much weaker than the bonds that form within molecules.

internal energy (p. 35) The sum of the potential and kinetic energies of an object's component particles.

International System of Units (SI) (p. 50) The modern standard metric system used by research scientists worldwide; in French, *Système International d'Unités*.

International Union of Pure and Applied Chemistry (IUPAC) (p. 111) The body responsible for the standardization of chemical nomenclature and usage.

ion (p. 98) A charged atom; an atom or molecule that has gained or lost electrons, thus producing an imbalance between the number of protons and electrons in the particle.

ionic bond (p. 141) The electrostatic attraction between two oppositely charged ions in a solid.

ionic compound (p. 190) A type of compound that consists of positive and negative ions whose electrical charges hold them together while neutralizing each other.

ionic equation (p. 206) An equation that represents all the substances present during a reaction, including the spectator ions, non-ionic products, and insoluble precipitates; written only for reactions taking place in a solution.

ionization (p. 302) A process by which covalent compounds form ions when they dissolve in a solution.

ionization constant of water (K_w) (p. 402) In self-ionization, the product of the concentrations of the hydronium ion (H_3O^+) and the hydroxide ion (OH^-); 1.0×10^{-14} at 25 °C; also called the *dissociation constant of water*.

irreversible reaction (p. 372) A reaction that proceeds in only one direction. The reactants change into products but not vice versa.

isotope (p. 93) One of two or more atoms of the same element with the same number of protons (atomic number) but with different numbers of neutrons.

isotopic notation (p. 93) A naming convention that includes the symbol, atomic number, and mass number of an element. It specifies the exact composition of an atom.

IUPAC (p. 111) *See* **International Union of Pure and Applied Chemistry**.

J

joule (J) (pp. 38, 327) The SI unit of energy and work; equal to $1 \text{ kg} \cdot \text{m}^2/\text{s}^2$.

K

Kelvin scale (p. 37) A temperature scale that divides the range from the freezing point of water (273 K) to the boiling point of water (373 K) into 100 increments and labels absolute zero (0 K) as its zero point; the absolute scale. One kelvin is the same size unit as one degree Celsius.

ketone (p. 458) Any organic compound of the general form

$$R-\underset{\underset{\displaystyle }{}}{\overset{\overset{\displaystyle O}{\|}}{C}}-R'$$

in which an interior carbon forms a double bond with an oxygen atom.

kilocalorie (kcal) (p. 38) Equivalent to 1000 calories.

kinetic energy (p. 32) Energy produced from motion.

kinetic-molecular theory (p. 39) A theory stating that the particles of matter are in constant motion and that the properties of matter are consequences of that motion; usually used in reference to gases and their properties but sometimes applied to solids and liquids.

kinetics (p. 353) In chemistry, the study of the rates of reactions and the steps by which they occur.

L

lanthanide series (p. 112) A portion of the sixth series of the periodic table that includes the inner transition metals from lanthanum to lutetium.

laser (p. 134) An acronym for *l*ight *a*mplification by *s*timulated *e*mission of *r*adiation.

latent heat (p. 328) Heat that produces a phase change while no temperature change is observed.

lattice energy (p. 284) The energy released when gaseous particles form crystals; expressed as a negative number.

law (p. 17) A description (often mathematical) of the behavior of matter and energy based on the results of many experiments; describes a recognizable, repeating pattern in nature.

law of chemical equilibrium (p. 375) The law stating that at a fixed temperature, a chemical system may reach a point at which the ratio of the concentration of the products to the reactants is constant.

law of combining volumes (p. 258) A law of gas behavior formulated by Gay-Lussac stating that, under equivalent conditions, the volumes of reacting gases and their gaseous products are expressed in ratios of small whole numbers (moles of reactants to moles of products).

law of conservation of mass-energy (p. 33) The physical law stating that during ordinary physical and chemical processes, mass is neither created nor destroyed, only converted from one form to another; a corollary to the first law of thermodynamics.

law of definite composition (p. 74) A law stating that every compound is formed of elements combined in ratios by mass that are unique for that compound.

law of mass conservation (p. 196) *See* **law of conservation of mass-energy**.

Le Châtelier's principle (p. 379) The principle stating that when a reversible process is disturbed, it will proceed in the direction that relieves the stress.

Lewis acid (p. 402) Any substance that can accept a pair of electrons.

Lewis base (p. 402) Any substance that can donate a pair of electrons.

Lewis structure (p. 144) A two-dimensional diagram that uses element symbols and dots to show the bonds between different atoms.

limiting reactant (p. 233) A reactant that is used up before the other reactants in a reaction, preventing more products from forming.

linear (p. 167) A structural arrangement in which particles are positioned in a straight line.

linear polymer (p. 486) A long chain of polymer molecules without branches or cross-linked polymers; results in increased rigidity, for example, high-density polyethylene.

line spectrum (p. 78) A spectrum showing only certain colors or wavelengths of light.

lipid (p. 468) A member of the large class of biological molecules that are characterized by a high proportion of C–H bonds, causing them to be insoluble in water, for example, fat, oil, wax, steroids, and fatty acids.

liquid (p. 39) A state of matter in which the particles have enough energy to partially overcome the attractive forces; the liquid conforms to the shape of its container.

literal equation (p. 64) Statement of a scientific law that relates quantities, whether variables or constants, usually in letters or symbols.

localized electron theory (p. 161) *See* **valence bond theory**.

M

magic number (p. 511) The number of nucleons (either protons or neutrons) in a completed nuclear shell, according to the nuclear shell model.

magnetic quantum number (m) (pp. 83–84) The number that defines the spatial orientation of the orbitals within the sublevel; sometimes called the *orbital quantum number*.

malleability (p. 25) A physical property of matter indicating the capability of matter to be shaped by pounding.

mass (p. 8) A measure of the amount of matter in a given substance.

mass defect (p. 516) The difference between the mass of a nucleus and the sum of the masses of the particles from which it was formed.

mass number (A) (p. 93) A number equal to the sum of the number of protons and neutrons in the nucleus of an atom.

matter (p. 8) Anything that occupies space and has mass.

measurement (p. 49) The act of comparing an unknown quantity to a standard unit; a number that indicates quantity and is followed by a unit.

mechanical energy (p. 32) Energy possessed by objects that are moving or have the potential to move.

melting (p. 42) A physical change from the solid to the liquid state.

melting point (p. 281) The temperature at which a pure solid turns into a liquid at 1 atm; the same temperature as the freezing point.

meniscus (p. 288) The curved upper surface of a column of liquid that results from intermolecular attractions within the liquid and between the liquid and the container.

metabolites (p. 365) The intermediates or products in the metabolic process.

metal (p. 111) An element located to the left of the heavy stair-step line in the periodic table; an element that is typically solid, malleable, ductile, lustrous, that conducts electricity and heat well, and that forms positive ions when it gives away its few valence electrons.

metallic bond (p. 141) A communal sharing of electrons between metal atoms. *See* **electron-sea theory**.

metalloid (p. 112) An element whose properties lie between those of metals and nonmetals; a compound found touching the heavy stair-step line in the periodic table.

metallurgy (p. 9) The process of extracting metals from their ores and adapting them for commercial use.

metric system (p. 49) Any measuring system in which different-sized units are related to each other by multiples of ten.

millimeters of mercury (mm Hg) (p. 246) A standard unit of pressure derived from the fact that normal atmospheric pressure at sea level can support 760 mm Hg in a column; 1 mm Hg = 1 torr.

miscible (p. 300) A term describing two liquids that are completely soluble in each other.

mixture (p. 27) Two or more pure substances physically combined with no definite proportions that can be separated by physical means.

molal boiling-point constant (K_b) (p. 313) A number that relates the change in boiling point of a particular solvent to the concentration of solute particles.

molal freezing-point constant (K_f) (p. 315) A number that relates the change in freezing point of a particular solvent to the concentration of solute particles.

molality (m) (p. 312) A quantitative measure of concentration equal to the number of moles of solute per kilogram of solvent; m = mol solute/kg solvent.

molar enthalpy of combustion ($\Delta H°_c$) (p. 333) The energy released as heat by the complete burning of one mole of a substance.

molar enthalpy of formation (ΔH_f) (p. 332) The change in enthalpy that occurs when one mole of a compound is formed from its elements.

molar enthalpy of fusion (ΔH_{fus}) (p. 328) The quantity of heat required to melt one mole of a solid to a liquid with no temperature change; usually expressed in units of kilojoules per mole. Also called the *heat of fusion*.

molar enthalpy of vaporization (ΔH_{vap}) (p. 329) The quantity of heat required to convert one mole of a liquid at its boiling point to its vapor at the same temperature. Also called the *heat of vaporization*.

molarity (M) (p. 311) The most common quantitative measure of concentration equal to the number of moles of solute per liter of solution; M = mol solute/L solution.

molar mass (p. 216) The mass of one mole of any pure substance; expressed in grams per mole (g/mol).

molar volume of a gas (p. 259) The volume that a mole of gas occupies if it is at standard temperature and pressure; 1 mole of gas at STP = 22.4 L.

mole (mol) (p. 216) The amount of substance contained in 6.022×10^{23} particles.

molecular (p. 39) Pertaining to molecules.

molecular formula (p. 219) A formula that shows the types of atoms involved and the exact composition of each molecule. For example, C_2H_4 is the molecular formula for ethene.

molecular orbital theory (p. 164) A bonding theory that suggests that the orbitals of a molecule's atoms are replaced by totally new orbitals when a molecule forms. These orbitals, both bonding and non-bonding, are linear combinations of the atomic orbitals.

molecule (p. 29) Two or more covalently bonded atoms found as a separate, distinct, independent unit.

mole ratio (p. 222) The ratio between the moles of one substance and the moles of another substance as indicated by the coefficients in the balanced equation; also the ratio between elements within a compound as indicated by the subscripts.

monatomic element (p. 28) An element whose atoms exist independently, for example, noble gases.

monomers (p. 484) Simple molecules that can bond to form very large molecules called polymers, for example, glucose.

monoprotic acid (p. 410) An acid that can donate only one proton.

monosaccharide (p. 466) A three- to six-carbon carbohydrate with attached hydroxyl groups and either an aldehyde or ketone group that cannot be hydrolyzed into simpler compounds; a simple sugar, for example, glucose.

N

nanotechnology (p. 492) The branch of science that combines the disciplines of chemistry, engineering, and materials science, offering the ability of manipulating atomic or molecular substances for the purpose of creating useful new products to benefit mankind.

natural experiment (p. 16) An experiment in which the conditions cannot be controlled.

naturalistic worldview (p. 5) The assumption that only matter exists and that there is nothing supernatural.

net ionic equation (p. 206) An equation that shows only the substances actually involved in a reaction and excludes spectator ions.

network covalent substance (p. 153) Atoms covalently bonded into a continuous three-dimensional network, for example, diamond.

neutralization reaction (p. 398) The reaction of an acid and a base to produce a neutral (pH = 7) solution of water and a salt.

neutral solution (p. 404) A solution with equal numbers of H_3O^+ and OH^- ions, resulting in a pH of 7.

neutron (N) (p. 78) A neutral particle within the nucleus; has a mass of 1.0087 u, slightly greater than the mass of a proton.

noble gas (p. 132) A Group 18 (8A) element; has a full outer energy level, very stable and thus essentially inert.

nomenclature (p. 188) A system of naming that follows a standardized set of rules. The International Union of Pure and Applied Chemistry developed a nomenclature for compounds.

non-electrolyte (p. 434) A substance that will not conduct electricity when melted or dissolved because it does not release ions.

nonmetal (p. 112) An element located to the right of, but not touching, the heavy stair-step line in the periodic table; an element that is generally a gas or a soft, crumbly solid, that is nonductile, nonmalleable, and nonconducting, and that usually forms negative ions because it has a strong attraction for its numerous valence electrons.

normal boiling point (p. 290) The temperature at which the vapor pressure equals 760 torr; the boiling point temperature of a liquid at STP.

North American Convention Periodic Table (p. 111) A combination of Arabic numerals and letters are placed above each column to indicate the valence electron structure of the elements in each column.

nuclear (p. 39) Pertaining only to the nucleus and its major components, for example, protons and neutrons.

nuclear binding energy (p. 516) The energy required to separate all the protons and neutrons in a specific nucleus from each other; the energy equivalent of the nucleus's mass defect.

nuclear bombardment reaction (p. 512) A non-spontaneous nuclear reaction that is forced to occur when the nucleus is struck by a high-energy particle or another nucleus.

nuclear chemistry (p. 501) The study of radioactivity, the nucleus, and the changes the nucleus undergoes.

nuclear energy (p. 32) The potential energy stored in an atom's nucleus that is released or absorbed when an atom experiences nuclear fission or fusion.

nuclear equation (p. 505) An equation that describes what occurs when nuclei split, fuse, or release radiation.

nuclear fission (p. 512) The process of splitting a massive nucleus, usually with the release of great amounts of energy and two large fragments of comparable mass, to form smaller and more stable nuclei.

nuclear fusion (p. 513) The transmutation process of combining two or more smaller nuclei into one larger, more stable nucleus, releasing great amounts of energy; the opposite of nuclear fission.

nuclear reaction (p. 501) A reaction that takes place within the nucleus of an atom.

nuclear shell model (p. 511) A model stating that the protons and neutrons in an atomic nucleus exist in different shells within the nucleus.

nuclear wastes (p. 518) Spent radioactive control rods and byproducts of nuclear fission.

nucleic acid (p. 471) A large molecule that stores and translates genetic information in living cells and consists of sugar units, nitrogenous bases, and phosphate groups.

nucleon (p. 509) A proton or neutron found in the nucleus of an atom.

nucleotide (p. 471) The "building block" of a DNA or RNA molecule. Each block is made of a sugar unit, a phosphate group, and a nitrogenous base (adenine, guanine, cytosine, thymine, or uracil).

nucleus (p. 76) The dense central part of an atom made up of protons and neutrons. The nucleus contains virtually all of the atom's mass but only a small portion of the atomic volume.

nuclide (p. 505) A unique atom of an element expressed with the isotopic notation $^A_Z X$, where X is the element's symbol, A is the atomic mass, and Z is the atomic number. The term is used instead of *isotope* when discussing the nuclear characteristics of different elements.

O

objective (p. 13) Unaffected by the observer's personal biases and presuppositions.

observation (p. 13) Using our senses of taste, touch, hearing, sight, and smell to learn about the natural world.

octet rule (p. 140) A rule stating that an atom tends to gain, lose, or share electrons until its outer level s and p orbitals are filled with eight electrons; this gives an element the electron configuration of a noble gas.

oil (p. 468) A molecule formed from glycerol and three carboxylic acids in which the carbon chains have a high degree of unsaturation; it differs from a fat in that it is liquid at room temperature, while a fat is solid.

orbital (p. 81) A four-dimensional region of space in which as many as two electrons may exist; sections of the sublevels.

orbital notation (p. 89) A diagrammatic representation that uses dashes and arrows to show the principal energy levels, sublevels, and orbitals for all the electrons in an atom.

organic compound (p. 445) Covalently bonded carbon compounds, with the exception of carbonates, carbon oxides, and carbides.

osmosis (p. 318) Diffusion of pure solvent molecules, such as water, through a semipermeable membrane.

osmotic pressure (p. 318) The amount of pressure required to prevent osmosis.

oxidation (p. 425) A chemical process in which electrons are lost and an oxidation number increases.

oxidation number (p. 182) A number that represents the number of electrons (the charge) that an atom in a compound must gain or lose to return to its neutral state. Also called an *oxidation state*.

oxidation-reduction reaction (p. 425) Any chemical reaction in which electrons transfer or shift; the shift of electrons is shown by changes in oxidation numbers. Also called a *redox reaction*.

oxide (p. 130) A binary compound in which the oxidation number of oxygen is –2, for example, Li_2O.

oxidizing agent (p. 427) The atom or ion that receives electrons during a redox reaction; the substance that causes other substances to be oxidized; the substance that is reduced.

oxyanion (p. 190) An anion that contains oxygen and one other element, for example, SO_4^{2-}.

P

paramagnetism (p. 124) A weak attraction of a substance by a magnetic field, usually as a result of unpaired electrons.

particle model of matter (p. 35) The theory that all matter is constructed from particles (atoms, molecules, or combinations of both) and that the particles are in constant motion.

particle molality (p. 314) The molality of a solution based on the number of solute particles in the solution. For example, a 1-molal solution of $AlCl_3$ has a particle molality of 4 (three Cl^- ions and one Al^+ ion).

pascal (Pa) (p. 246) The pascal is the SI unit of pressure, defined as the pressure of one newton acting on an area of one square meter.

Pauli exclusion principle (p. 84) The rule that an orbital can hold only two electrons with opposite spin. Therefore, no two electrons in the same atom can have the same set of four quantum numbers.

peptide bond (p. 470) The bond between an amino group of one amino acid and a carboxyl group of another.

percent by mass (p. 310) A quantitative measure of concentration in which the mass of the solute is compared to the mass of the solution; mass of solute/mass of solution × 100%.

percent by volume (p. 310) A quantitative measure of concentration in which the volume of the solute is compared to the volume of the solution; volume of solute/volume of solution × 100%

percent composition (p. 220) A percent that gives the relative amount (based on mass) of each element present. For example, there is 52.9% Al and 47.1% O by mass in Al_2O_3.

percent error (p. 56) A quantity derived from a method that compares the magnitude of the measurement error with the size of the measurement.

percent yield (p. 235) The percentage of the theoretical yield that was actually produced.

period (p. 111) A horizontal row of elements in the periodic table; also called a *series*.

periodic law (p. 107) The law stating that the properties of elements vary with their atomic numbers in a periodic way.

periodic table of the elements (p. 105) A table of the chemical elements arranged to display their periodic properties in relation to their atomic numbers.

permeability (p. 245) A qualitative or quantitative measure of how easily a fluid can move through the spaces between particles in a substance.

pH (p. 403) A measure of the hydronium ion (H_3O^+) concentration; the negative logarithm of the molar H_3O^+ ion concentration: $pH = -\log [H_3O^+]$.

pharmacokinetics (p. 365) The many branches of study of how the human body processes medication; includes absorption, distribution, metabolism, and excretion.

phase (p. 27) A homogeneous region of a heterogeneous mixture.

phase diagram (p. 291) A graphical diagram summarizing the temperature and pressure conditions under which a pure substance exists as a solid, liquid, or gas.

phlogiston (p. 14) A nonexistent substance once thought to be the volatile component of all combustible materials.

photon (p. 80) A packet of electromagnetic energy; an elementary nonmaterial "particle" that transmits the electromagnetic force in the standard model of matter.

physical change (p. 25) A change that alters the physical properties of a substance (state, shape, size) but that does not change the substance's identity.

physical property (p. 24) A property of matter that results from the position and characteristics of its particles and that can be measured without causing a change in the identity of the material.

pi (π) bond (p. 162) A bond in which there is a side-by-side overlap of the orbitals.

plasma (p. 40) The most abundant form of matter in the universe, consisting of a gaseous sea of high-velocity electrons, ions, and neutral atoms.

plastics (p. 489) Polymers that can be formed into different shapes, usually when heated or pressurized.

pOH (p. 404) A measure of the hydroxide ion (OH^-) concentration; the negative logarithm of the molar OH^- ion concentration: $pOH = -\log [OH^-]$.

polar covalent bond (p. 172) A chemical bond that has partially positive and partially negative ends because of unevenly shared electrons. All bonds between nonidentical atoms are polar.

polarity (p. 141) The tendency of an object to form two localized regions of opposite character; in chemistry, polarity relates to electrical charge.

polyatomic element (p. 28) An element whose atoms bond in units of three or more identical atoms, for example, S_8.

polyatomic ion (p. 148) A group of atoms that maintains a constant electrical charge while existing as a unit in a wide variety of chemical reactions, for example, SO_4^{2-}.

polyatomic ionic compound (p. 192) Ionic compounds that contain polyatomic ions.

polycyclic aromatic hydrocarbons (PAHs) (p. 454) Large, platelike molecules of fused benzene rings; composed of only hydrogen and carbon.

polyhydroxy alcohol (p. 457) An alcohol that contains more than one –OH group.

polymer (p. 464) A substance consisting of huge molecules that have repeating simpler structural units.

polymorphous (p. 284) A term describing substances (either elements or compounds) that can occur in more than one type of crystalline form.

polypeptide (p. 470) A series of many amino acids joined by peptide bonds.

polyprotic acid (p. 410) An acid that can donate more than one proton per molecule.

polysaccharide (p. 477) A carbohydrate composed of many monosaccharide units.

positron ($^{0}_{+1}\beta$) (p. 507) An electron with a positive charge; an anti-electron.

positron emission (p. 507) A reaction in which a proton is converted to a neutron and a positron.

post-transition metal (p. 125) A metal found in families 3–5 in the periodic table.

potential energy (PE) (p. 32) The energy of a system due to its position or condition; a form of mechanical energy.

pounds per square inch (psi) (p. 246) A unit of pressure equal to 6.895×10^3 Pa or 51.71 torr.

precipitate (p. 198) A solid that separates from a solution as the result of physical or chemical changes.

precision (p. 56) The agreement between two or more measurements; a quantitative evaluation of the exactness of a measurement or a measuring instrument.

premise (p. 13) A statement of assumption or fact; the basis of deductive reasoning.

pressure (p. 246) The average force exerted per unit area when molecules collide with a boundary.

presupposition (p. 5) An idea assumed true without proof, often used as a basis for understanding or proving other ideas.

principal energy level (p. 80) A region around the nucleus containing a specified group of electrons in sublevels and orbitals.

principal quantum number (n) (p. 83) The number indicating the main or principal energy level in which an electron is found.

product (p. 196) An element or compound that is produced from a chemical reaction and is usually written to the right of the arrow in a chemical equation.

protein (p. 470) A complex polymer structure of many amino acids that is joined by peptide bonds and has a molecular mass greater than 10,000 amu.

proton (p. 78) A particle in the nucleus of an atom with a +1 charge and a mass of 1.0073 u.

protonation (p. 399) The process of gaining a proton; example: $H_2O + H^+ \longrightarrow H_3O^+$

pure science (p. 12) Studying nature simply to learn new things about the universe we live in.

pure substance (p. 26) A substance that is made up of only one kind of particle and has uniform composition.

pyramidal (p. 169) A structural arrangement in which three particles and an unshared electron pair surrounding a central particle are oriented toward the corners of a four-sided pyramid.

Q

qualitative data (p. 13) Observations such as texture or color that use non-numerical data.

quantitative data (p. 13) Observations such as weight or mass that use numerical data.

quantized (p. 80) A term describing something that has separate, discrete values.

quantum model (p. 80) The current atomic model developed during the twentieth century in which the tiny, dense atomic nucleus is surrounded by a "cloud" of electrons occupying three-dimensional orbitals according to their energies.

quantum numbers (pp. 82–84) Four numbers that describe the location of an electron in an atom: the first number identifies the relative size of the principal energy level, the second describes the type of sublevel, the third indicates the direction of the orbital in space, and the fourth describes the spin of the electron.

quark-gluon plasma (QGP) (p. 41) A high-temperature state of matter that forms when nuclei collide with such energy that their protons and neutrons are broken down into a "soup" of elementary particles called quarks and gluons.

R

rad (p. 504) A *r*adiation *a*bsorbed *d*ose; a measurement of radiation equivalent to 0.01 gray.

radioactive decay series (p. 508) A series of sequential reactions of alpha and beta emissions that change larger, unstable nuclides to smaller, stable nuclides.

radioactivity (p. 501) The spontaneous emission of penetrating rays from nuclei.

radioisotope (p. 502) Any radioactive isotope of an element.

rate-determining step (p. 361) The slowest elementary step in the reaction; the speed-limiting step.

rate law (p. 363) An equation that, by using a numerical constant, mathematically describes how fast a reaction occurs.

reactant (p. 196) An element or compound that undergoes chemical change and is usually written to the left of the arrow in a chemical equation.

reaction mechanism (p. 361) The series of steps that make up a reaction.

reaction order (p. 362) An indicator of how the rate of a reaction is affected by that specific reactant's concentration.

reaction rate (p. 357) The speed at which reactants disappear or products appear in a chemical reaction.

redox reaction (p. 425) *See* **oxidation-reduction reaction**.

reducing agent (p. 426) The atom or ion that supplies electrons during a redox reaction; the substance that causes other substances to be reduced; the substance that is oxidized in a redox reaction.

reduction (p. 425) A chemical process in which electrons are gained and an oxidation number decreases.

representative group (p. 111) Elements in the *s* and *p* blocks of the periodic table; designated with *A* suffixes in the North American Convention Periodic Table.

resonance (p. 163) A molecular property where, instead of oscillating between two different bond regions, the bonds take on an intermediate character.

reversible reaction (p. 198) A reaction in which the products can change back into the original reactants so that an equilibrium is reached; a reaction that can occur in the reverse direction as well as the forward direction.

roentgen (R) (p. 504) A measure of dosage for radioactivity; 1 R = 0.0096 gray.

Roman numeral system (p. 192) *See* **Stock system**.

rotational motion (p. 337) Particle motion occurring as a spinning action.

S

salt (p. 413) A compound formed from the cation of a base and the anion of an acid.

salt bridge (p. 437) A tube of electrolytic gel that connects the two half-cells of a voltaic (galvanic) cell and allows the flow of ions.

saturated (pp. 304, 448) A term describing a solution that contains the maximum amount of solute possible at a given set of conditions; a term describing an organic compound that contains the maximum possible number of hydrogens; the compound has no double or triple bonds.

scale (p. 49) A means of measuring. A scale is subdivided into measuring units.

science (p. 12) The systematic study of nature based on observations and the collection of knowledge that results from that study.

scientific method (p. 17) A logical method of problem-solving that starts with observations and is based on inductive reasoning.

scientific model (p. 4) A simplified representation of phenomena.

scientific questions (p. 13) Questions which direct and stimulate scientific inquiry.

scientific survey (p. 16) A process that involves randomly selecting representative samples from a larger population to learn about the characteristics of the population as a whole.

scientism (p. 5) Extreme faith in human reason and science as the only reliable source of truth.

second law of thermodynamics (pp. 34, 347) The physical law that applies to natural processes unaffected by outside manipulation by any intelligence and states that during any energy transformation, some energy goes to an unusable form. Also called the *law of increasing entropy*.

self-ionization (p. 402) A process in an acid-base reaction in which one molecule donates a proton to another molecule of the same substance. Also called *auto-ionization* or *autoprotolysis*.

semiconductor (p. 125) A substance with an electrical conductivity intermediate between a conductor and an insulator; can act as either a conductor or an insulator depending on the circumstances.

sensible heat (p. 328) Heat energy that, when applied to a substance, produces a temperature change in the substance.

series (p. 111) *See* **period**.

serum (p. 58) The clear, yellowish fluid derived from blood plasma.

sievert (Sv) (p. 504) A unit of measure of the biological effects of different types of radiation on man; named for Rolf Sievert.

sigma (σ) bond (p. 161) A hybrid bond in which an *s* and *p* orbital overlap end to end.

significant digits (SD) (p. 58) The digits in the numerical value of a measurement that indicate its precision. The significant digits consist of all the certain digits plus one estimated digit.

single replacement reaction (p. 204) A chemical reaction of the general form A + BZ ⟶ B + AZ, in which a reactive element replaces a less reactive element in a compound; also called *displacement* or *substitution reactions*.

sinter (p. 482) The process of causing materials to cohere by heating at high temperatures without melting the substances.

sol-gel process (p. 482) The most common method used to produce advanced ceramics that are simple, economic, and of high quality.

solid (p. 39) A state of matter in which the particles have relatively little energy and cannot overcome the attractive forces; the particles of a solid remain in fixed positions with set distances between them.

solubility (p. 303) The maximum amount of solute that can dissolve in a specific solvent under specific conditions, such as temperature and pressure.

solubility product constant (K_{sp}) (p. 388) The equilibrium constant for the dissolving of a slightly soluble salt.

soluble (p. 303) The ability of a solute to dissolve in a certain solvent.

solute (p. 300) One of the least abundant substances in a solution; the substance that is dissolved.

solution (pp. 27, 300) A homogeneous mixture of two or more substances in a single phase.

solvation (p. 302) The process in which solvent particles surround and interact with solutes; the dissolving process in solid-in-liquid solutions, for example, a sugar-water solution.

solvent (p. 300) The most abundant substance in a solution; the substance that does the dissolving.

sound energy (p. 32) The transmission of energy through matter by the periodic motion of particles (mechanical waves).

specific heat (c_{sp}) (p. 329) The amount of heat required to raise the temperature of 1 g of the substance by 1 °C.

specific rate constant (k) (p. 363) The experimentally determined rate for each individual reaction, for example, for the reaction $H_2 (g) + I_2 (g) \longrightarrow 2HI (g)$, rate = $k[H_2][I_2]$.

spectator ion (p. 206) An ion present on both sides of an ionic equation; it does not actually participate in the reaction.

spectroscopy (p. 78) The analysis of light emitted or absorbed by matter.

spontaneous combustion (p. 356) The ignition of a substance without an outside source of heat; caused by an exothermic reaction between materials.

standard molar enthalpy of formation ($\Delta H°_f$) (p. 332) The enthalpy change for the reaction that produces one mole of a compound in its standard state from its elements in their standard states. Also called the *heat of formation*.

standard solution (p. 415) In a titration, the solution with a precisely known concentration; also called the *titrant*.

standard state (p. 332) In thermodynamics, 25 °C (298 K) and 1 atm of pressure.

standard temperature and pressure (STP) (p. 249) A value used when measuring gases; standard temperature is 0 °C or 273 K, and standard pressure is 760 torr, 101 325 Pa, or 1 atm.

starch (p. 477) A mixture of straight and branched polymers of glucose that serves to store energy for plants.

steroid (p. 469) A lipid that contains a set of three six-membered rings and one five-membered ring, for example, cholesterol.

Stock system (p. 192) A convention used to show the oxidation state of a metal ion. For example, the Stock system nomenclature for $CuCl_2$ is copper (II) chloride. This system is sometimes referred to as the *Roman numeral system*.

stoichiometry (p. 227) The measurement and calculation of the mass and molar relationships between reactants and products in chemical reactions.

strong acid (p. 406) A substance that gives up protons very easily and ionizes completely in an aqueous solution.

strong base (p. 406) A substance that readily accepts protons.

strong nuclear forces (p. 509) A term used to describe the forces that hold a nucleus together.

structural formula (p. 219) A formula that shows the types of atoms involved, the exact composition of a molecule, and the location of chemical bonds. Example: The structural formula of C_2H_4 is $-\overset{|}{C}=\overset{|}{C}-$.

structural isomers (p. 449) Compounds that have the same molecular formula but different structural formulas.

subatomic (p. 39) Pertaining to particles making up atoms, for example, protons, neutrons, and electrons.

subcritical mass (p. 513) A measure of the mass of uranium too small to sustain a chain reaction.

sublevel (s, p, d, f) (pp. 85–86) A portion of a principal energy level made up of one or more orbitals.

sublimation (pp. 42, 282) A physical change from the solid directly to the gaseous state. For example, solid CO_2 (dry ice) sublimates to gaseous CO_2.

substitution reaction (p. 464) A reaction in which one atom or group replaces another atom or group in a molecule.

sulfide (p. 130) A binary compound in which the oxidation number of sulfur is –2, for example, H_2S.

supercooled liquid (p. 281) A liquid cooled so fast that its particles may not have had time to form a preferred crystalline pattern; therefore, the molecules are locked in random positions; an amorphous solid.

supercritical mass (p. 513) The amount of fissionable material that can support an explosion; an uncontrolled chain reaction.

supersaturation (p. 305) A term describing a solution that contains more dissolved solute than at equilibrium.

surface tension (p. 287) The elastic "skin" that forms the surface of liquids; produced by the imbalance of intermolecular forces on the surface of a liquid.

surfactant (p. 287) A substance added to a liquid that acts to reduce the surface tension of that liquid by interfering with hydrogen bonds.

suspension (p. 319) A heterogeneous mixture containing solid particles that are large enough to eventually yield a precipitate.

synthesis reaction (p. 202) A chemical reaction of the general form A + B ⟶ AB in which two or more reactants combine into a single product.

system (p. 34) A portion of the universe under study.

T

temperature (p. 35) A measure of the average kinetic energy of the atoms, molecules, or ions in matter (measured in degrees Celsius or in kelvins).

ternary acid (p. 194) An oxyacid; an acid molecule that contains three different elements—hydrogen, oxygen, and another nonmetal. The oxygen and other nonmetal are often combined in a polyatomic ion, for example, $HClO_4$.

tetrahedral (p. 166) A structural arrangement in which four areas of electron density surrounding a central atom are oriented toward the corners of a four-sided pyramid.

theoretical yield (p. 235) The maximum amount of product that could be created from a given amount of reactant; normally expressed in grams of the product.

theory (p. 17) An overarching explanation of scientific observations. Theories are valuable for guiding future study and making predictions.

thermal energy (p. 32) The measure of the total kinetic energy of the molecules or ions in matter.

thermochemical equation (p. 332) A chemical equation that shows the reactants, products, and amount of energy that is released or absorbed as heat.

thermochemistry (p. 327) The branch of science that studies the transfer of energy during chemical reactions or phase changes.

thermodynamics (p. 32) The study of energy transformations in chemical and physical processes.

thermonuclear reaction (p. 514) *See* **nuclear fusion**.

thermoplastic material (pp. 489–90) Plastics that can be heated and then reshaped into other forms.

thermosetting material (p. 490) Polymers that are created by irreversible chemical reactions that make them difficult or impossible to melt and reshape; also called *thermosets*.

third law of thermodynamics (p. 37) The physical law stating that it is impossible to reach absolute zero.

titrant (p. 415) *See* **standard solution**.

titration (p. 415) A controlled reaction (often neutralization) conducted to measure the capacity of a solution of unknown concentration to react with one of known concentration.

titration curve (p. 416) A graph that plots the pH change of a solution versus the volume of added acids or bases.

torr (p. 246) *See* **millimeters of mercury**.

transition interval (p. 411) An indicator's range of pH values over which the color change occurs.

transition metal (p. 123) An element in the *d* block groups (3–12) of the periodic table.

transition state (p. 355) *See* **activated complex**.

translational motion (p. 35) Particles of matter moving in straight lines.

transmutation (p. 505) Any process that converts one element into another by a change in the number of protons, for example, the changing of lead into gold.

transuranium elements (p. 107) An element with an atomic number higher than ninety-two.

trigonal planar (p. 167) A structural arrangement in which three particles surrounding a central particle are oriented toward the corners of a flat triangle.

triple point (p. 291) The temperature and pressure at which the solid, liquid, and gaseous states of a substance exist in equilibrium.

triprotic acid (p. 411) An acid that can donate three protons.

Tyndall effect (p. 320) The scattering of light by particles in a colloidal dispersion.

U

unified atomic mass unit (u or amu) (p. 95) A small unit of mass equal to one-twelfth of the mass of the nucleus of a carbon-12 atom, or approximately the same size as a proton or a neutron. This unit is used to represent the mass of atoms and molecules.

unit (p. 49) A label, such as "inches" or "meters," used to specify the terms in which a measurement is being reported.

unit cell (p. 282) The basic building block for any type of crystal; the section of a crystal lattice that contains one formula unit of the compound or one repeating segment of the overall lattice structure.

unit conversion (p. 53) Multiplication of a measurement by a conversion factor.

universal gas constant (R) (p. 266) The constant R in the ideal gas law ($R = PV/nT$), whose value and units depend on the units used for P, V, n, and T.

universal indicators (p. 412) Carefully chosen combinations of indicators used for more accurate values.

unsaturated (pp. 304, 450) A term describing a solution that contains less than the maximum amount of solute at a given set of conditions; a term describing organic compounds that contain less than the maximum possible number of hydrogen atoms because they have at least one double or triple bond between the carbon atoms.

V

valence bond theory (p. 161) The concept that covalent bonds are formed when orbitals of different atoms overlap; also called the *localized electron theory*.

valence electrons (p. 96) The most loosely bound electrons, which are usually found in the highest or outermost energy level of a neutral atom.

valence shell electron pair repulsion (VSEPR) theory (p. 166) A theory stating that because of electron-electron repulsion, the electron orbitals in molecules are arranged so that they are as far apart as possible around the central atom.

vaporization (pp. 42, 288) A physical change from a liquid state to the vapor or gaseous state.

vapor pressure (pp. 255, 289) The pressure exerted by a vapor in equilibrium with its solid or liquid state at a specified temperature.

velocity (v) (p. 35) The rate of displacement of a system. It is always a vector quantity.

vibrational motion (p. 337) Particle motion occurring as small, random vibrations.

viscosity (p. 288) The ability of a liquid to resist flowing; the amount of internal resistance.

volatile (pp. 176, 313) Having the ability to vaporize (become a gas) at a relatively low temperature; the ability to release molecules into the air.

voltaic cell (p. 434) An electrochemical cell in which a spontaneous redox reaction produces electricity; also called a *galvanic cell*.

volume (p. 24) An amount or quantity, sometimes expressed in cubic units.

volumetric flask (p. 311) Carefully calibrated glass apparatus used by chemists to yield exact amounts of solutions.

vulcanization (p. 486) A chemical reaction perfected by Charles Goodyear, heating natural rubber with sulfur to high temperatures causing the formation of cross links within the substance, strengthening the rubber for practical purposes.

W

water displacement (p. 71) A method to measure volume by completely immersing the object in water in a graduated cylinder or an overflow container. The object's volume is measured by either directly reading the volume change on the graduated cylinder or by collecting and measuring the water from the overflow container in a graduated cylinder.

water of hydration (p. 193) The water molecules held in some definite molar ratio to the rest of the substance.

weak acid (p. 406) A substance that does not give up protons easily and does not ionize completely in an aqueous solution; a weak electrolyte.

weak base (p. 406) A substance that is a poor proton acceptor.

word equation (p. 196) An expression using the chemical names instead of symbols. Such equations list all the substances involved in a chemical reaction but contains no information about the quantities of the materials involved.

workability (p. 15) The characteristic that makes information useful or valuable in applying to other situations; the most important characteristic of a scientific model.

worldview (p. 2) The perspective from which one sees and interprets all of life.

Index

A

absolute zero, 37, 339
absorption, 359, 365
accuracy, 55, 68
acid: Arrhenius, 398; binary, 194; boric, 408–9; Brønsted-Lowry, 399, 410; citric, 459; conjugate, 400; hydrochloric, 414–15; Lewis, 401; phosphoric, 227; properties of, 397; sulfuric, 215, 236; ternary, 194
acid-base reactions, 402
acidic solution, 403
acid-ionization constant (K_a), 407
acid reflux, 414–15
actinide series, 112, 124
activated complex, 355
activation energy (E_a), 355
active site, 360
activity series, 205
actual yield, 235
addition polymers, 485
addition reaction, 464
adenine, 471–72
adhesion, 287
aerosol, 320
airbag, 267
alchemy, 10
alcohol, 291, 310, 419, 456
aldehyde, 458
aliphatic compound, 447
alkali, 120
alkali metals, 120, 183
alkaline battery, 438
alkaline-earth metals, 120, 122, 183
alkaline solution, 404
alkane, 447–49
alkene, 450
alkyl group, 448
alkyl halide, 455
alkyne, 449–50
allergy, 48, 57
allotrope, 284
allotropic element, 284
alloy, 126, 154
alpha decay, 505–6
alpha particle (α), 76, 502, 511
aluminum, 126, 434
amalgam, 301
amide, 461–62
amine, 461
amino acid, 469–70
ammonia, 120, 169, 173, 384, 386, 404, 433
ammonium nitrate, 336
amorphous carbon, 127
amorphous solid, 280
amphoteric substance, 410
analog instrument, 57
analytical chemist, 262, 281
anesthetist, 274
angular momentum equations, 83
anhydrous, 194
anion, 98, 432
anode, 434, 436
antacid, 414–15
antibonding orbital, 165
antifreeze, 317, 457
antimony, 129
Apollo space program, 139, 238
apothecary, 9
applied science, 12
Arabic science, 10
area, 51
argon, 133, 269
Aristotle, 10
aromatic compound, 447, 452–53
array, 29
Arrhenius acid, 398
Arrhenius model, 398, 400
Arrhenius, Svante, 398
arsenic, 129
artificial elements, 107
aryl group, 455
aryl halide, 455
astatine, 131
asthma, 48, 57
atmosphere, 128, 130, 255, 269, 433
atmosphere (atm), 246
atmospheric pressure, 255, 289
atom, 28, 44, 73, 77
atomic crystal, 285
atomic mass, 106–7
atomic model, 4, 73
atomic number (Z), 78, 89, 93
atomic radius, 113
atomism, 10, 73, 100
Aufbau principle, 89
autoclave, 289
auto-ionization of water, 402
autoprotolysis of water, 402
Avogadro's law, 259
Avogadro's number (N_A), 216
azimuthal quantum number (l), 83

B

balanced chemical equation, 197
balancing redox reactions, 431
band of stability, 510
barium x-ray, 389
barometer, 246
base: Arrhenius, 398; Brønsted-Lowry, 399, 410; conjugate, 400; Lewis, 401; properties of, 397–98
base-ionization constant (K_b), 407
base unit, 50
basic solution, 404
battery, 436, 438–40
becquerel (Bq), 504
Becquerel, Henri, 501
bent molecule, 170, 174
benzene, 280, 452–53
Berzelius, Jons Jakob, 129, 484
beta decay, 506
beta particle (β), 503
bias, 5, 19
big bang, 121
binary acid, 194
binary covalent compound, 188
binary ionic compound, 189–90
binding energy, 517
biochemistry, 11, 126, 445, 465
biopolymer, 494
bleach, 427
blood, 417–18
blood alcohol content (BAC), 419
Bohr, Niels, 78
Bohr's atomic model, 78, 80
boiling, 25, 289
boiling point, 290
boiling-point elevation (ΔT_b), 313
bond: chemical, 337–38; covalent, 141, 143–48; ionic, 141, 149; metallic, 141, 151; polar covalent, 172
bond axis, 161
bonding orbital, 165
bonding pair, 142–43
bonding theories, 161
boric acid, 408–9
boron, 125
boron nitride, 481
Bose-Einstein condensate (BEC), 40
Boyle, Robert, 10, 249
Boyle's law, 249
branched polymer, 486
bridge notation, 53–54
brimstone, 130
brine, 435–36
bromine, 131
Brønsted, Johannes, 399
Brønsted-Lowry acid, 399, 410
Brønsted-Lowry base, 399, 410
Brønsted-Lowry theory, 400
bronze, 128
Brookhaven Relativistic Heavy Ion Collider, 41
Brownian motion, 35, 321

Brown, Robert, 35, 320–21
buckminsterfullerene, 492
buckyball, 492
buckytube, 493
buffer, 417–18
Bunsen, Robert, 77
buret, 416

C

calcium chloride, 317
calcium fluoride, 150
calcium hydroxide, 207
calibrate, 49
calorie (cal), 38
Calorie (kcal), 38
calorimeter, 327
cancer, 444, 473, 500
candy, 309
capillary action, 288
carbohydrate, 465–68
carbon, 126, 445, 492–493
carbon-14 dating, 519
carbonation, 305
carbon dioxide, 14, 42, 238, 269, 282
carbon nanotube, 127, 493
carbon tetrachloride, 166
carbonyl group, 458–59
carboxyl group, 459
carboxylic acid, 459
careers in chemistry, 7, 19, 43, 138, 214, 274, 348, 443, 496
Carothers, Wallace, 488
catalysis, 359
catalyst, 199, 359
catalytic converter, 360
cathode, 434
cathode ray, 75
cathode-ray tube, 275
cation, 98, 432
Cavendish, Henry, 119
cellulose, 467–68
Celsius scale, 36
centrifuge, 198
ceramic-based composites, 482–84
ceramics, 481–82
Chadwick, James, 78
Chadwick's model of the atom, 78
chain reaction, 513
Challenger (space shuttle), 5
changes, nuclear, 121
charcoal, 127
Charles, Jacques, 250
Charles's law, 250
chemical bond, 140, 337–38
chemical change, 25, 44, 196
chemical energy, 32
chemical equation, 196
chemical equilibrium, 372: catalyst, 381–82; concentration, 379; law of, 375; pressure, 380; temperature, 381

chemical formula, 29, 44
chemical property, 25
chemical subscript, 29
chemical symbol, 28, 44
chemical warfare, 385
chemist, analytical, 262
chemistry, 8, 19: biochemistry, 11, 126, 445, 465; branches of, 11, 19; history of, 8–12, 19; inorganic, 11; nuclear, 11; organic, 11, 127; physical, 11
chemoreception, 176
Chernobyl (Ukraine), 518
chitin, 466
chitosan, 176
chlorine, 132, 207
chlorophyll, 454
cholesterol, 469
Christian worldview, 6, 19
chromosome, 471
Chunnel, 104, 133
citric acid, 459
clay, 481
closed system, 347
coagulation, 321
coal, 356
coefficient, 30, 197
cohesion, 287
colligative property, 313
collision, 244: theory of, 354
colloid, 319–20
combined gas law, 254
combining volumes, law of, 258–59
combustion, 10, 25: spontaneous, 356
commodity chemicals, 215
common-ion effect, 390
complex reaction, 362
compound, 29: binary ionic, 189–90; ionic, 150, 276; naming flow chart, 189; polyatomic ionic, 190
compressibility, 245
concentration, 308; of reactant, 357–58, 362–65
condensation, 42, 289: polymers, 488; reaction, 464
conductivity, 25, 154, 432
conjugate acid, 400
conjugate base, 400
conjugate pair, 399
contact process, 236
continuous spectrum, 78
continuous theory of matter, 73
conversion factor, 53
copolymers, 484
copper, 28
corrosion, 429–30
covalent bond, 141, 143–48
covalent molecular crystal, 285
covalent network crystal, 285
Creation, 6, 347
Creation Mandate, 3, 19, 49
criss-cross method, 184

critical mass, 513
critical pressure (P_c), 294
critical temperature (T_c), 292
cross-linked polymer, 486
crude oil, 448–49
cryogenics, 292–93
cryonics, 293
cryosurgery, 293
crystal, 280, 285–86
crystal lattice, 151, 282
crystalline solid, 280
crystallization, 305
crystals, classes of, 283
cubic, 283
curie (Ci), 504
Curie, Marie Sklodowska, 501
Curie, Pierre, 501
current, electrical, 50
Curse, 6
cyanide, 385
cyclic aliphatic compound, 451–52
cycloalkane, 452
cycloalkene, 452
cytosine, 471–72

D

Dalton, John, 74, 105, 255
Dalton's law of partial pressures, 255
Dalton's particle model of the atom, 74
Damadian, Raymond, 91
data, 13, 58
Davy, Sir Humphrey, 120
de Broglie, Louis, 81
de Broglie's hypothesis, 81
decomposition reaction, 203
deductive reasoning, 13
definite composition, law of, 74
delocalized electron, 151, 154, 156, 453
delta, 142
Democritus, 10, 73, 100
density, 24, 61
deoxyribonucleic acid (DNA), 454, 471–73
deposition, 42, 282
deprotonation, 399, 406
derived unit, 50
descriptive chemistry, 118
diagonal rule, 86
diamond, 127, 153–55, 285
diamond, Jonker, 154–55
diatomic element, 28
diatomic molecule, 116, 141, 144
diffraction grating, 77
diffusibility, 288
diffusion, 34, 244, 302
digital instrument, 57
dimension, 49, 68
dipeptide, 470
dipole-dipole force, 276
dipole moment (μ), 173–74

diprotic, 410
disaccharide, 467
dispersed phase, 319
dispersing medium, 319
dispersion (London) force, 277–79
displacement, water, 255
dissociation, 302
dissolve, 300
distillation, 291: fractional, 291
DNA, 277
DNA replication, 444
Döbereiner, Johann, 105
Döbereiner's triads, 105
dominion science, 3, 19
double replacement reaction, 206
driving, 396
dry ice, 42, 282
ductility, 25, 156
dynamic equilibrium, 289, 304

E

effervescence, 308
effusion, 244: Graham's law of, 245
Egypt, 9–10
Einstein, Albert, 31
elastomer, 489
electrical current, 50
electrical energy, 32
electricity, 432
electrochemical cell, 434
electrochemistry, 432
electrode, 434
electrodeposition, 436
electrolysis, 119, 434
electrolyte, 397, 432
electrolytic cell, 434
electromagnetic energy, 32
electromagnetic radiation, 77, 79
electromagnetic spectrum, 79
electron, 28, 75, 80: delocalized, 151, 154, 156; gain, 425–26; loss, 425; valence, 96
electron affinity, 115
electron capture, 507
electron configuration, 89
electron-dot notation, 97–98
electronegativity, 116, 142, 182
electronegativity scale, Pauling's, 117
electron pair, 401
electron-sea theory, 151, 154
electron-spin quantum number (m_s), 84
electroplating, 436
electrostatic force, 144
element, 10, 27
elementary steps, 361
elements: artificial, 107; periodic table of, 88, 105–13, 529–30; super-heavy, 110
empirical, 16

empirical formula, 219, 222–23, 225
emulsion, solid, 320
endothermic reaction, 36, 332, 336, 340
end point, 416
energy, 31: activation (E_a), 355; kinetic, 244, 327; thermal, 302, 329, 336; types, 32
energy changes, 353
energy diagram, 353–54
energy transfer, 327
English Channel, 104
enthalpy (H), 327: (heat) of solution (ΔH_{soln}), 302, 336; of condensation, 329; of reaction (ΔH), 331
entropy (S), 34, 339–48
environmentalist, 8
enzyme, 360
Epsom salts, 187
equation: chemical, 196; ionic, 206; net ionic, 206; thermochemical, 332; word, 196
equilibrium: heterogeneous, 376; homogeneous, 376; ionic, 387
equilibrium constant (K_{eq}), 375
equivalence point, 416
Escherichia coli, 472
essential amino acid, 469–70
ester, 384, 461
esterification, 464
ethanol, 419
ether, 457
evaporation, 289
evolution, 5, 121, 518: of life, 433, 454
excess reactant, 233
ex nihilo, 24
exothermic, 336
exothermic processes, 36
exothermic reaction, 157, 332, 340
expansibility, 245
experiment, 15: controlled, 16; natural, 16
explosion, 353

F

Fahrenheit scale, 37
faith, 4
Fall, 6, 347
family, 111
Faraday, Michael, 452
fat, 460, 468: saturated, 460, 468; unsaturated, 460, 468
fatty acid, 459, 468
Feynman, Richard, 4
firefly, 425
fireworks, 122
first commandment, 2
first ionization energy, 115
first law of thermodynamics, 33, 428
flask, volumetric, 311

flavor, fruit, 461
flocculation, 207, 321
fluid, 245
fluidity, 245
fluorescent light, 133
fluorine, 116, 131–32
foam, 320
force, 51: intermolecular, 276
formaldehyde, 146, 458
formula: chemical, 29, 44; empirical, 219, 225; molecular, 219, 225; structural, 219, 446
formula unit, 29, 150
fractional distillation, 291
fractions, 291
fragrance, flower, 461
free energy (Gibbs free energy) (G), 340, 353
free-energy change (ΔG), 340–43
free radical, 164
freezing, 42
freezing-point depression (ΔT_f), 315
frequency, 51
frog, poison dart, 461
fuel cell, 439
Fuller, R. Buckminster, 492
fullerene, 493
functional group, 455

G

galvanization, 429–30
gamma emission, 507
gamma ray (γ), 505
Garden of Eden, 9
gas, 39, 244: density, 244, 261; ideal, 264; molar volume, 259
gasoline, 457
gas pressure, 247
gastroesophageal reflux disease (GERD), 414–15
gas volume, 247
gas warfare, 384
Gay-Lussac, Joseph Louis, 131, 252, 258
Gay-Lussac's law, 252–54
Geiger, Hans, 76
gel, 320
gene, 471
Genesis Flood, 24
Germany, 384–85
Gibbs, J. Willard, 340
globular, 281
glucose, 466–67
glycerol, 468
God: as Creator, 347; glory of, 2, 19; image of, 2
goiter, 132
gold, 10
gold foil experiment, Hans Geiger's, 76
Goodyear, Charles, 486–87

Graham, Thomas, 244–45
Graham's law of effusion, 245
graphite, 127
gray (Gy), 504
Greek philosophy, 9–11
Greek prefix system, 188
groundskeeper, 443
ground state, 80, 84, 86
group, 111
guanine, 471

H

Haber, Fritz, 384–85
Haber process, 384, 386
Haldane, John, 433
half-life, 508–9
Hall-Héroult process, 435
halogen, 131
hard water, 207
hay, 356
hazardous waste, 181, 201
HDPE, 485
heartburn, 414–15
heat, 36, 327: latent, 328; sensible, 328
heat transfer, 36
Heisenberg uncertainty principle, 81
Heisenberg, Werner, 81
helium, 77, 121, 132, 244, 269, 519
hemoglobin, 383, 454
Henry's law, 307
Hess's law, 335
heterogeneous alloys, 301
heterogeneous catalyst, 359
heterogeneous equilibrium, 376
heterogeneous mixture, 27, 44
heterogeneous reaction, 359
hexagonal, 283
Hiroshima (Japan), 501
hoarfrost, 282
homogeneous catalyst, 359
homogeneous equilibrium, 376
homogeneous mixture, 27, 44
homogeneous reaction, 362
Hund, Friedrich, 90
Hund's rule, 90
HVAC technician, 348
hybridization, 168
hydrate, 193, 204
hydration, 302
hydration, water of, 193
hydraulic pressure, 286–87
hydrocarbon, 119, 278, 447
hydrochloric acid, 414–15
hydrogen, 119, 121: isotopes, 94
hydrogenation, 460
hydrogen bond, 276
hydrogen chloride, 171
hydrogen fluoride, 170
hydrogen sulfide, 177
hydronium ion, 402–3
hydroxide ion, 402–3
hypertonic, 318
hypothesis, 16
hypotonic, 318
hypoxia, 371

I

ideal gas, 264
ideal gas law, 265–66
Iijima, Sumio, 493
immiscible, 300
immunotherapy, 57
indicator, 411, 416
indicator, universal, 412
inductive reasoning, 13
Industrial Revolution, 32
inhibitor, 360
inner transition metal, 111–12, 124
insoluble, 304
inspection, 198
instant ice pack, 336
instrument, 49
intermediate, 362
intermolecular attraction, 329
intermolecular force, 276
internal energy, 35
International System of Units (SI), 38
International Union of Pure and Applied Chemistry (IUPAC), 107, 110
intoxication, 396
intravenous fluid, 318
iodine, 132, 282
ion, 28, 98: polyatomic, 148; spectator, 414
ionic bond, 141, 149
ionic bonding, 149–50
ionic compound, 150, 276
ionic crystal, 285
ionic equation, 206
ionic radius, 114
ionization, 302
ionization constant of water (K_w), 402
ionization energy, first, 115
iris, 458
iron, 429–30
iron oxide, 429–30
irreversible reaction, 372
isotonic, 318
isotope, 93
isotopic notation, 93, 505
IUPAC (International Union of Pure and Applied Chemistry), 29, 188

J

joule (J), 38, 44, 327
Joule, 32
Joule, James Prescott, 38

K

Kekule, August, 452
Kelvin, Lord, 37
Kelvin scale, 36–37
keratin, 277
ketone, 458–59
kilocalorie, 38
kilogram, 50
kilojoule (kJ), 327
kinetic energy (KE), 32, 35, 244, 327, 339
kinetic-molecular theory, 35, 39, 44, 264, 279, 286
kinetics, 353
Kirchoff, Gustav, 77
Kroto, J. F., 492
krypton, 133

L

lactose, 467
Lake Nyos, Cameroon, 261
landfill, 480, 490
lanthanide series, 112, 124
laser, 133–34
latent heat, 328
lattice energy, 284
Lavoisier, Antoine, 10, 14, 19, 105, 130
law, 17: of chemical equilibrium, 375; of combining volumes, 258–59; of conservation of mass-energy (energy conservation), 33; of definite composition, 74, 221; of mass conservation, 196; of octaves, Newlands's, 105
laws of thermodynamics, 327
LDPE, 486
lead, 281, 429–30
lead-acid storage battery, 438
lead oxide, 487
Le Châtelier, Henri, 379
Le Châtelier's principle, 379
length, 50
Lewis acid, 401
Lewis base, 401
Lewis, Gilbert, 144, 400
Lewis structures, 144–48, 161
Lewis theory, 400
light, fluorescent, 133
light intensity, 50
lightning, 130
limiting reactant, 233
linear, 167
linear polymer, 486
line spectrum, 78
lipid, 468
liquefaction, 264, 281
liquid, 39
liquid crystal, 294
liquid crystal display (LCD), 294

liquid gas, 292
liquid nitrogen, 252
liquid oxygen, 164
liquid, supercooled, 281
literal equation, 65
lithium hydroxide, 238
litmus, 397–98
localized electron theory, 161
logarithms, 403–4
London, Fritz, 277
Long, Crawford, 457
lower esophageal sphincter, 414–15
Lowry, Thomas, 399
luster, metallic, 120, 123

M

magic number, 511
magnetic field, 32
magnetic (orbital) quantum number (m), 83
magnetic precession, 91
magnetic resonance imaging (MRI), 91
malleability, 25, 156
maltose, 467
Manhattan Project, 5
mass, 24, 50: atomic, 106–7
mass composition, 220
mass defect, 516
mass density, 51
mass number (A), 93
mass-to-mass conversion, 231–33
mass-to-mole conversion, 230–31
mathematical formula, 61
math rules (for measured data), 62–63
matter, 8, 19, 24, 44
measurement, 49, 68
measurement, uncertainty, 55
mechanical energy, 32
mechanism, reaction, 361
medication, 352
melting, 42
melting point, 281
melting point apparatus, 281
membrane, semipermeable, 318
Mendeleev, Dmitri, 106
meniscus, 288
mercury, 246
metabolism, drug, 365
metabolite, 365
metal, 111, 434
metal fatigue, 23, 26
metal hydroxide, 203
metallic bond, 141, 151
metallic crystal, 286
metallic luster, 120, 123
metalloid, 112, 125
metallurgist, 138
metallurgy, 9, 138
metals: alkali, 120, 183; alkaline-earth, 120, 122, 183; inner transition, 111–12, 124; post-transition, 125; precious, 123; transition, 106, 123
methane, 168–69
metric system, 49, 68
milk, 358
Miller, Stanley, 433
millimeters of mercury (mm Hg), 246
miscibility, 300, 303
mixture, 27
model, 15, 264: atomic, 4, 73
molal boiling-point constant (K_b), 313
molal freezing-point constant (K_f), 316
molality (m), 312
molar enthalpy: of combustion ($\Delta H°_c$), 333; of formation ($\Delta H°_f$) (or heat of formation), 332; of fusion (ΔH_{fus}) (or heat of fusion), 328; of vaporization (ΔH_{vap}) (or heat of vaporization), 329
molarity (M), 311, 362, 375
molar mass, 216
molar volume of a gas, 259
mole, 216–27
molecular-bonding model, 4
molecular formula, 219, 225
molecular orbital theory, 164
molecule, 28, 44
mole ratio, 222–24, 229, 258
mole-to-mole conversion, 227–30
monatomic element, 28
monoclinic, 283
monomers, 484
monoprotic, 410
monosaccharide, 466–67
Moseley, Henry Gwyn Jeffreys, 107
mothball, 282
motion, Brownian, 321
motion, types, 35
mucus, 176, 414–15
multi-walled nanotube, 493–94
mustard gas, 385

N

nanometer, 492
nanoparticles, 495
nanorobot, 494
nanotechnology, 492: risks, 495; timeline, 492
nanotube, carbon, 127
naphthalene, 282
NASA, 139, 156, 238
native minerals, 123
natural experiment, 16
naturalism, 33
naturalistic worldview, 5, 19
neon, 133
net ionic equation, 206
network covalent substance, 153
neutralization reaction, 398, 413–14
neutral solution, 404
neutron, 28, 78, 509–10
Newlands, John, 105
Newlands's law of octaves, 105
nitrogen, 128, 269, 384, 454: liquid, 252
noble gases, 131–32
noble gas notation, 90
nomenclature, 187: of organic compounds, 448
nonelectrolyte, 434
nonmetal, 112, 143
non-oxides, 482
normal boiling point, 290
North American Convention Periodic Table, 111
North American Society of the Plastics Industry, 490
notation: electron-dot, 97–98; isotopic, 93; noble gas, 90; orbital, 90
nuclear binding energy, 517
nuclear bombardment reaction, 513
nuclear changes, 121
nuclear chemistry, 501
nuclear energy, 32
nuclear equation, 506–7
nuclear fission, 513
nuclear fusion, 40, 514–15
nuclear medicine, 501
nuclear physics, 110
nuclear power, 501
nuclear reaction, 501
nuclear shell model, 511
nuclear stability, 509–10
nuclear wastes, 519
nuclear weapon, 501
nucleic acid, 471–72
nucleon, 509
nucleotide, 471
nucleus, 76
nuclide, 506
number: atomic (Z), 78, 93; Avogadro's (N_A), 216; mass (A), 93; oxidation, 182
nylon, 488

O

objective, 13
observations, 13
Ockham's razor, 16
octet, 163
octet rule, 140
oil, 468
Oparin, A. I., 433
orbital, 81, 85, 161: antibonding, 165; bonding, 165

orbital notation, 89
organic compound, 444
orthorhombic, 283
osmosis, 318
osmotic pressure, 318
oxidation, 182, 425
oxidation number, 182, 430
oxidation number method of balancing equations, 430
oxidation number rules, 182–83
oxidation-reduction reaction, 425, 462: balancing, 428–31
oxidation state, 182
oxide, 130, 203
oxides, 482
oxidizing agent, 427
oxyacid, 203
oxyanion, 190–91
oxygen, 10, 14, 129–30, 269: liquid, 164
oxygen bridge, 457, 468
oxygine, 129
oxyhemoglobin, 383
ozone, 130, 163

P

PAH world hypothesis, 454
paramagnetism, 124, 165
particle, alpha, 76
particle model of matter, 35
particle molality, 313
particle theory of matter, 73
pascal (Pa), 246
Pascal, Blaise, 246
passivation, 428
patent attorney, 496
pathogen, 321
Pauli exclusion principle, 84
Pauling, Linus, 116, 142
Pauling's electronegativity scale, 117
Pauling unit, 117
peer review, 17
peptide bond, 470
percent by mass, 310
percent by volume, 310
percent composition, 220
percent error, 56: absolute, 56
percent yield, 235
period, 111
periodicity, 105
periodic law, 106–7
periodic table, 88, 105–13, 529–30
periodic trends, 113
permeability, 245, 288
peroxide ion, 183–84
PET, 488–89
pH, 403
pH, measurement, 411–13
pH meter, 412, 414–15
pharmacokinetics, 365

phase, 27
phase diagram, 291
phenolphthalein, 411
phlogiston theory, 14
phosphoric acid, 227
phosphorus, 128
photography, 426
photon, 80, 134
photosynthesis, 347
physical change, 25, 44
physical property, 24
pi (π) bond, 162
plasma, 40–41
plastic, 480, 489–491, 494–95: PVC, 455
plutonium, 125
pOH, 404
polar covalent bond, 172
polarity, 141, 276
polar molecules, 174
polonium, 129, 131
polyatomic element, 28
polyatomic ion, 148
polyatomic ionic compound, 190
polyatomic ions, 185
polycyclic aromatic hydrocarbons (PAHs), 454
polyester, 488–89
polyethylene, 485–86
polyhydroxy alcohol, 457
polyhydroxyalkanoates (PHA), 494
polylactide (PLA), 494
polymer, 464
polymer, glucose, 466
polymorphous, 284
polypeptide, 470–71
polyprotic, 410–11
polysaccharide, 467–68
polystyrene, 486
porcelain, 482
positron, 507
positron emission, 507
post-transition metals, 125
potassium-40–argon-40, 519
potassium uranyl sulfate, 501
potential energy, 32
pounds per square inch (psi), 246
precious metals, 123
precipitate, 198
precipitation, 390
precision, 56, 68
prefix, metric, 51–52
premise, 13
pressure, 51, 246: atmospheric, 289; critical (P_c), 294; gas, 247; osmotic, 318; vapor, 255–56, 289
pressure cooker, 248
presupposition, 5
Priestley, Joseph, 10, 14, 129
principal energy level, 80
principal quantum number (n), 83

problem-solving steps, 64–65
product, 196
propane, 264
protein, 470–71
proton, 28, 78
protonation, 399, 406
pure science, 12
pure substance, 26, 44
PVC plastic, 455
pyramidal, 169, 174
pyroxylin, 356

Q

qualitative data, 13
quantitative data, 13
quantize, 80
quantum model of the atom, 80
quantum number, 82, 84. See also listings for specific quantum numbers.
quark, 41
quark-gluon plasma, 41
quartz, 153

R

rad, 504
radiation: effects, 504; electromagnetic, 76, 79; measurement, 504; ultraviolet, 433
radioactive, 125
radioactive dating methods, 518–19
radioactive decay, 51
radioactive decay series, 507–8
radioactivity, 501
radiohalos, 519
radioisotope, 502
Radioisotopes and the Age of The Earth (RATE) Project, 519
radius: atomic, 113; ionic, 114
radon, 133
radon-222, 511
radon detection device, 511–12
Ramsay, Sir William, 132
rate-determining step, 361
rate law, 363
ratio, mole, 222–24, 229
Rayleigh, Lord (John William Strutt), 132
reactant, 196: excess, 233; limiting, 233
reactant concentration, 357–58, 362–65
reaction: addition, 464; complex, 362; condensation, 464; decomposition, 203; double replacement, 206; endothermic, 332, 336, 340; exothermic, 157, 332, 340; homogeneous, 362; irreversible, 372; neutralization, 413–14; oxidation-reduction, 425; precipitation, 390–91; reversible, 198, 372; single replacement, 204–6;

spontaneous, 338; substitution, 464; synthesis, 202; types, 202–8
reaction mechanism, 361
reaction order, 362
reaction rate, 357
recycling, 480, 490–91
Redemption, 6
redox reaction, 425
reducing agent, 426
reduction, 425
replication, 472
representative group, 111
resonance, 163
reversible reaction, 198, 372
rhombohedral, 283
ribonucleic acid (RNA), 454, 471–72
rocket fuel, solid, 156
rock salt, 316
roentgen (R), 504
Roentgen, Wilhelm, 75
Roman numeral system, 192
Rosenberg, Barnett, 472
rotation, 337
rotational motion, 35
rubber, 486–87
rubbing alcohol, 310
Rush, Benjamin, 11, 19
rust, 424, 428–30
Rutherford, Daniel, 128
Rutherford, Ernest, 76, 515
Rutherford's nuclear model of the atom, 76

S

sacrificial anode, 436
Sagan, Carl, 5–6
salt, 203, 413
salt bridge, 437
salts, 132
saturated, 304, 448
saturation, 390
scale, 49
scanning tunneling microscope (STM), 73
science, 12–13, 19: applied, 12; pure, 12
science illustrator, 214
scientific method, 17, 20
scientific model, 4, 19
scientific notation, 60
scientific problems, 13, 19
scientific questions, 13
scientific survey, 16
scientism, 5
second commandment, 2
second law of thermodynamics, 34, 140, 339, 347
selenium, 131
self-ionization of water, 402
semiconductors, 125

semipermeable membrane, 318
sensible heat, 328
series, 111
serum, 58
shot (laser) peening, 26
SI, 50, 68
sievert (Sv), 504
sigma (Σ), 335
sigma (σ) bond, 161
significant digit (SD), 58, 68: rules, 59–60, 68
significant digits, logarithms, 405
silicon, 492
silicon dioxide, 127
silver, sterling, 436
single replacement reaction, 204–6
single-walled nanotube, 493
Smalley, R. E., 492–93
smelling, 176–77
soap, 287
sodium chloride, 132, 150, 279, 283, 317, 390
sol, 320
sol-gel process, 482
solid, 39, 279: amorphous, 280; crystalline, 280; density, 280
solid emulsion, 320
solid-in-liquid solution, 301
solid rocket fuel, 156
solubility, 303, 306
solubility product constant, 388
solubility rules, 304
soluble, 303
solute, 300
solution, 27, 300: acidic, 404; alkaline, 404; basic, 404; gaseous, 301; neutral, 404; solid-liquid, 300; solid-solid, 300
solvation, 302, 336
solvent, 300
sound (acoustic) energy, 32
space shuttle, 156–57, 238
specific heat (c_{sp}), 329–31
specific rate constant (k), 363
spectator ion, 206, 414
spectroscopy, 77–78
spectrum: continuous, 78; line, 78
spontaneous combustion, 356
spontaneous generation, 433
stability, 140
standard molar enthalpy of formation ($\Delta H°_f$) (heat of formation), 332
standard solution, 415
standard state, 332
standard temperature and pressure (STP), 249
Standard Thermodynamic Property Values Table, 343
starch, 467
stars, 121

stereochemical theory, 176–77
sterilization, 207
steroid, 1, 469
Stock, Alfred, 192
Stock system, 192
stoichiometric conversion flow chart, 233
stoichiometry, 227–36, 415: gas, 258, 263
stroke, 72
strong acid, 406
strong base, 406
strong nuclear force, 509
structural formula, 219, 446
structural isomer, 449–50
subcritical mass, 513
sublevel (s, p, d, f), 85, 87
sublimation, 42, 282
substituted hydrocarbons, 455
substitution reaction, 464
sucrose, 309, 467
sugar, 309
sulfide, 130
sulfur, 130, 487
sulfuric acid, 215, 236
Sumerians, 9
sun, 514
supercooled liquids, 281
supercritical mass, 514
super-heavy elements, 110
supersaturation, 305, 309
surface area, 358–59
surface tension, 287
surfactant, 287
suspension, 27, 319
symbol, 107: chemical, 28, 44
symbols in chemical equations, 198–99
synthesis reaction, 202
synthetic elements, 515
system, 34
system, closed, 347

T

teacher, chemistry, 43
technology, electronic, 481, 483
tellurium, 129
temperature, 35, 50–51, 327, 358: critical (T_c), 292
temperature conversion, 37
ternary acid, 194
tetragonal, 283
tetrahedral, 166, 174
theoretical yield, 235
theory, 17: of matter, 73; of relativity, 513
thermal energy, 32, 35, 44, 329
thermochemical equation, 332
thermochemistry, 327
thermodynamics, 32, 44, 353:

first law of, 33, 428; laws of, 327; second law of, 34, 140, 339, 347; third law of, 37
thermometer, 36
thermonuclear reaction, 515
thermoplastic polymer, 490
thermoplastic starch (TPS), 494
thermosetting polymer (thermoset), 490
third law of thermodynamics, 37
Thomson, Joseph John, 75
Thomson's plum-pudding model of the atom, 75
thymine, 471–72
time, 50–51
tin, 429–30
titrant, 415
titration, 415
titration curve, 416
torr, 246
trans fats, 460
transition interval, 411
transition metals, 106, 123
transition state, 355
translation, 337
translational motion, 35
transmutation, 505: induced, 515
transuranium elements, 107, 124, 515
triad, 105
triclinic, 283
trigonal planar, 167
triple point, 291: of water, 37
triprotic, 411
truth, 5, 19, 58
Tubal-Cain, 9
twisted nematic phase, 294
Tyndall effect, 320
Tyndall, John, 320

U

unified atomic mass unit (u), 95
unit, 49
unit cell, 282–83
unit conversion, 53
universal gas constant (R), 266
universe, 347
unsaturated, 304, 450
ununhexium, 129
ununoctium, 133
uracil, 471
uranium, 125, 501
uranium-235, 513
uranium-lead method, 518–19
urea, 445
Urey, Harold, 433

V

vacuum pump, 74
valence bond theory, 161
valence electron, 96
valence shell, 140, 182
valence shell electron pair repulsion (VSEPR) theory, 166
vanillin, 177
vaporization, 42, 288
vapor pressure, 255–56, 289, 313
vapor pressure table, 256
velocity, 35
vibration, 337
vibrational motion, 35
vinegar, 310
viscosity, 288
volatility, 176, 313
voltaic cell, 434, 436–37
volume, 24, 51
volumetric flask, 311
vulcanization, 486–87

W

warming curve, 328
Washington Monument, 435
waste: hazardous, 181, 201; nuclear, 509, 519
water, 161, 170, 173, 276–77, 318: auto-ionization, 402; autoprotolysis, 402; density, 280; drinking, 299; percent composition, 220–21; self-ionization, 402
water displacement, 255
water of hydration, 193
water purification, 321
water softener, 207
water treatment, 207
water vapor, 269
wave-particle duality, 81
wax, 468
weak acid, 406
weak base, 406
weighted average, 95
Wöhler, Friedrich, 11, 19, 445
word equation, 196
work, 31
workability of models, 4, 12, 15, 19
worldview, 2, 41, 43, 49, 55, 80, 496

X

xenon, 133
x-ray, 107, 501
x-ray diffraction, 113

Y

yield: actual, 235; percent, 235; theoretical, 235

Z

zero, absolute, 339
zinc, 429–30
zinc-carbon dry cell battery, 438
zircon crystals, 519

Photograph Credits

The following agencies and individuals have furnished materials to meet the photographic needs of this textbook. We wish to express our gratitude to them for their important contribution.

Agence France-Press (AFP)
Air Products
Alamy
American Geological Institute (AGI)
American Institute of Physics (AIP)
American Pharmacists Association Foundation (APhA)
Art Resource
Associated Press (AP)
Baron
Lynn Betts
Blue Bird Corporation
Boeing
Brookhaven National Laboratory
Brymill Cryogenic Systems
Jeremy Burgess, MD
Burndy Library
Dr. Richard Busch
Cambridge University Press
Camp Chef
Jean-Loup Charmet
Chemical Heritage Foundation
Martyn F. Chillmaid
Cole-Parmer Instrument Company
Communications Network Research Institute (CNRI)
Correct Building Products, LLC
Dr. Stewart Custer
Digital Vision
Earth Science World
Eastman Chemicals Division
Chuck Eckert
Wim van Egmond
Dr. Tim Evans
Kenneth Eward
Charles Falco
Fermilab
Sabrina Fletcher
James A. Finley
Arnold Fisher
Formica Corporation
Fotolia
Prof. Peter Fowler
Getty Images
Peggy Greb
Klaus Guldbrandsen
Erwin Harrison
Harvard University
Hulton Archive
Humco
Imagno
iStockphoto
Jet Propulsion Laboratory (JPL)
JupiterImages Corporation
Andrew Lambert Photography
Joyce Landis
Lawrence Livermore National Laboratory
Library of Congress
Living Art Enterprises
Jack Lockwood
Fred Mang, Jr.
Steve McCutcheon
Carolyn A. McKeone
Lennox McLendon
Garry D. McMichael
Meggers Gallery
The Metropolitan Museum of Art
Dr. G. Moscoso
Nano-Tex
National Aeronautics and Space Administration (NASA)
National Institute for Materials Science, Japan
National Institute of Standards and Technology (NIST)
National Nuclear Security Administration
Natural Resources Conservation Service (NRCS)
NEC Corporation
New York Public Library
North Carolina State University (NCSU)
Novartis Health, Inc.
Ohaus
Ormco Corp.
Oxford Physics Media Services Unit
Susan Perry
Photofusion Picture Library
Photographische Gesellschaft
Photo Researchers, Inc.
Robert Rathe
Rebreather Lab
Dr. H. C. Robinson
The Royal Society
Schaeffler KG
Cliff Schiappa
Mark A. Schneider
Erich Schrempp
Science Photo Library (SPL)
Science Source
The Scotts Company

Emilio Segrè Visual Archives
Solar and Heliospheric Observatory (SOHO)
Stockbyte Platinum
Thomas Tegge
Sheila Terry
Del Thompson
Rollin Thurlow
United States Department of Agriculture (USDA)
United States Department of Defense (DOD)
United States Department of Energy (DOE)
United States Fish and Wildlife Service (USFWS)
United States Geological Survey (USGS)
United States Postal Service (USPS)
University of California
University of Cambridge
University of Florida Institute of Food and Agriculture Sciences
University of Manchester
University of Pennsylvania Library
Unusual Films
Vernier Software & Technology
Visuals Unlimited
Ward's Natural Science Establishment, Inc.
M. E. Warren
Keith Weller
Wikimedia
R. E. Wilcox
Charles D. Winters
Ray Witlin
World Bank
Yellowstone National Park

Cover
Getty Images/Dorling Kindersley/Clive Streeter (back)

Chapter 1
King James Bible, Concord Wide-Margin Reference Edition published by Cambridge University Press/Photo by Unusual Films 1; Unusual Films 1, 2, 8; © 2008 JupiterImages Corporation 3 (top), 9 (middle), 12; © Dr. Tim Evans/Photo Researchers, Inc. 3 (bottom); www.istockphoto.com/Nikolay Staykov 4; Photo Courtesy of National Nuclear Security Administration/Nevada Site Office 5; AP Photo/Lennox McLendon 6; Susan Perry 7 (top); www.istockphoto.com/Andrei Malov 7 (middle); www.istockphoto.com/Frances Twitty 7 (bottom); Printed with permission of American Pharmacists Association Foundation. Copyright 2007 APhA Foundation 9 (top); www.istockphoto.com/Adam Korzekwa 9 (bottom); Courtesy of the Chemical Heritage Foundation Collections 10; USDA/Keith Weller 13 (top); © CNRI/Photo Researchers, Inc. 13 (bottom); © The Metropolitan Museum of Art/Art Resource/NY 14; NASA/JPL 16 (top); AP Photo/James A. Finley 16 (bottom)

Chapter 2
www.istockphoto.com/Glen Jones 23; Digital Vision/Getty Images 24; Unusual Films 26 (top), 27, 42 (bottom); Richard Lake/Yellowstone National Park 26 (bottom); U.S. Navy photo by General Dynamics Electric Boat 32; Joyce Landis 33; Copyright © Boeing. All rights reserved. 34; Courtesy Rollin Thurlow 35; www.istockphoto.com/Javier Lacuey 37 (top); www.istockphoto.com/G. W. Mullis 37 (bottom); www.istockphoto.com/Ed Fuentes 39; NIST/JILA/CU-Boulder 40 (left); Courtesy of Brookhaven National Laboratory 40 (right), 41; © 2008 JupiterImages Corporation 42 (top left, top right); SOHO (ESA & NASA) 42 (middle); PhotoDisc/Getty Images 43

Chapter 3
www.istockphoto.com/Amanda Rohde 48; Digital Vision/Getty Images/Frank Gaglione 49; © Robert Rathe 50; www.istockphoto.com/Mitja Mladkovic 54; NIST 55; Ohaus 56, 57 (top left); Unusual Films 57 (right), 59, 65; Photograph copyright © by Cole-Parmer Instrument Company, used with permission 57 (bottom left); www.istockphoto.com/Sean Warren 58; PhotoDisc/Getty Images 61

Chapter 4
© Gibson Stock Photo/JupiterImages Corporation 72; Cavendish Laboratory, University of Cambridge 74; Burndy Library, courtesy AIP/Emilio Segrè Visual Archives 75; © Prof. Peter Fowler/Photo Researchers, Inc. 76 (top); © AIP/Photo Researchers, Inc. 76 (bottom); © Sheila Terry/Photo Researchers, Inc. 77 (top); © 2008 JupiterImages Corporation 77 (left); Photographische Gesellschaft, Berlin, courtesy AIP/Emilio Segrè Visual Archives, W. F. Meggers Collection, Brittle Books Collection, Harvard University Collection 77 (right); University of Manchester 78 (top); Photo by Baron/Getty Images 78 (bottom); © Meggers Gallery/AIP/Photo Researchers, Inc. 81 (top); Imagno/Getty Images 81 (bottom); Unusual Films 87; Getty Images/DAJ 92

Chapter 5
SOHO (ESA & NASA) 104; © Photo Researchers, Inc. 105; © Dr. H. C. Robinson/Photo Researchers, Inc. 106; © SPL/Photo Researchers, Inc. 107 (top); Oxford Physics Media Services Unit 107 (bottom); Sabrina Fletcher and Thomas Tegge/Lawrence Livermore National Laboratory 110; © Andrew Lambert Photography/Photo Researchers, Inc. 111; © Klaus Guldbrandsen/Photo Researchers, Inc. 112; Library of Congress 116; Lynn Betts/USDA NRCS Photo 119; Unusual Films 120, 123 (right), 127 (top right), 129 (middle), 133 (bottom left); Fred Mang, Jr. 122 (top left); © 2008 JupiterImages Corporation 122 (top right), 126 (right), 127 (top left), 132 (middle); Copyright © Boeing 122 (bottom); www.istockphoto.com/Timothy Babasade 123 (top left); PhotoDisc/Getty Images 123 (bottom left); DOE 124 (left), 134; www.istockphoto.com/Alex Hinds 124 (right);

www.istockphoto.com/George Argyropoulos 126 (left); © Dr. Tim Evans/Photo Researchers, Inc. 127 (bottom left); © Jeremy Burgess, MD/Photo Researchers, Inc. 127 (bottom right); www.istockphoto.com/Greg Nicholas 129 (top left); The Scotts Company 129 (top right); © Dr. Richard Busch/Image courtesy AGI Image Bank/www.earthscienceworld.com 129 (bottom left), 130 (middle right); © Charles D. Winters/Photo Researchers, Inc. 129 (bottom right); © Arnold Fisher/Photo Researchers, Inc. 130 (top); www.istockphoto.com/Moritz von Hacht 130 (middle left); www.istockphoto.com/Don Wilkie 130 (bottom); www.istockphoto.com/Christine Balderas 132 (top left); www.istockphoto.com/Lisa McDonald 132 (bottom left); Humco, 7400 Alumax Drive, Texarkana, TX 75501, 1-800-662-3435 132 (right); www.istockphoto.com/Charles Shapiro 133 (top left); DOD 133 (right); © M. E. Warren/Photo Researchers, Inc. 138

Chapter 6
Courtesy of the Bancroft Library, University of California, Berkeley 144; © Charles Falco/Photo Researchers, Inc. 151; © 2008 JupiterImages Corporation 154; NASA 157

Chapter 7
© Garry D. McMichael/Photo Researchers, Inc. 160; PhotoDisc/Getty Images/Patrick Clark 163 (top); PhotoDisc/Getty Images/Emanuele Taroni 163 (bottom); www.istockphoto.com/Randy Harris 164 (top); Photo courtesy of NCSU Physics Demons 164 (bottom); Unusual Films 166, 173; Photographer: Roger Ingebo, Copyright 2007 Rebreather Lab 167; © Dr. G. Moscoso/Photo Researchers, Inc. 176

Chapter 8
© 2008 JupiterImages Corporation 181, 196, 207 (both); Unusual Films 183, 193, 195, 200, 204 (both), 206; © PHOTOTAKE Inc./Alamy 188; www.istockphoto.com/Tina Lorien 194; www.istockphoto.com 198; www.istockphoto.com/Jean Frooms 203; © Charles D. Winters/Photo Researchers, Inc. 205; Del Thompson 214

Chapter 9
© Erich Schrempp/Photo Researchers, Inc. 215; Unusual Films 217, 228, 234 (both); Eastman Chemicals Division 223; NASA 238 (top); NASA/JSC 238 (bottom)

Chapter 10
© 2008 JupiterImages Corporation 243; Unusual Films 245, 248, 252 (all), 255; PhotoDisc/Getty Images 246, 274; Susan Perry 253; Fotolia.com/Norman Chan 256; R. E. Wilcox, U.S. Geological Survey 258; Photo by Jack Lockwood, 1986 (U.S. Geological Survey) 261 (top); AFP/Getty Images 261 (bottom); Camp Chef 264

Chapter 11
www.istockphoto.com/Moritz von Hach 275; Unusual Films 280 (top), 281 (all), 282 (left both), 287 (top left both), 288 (bottom); Ward's Natural Science Establishment, Inc. 280 (bottom), 284 (bottom left); © 2008 JupiterImages Corporation 282 (right); © Mark A. Schneider/Photo Researchers, Inc. 284 (top left); Wikimedia 284 (top middle); © Dr. Richard Busch/American Geological Institute/www.earthscienceworld.com 284 (top right); Dr. Stewart Custer/Unusual Films 284 (bottom right); www.istockphoto.com/Glenn Young 285; www.istockphoto.com/Mario Hornik 287 (top right); PhotoDisc/Getty Images 287 (bottom right); www.istockphoto.com/Greg Nicholas 288 (top); © Charles D. Winters/Photo Researchers, Inc. 289 (top); © Photofusion Picture Library/Alamy 289 (bottom); Air Products 292; Brymill Cryogenic Systems 293 (left); Peggy Greb/USDA 293 (right); © Chuck Eckert/Alamy 294

Chapter 12
www.istockphoto.com/Tammy Peluso 299; www.istockphoto.com/Jeff Martinez 300 (left); © Wim van Egmond/Visuals Unlimited 300 (right); www.istockphoto.com/Martin McCarthy 301; Unusual Films 305 (top), 310 (bottom), 311 (top), 314 (both), 316, 320, Susan Perry 305 (bottom), 309; Photo courtesy of Vernier Software & Technology 310 (left); PhotoDisc/Getty Images 311 (bottom), 319; www.istockphoto.com/Jane Norton 317; Stockbyte Platinum/Getty Images/Paul Tearle 318; Ray Witlin/World Bank 321

Chapter 13
www.istockphoto.com/Tanja Miro 326; www.istockphoto.com/Joe Potato 329; Susan Perry 333; © Carolyn A. McKeone/Photo Researchers, Inc. 336; Unusual Films 338; © 2004 USPS/Unusual Films 340; www.istockphoto.com/Sheryl Griffin 347; www.istockphoto.com/Frances Twitty 348

Chapter 14
Unusual Films 352, 358 (top both); PhotoDisc/Getty Images 356 (left); © 2008 JupiterImages Corporation 356 (right); © Martyn F. Chillmaid/Photo Researchers, Inc. 358 (bottom); AP/CLIFF SCHIAPPA 359

Chapter 15
www.istockphoto.com/Dave Hughes 371; © Charles D. Winters/Photo Researchers, Inc. 375, 377 (top); Unusual Films 377 (bottom), 390; Edgar Fahs Smith Collection, University of Pennsylvania Library 379 (top); www.istockphoto.com/Chad McDermott 383 (top); © Kenneth Eward/Photo Researchers, Inc. 383 (middle); © Science Source/Photo Researchers, Inc. 383 (bottom); Hulton Archive/Getty Images 384, 385 (bottom); © 2008 JupiterImages Corporation 385 (top left); www.istockphoto.com/Nathan Maxfield 385 (top right); www.istockphoto.com/Amanda Rohde 385 (middle);

© Steve McCutcheon/Visuals Unlimited 386; © Living Art Enterprises, LLC/Photo Researchers, Inc. 389

Chapter 16
Unusual Films 396; © 2008 JupiterImages Corporation 397 (left), 403 (top), 413 (bottom); © Charles D. Winters/Photo Researchers, Inc. 397 (right), 413 (top), 416 (top); © Andrew Lambert Photography/Photo Researchers 398 (top), 407; Elliot & Fry, courtesy AIP/Emilio Segrè Visual Archives 398 (bottom); courtesy AIP/Emilio Segrè Visual Archives 399 (left); The Royal Society 399 (right); www.istockphoto.com/Branislav Ostojic 409 (bottom); Unusual Films, used by permission: Novartis Health, Inc. 418; © L. O'Shaughnessy/Visuals Unlimited 419

Chapter 17
PhotoDisc/Getty Images 424; Photos by Jim Lloyd, University of Florida Institute of Food and Agriculture Sciences 425; PHOTOTAKE Inc./Alamy 427 (top right), 439; www.istockphoto.com/Darren Baker 427 (bottom left); www.istockphoto.com/Jorge Farres Sanchez 429 (top right); www.istockphoto.com/Andreas Remiorz 429 (middle left); geogphotos/Alamy 429 (bottom right); Unusual Films 432; Library of Congress 435; www.istockphoto.com/Jim Jurica 443

Chapter 18
Getty Images/Gary Pearl 444; Edgar Fahs Smith 445; Image of Blue Bird Vision® Propane-Powered School Bus provided by Blue Bird Corporation; photograph by Erwin Harrison 448; USFWS 449; © 2008 JupiterImages Corporation 450, 457, 458 (top), 464 (top), 467 (top); © Charles D. Winters/Photo Researchers, Inc. 451; © Sheila Terry/Photo Researchers, Inc. 452; www.istockphoto.com/Manfred Konrad 454; Fotolia.com/Joe Gough 455 (left); Unusual Films 456, 468; www.istockphoto.com/Oleksandr Staroseltsev 459; www.istockphoto.com/Aleksandr Stennikov 460; www.istockphoto.com/Christine Balderas 461 (top); Digital Vision 461 (bottom); © Andrew Lambert Photography/Photo Researchers, Inc. 464 (bottom both); www.istockphoto.com/Laura Callaghan 466; www.istockphoto.com/Mike Dabell 467 (bottom)

Chapter 19
www.istockphoto.com/PeskyMonkey 480; www.istockphoto.com/Lynn Watson 481 (top right); Eric Lessing/Art Resource, NY 481 (bottom left); Courtesy of National Institute for Materials Science, Japan 482; Schaeffler KG 483 (top); www.istockphoto.com/Jon Meier 483 (middle); www.istockphoto.com/Lee Pettet 483 (bottom); Inspire ICE™ image courtesy of Ormco Corp. 484; © Charles D. Winters/Photo Researchers, Inc. 485; Unusual Films 486 (top), 489 (left), 490 (all); www.istockphoto.com/Marco Rametta 486 (bottom); www.istockphoto.com/Denis Vorob'yev 487 (top); © Humanities and Social Sciences Library/New York Public Library/Photo Researchers, Inc. 487 (bottom); www.istockphoto.com/Susan Trigg 488; Photo courtesy of Formica Corporation 489 (right); Copyright Correct Building Products, LLC www.correctdeck.com 491; Dr. Sumio Iijima discovered the carbon nanotube in 1991 at NEC. Photo courtesy of NEC Corporation 493; Wikimedia/GNU Free Documentation License 494 (top); Photo courtesy of Nano-Tex 494 (bottom); PhotoDisc/Getty Images 495; Courtesy of DOE/NREL, Credit—Warren Gretz 496

Chapter 20
Getty Images/James Hardy 500; © Jean-Loup Charmet/Photo Researchers, Inc. 501; © SPL/Photo Researchers, Inc. 502 (left); www.istockphoto.com/Rade Lukovic 504; © Michael Melford/Getty Images 509; U.S. Geological Survey 511 (top); Getty Images/Steve Cole 511 (bottom); Fermilab/Reidar Hahn 512; Courtesy of National Nuclear Security Administration/NV Office

Images generated using Spartan software 28, 30, 139, 142, 167, 171, 172, 173, 177, 365, 446, 455, 458, 461, 469, 472

Periodic Table

Periodic Table of Elements

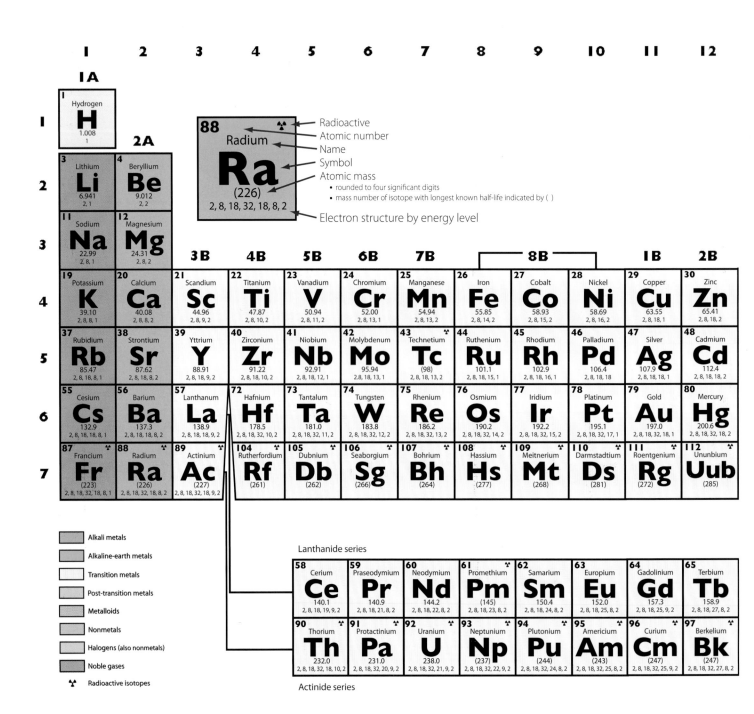

13	14	15	16	17	18
					8A
					2 Helium **He** 4.003 2
3A	4A	5A	6A	7A	
5 Boron **B** 10.81 2, 3	6 Carbon **C** 12.01 2, 4	7 Nitrogen **N** 14.01 2, 5	8 Oxygen **O** 16.00 2, 6	9 Fluorine **F** 19.00 2, 7	10 Neon **Ne** 20.18 2, 8
13 Aluminum **Al** 26.98 2, 8, 3	14 Silicon **Si** 28.09 2, 8, 4	15 Phosphorus **P** 30.97 2, 8, 5	16 Sulfur **S** 32.07 2, 8, 6	17 Chlorine **Cl** 35.45 2, 8, 7	18 Argon **Ar** 39.95 2, 8, 8
31 Gallium **Ga** 69.72 2, 8, 18, 3	32 Germanium **Ge** 72.64 2, 8, 18, 4	33 Arsenic **As** 74.92 2, 8, 18, 5	34 Selenium **Se** 78.96 2, 8, 18, 6	35 Bromine **Br** 79.90 2, 8, 18, 7	36 Krypton **Kr** 83.80 2, 8, 18, 8
49 Indium **In** 114.8 2, 8, 18, 18, 3	50 Tin **Sn** 118.7 2, 8, 18, 18, 4	51 Antimony **Sb** 121.8 2, 8, 18, 18, 5	52 Tellurium **Te** 127.6 2, 8, 18, 18, 6	53 Iodine **I** 126.9 2, 8, 18, 18, 7	54 Xenon **Xe** 131.3 2, 8, 18, 18, 8
81 Thallium **Tl** 204.4 2, 8, 18, 32, 18, 3	82 Lead **Pb** 207.2 2, 8, 18, 32, 18, 4	83 Bismuth **Bi** 209.0 2, 8, 18, 32, 18, 5	84 Polonium **Po** (209) 2, 8, 18, 32, 18, 6	85 Astatine **At** (210) 2, 8, 18, 32, 18, 7	86 Radon **Rn** (222) 2, 8, 18, 32, 18, 8
113 Ununtrium **Uut** (284)	114 Ununquadium **Uuq** (289)	115 Ununpentium **Uup** (288)	116 Ununhexium **Uuh** (291)		118 Ununoctium **Uuo** (294)

The names given to elements 112–116 and 118 represent the Latin and Greek names for their Arabic numbers.

| 66 Dysprosium **Dy** 162.5 2, 8, 18, 28, 8, 2 | 67 Holmium **Ho** 164.9 2, 8, 18, 29, 8, 2 | 68 Erbium **Er** 167.3 2, 8, 18, 30, 8, 2 | 69 Thulium **Tm** 168.9 2, 8, 18, 31, 8, 2 | 70 Ytterbium **Yb** 173.0 2, 8, 18, 32, 8, 2 | 71 Lutetium **Lu** 175.0 2, 8, 18, 32, 9, 2 |
| 98 Californium **Cf** (251) 2, 8, 18, 32, 28, 8, 2 | 99 Einsteinium **Es** (252) 2, 8, 18, 32, 29, 8, 2 | 100 Fermium **Fm** (257) 2, 8, 18, 32, 30, 8, 2 | 101 Mendelevium **Md** (258) 2, 8, 18, 32, 31, 8, 2 | 102 Nobelium **No** (259) 2, 8, 18, 32, 32, 8, 2 | 103 Lawrencium **Lr** (262) 2, 8, 18, 32, 32, 9, 2 |

Element	Number	Element	Number
Actinium	89	Neodymium	60
Aluminum	13	Neon	10
Americium	95	Neptunium	93
Antimony	51	Nickel	28
Argon	18	Niobium	41
Arsenic	33	Nitrogen	7
Astatine	85	Nobelium	102
Barium	56	Osmium	76
Berkelium	97	Oxygen	8
Beryllium	4	Palladium	46
Bismuth	83	Phosphorus	15
Bohrium	107	Platinum	78
Boron	5	Plutonium	94
Bromine	35	Polonium	84
Cadmium	48	Potassium	19
Calcium	20	Praseodymium	59
Californium	98	Promethium	61
Carbon	6	Protactinium	91
Cerium	58	Radium	88
Cesium	55	Radon	86
Chlorine	17	Rhenium	75
Chromium	24	Rhodium	45
Cobalt	27	Roentgenium	111
Copper	29	Rubidium	37
Curium	96	Ruthenium	44
Darmstadtium	110	Rutherfordium	104
Dubnium	105	Samarium	62
Dysprosium	66	Scandium	21
Einsteinium	99	Seaborgium	106
Erbium	68	Selenium	34
Europium	63	Silicon	14
Fermium	100	Silver	47
Fluorine	9	Sodium	11
Francium	87	Strontium	38
Gadolinium	64	Sulfur	16
Gallium	31	Tantalum	73
Germanium	32	Technetium	43
Gold	79	Tellurium	52
Hafnium	72	Terbium	65
Hassium	108	Thallium	81
Helium	2	Thorium	90
Holmium	67	Thulium	69
Hydrogen	1	Tin	50
Indium	49	Titanium	22
Iodine	53	Tungsten	74
Iridium	77	Ununbium	112
Iron	26	Ununhexium	116
Krypton	36	Ununoctium	118
Lanthanum	57	Ununpentium	115
Lawrencium	103	Ununquadium	114
Lead	82	Ununtrium	113
Lithium	3	Uranium	92
Lutetium	71	Vanadium	23
Magnesium	12	Xenon	54
Manganese	25	Ytterbium	70
Meitnerium	109	Yttrium	39
Mendelevium	101	Zinc	30
Mercury	80	Zirconium	40
Molybdenum	42		